RG

Edited by
Herbert Waldmann and
Petra Janning

Concepts and Case Studies in Chemical Biology

Related Titles

Trabocchi, A. (ed.)

Diversity-Oriented Synthesis

Basics and Applications in Organic Synthesis, Drug Discovery, and Chemical Biology

2013

Print ISBN: 978-1-118-14565-4, also available in digital formats

Sierra, M.A., de la Torre, M.C., Cossio, F.P.

More Dead Ends and Detours

En Route to Successful Total Synthesis

2013

Print ISBN: 978-3-527-32976-2

Christmann, M., Bräse, S. (eds.)

Asymmetric Synthesis II

More Methods and Applications

2012

Print ISBN: 978-3-527-32900-7, also available in digital formats

Civjan, N. (ed.)

Chemical Biology

Approaches to Drug Discovery and Development to Targeting Disease

2012

Print ISBN: 978-1-118-10118-6, also available in digital formats

Luisi, P.P. (ed.)

Chemical Synthetic Biology

2011

Print ISBN: 978-0-470-71397-6, also available in digital formats

Waldmann, H., Janning, P. (eds.)

Chemical Biology

Learning through Case Studies

2009

Print ISBN: 978-3-527-32330-2

Edited by Herbert Waldmann and Petra Janning

Concepts and Case Studies in Chemical Biology

WILEY-VCH
Verlag GmbH & Co. KGaA

The Editors

Prof. Dr. Herbert Waldmann
MPI of Molecular Physiology
Otto-Hahn-Str. 11
44227 Dortmund
Germany

Dr. Petra Janning
MPI of Molecular Physiology
Otto-Hahn-Str. 11
44227 Dortmund
Germany

Cover

fotolia.com © Paolo Toscani

We thank Claudia Pieczka, MPI of Molecular Physiology, Dortmund, Germany for preparing the cover illustration.

Library of Congress Card No.: applied for

British Library Cataloguing-in-Publication Data
A catalogue record for this book is available from the British Library.

Bibliographic information published by the Deutsche Nationalbibliothek
The Deutsche Nationalbibliothek lists this publication in the Deutsche Nationalbibliografie; detailed bibliographic data are available on the Internet at <http://dnb.d-nb.de>.

Print ISBN: 978-3-527-33611-1
ePDF ISBN: 978-3-527-67598-2
ePub ISBN: 978-3-527-67600-2
Mobi ISBN: 978-3-527-67599-9

Cover-Design Adam Design Weinheim, Germany; Claudia Pieczka, MPI of Molecular Physiology, Dortmund
Typesetting Laserwords Private Limited, Chennai, India
Printing and Binding Markono Print Media Pte Ltd, Singapore

Printed on acid-free paper

Contents

List of Contributors

Hans Aerts
University of Amsterdam
Department of Medical Biochemistry
Academic Medical Centre
Amsterdam
The Netherlands

Albert A. Antolín
Universitat Pompeu Fabra
Systems Pharmacology
Research Program on Biomedical Informatics
IMIM Hospital del Mar Medical Research Institute
Doctor Aiguader 88
08003 Barcelona
Spain

Thomas Beenakker
Leiden University
Leiden Institute of Chemistry
Einsteinweg 55
CC 2333 Leiden
The Netherlands

Rolf Boot
University of Amsterdam
Department of Medical Biochemistry
Academic Medical Centre
Amsterdam
The Netherlands

Gerjan de Bruin
Leiden University
Leiden Institute of Chemistry
Einsteinweg 55
CC 2333 Leiden
The Netherlands

Atwood K. Cheung
Novartis Institutes for BioMedical Research, Inc.
Global Discovery Chemistry
250 Massachusetts Avenue
Cambridge, MA 02139
USA

Jeroen Codée
Leiden University
Leiden Institute of Chemistry
Einsteinweg 55
CC 2333 Leiden
The Netherlands

Feng Cong
Novartis Institutes for BioMedical Research, Inc.
Developmental and Molecular Pathways
250 Massachusetts Avenue
Cambridge, MA 02139
USA

Alexander Deiters
University of Pittsburgh
Department of Chemistry
Chevron Science Center
219 Parkman Avenue
Pittsburgh, PA 15260
USA

Frank J. Dekker
Groningen University
Department of Pharmaceutical
Gene Modulation
Antonius Deusinglaan 1
9713 av Groningen
Netherlands

Hans van den Elst
Leiden University
Leiden Institute of Chemistry
Einsteinweg 55
CC 2333 Leiden
The Netherlands

Bogdan Florea
Leiden University
Leiden Institute of Chemistry
Einsteinweg 55
CC 2333 Leiden
The Netherlands

Yushi Futamura
RIKEN
Antibiotics Laboratory
2-1 Hirosawa
Wako
Saitama 351-0198
Japan

Warren R. J. D. Galloway
University of Cambridge
Department of Chemistry
Lensfield Road
Cambridge CB2 1 EW
UK

Malte Gersch
Technische Universität München
Department of Chemistry
Lichtenbergstraße 4
85748 Garching
Germany

Paul Geurink
The Netherlands Cancer Institute
(NKI)
Division of Cell Biology
Plesmanlaan 121
CX 1066 Amsterdam
The Netherlands

Roger S. Goody
Max Planck Institute of
Molecular Physiology
Department of Physical
Biochemistry
Otto-Hahn-Straße 11
44227 Dortmund
Germany

Tom N. Grossmann
Chemical Genomics Centre of
the Max Planck Society
Otto-Hahn-Straße 15
44227 Dortmund
Germany

Kathy Hadje-Georgiou
University of Cambridge
Department of Chemistry
Lensfield Road
Cambridge CB2 1 EW
UK

Christian Hedberg
Max Planck Institute of
Molecular Physiology
Department of Chemical Biology
Otto-Hahn-Straße 11
44227 Dortmund
Germany

Andreas Herrmann
Humboldt University Berlin
Department of Biology
Invalidenstrasse 42
10115 Berlin
Germany

Glenn R. Hicks
University of California
Riverside
Center for Plant Cell Biology and
Department of Botany and Plant
Sciences
900 University Avenue
Riverside, CA 92521
USA

Birgit Hoeger
European Molecular Biology
Laboratory (EMBL)
Genome Biology Unit
Meyerhofstrasse 1
69117 Heidelberg
Germany

Felix Hövelmann
Humboldt University Berlin
Department of Chemistry
Brook-Taylor-Straße 2
12489 Berlin
Germany

Barbara Imperiali
Massachusetts Institute of
Technology
Departments of Biology and
Chemistry
68-380, 77 Massachusetts
Avenue
Cambridge, MA 02139
USA

Aymelt Itzen
Technische Universität München
Center of Integrated Protein
Science Munich
Department Chemie
AG Proteinchemie
Lichtenbergstraße 4
85748 Garching
Germany

Jianbing Jiang
Leiden University
Leiden Institute of Chemistry
Einsteinweg 55
CC 2333 Leiden
The Netherlands

Markus Kaiser
Universität Duisburg-Essen
Zentrum für Medizinische
Biotechnologie
Fakultät für Biologie
Universitätsstrasse 2
45117 Essen
Germany

Wouter Kallemeijn
University of Amsterdam
Department of Medical
Biochemistry
Academic Medical Centre
Amsterdam
The Netherlands

Stefan Knapp
University of Oxford
Nuffield Department of Clinical
Medicine
Structural Genomics
Consortium and Target
Discovery Institute
Roosevelt Drive
Oxford OX3 7FZ
UK

Andrea Knoll
Humboldt University Berlin
Department of Chemistry
Brook-Taylor-Straße 2
12489 Berlin
Germany

Maja Köhn
European Molecular Biology
Laboratory (EMBL)
Genome Biology Unit
Meyerhofstrasse 1
69117 Heidelberg
Germany

Yasumitsu Kondoh
RIKEN Center for Sustainable
Resource Science (CSRS)
Chemical Biology Research
Group
2-1 Hirosawa
Wako
Saitama 351-0198
Japan

and

RIKEN
Antibiotics Laboratory
2-1 Hirosawa
Wako
Saitama 351-0198
Japan

Kamal Kumar
Max Planck Institute of
Molecular Physiology
Department of Chemical Biology
Otto-Hahn-Straße 11
44227 Dortmund
Germany

Susann Kummer
Universitätsklinikum Heidelberg
Department of Infectiology
Im Neuenheimer Feld 324
69120 Heidelberg
Germany

Luca Laraia
University of Cambridge
Department of Chemistry
Lensfield Road
Cambridge CB2 1 EW
UK

Kah-Yee Li
Leiden University
Leiden Institute of Chemistry
Einsteinweg 55
CC 2333 Leiden
The Netherlands

Nan Li
Leiden University
Leiden Institute of Chemistry
Einsteinweg 55
CC 2333 Leiden
The Netherlands

Hannah Lingard
University of Oxford
Nuffield Department of Clinical
Medicine
Structural Genomics
Consortium and Target
Discovery Institute
Roosevelt Drive
Oxford OX3 7FZ
UK

Wouter van der Linden
Standford University
Department of Pathology
School of Medicine
300 Pasteur Drive
Stanford, CA 94305-5324
USA

Qingyang Liu
North Carolina State University
Department of Chemistry
2620 Yarbrough Drive
Raleigh, NC 27695-8204
USA

Gijs van der Marel
Leiden University
Leiden Institute of Chemistry
Einsteinweg 55
CC 2333 Leiden
The Netherlands

Jordi Mestres
Universitat Pompeu Fabra
Systems Pharmacology
Research Program on Biomedical
Informatics
IMIM Hospital del Mar Medical
Research Institute
Doctor Aiguader 88
08003 Barcelona
Spain

Susanne Müller
University of Oxford
Nuffield Department of Clinical
Medicine
Structural Genomics
Consortium and Target
Discovery Institute
Roosevelt Drive
Oxford OX3 7FZ
UK

Makoto Muroi
RIKEN Center for Sustainable
Resource Science (CSRS)
Chemical Biology Research
Group
2-1 Hirosawa
Wako
Saitama 351-0198
Japan

and

RIKEN
Antibiotics Laboratory
Hirosawa 2-1
Wako
Saitama 351-0198
Japan

Feilin Nie
University of Cambridge
Department of Chemistry
Lensfield Road
Cambridge CB2 1 EW
UK

Julian Oeljeklaus
Universität Duisburg-Essen
Zentrum für Medizinische
Biotechnologie
Fakultät für Biologie
Universitätsstrasse 2
45117 Essen
Germany

Hiroyuki Osada
RIKEN Center for Sustainable Resource Science (CSRS)
Chemical Biology Research Group
Hirosawa 2-1
Wako
Saitama 351-0198
Japan

and

RIKEN
Antibiotics Laboratory
Hirosawa 2-1
Wako
Saitama 351-0198
Japan

Herman Overkleeft
Leiden University
Leiden Institute of Chemistry
Einsteinweg 55
CC 2333 Leiden
The Netherlands

Guillem Paniagua
Leiden University
Leiden Institute of Chemistry
Einsteinweg 55
CC 2333 Leiden
The Netherlands

Laura B. Peterson
Massachusetts Institute of Technology
Departments of Biology and Chemistry
68-380, 77 Massachusetts Avenue
Cambridge, MA 02139
USA

Natasha V. Raikhel
University of California Riverside
Center for Plant Cell Biology and Department of Botany and Plant Sciences
900 University Avenue
Riverside, CA 92521
USA

Daniel Rauh
Technische Universität Dortmund
Fakultät für Chemie und Chemische Biologie
Otto-Hahn-Straße 6
44227 Dortmund
Germany

Shin-ichi Sato
Kyoto University
Institute for Integrated Cell-Material Sciences (WPI-iCeMS)
Kyoto 611-0011
Japan

Tilmann Schneider-Poetsch
RIKEN
Chemical Genetics Laboratory
Hirosawa 2-1
Wako
Saitama 351-0198
Japan

Oliver Seitz
Humboldt University Berlin
Department of Chemistry
Brook-Taylor-Straße 2
12489 Berlin
Germany

Stephan A. Sieber
Technische Universität München
Department of Chemistry
Lichtenbergstraße 4
85748 Garching
Germany

Jeffrey R. Simard
Amgen, Inc.
360 Binney St.
Cambridge, MA 02142
USA

David R. Spring
University of Cambridge
Department of Chemistry
Lensfield Road
Cambridge CB2 1 EW
UK

Mario van der Stelt
Leiden University
Leiden Institute of Chemistry
Einsteinweg 55
CC 2333 Leiden
The Netherlands

Gemma Triola
Spanish National Research
Council (CSIC)
Institute of Advanced Chemistry
of Catalonia (IQAC)
Department of Biomedicinal
Chemistry
Jordi Girona 18-26
08034 Barcelona
Spain

Nachiket Vartak
Max Planck Institute of
Molecular Physiology
Department of Chemical Biology
Otto-Hahn-Straße 11
44227 Dortmund
Germany

Motonari Uesugi
Kyoto University
Institute for Integrated
Cell-Material Sciences
(WPI-iCeMS)
Kyoto 611-0011
Japan

and

Kyoto University
Institute for Chemical Research
Uji, Kyoto 611-0011
Japan

Gregory L. Verdine
Harvard University
Departments of Stem Cell &
Regenerative Biology
Chemistry & Chemical Biology,
and Molecular & Cellular Biology
Cambridge, MA 02138
USA

Martijn Verdoes
Radboud University
Department of Tumor
Immunology
Nijmegen Medical Centre
Geert Grooteplein 26/28
GA 6525 Nijmegen
The Netherlands

Bridget K. Wagner
Broad Institute
Center for the Science of
Therapeutics
7 Cambridge Center 3027
Cambridge, MA 02142
USA

Marthe Walvoort
Massachusetts Institute of Technology
Department of Biology
77 Massachusetts Avenue
Cambridge, MA 02139
USA

Yansong Wang
European Molecular Biology Laboratory (EMBL)
Genome Biology Unit
Meyerhofstrasse 1
69117 Heidelberg
Germany

David Wilcke
University of Cambridge
Department of Chemistry
Lensfield Road
Cambridge CB2 1 EW
UK

Lianne Willems
Leiden University
Leiden Institute of Chemistry
Einsteinweg 55
CC 2333 Leiden
The Netherlands

Martin Witte
University of Groningen
Stratingh Institute of Chemistry
Bio-Organic Chemistry
Nijenborgh 7
AG 9747 Groningen
The Netherlands

Yaowen Wu
Chemical Genomics Centre of the Max Planck Society
Otto-Hahn-Straße 15
44227 Dortmund
Germany

Bo-Tao Xin
Leiden University
Leiden Institute of Chemistry
Einsteinweg 55
CC 2333 Leiden
The Netherlands

Minoru Yoshida
RIKEN
Chemical Genetics Laboratory
Hirosawa 2-1
Wako
Saitama 351-0198
Japan

Chunhua Zhang
University of California Riverside
Center for Plant Cell Biology and Department of Botany and Plant Sciences
900 University Avenue
Riverside, CA 92521
USA

Lei Zhao
Chemical Genomics Centre of the Max Planck Society
Otto-Hahn-Straße 15
44227 Dortmund
Germany

Slava Ziegler
Max Planck Institute of Molecular Physiology
Department of Chemical Biology
Otto-Hahn-Straße 11
44227 Dortmund
Germany

Introduction and Preface

"Chemical Biology may be defined as the application of chemical methods and techniques to the study of biological phenomena, that is, chemical biology research seeks new insights into biology by means of an approach originating from an enabling chemistry tool box.

The chemical biological approach often starts with the analysis of a biological phenomenon in order to deduce structural information, for instance, about biomacromolecules or small molecules that interact with them. On the basis of this information, unsolved chemical problems are identified and the ability of the synthetic chemist to design and prepare tailor-made reagents and tool compounds, that is, proteins equipped with reporter groups and tags or potent and selective small molecule modulators of protein function, is employed as key enabling technology for subsequent research. Very frequently, the biochemical and biophysical properties of these reagents need to be determined for the proper design and execution of biological experiments, giving new insights into the originally motivating biological phenomenon. The results gleaned thereby may then lead to a better understanding of biology and fuel additional cycles of chemical biology research following the same logic Figure 1 illustrates the cycle of chemical biology research.

By its very nature, chemical biology is multidisciplinary and needs to bridge the approaches and cultures of the neighboring sciences, chemistry, biology, and physics, within a given research group or in collaborations between groups of complementary expertise (as is very frequently the case).

Thus, education in chemical biology requires training in these disciplines on a more or less advanced level. A chemical biology textbook, planned and organized similar to that of established textbooks of the individual disciplines mentioned, would have to face the challenge of not growing too large to be readable but at the same time be sufficiently comprehensive to cover the individual disciplines in the required scientific depth.

An alternative, and probably more efficient and appropriate, approach to chemical biology education may be to resort to the well-established, proven textbooks of chemistry, biology, and physics for in-depth courses and to complement them by lecture series, seminars, and practical courses that demonstrate the combination

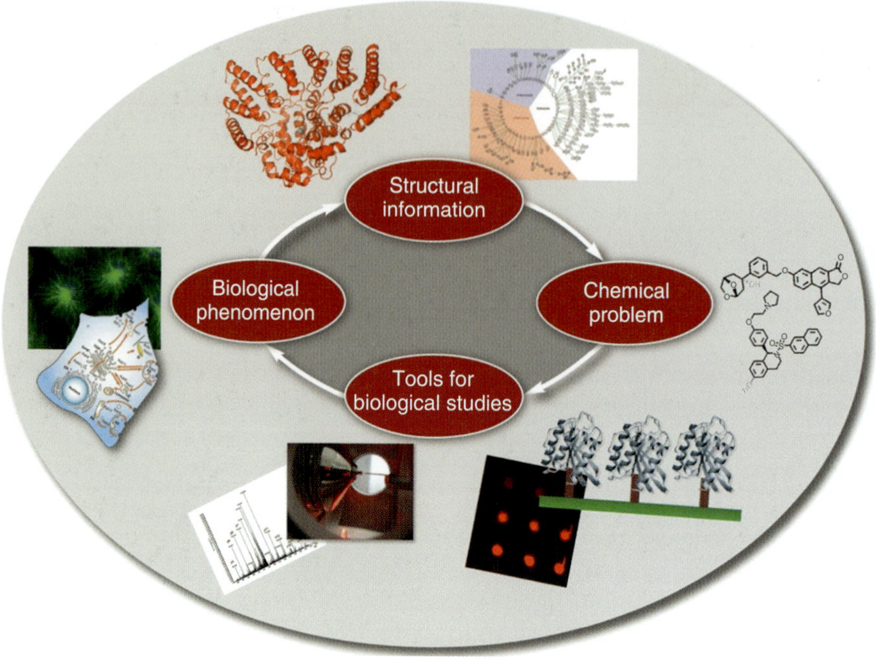

Figure 1 Illustration of the cycle of chemical biology research.

of and the interplay between these sciences and the corresponding experimental techniques in chemical biology." [1]

With this goal in mind, we planned and prepared our previous book *Chemical Biology: Learning through Case Studies* [1] in 2009. By concentrating on a series of individual successful cases of chemical biology research, it highlighted the combination of the different sciences involved in gaining new insights into biological phenomena with approaches originating from chemistry and integrating biophysics, biochemistry, and other disciplines whenever required.

The same concept has been chosen for this new book, entitled *Concepts and Case Studies in Chemical Biology*. It covers 27 new case studies in Chemical Biology, reflecting the rapid growth in this interdisciplinary topic since 2009.

Again, in each chapter, initially a biological problem is presented. To address this problem, a chemical approach is described and both together lead to chemical biology research. Following this line, for several different examples the reader is introduced into thinking and research in Chemical Biology, arriving at important scientific results and techniques and methods used in this field at the same time.

In contrary to the previous *Learning through Case Studies* book, we asked the researchers themselves to write a case study that has the origin in their own lab, rather than writing the chapters based on literature reports only.

We hope the book will be a valuable source of information for advanced students, postdoctoral researchers, and researchers working on the borderline between chemistry, biology, and biochemistry.

We are grateful to all authors of the individual chapters for their excellent work and trust in the concept. We are also grateful to Bernadette Gmeiner and Dr Anne Brennführer from Wiley-VCH for their editorial help and encouragement.

Dortmund, February 2014

Petra Janning
Herbert Waldmann

References

1. Waldmann, H. and Janning, P. (eds) (2009) *Chemical Biology – Learning through Case Studies*, Wiley-VCH Verlag GmbH, Weinheim.

Abbreviations

11ß-HSD	11ß-hydroxysteroid dehydrogenase
2D	Two dimensional
2D-DIGE	2-Dimensional difference gel electrophoresis
3D	Three dimensional
3MA	3-Methyladenine
4-DMN	4-*N*,*N*-dimethylamino-1,8-naphthalimide
AB	Aminobenzamide
ABA	Abscisic acid
ABC transporter	ATP-binding cassette
ABF	ABRE binding factor
ABG	Almond retaining beta-exoglucosidase
abi mutant	ABA insensitive mutant
Abl kinase	Abelson murine leukemia viral oncogene homolog kinase
ABP	Activity-based probe
ABPP	Activity-based protein profiling
ABRE	ABA-responsive promoter element
ACC	Acetyl-CoA carboxylase
AChE	Acetylcholinesterase
ACN	Acetonitrile
AcOH	Acetic acid
AD	Activating domain
Adda	β-(2s,3s,8s,9s)-3-amino-9-methoxy-2,6,8-trimethyl-phenyldeca-4,6-dienoic acid
ADEP	Acyldepsipeptide
ADP	Adenosine diphosphate
Aeg	Aminoethyl glycine
AFM	Atomic force microscopy
AFP	*Aequorea victoria* fluorescent proteins
AIBN	2,2′-Azobis(2-methylpropionitril)
AIC	Anthocyanin-inductive condition
Akt	AKT8 virus oncogene cellular homolog, AKT8 virus oncogene cellular homolog, v-akt thymoma viral oncogene homolog, protein kinase B, PKB

ALL	Acute lymphoid leuk emia
Alloc	Allyloxycarbonyl
AlphaScreen	Amplified luminescence proximity homogeneous assay
AML	Acute myeloid leukaemia
AMNB	5-Aminomethyl-2-nitrobenzyl
AMP	Adenosine monophosphate
AmR	Aminonitrile retinoid
APC	Adenomatous polyposis coli
APM-DNM	Adamantanepentyloxydeoxynojirimycin
APT	Acyl protein thioesterase
AR	Atypical retinoid
ARTD	ADP-ribosyltransferase
ASF	Anti-silencing function protein
ATase	Adenylyltransferase
AtCPY	Arabidopsis CPY
atg	Autophagy-related gene
ATP	Adenosine triphosphate
ATPase	Adenosine triphosphatase
ATRA	All-*trans* retinoic acid
Baf	Bafilomycin
BAmR	Boron-aminonitrile retinoid
BAPTA-AM	1,2-Bis(*O*-aminophenoxy)ethane-*N*,*N*,*N*',*N*'-tetraacetic acid tetra(acetoxymethyl) ester
Bcl	B-cell lymphoma
Bcr-Abl	Breakpoint cluster region – Abelson murine leukemia viral oncogene homolog
BD	Binding domain
BET	Bromo and extra terminal
Bhoc	Benzhydryloxycarbonyl
BHQ	Bromohydroxyquinoline
BID	Intra-peritoneal injection twice a day
BIOS	Biology oriented synthesis
BLAST	Basic Local Alignment Search Tool
BLI	Biolayer interferometry
BMMs	Bone morrow-derived macrophages
BO	Pyridinium benzothiazole
Boc	*N*-*tert*-butoxycarbonyl
BODIPY	Boron dipyrromethene difluoride
bp	Base pair
Bpa	L-4-Benzoylphenylalanine
BPS	Branch point sequence
BRD	Bromodomain containing proteins
BSA	Bovine serum albumin
BSO	Buthionine sulfoximine
BTF	β-turn focused sensors

BZD	Benzodiazepine
BzT	Benzotriazepine
C/M precipit.	Chloroform/methanol precipitation
Cal-A	Calyculin A
cAMP	Cyclic adenosine monophosphate
CaN	Calcineurin
CaNAR1	Calcineurin activity reporter 1
cat2	ß-catenin 2
CBD	ß-catenin binding domain
CBP	CREB-binding protein
CbZ	Carboxybenzyl
CCD	Charge-couple device
CD	Circular dichroism
Cdc	Cell division control protein
Cdc42	Cell division cycle 42
CDK	Cyclin dependant kinase
cDNA	Complementary DNA
CENP	Centromeric protein
CFP	Cyan fluorescent protein
CHEF	Chelation enhanced fluorescence
CID	Collision-induced dissociation
CK	Casein kinase
Clk	cdc2-like kinase
CLSM	Confocal laser scanning microscope
CLV3	CLAVATA3
CMA	Chaperone-mediated autophagy
CML	Chronic myelogenous leukemia
cMO	Caged MO
CNE	2-Cyanothyl
Col	wt *Arabidopsis* (Col-0)
Cos7 cells	Cells being CV-1 (simian) in Origin, and carrying the SV40 genetic material
COX	Cyclooxygenase
Cpm	Counts per minute
CPY	Carboxypeptidase Y
CRBN	Cereblon
CSA	Camphorsulfonic acid
CTAB	Cetyl trimethylammonium bromide
CTP	Cytidine triphosphate
CTPP	Carboxyl-terminal propeptide
Ctr	Control
ctrl	Control
Cy3	Cyanine dye 3
Cy5	Cyanine dye 5
Dans	Dansyl

DAPI	4,6-Diamindino-2-phenylindole
DBU	1,8-Diazabicyclo[5.4.0]undec-7-ene
DDB1	DNA binding protein 1
DFG motif	Asp-Phe-Gly motif
DIAD	Diisopropyldiazadicarboxylate
DIC	Differential interference contrast
DiFMUP	6,8-Difluoro-4-methylumbelliferyl phosphate
DIPEA	*N,N*-diisopropylethylamine
DLS	Dynamic light scattering
DMACA	*p*-Dimethylaminocinnamaldehyde
DMEM	Eagle's minimal essential medium
DMF	*N,N*-dimethylformamide
DMNB	Dimethoxynitrobenzyl
DMSO	Dimethyl sulfoxide
DNA	Deoxyribonucleic acid
DOS	Diversity oriented synthesis
DPBS	Dulbecco's phosphate buffered saline
DPD	Compound under investigation
dpf	Days post fertilization
DPPE	1,2-Dipalmitoyl-*sn*-glycero-3-phosphoethanolamine
DrrA	Defects in Rab1 recruitment protein A
DSF	Differential scanning fluorimetry
Dsh	Dishevelled
DTF	Downstream transcription factor
DTT	Dithiothreithol
DUSP	Dual specificity phosphatase
EC_{50}	Effective concentration 50
EDC	1-Ethyl-3-(3-dimethylaminopropyl)carbodiimide
EEDQ	Ethoxycarbonyl-ethoxy-dihydroquinoline
EGDE	Ethyleneglycoldiglycidylether
EGF	Epidermal growth factor
EGFP	Enhanced green fluorescent protein
EGFR	Epidermal growth factor receptor
EGTA	Ethylene glycol tetraacetic acid
EHEC	Enterohemorrhagic *Escherichia coli*
EJC	Exon junction complex
ELISA	Enzyme linked immunosorbent assay
EMS	Ethyl methanesulfonate
EPL	Expressed protein ligation
ER	Endoplasmic reticulum
Erk	Extracellular signal-regulated kinase
ESE	Exonic splicing enhancer
ESI	Electrospray ionization
ESI-MS	Electrospray mass spectrometry
ET	Extra terminal

Et_3N	Triethylamine
ETD	Electron transfer dissociation
EtOAc	Acetic acid ethylester
EtOH	Ethanol
etv2	ETS1-related protein
EYFP	Enhanced yellow fluorescent protein
FACS	Fluorescence-activated cell sorting
FAM	5-Carboxyfluorescein
FANA	2′-Fluoro-arabino nucleic acid
far	Farnesyl
FDA	Food and Drug Administration (USA)
fgf	Fibroblast growth factor
FIT	Forced intercalation (of thiazole orange)
FITC	Fluorescein isothiocyanate
FKB12	FK506-binding protein 12
FKBP	FK506-binding protein
FLAP	Fluorescence loss after photo-activation
flh	Floating head
FLiK	Fluorescence labels in kinases
FLIM	Fluorescence lifetime imaging microscopy
FliP	Fluorescent labels in phosphatases
Fmoc	Fluorenylmethoxycarbonyl
FOSL	FOS-like
FP	Fluorescence polarization (Chapter 16, Chapter 25)
FP	Fluorescent protein (Chapter 6)
FPD	Feature-pair distribution
FPP	Farnesylpyrophosphate
FRAP	Fluorescence recovery after photobleaching
FRET	Förster/fluorescence resonance energy transfer
GABARAP	γ-Aminobutyric acid type A receptor associated protein
GalT	Galactosyltransferase
GAP	GTPase activating protein
GATE-16	Golgi-associated ATPase enhancer of 16 kDa
GBA	Glucosidase, beta, acid
GDF	GDI displacement factor
GDI	GDP dissociation inhibitor
GdnHCl	Guanidine hydrochloride
GDP	Guanosine diphosphate
GEF	Guanine nucleotide exchange factor
ger	Geranyl
GFP	Green fluorescent protein
GGPP	Geranylgeranylpyrophosphate
GGTase	Geranylgeranyl transferase
GIST	Gastrointestinal stromal tumor
GLO1	Glyoxalase 1

GMA	Glycidylmethacrylate
GMP	Guanosine monophosphate
GppNHp	Guanosine-5′-*O*-[(β,γ)-imido]-triphosphate
G-protein	GTP binding protein
GR	Guanidine retinoid
GSH	Glutathione
GSK	Glucogen synthase kinase
GST	Glutathione S-transferase
GTP	Guanosine triphosphate
GTPase	Guanosine triphosphatase
HA	Human influenza hemagglutinin
HAD	Haloacid dehalogenase
HATU	*O*-(7-Azabenzotriazole-1-yl)-1,1,3,3-tetramethyluronium hexafluorophosphate
HCA	High content analysis
HCD	High-energy collision dissociation
HCS	High content screen
HCTU	2-(6-Chloro-1-*H*-benzotriazole-1-yl)-1,1,3,3-tetramethylaminium hexafluorophosphate
HD	Hintington disease
HDAC	Histone deacetylase
heg	Heart of glass
HeLa cells	Cells from Henrietta Lacks
HEPES	2-[4-(2-Hydroxyethyl)piperazin-1-yl]ethanesulfonic acid
HMG-CoA	3-Hydroxy-3-methylglutaryl-coenzyme A
hnRNP	Heterologous nuclear ribonuclear particles
HOBT	Hydroxybenzotriazole
hpf	Hours post fertilization
HPLC	High performance liquid chromatography
HRP	Horseradish peroxidase
HRV	Human rhinovirus
Hsp	Heat shock protein
HSQC	Heteronuclear single quantum coherence
HTS	High throughput screen/screening
HVR	Hypervariable region
I2	Inhibitor-2
IC_{50}	Inhibitor concentration 50
iFLiK	Interface-FliK
IGF	Insulin-like growth factor
IM	Isolation membrane
IMPACT	Intein mediated purification with an affinity chitin binding tag
IMPACT-TWIN	Intein mediated purification with an affinity chitin binding tag-two intein

IP	Immunoprecipitation
IP_3R	Inositol-1,4,5-trisphosphate receptor
IR	Insulin receptor
IRRA	Infrared reflection absorption
ISE	Intronic splicing enhancer
ITC	Isothermal titration calorimetry
iTRAQ	Isobaric tags for quantitation
JAK2	Janus kinase 2
JNK	c-jun N-terminal kinase
lacZ gene	Gene coding ß-galactosidase
LAMP	Lysosome-associated membrane protein type
LC	Liquid chromatography
LC/MS	Liquid chromatography coupled to mass spectrometry
LC3	Microtubule-associated protein 1 (MAP1) light chain 3
LC-MS/MS	Liquid chromatography coupled with tandem mass spectrometry
LDL	Low-density lipoprotein
LDLR	Low-density lipoprotein receptor
LEF	Lymphoid enhancer factor
Ler	wt Arabidopsis Landsberg erecta ecotype
LNA	Locked nucleic acid
LRP	Low-density lipoprotein-related receptor
Luc	Luciferase
MALDI-TOF	Matrix assisted laser desorption ionization – time of flight
mant	2′,3′-*N*-methylanthraniloyl
MAP	Mitogen activated protein
MAP1	Microtubule-associated protein 1
MAPK	Mitogen activated protein kinase
MATE	Multidrug and toxic compound extrusion
MBP	Maltose-binding protein
mCFP	Monomeric CFP
mCh	mCherry
mCherry	Membrane localized cherry red fluorescent protein
mCit	mCitrine
mCitrine	Monomeric yellow fluorescent protein
m-CPBA	*meta*-Chloroperbenzoic acid
M-CSF	Macrophage colony-stimulating factor
MDCK cells	Madine-darby canine kidney cells
MeCN	Acetonitrile
MeI	Methyl iodide
MEK	MAPK/Erk kinase
MeNHPh	*N*-methyl-phenylamine
MeOH	Methanol
MESNA	2-Mercaptoethanesulfonate
MFP	Membrane fusion protein

MG	Methylglyoxal
M-GFN	Methyl-gerferin
MHCI	Major histocompatibility class I
MIC	Minimal inhibitory concentration
MICAL3	Microtubule associated monooxygenase, calponin, and LIM domain containing 3
MK2	(MAPK)-activated *protein* kinase 2
MLL	Mixed linage leukemia
MO	Morpholino
MOA	Mechanism of action
MPAA	(4-Carboxylmethyl)thiophenol
Mps1	Monopolar spindle protein 1
mRFP	Monomeric red fluorescent protein
mRNA	Messenger RNA
MRSA	Methicillin-resistant *Staphylococcus aureus*
MS	Mass spectrometry
MS/MS	Tandem mass spectrometry
mTOR	Mammalian target of rapamycin
MTT	Methylthiazol tetrazolium
Myc	Myelocytomatosis oncogene cellular homolog
NA	Neuraminidase
NAD	Nicotinamide adenine dinucleotide
NBD	Nitrobenzoxadiazole
NCL	Native chemical ligation
NEt_3	Triethylamine24
NF-κB	Nuclear factor kappa B
NHS	*N*-hydroxysuccinimide
NIH	National Institutes of Health
NIPP1	Nuclear Inhibitor of PP1
Nle	Norleucine
NMC	NUT midline carcinoma
NMD	Nonsense-mediated mRNA decay
NMM	*N*-methyl morpholine
NMO	*N*-methylmorpholine-*N*-oxide
NMP	*N*-methylpyrrolidin-2-one
NMR	Nuclear magnetic resonance
NPDepo	Natural products depository
NPM	Nucleophosmin
NPOM	6-Nitropiperonyloxymethyl
NRSB	Neutral red staining bodies
nt	Nucleotide
NTC	No template control
ntla	No tail a
NTPP	Amino-terminal propeptide
NUT	Nuclear protein in testis

OA	Ocadaic acid
OAc	Acetate
OD	Optical density
OMFP	3-*O*-Methylfluorescein phosphate0
ONB	*o*-Nitrobenzyl
OTf	Trifluoromethanesulfonate, triflate
p.i.	Post infection
PA	Photoactivatable
PAGE	Polyacrylamide gel electrophoresis
PalFar	Palmitoylated and farnesylated
PARP	Poly(ADP-ribose) polymerase
PARsylation	Poly-ADP-ribosylation
PAT	Protein acyl transferase
PCA	Principal component analysis
PCR	Polymerase chain reaction
PDAC	Pancreatic ductal adenocarcinoma
PDB	Protein data base
PDEδ	Phosphodiesterase δ
PDP	PP1-disrupting peptide
PE	Phosphatidylethanolamine
PEG	Polyethylene glycol
pErk	Phosphorylated ERK
PhMe	Toluene
PHRAG	Parmacophoric fragment
pI	Isoelectric point
PI(3,4,5)P3	Phospoinositide-3,4,5-triphosphate
PI3K	Phosphoinositide-3-kinase
PI3P	Phosphatidyilinositol-3-phosphate
PKA	Protein kinase A
PKB	Protein kinase B
PKC	Protein kinase C
Plk	Polo-like kinase
P-loop	Glycine-rich loop
PNA	Peptide nucleic acid
pNPP	*para*-Nitrophenol phosphate
PNT	N-terminal pointed domain
PP	Protein phosphatase
PPI	Protein-protein interaction
PP_i	Pyrophosphate
PPM	Metal-dependent protein phosphatase
PPP	Phosphoprotein phosphatase
PPT	Polypyrimidine tract
PPTS	Pyridinium *para*-toluene sulfonate
Pra1	Prenylated Rab acceptor 1
PS DNA	Phosphorothioate DNA2

PSCC	Protein structure similarity clustering
PSTK	Protein Ser/Thr kinase
PSTP	Protein Ser/Thr phosphatase
P-TEF	Positive transcription elongation factor
ptfl	Pancreas transcription factor
PTM	Post-translational modification
PTP	Protein tyrosine phosphatase
PTSA	*p*-Toluenesulfonic acid
PVDF	Polyvinylidene difluoride
PyBOP	Benzotriazole-1-yl-oxy-tris-pyrrolidino-phosphonium hexafluorophosphate
PYL	PYR like
PyMPO	1-(2-Maleimidylethyl)-4-(5-(4-methoxyphenyl)oxazol-2-yl)pyridinium methanesulfonate
PYR	Pyrabactin resistance
Q	Quadrupole mutant
QD	Intra-peritoneal injection once a day
qPCR	Quantitative PCR
qRT-PCR	Quantitative RT-PCR
r.t.	Room temperature
Rab	Ras related in brain
Rac	Ras-related C3 botulinum toxin substrate
RANKL	Receptor activator of NF-κB ligand
RAR	Retinoic acid receptor
Ras	Rat sarcoma
RCAR	Regulatory component of ABA receptors
RDF	Recognition-domain focused sensors
REP	Rab escort protein
rheb	Ras homolog enriched in brain
Rho	Ras homologous
Rhod	Rhodamine B
RIKEN	Rikagaku Kenkyusho (Institute of Physical and Chemical Research, Japan)
RMS	Root mean square
RNA	Ribonucleic acid
RNAi	RNA interference
RNAPII	RNA polymerase II
RNase	Ribonuclease
RNA-seq	RNA sequencing
ROS	Reactive oxygen species
RT-PCR	Real time PCR
S1P	Sphingosine-1-phosphate
salph	*N,N-o*-bis(3,5-di-*tert*-butylsalicylidene) – 1,2-phenylenediamine
SAM	Simplest active molecule

SAR	Structure activity relationship
SAS	Simplest active subgraph
SDS	Sodium docecylsulfate
SF	Splicing factor
SFV	Semliki forest virus
SHED	Shannon entropy descriptor
shRNA	Small hairpin RNA
SILAC	Stable isotope labeling by amino acids in cell culture
siRNA	Small interfering RNA
SLO	Streptolysine O
SMER	Small molecule enhancers of rapamycin
snRNP	Small ribonuclear particles
Sortin	SORTing inhibitor
sox	Sry-box transcription factor
SPPS	Solid phase peptide synthesis
SPR	Surface plasmon resonance
SR	Serine rich
Src	Rous sarcoma oncogene cellular homolog
SRSF	Serine/arginine splicing factor
SSA	Spliceostatin A
ssDNA	Single stranded DNA
StAR protein	Steroidogenic acute regulatory protein
START protein	StAR-related lipid transfer protein
STAT3	Signal transducer and activator of transcription 3
StAx	Axin-derived stapled peptide
STF	Super-topflash
SylA	Syringolin A
TAMRA	Tetramethyl-6-carboxyrhodamine
TBTA	Tris-(benzyltriazolylmethyl)amine
TBTU	*O*-Benzotriazole-1-yl-1,1,3,3-tetramethyluronium tetrafluoroborate
*t*Bu	*tert*-Butyl
TCEP	Tris(2-carboxyethyl)phosphine
TCF	T-cell factor
TCPTP	T-cell PTP
TEF	Transcription elongation factor
TEV	Tobacco etch virus
TFA	Trifluoro acetic acid
TGF	Transforming growth factor
TGN	Trans-Golgi network
THF	Tetrahydrofurane
tmRNA	Transfer-messenger RNA
TNF	Tumor necrosis factor
TNKS	Tankyrase
TNT	Trinitrotoluene

TO	Thiazole orange
TOP2	Topoisomerase II
TOS	Target oriented synthesis
TosOMe	*para*-Toluenesulfonic acid methyl ester
TRAP+	Tartrate-resistant acid phosphatase-positive
TrCP	Transducing repeat-containing protein
Trt	Trityl
TSS	Transcriptional start site
UAS	Upstream activating sequence
UPS	Ubiquitin proteasome system
UTP	Uridine triphosphate
UV	Ultra violet
V-ATPase	Vacuolar ATPase
VCP	Valosin-containing protein
VEGF	Vascular endothelial growth factor
VopS	Vibrio outer protein S
Vps34	Vacuolar protein sorting 34
vRNP	Viral ribonucleoprotein
VSR	Vacuolar sorting receptor
Wnt	Wingless and INT
WPD-loop	Trp-Pro-Asp loop
wt	Wild type
XN	Xanthohumol
Y2H	Yeast-2-hybrid
YFP	Yellow fluorescent protein
Yip3	YPT-interacting protein 3
YPT	Yeast protein transport
δTIP	Delta-tonoplast intrinsic protein

1 Real-Time and Continuous Sensors of Protein Kinase Activity Utilizing Chelation-Enhanced Fluorescence

Laura B. Peterson and Barbara Imperiali

1.1 Introduction

Protein kinases, the enzymes responsible for phosphoryl transfer from a chemical donor such as adenosine triphosphate (ATP) to a peptide or a protein acceptor, are integral enzymes in signaling cascades, play crucial roles in numerous cellular processes, and are of fundamental importance in systems biology. In addition, aberrant kinase activities are commonly associated with disease states, making kinases important therapeutic targets in current drug development initiatives. Therefore, understanding kinase activation dynamics is of utmost biological and clinical importance. Accurate and physiologically relevant methods to quantify kinase activities are needed to understand the intricate dynamics of kinase activation and inactivation. This chapter describes the design, evolution, and application of fluorescent-based Ser/Thr/Tyr kinase activity sensors that take advantage of chelation-enhanced fluorescence (CHEF). These sensors are compatible with physiological conditions, are selective for specific protein kinases, and provide real-time kinetic information regarding kinase activity.

1.2 The Biological Problem

Phosphorylation, or the attachment of a phosphate group to amino acid side chains, is one of the most abundant posttranslational modifications (PTMs) of proteins. Phosphorylation reactions are mediated by phosphotransferase enzymes, termed *kinases*, with ATP as the typical source of the transferred phosphoryl group. Ser, Thr, and Tyr are the most commonly phosphorylated residues in eukaryotes, while His and Asp phosphorylation has also been observed, predominantly in prokaryotes. Protein activity, localization, and structure as well as protein–protein interactions are all affected by protein phosphorylation [1, 2]. As kinases play integral roles in cellular signaling, dysregulated kinase function has emerged as a driver for many different disease states, including

Concepts and Case Studies in Chemical Biology, First Edition. Edited by Herbert Waldmann and Petra Janning.

cancer, neurodegenerative diseases, and metabolic disorders [3, 4]. Accordingly, much effort has been put forth toward understanding kinase structure, function, and activity as well as toward the clinical development of kinase inhibitors for the treatment of human disease.

Of considerable value to the scientific community are methods to study kinase activity, providing a means to evaluate kinase activity dynamics, inhibitor activities, and roles in cell signaling. Traditional assays for monitoring kinase activity utilize antibodies specific for the phosphorylated (activated) kinase, which is a common proxy for kinase activation, or rely on radioactivity-based measurements by monitoring the transfer of the radioactive γ-phosphoryl group from [γ-^{32}P]ATP to a substrate protein or peptide. Although the use of phosphopeptide/protein-specific antibodies is widely accepted as a useful detection method of kinase activity, antibodies may not take into account other factors affecting kinase activity, including kinase or substrate localization or additional PTMs that may also modulate activity. Radioactivity-based assays are limited in throughput, are inherently noncontinuous, and radioactive reagents require special handling. Mass spectrometry-based methods have also been developed and rely on the detection of phosphopeptides after enzymatic degradation. Fluorescence-based approaches represent valuable alternative methods for monitoring kinase activity.

Many strategies using fluorescence have been employed to detect kinase activity. Generally, these kinase sensors manifest increased fluorescence emission upon phosphorylation, while both dual fluorophore (fluorescence resonance energy transfer, FRET, Box 1.1) sensors and single fluorophore-containing sensors have been developed.

Box 1.1 Förster Resonance Energy Transfer (FRET)

FRET is the process by which one fluorophore, "the donor," transfers energy to a second fluorophore, "the acceptor." When both chromophores are fluorescent, FRET occurs. In the case of FRET between fluorophores, the emission spectrum of the donor fluorophore must overlap with the absorption spectrum of the acceptor. In this case, the emission from the donor excites the acceptor causing it to emit light (fluoresce). The efficiency of FRET depends on the distance between the two fluorophores, the spectral overlap, and the relative orientations between the donor emission dipole and the acceptor absorption dipole.

FRET-based sensors rely on conformational changes that often accompany phosphorylation, which alter the FRET efficiency. Many FRET sensors are plagued by small changes in fluorescence upon phosphorylation, rely on the bulky *Aequorea victoria* fluorescent proteins (AFP), are not compatible with high-throughput methods, and/or require genetic manipulation to incorporate the sensor into the system of choice [5]. Therefore, peptide-based fluorescent sensors provide an alternative approach to kinase sensing. Ideal kinase activity sensors should manifest high fluorescence changes upon phosphorylation, should provide a quantitative measure of catalytic activity, and be amenable to the establishment of continuous assays, ideally in a high-throughput format. They should also be selective for the kinase of interest, readily prepared, and the design should be generalizable to the diverse families of kinases that comprise the kinome. In addition, the sensors should be operationally compatible with endogenous concentrations of the ATP cosubstrate. This chapter describes the design, development, and application of fluorescent sensors for kinase activity that are based on the principle of CHEF using 8-hydroxyquinoline fluorophores.

1.3 The Chemical Approach

Kinase substrate peptides represent ideal platforms for sensor design. Peptides are readily prepared by solid-phase peptide synthesis (SPPS), can be chemically modified with fluorescent probes or other small molecules, and retain recognition elements contained within endogenous kinase substrates. Commonly, kinases recognize a consensus sequence of four or five amino acids flanking the phosphorylated residue. Each kinase recognizes a unique consensus sequence and uses this molecular interaction as one level of substrate selectivity. Several methods for consensus sequence determination exist and have allowed for the generation of kinase-specific substrate peptides. Further modifications of substrate peptides with fluorophores have inspired new methods for kinase activity sensing that offer many advantages over conventional methods.

Fluorophores capable of CHEF and other types of environment-sensitive (or solvatochromic) fluorophores have been utilized in sensors for kinase activity. Chelation-sensitive fluorophores manifest altered fluorescent properties upon chelation of various metal ions, while environment-sensitive fluorophores exhibit altered excitation and emission properties with changing environment, such as solvent polarity (Figure 1.1).

One of the first reported examples of a fluorescence-based kinase activity sensor exploits the native fluorescence of Trp and a change in the local environment of the indole fluorophore upon phosphorylation. A peptide-substrate-based sensor for cyclic adenosine monophosphate (cAMP)-dependent kinase, containing a Trp-Ser motif, was prepared and manifested a 20% increase in fluorescence upon phosphorylation (**1**, Figure 1.2a) [6]. Additional examples include peptide- and protein-based sensors with appended environment-sensitive fluorophores, although these sensors are generally plagued by small signal changes resulting in low sensitivity [7–9].

1.3.1 Chelation-Enhanced Fluorescence

CHEF was originally exploited for the detection and quantification of metal ions, such as those developed for the detection and quantification of Ca^{2+} [10]. Given

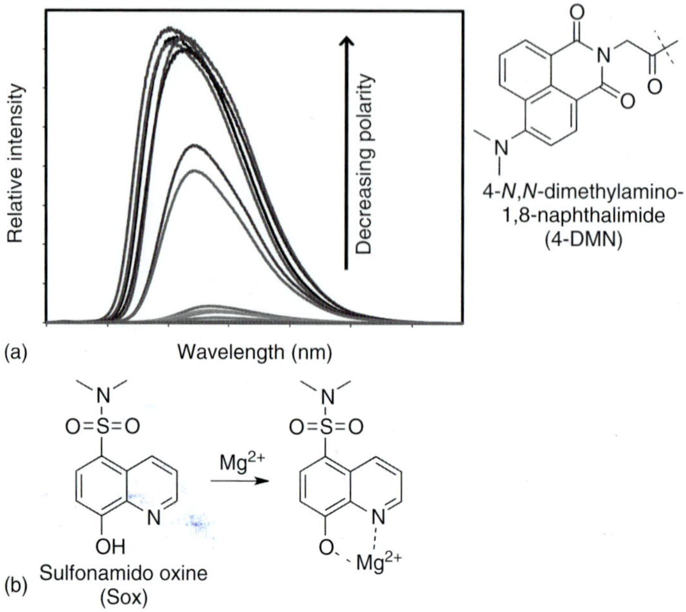

Figure 1.1 *Environment-sensitive fluorescence.* (a) Fluorescence emission spectrum of the environment-sensitive fluorophore, 4-DMN, in solvents of various polarities and (b) structure of the Sox fluorophore demonstrating chelation-enhanced fluorescence.

the ability of alkyl and aryl monophosphate esters to also chelate metal ions, it was soon after realized that chelation-sensitive fluorophores could be incorporated proximal to the phosphorylatable Ser/Thr/Tyr residue into peptide substrates to provide robust fluorescence readout upon phosphorylation.

One early example, modeled on the aforementioned Ca^{2+} sensors, included a carboxylate-containing fluorophore proximal to a Ser residue in a substrate peptide for protein kinase C (PKC) (**2**, Figure 1.2b). Upon phosphorylation, the Ser-phosphate and fluorophore carboxyl groups chelate Ca^{2+} resulting in a twofold enhancement of fluorescence [11]. This sensor relied on chemical modification of the peptide substrate following peptide synthesis. An advantageous alternative is the use of modified synthetic amino acid building blocks, wherein the fluorescent reporting moiety can be directly incorporated during peptide synthesis.

A class of zinc ion (Zn^{2+}) sensors, which utilized an unnatural amino acid that included the chelation-sensitive fluorophore, 8-hydroxy-4-(*N*,*N*-dimethylsulfonamido)-2-methylquinoline (sulfonamido oxine (Sox), Figure 1.3a), provided inspiration for a second class of kinase activity sensors [12–14]. The Sox amino acid was prepared via asymmetric synthesis and converted to the fluorenylmethyloxycarbonyl (Fmoc)-protected derivative and incorporated via SPPS into a peptide containing a proline-mediated β-turn sequence [13]. The β-turn was flanked by both Zn-chelating amino acids and the Sox fluorophore (Figure 1.3b). In this case, the β-turn was included to provide preorganization

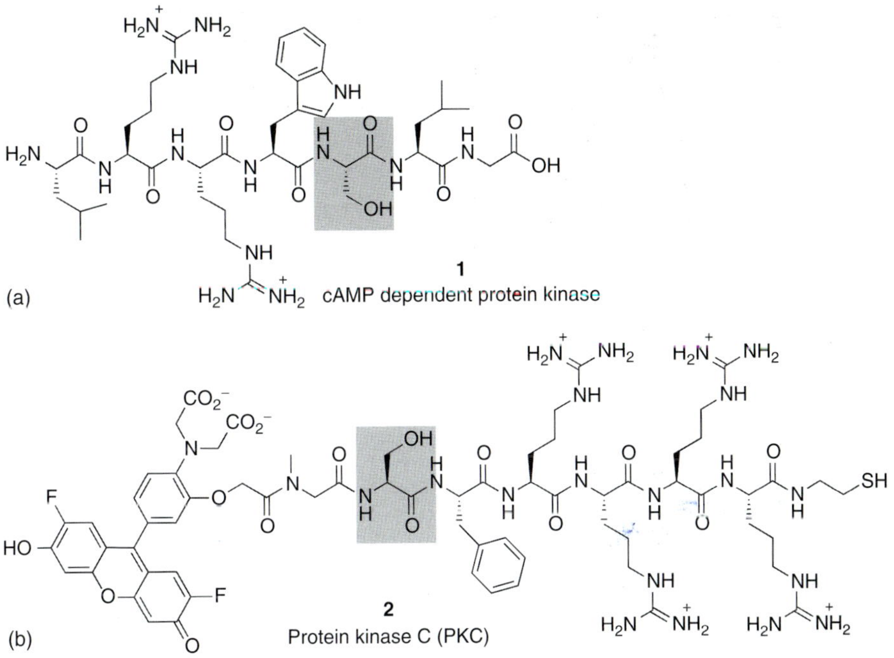

Figure 1.2 *Structures of (a) tryptophan-based and (b) fluorescein-based kinase activity sensors.* Phosphorylated residue (Ser) highlighted for clarity.

for the Zn^{2+} chelation event. In the presence of Zn^{2+}, the flanking residue (for example, His, Cys, Glu, or Asp) and the Sox fluorophore chelate Zn^{2+}, while metal ion binding to the Sox moiety results in an increased fluorescence signal due to CHEF (Figure 1.3c,d, Box 1.2).

Box 1.2 Chelation-Enhanced Fluorescence of 8-hydroxyquinolines

The 8-hydroxyquinoline chromophore manifests weak fluorescence in the absence of metal ions in aqueous solution. However, in the presence of various metal ions (e.g., Zn^{2+} and Mg^{2+}), 8-hydroxyquinoline becomes strongly fluorescent through CHEF. One plausible mechanism to describe the observed CHEF of 8 hydroxyquinolines involves the change in the lowest energy excited state. The lowest energy transition of unbound 8-hydroxyquinoline is the n to π^* transition, rapid intersystem crossing prevents this transition from producing fluorescence. However, upon metal chelation, the lowest energy transition becomes the π to π^* transition, which does not undergo intersystem crossing and is fluorescent [15]. A second theory to describe CHEF involves a photoinduced proton transfer from the phenol to the quinoline ring nitrogen, which upon excitation quenches fluorescence. Metal chelation promotes deprotonation of the phenol, preventing fluorescence quenching upon excitation [16].

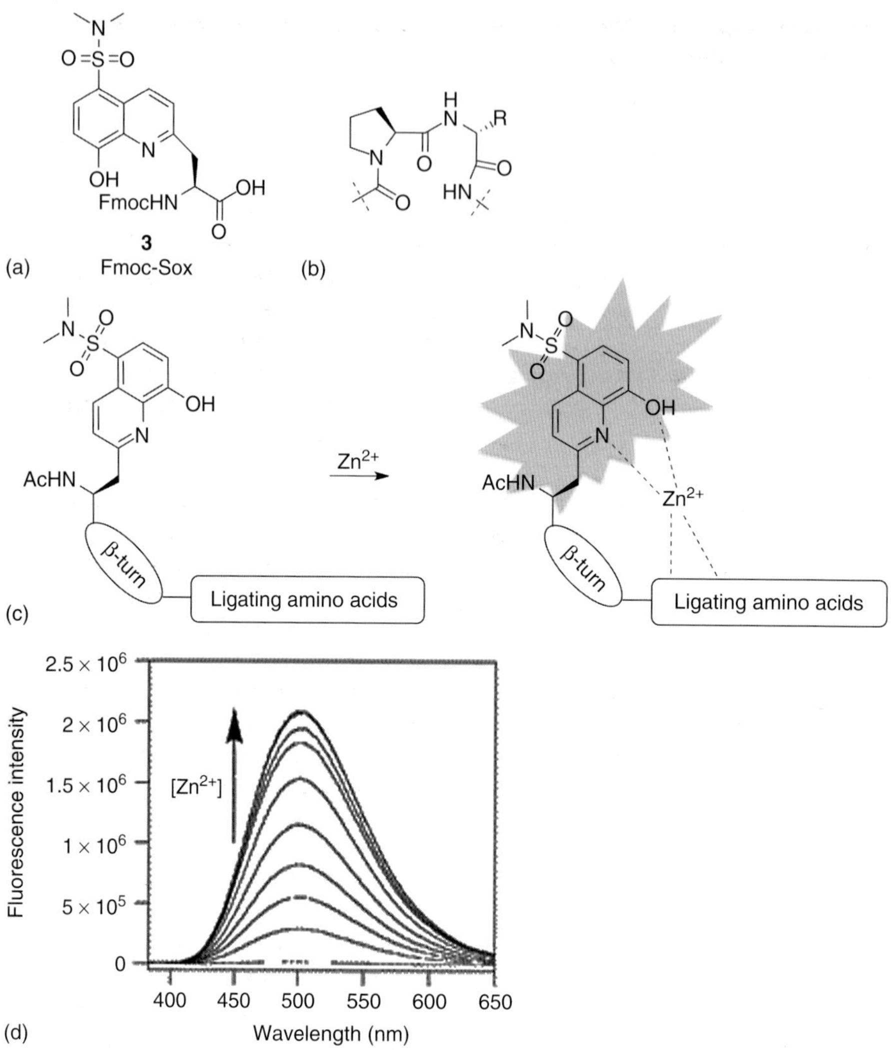

Figure 1.3 *CHEF-based sensors for divalent zinc.* (a) Structure of Fmoc-Sox used in SPPS synthesis of Zn^{2+} sensors; (b) depiction of the β-turn motif; (c) schematic of a peptidic CHEF-based Zn^{2+}; and (d) fluorescence emission spectrum of a peptide-based Zn^{2+} sensor with increasing concentrations of Zn^{2+} ($\lambda_{ex} = 360$ nm). (Reprinted with permission from Ref. [13]. Copyright 2003 American Chemical Society.)

Altering the flanking amino acids provides a means to fine-tune the binding affinity for Zn^{2+}, and thus the capacity to detect the divalent ion at different target concentrations.

Although this strategy was originally employed for the detection of Zn^{2+}, it was quickly realized that there would potentially be greater impact in the application of the quinolone fluorophore and CHEF for detecting kinase activity, as the phosphate transferred to Ser/Thr/Tyr could serve as the flanking group capable of metal chelation (Figure 1.4a).

1.3.2 β-Turn-Focused Kinase Activity Sensors

The first generation of Sox-containing kinase activity sensors, the β-turn-focused sensors (BTF), utilized a β-turn motif flanked by the Sox amino acid and either an N- or C-terminal kinase recognition motif, which includes the phosphorylatable residue (Figure 1.4b) [17–19]. The Sox amino acid is an ideal fluorophore in this context as it is relatively small in size, which prevents perturbation of native kinase–substrate interactions. In addition, the Sox fluorophore is relatively stable and resistant to photobleaching. Finally, Sox undergoes CHEF upon Mg^{2+} chelation, resulting in a robust increase in fluorescence ($\lambda_{ex} = 360$ nm; $\lambda_{em} = 485$ nm).

In this kinase sensor design, the role of the β-turn, made up of two amino acids, namely, XaaPro or ProXaa, is to preorganize the incipient Mg^{2+}-binding site comprised of Sox and the transferred phosphoryl group. The kinase recognition motif, typically based on an optimum peptide substrate or substrate consensus sequence, can be placed at either the C- or N-terminus of the peptide, relative to the β-turn/Sox motif. This modular design allows one to empirically determine the contribution of either the N- or C-terminal recognition elements and establish which may contribute optimally to kinase selectivity and/or enzyme turnover.

An essential feature of Sox-containing sensors is the differential binding affinity for Mg^{2+} between the substrate and product (phosphorylated) peptides. A 10- to 25-fold enhancement of binding affinity for Mg^{2+} (as measured by dissociation constant, K_D) is observed upon phosphorylation. Therefore, substrate peptides manifest low background fluorescence, while phosphorylation results in robust fluorescence increases (three- to eightfold) in the presence of Mg^{2+}. In the presence of Mg^{2+} and ATP, these sensors accurately report kinase activity, while providing kinetic detail. The kinetic parameters (K_M and V_{max}) of the BTF sensors are in agreement with the corresponding non-Sox-containing substrate peptides as determined by ^{32}P incorporation from radiolabeled ATP and scintillation counting [17]. Further application of these sensors is discussed in Section 1.4. Although the BTF sensors provided a reliable method to quantify kinase activity, second-generation Sox sensors addressed one shortcoming of the first-generation design and provided a means to incorporate both N- and C-terminal kinase recognition elements.

1.3.3 Recognition-Domain-Focused Kinase Activity Sensors

Given the multitude of kinases encoded in the human genome, substrate selectivity and specificity is of paramount importance when designing kinase activity sensors for application in complex unfractionated samples. For this reason, one

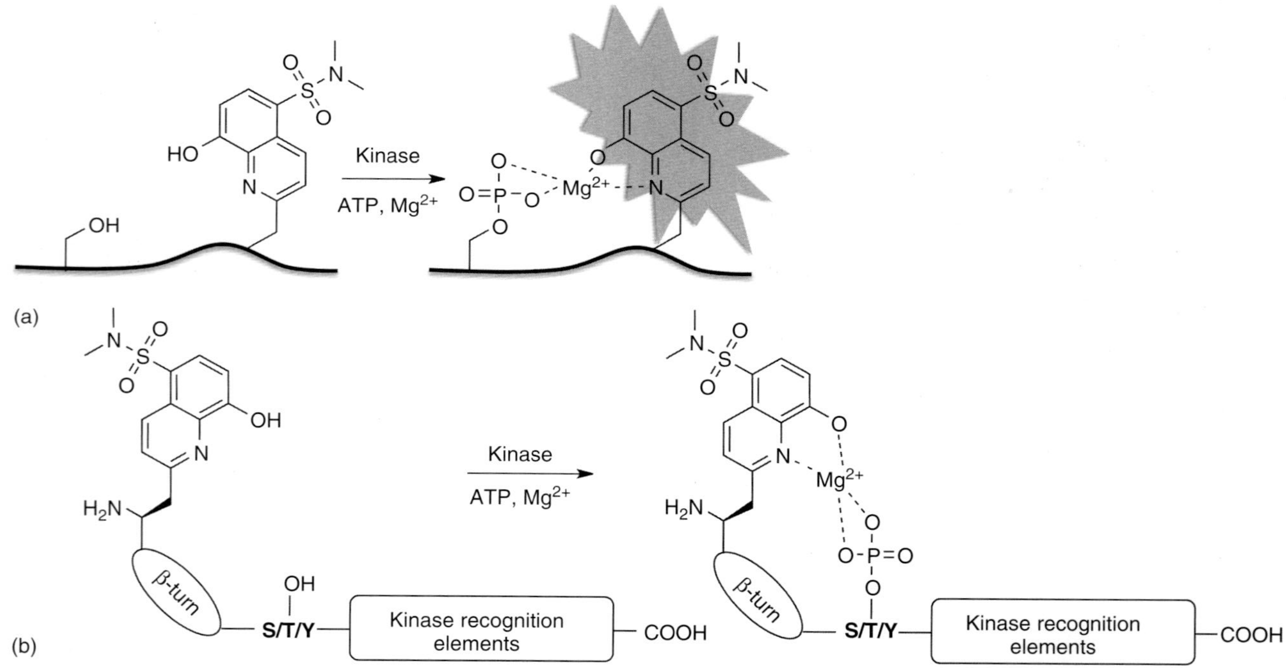

Figure 1.4 *Sox-containing kinase activity sensors.* (a) Schematic of Sox-based peptide sensors of kinase activity and (b) β-turn-focused kinase sensors including C-terminal recognition elements.

major disadvantage of the BTF sensors was the loss of either the N- or C-terminal substrate recognition determinants. Accordingly, strategies to mitigate this limitation were investigated, leading to the development of a recognition-domain focused (RDF) sensor design that utilized the more flexible Sox-containing unnatural amino acid, cysteine-Sox (C-Sox, Figure 1.5). Given the increased flexibility, the preorganizing β-turn motif became unnecessary.

C-Sox was able to coordinate the phosphate-bound Mg^{2+} without the predisposed β-turn-mediated conformational bias. Synthesis of both Fmoc-protected C-Sox (**4**) and Sox-Br (**5**) allowed for facile incorporation into peptides via either SPPS or cysteine-selective alkylation, respectively, (Figure 1.5) [20]. This increasingly versatile approach allowed recognition elements on both sides of the phosphorylatable residue to be included in the peptidic sensor, generally resulting in superior kinase specificity and selectivity.

The optimal C-Sox location was empirically determined to be at the +2/−2 location, relative to the phosphorylatable Ser/Thr/Tyr, in most cases. One notable exception being for the mitogen-activated protein kinases (MAPKs), which recognize either SerPro or ThrPro as the minimal consensus sequence. MAPK activity sensors included C-Sox juxtaposed to Pro at the −3 position. Using the RDF approach, sensors were prepared for a variety of different kinases (Table 1.1) that generally exhibited good fluorescence increases upon phosphorylation (3- to 10-fold) and showed significant improvements in kinetic parameters as compared to the first-generation BTF sensors.

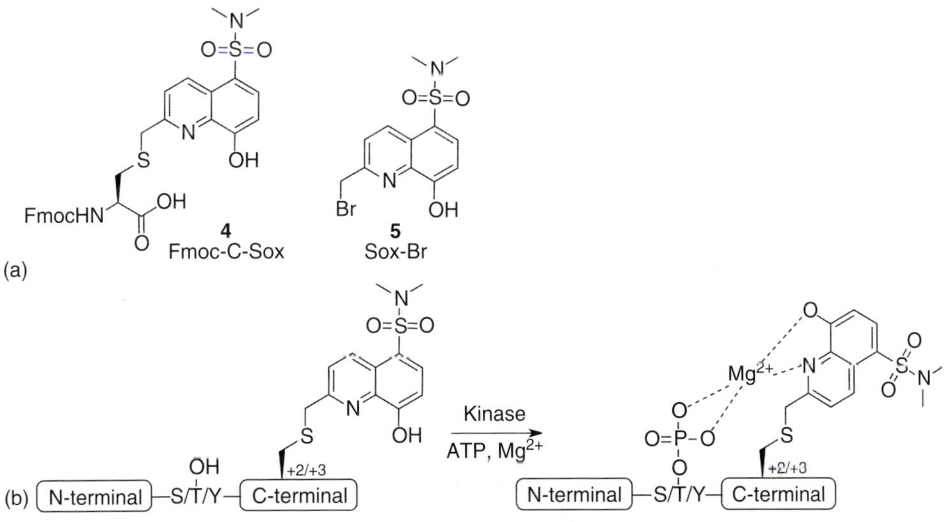

Figure 1.5 *Recognition-domain-focused kinase activity sensors.* (a) Structures of Fmoc-C-Sox and Sox-Br used in sensor synthesis and (b) RDF sensors with the C-Sox moiety placed in the +2/+3 position relative to the phosphorylated residue.

Table 1.1 C-Sox-containing kinase activity sensors.

Kinase	Substrate sequence	Fluorescence increase	K_M (μM)	Vmax (μmol mg^{-1} min^{-1})
PKC	Ac-RRR-CSox-GS*FRRR-$CONH_2$	3.5	0.1	1.8
Akt1	Ac-ARKRERAYS*F-CSox-HHA-$CONH_2$	3.9	0.69	2.5
MK2	Ac-AHLQRQLS*I-CSox-HH-$CONH_2$	4.4	1.2	1.3
Src	Ac-AEE-CSox-IY*GEFEAKKKK-$CONH_2$	2.2	7	3.4
Pim2	Ac-ARKRRRHPS*G-CSox-PTA-$CONH_2$	3.2	1.4	0.67
PKA	Ac-ALRRAS*L-CSox-AA-$CONH_2$	5	2.6	17.9
Abl	Ac-E-CSox-IY*AAPFAKKK-$CONH_2$	5.2	10.5	19.1
IRK	Ac-R-CSox-DY*-Nle-TMQIGKK-$CONH_2$	4.2	25.9	8.7

Fluorescent properties and kinetic parameters for RDF-based kinase activity sensors. Asterisk (*) denotes the phosphorylated residue, while underlined residues are those important for kinase recognition, Nle = norleucine.
Adapted with permission from Ref. [20]. Copyright (2008) American Chemical Society.

1.3.4 Chimeric Kinase Activity Sensors

Although RDF-based sensors provided improvement over BTF sensors, some kinase targets were particularly challenging targets for sensor design, in particular, those with short or ubiquitous consensus sequences. Although many kinases have linear consensus sequences comprising 8–10 amino acids, the MAPKs require only the presence of SerPro or ThrPro at the phosphorylation site. The MAPKs are involved in many important signaling pathways, making the development of activity sensors for individual MAPKs, such as epithelial growth factor-related kinase (ERK), c-jun N-terminal kinase (JNK), and p38 an important endeavor. In nature, MAPKs achieve high target specificity by interactions with a secondary recognition element proximal to the kinase active site where ATP and substrate peptides bind (Figure 1.6b).

This "docking" site is typified by an acidic cleft adjacent to a hydrophobic pocket. Kinases that are upstream in signaling pathways and substrate proteins include a complementary basic-hydrophobic motif that docks into this groove. This docking interaction provides a second layer of specificity for the MAPKs. Accordingly, incorporation of a docking motif into an MAPK activity sensor was proposed to enhance the selectivity for a given MAPK (Figure 1.6a).

In pursuit of an ERK activity sensor, a chimeric sensor was envisioned taking advantage of this secondary docking interaction [21]. The chimeric sensor comprised a docking motif (the N-terminal pointed domain (PNT) from an ERK substrate) [22] fused to a C-Sox-containing ERK substrate peptide (Figure 1.6c). The two parts were independently prepared, the substrate peptide via SPPS and the PNT domain via homologous expression, and joined via native chemical ligation (Figure 1.6c). The resulting ERK sensor displayed vastly improved kinetic

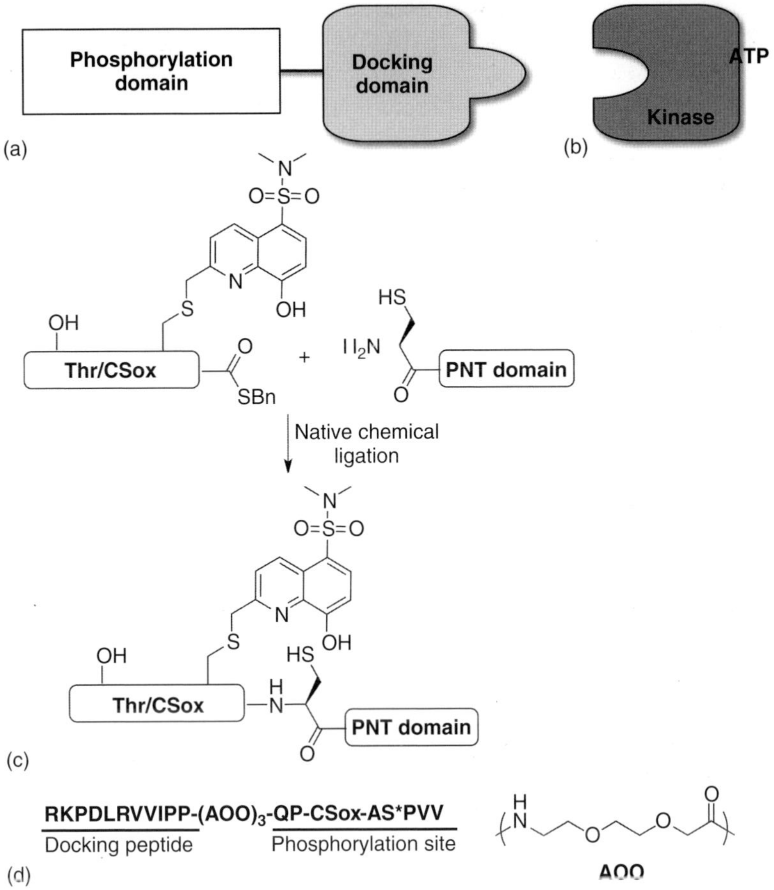

Figure 1.6 *Chimeric kinase activity sensors.* (a) Schematic of a chimeric sensor; (b) schematic of kinase illustrating docking groove relative to ATP binding site; (c) preparation of ERK activity sensor; and (d) sequence of p38 activity sensor.

parameters and selectivity as compared to the phosphorylation motif alone. This example demonstrates the utility and adaptability of C-Sox-containing kinase activity sensors. A similar approach facilitated the design of a p38α chemosensor [23]. One advantage of the p38α sensor is that the docking motif used was only 11 amino acids; its incorporation could be achieved through SPPS. The docking motif and C-Sox phosphorylation motif were connected via a flexible polyethyleneglycol (PEG) linker, which was also installed via SPPS using commercially available Fmoc-$(PEG)_X$-CO_2H units. The development of C-Sox-containing sensors illustrates how the CHEF principle can be combined with unnatural amino acids and exploited to generate valuable protein kinase sensors. These sensors accurately and robustly report kinase activity in recombinant systems, in cell lysates, and even in tissue homogenates. The next section addresses the evaluation and application of these sensors.

1.4
Chemical Biological Research/Evaluation

Sox-based kinase probes report activity in continuous assays, are compatible with cell lysates and tissue homogenates, and can be used for small-molecule inhibitor screening. Upon design and synthesis, kinetic parameters for each Sox-peptide are determined and kinase selectivity is addressed. Subsequently, sensors are evaluated in cell-lysate-based systems in the presence and absence of selective inhibitors to establish selectivity in complex target samples.

1.4.1
Kinetic Parameters

It is straightforward to determine the kinetic parameters, namely, K_M and V_{max}, for each kinase sensor as fluorescence emission (F_{em}) is monitored over time in a continuous format (Figure 1.7a). Reaction volumes are relatively small, while fluorescence can be monitored in microcuvettes or in multiwell plates (96-, 384-, or even 1536-well format). Initial slopes taken directly from the F_{em} versus time plots can be used to determine Michaelis constants. In order to determine V_{max}, one must have a way to convert F_{em} to units of product formation, which can be achieved by chemically synthesizing the corresponding product (phosphorylated) peptide assuming that total F_{em} is the sum of the product and substrate intensities. In general, both BTF and RDF sensors manifest kinetic parameters very similar to those determined by other means, such as [γ-^{32}P]ATP-based assays.

1.4.2
Assessing Kinase Selectivity

In order to determine the selectivity of each peptide sensor substrate, assays with a panel of recombinantly expressed, activated kinases can be performed (Figure 1.7b). In this case, the concentration of substrate peptide is held constant at two to three times the determined K_M. For example, the p38 sensor depicted in Figure 1.6 was incubated with various MAP (mitogen-activated protein) and non-MAP kinases and fluorescence emission was monitored. Figure 1.7b demonstrates the selectivity of the p38 sensor for the target kinase [23]. Following this initial screen, kinase selectivity can be addressed directly in cell lysates. Cells can be stimulated to activate the kinase of interest and kinase activity can be determined in the presence and absence of a selective inhibitor of the kinase of interest. Residual activity in the presence of inhibitor would indicate sensor cross talk with other kinases.

Alternatively, following stimulation, the kinase of interest can be immunodepleted using an antibody specific for the desired kinase (Figure 1.7c). In this case, residual kinase activity in the depleted lysate can be attributed to sensor cross talk. The chimeric ERK sensor was found to be selective for ERK in this manner.

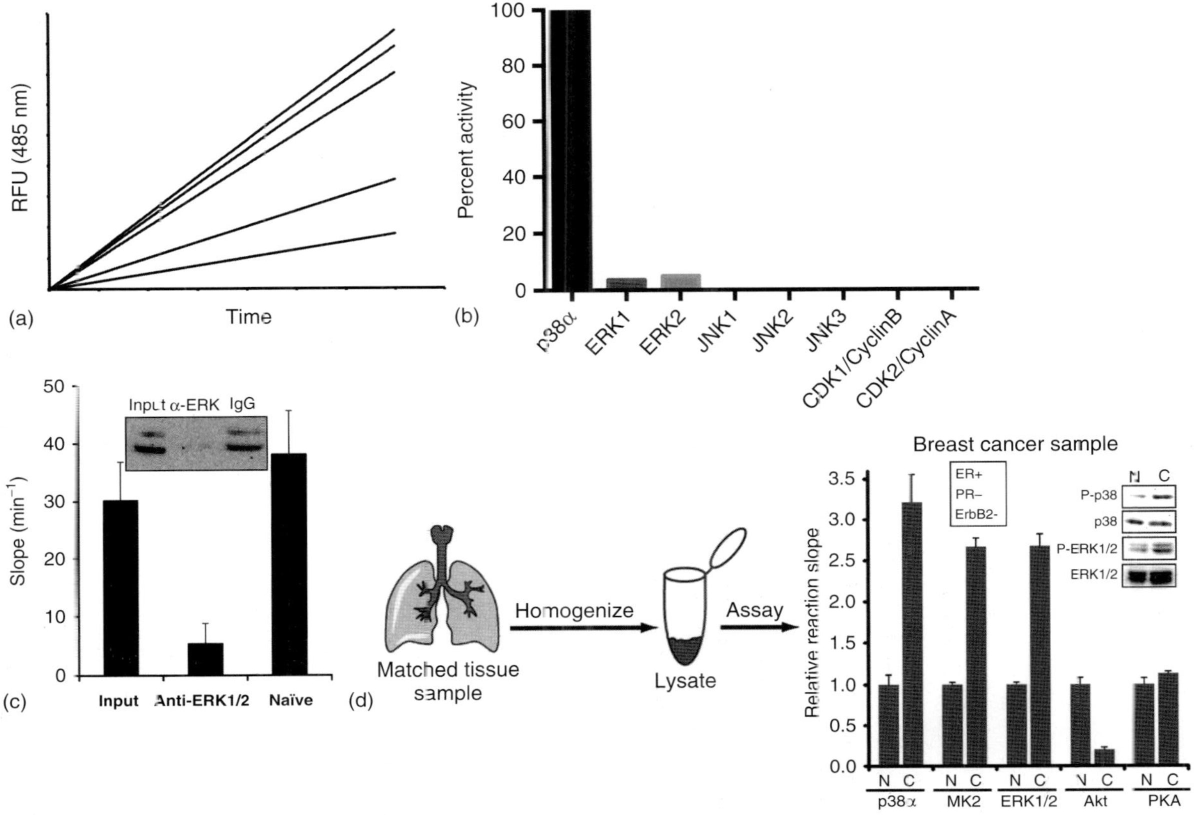

Figure 1.7 *Characterization and application of Sox-containing kinase activity sensors.* (a) Example data for a typical kinetic experiment; (b) kinase selectivity profile for p38 sensor shown in Figure 1.6d; and (c) ERK sensor activity in ERK-stimulated lysates following immunodepletion. (Reprinted with permission from Ref. [21]. Copyright (2009) Wiley-VCH Verlag GmbH & Co. KGaA, Weinheim.) (d) Diagram of kinase profiling in human samples and kinase activity profile comparing a normal "N" and breast cancer "C" sample. (Reprinted with permission from Ref. [24]. Copyright (2012) Cell Press.)

ERK activity was stimulated with EGF (epithelial growth factor) in HeLa cells. The cells were lysed and either incubated with no antibody, an antibody for ERK, or a control-naïve antibody. Following immunodepletion, ERK activity was assessed. As evident in Figure 1.7c, most of the ERK activity is absent following depletion of ERK, providing evidence that the sensor is selective for ERK. In the event that a kinase activity sensor is not completely selective, adding specific kinase inhibitors to the assay buffer can minimize activity resulting from "off-target" kinases. Once the selectivity of a kinase sensor has been determined, these sensors can be used to monitor kinase activity in cell lysates and/or tissue homogenates under a variety of different contexts.

1.4.3
Kinase Profiling in Cell Lysates and Tissue Homogenates

Sox-containing kinase activity sensors provide a means to directly quantify enzymatic activity in unfractionated cell lysates and tissue homogenates. Assays with lysates can be performed in multiwell plates, allowing one to monitor the activity of multiple kinases simultaneously. In one example, a panel of five activity sensors (MK2 ((MAPK)-activated protein kinase 2), p38α, ERK, Akt, PKA (protein kinase A)) was used to monitor kinase activation dynamics in a model of skeletal muscle differentiation [24]. For Sox-based assays, the amount of lysate required is comparable to the alternative method of Western blotting (10–40 μg total protein/replicate in a 96-well format). In another example, the same kinase sensors were used to quantitatively determine kinase activity in human cancer tissue samples and were compared to matched healthy tissue controls (Figure 1.7d). Experiments were validated by comparing results to traditional Western blot analysis. These types of experiments highlight the utility of this method of detecting kinase activity and provide the means to profile kinase activities under a variety of conditions.

1.5
Conclusions

This chapter details the development and application of kinase activity probes that utilize CHEF manifested by 8-hydroxyquinoline derivatives. These sensors provide significant advantages over traditional kinase-sensing protocols. Namely, Sox-based sensors provide a quantitative readout of kinase activity in a sensitive and continuous format. The approach is generalizable and has been applied to Ser, Thr, and Tyr kinases representing many diverse families of kinases. Owing to the importance of kinases in different diseases, the need for additional probes of this type is clear, as these probes provide valuable insight into kinase (in)activation dynamics as well as kinase inhibitor activities. The potential for these sensors in a systems biology platform is significant, as many kinases can be profiled in high throughput. Subsequent generations of sensors should address other difficult-to-target kinases, should expand the application of these probes to live-cell imaging, and should provide a means to multiplex the assay with modified Sox fluorophores.

References

1. Endicott, J.A., Noble, M.E.M., and Johnson, L.N. (2012) The structural basis for control of eukaryotic protein kinases. *Annu. Rev. Biochem.*, **81**, 587–613.
2. Manning, G., Whyte, D.B., Martinez, R., Hunter, T., and Sudarsanam, S. (2002) The protein kinase complement of the human genome. *Science*, **298**, 1912–1934.
3. Lahiry, P., Torkamani, A., Schork, N.J., and Hegele, R.A. (2010) Kinase mutations in human disease: interpreting genotype-phenotype relationships. *Nat. Rev. Genet.*, **11**, 60–74.
4. Blume-Jensen, P. and Hunter, T. (2001) Oncogenic kinase signalling. *Nature*, **411**, 355–365.
5. Rothman, D.M., Shults, M.D., and Imperiali, B. (2005) Chemical approaches for investigating phosphorylation in signal transduction networks. *Trends Cell Biol.*, **15**, 502–510.
6. Wright, D.E., Noiman, E.S., Chock, P.B., and Chau, V. (1981) Fluorometric assay for adenosine 3′,5′-cyclic monophosphate-dependent protein kinase and phosphoprotein phosphatase activities. *Proc. Natl. Acad. Sci. U.S.A.*, **78**, 6048–6050.
7. Bowman, B.F., Peterson, J.A., and Stull, J.T. (1992) Pre-steady-state kinetics of the activation of rabbit skeletal muscle myosin light chain kinase by Ca2+/calmodulin. *J. Biol. Chem.*, **267**, 5346–5354.
8. McIlroy, B.K., Walters, J.D., and Johnson, J.D. (1991) A continuous fluorescence assay for protein kinase C. *Anal. Biochem.*, **195**, 148–152.
9. Post, P.L., Trybus, K.M., and Taylor, D.L. (1994) A genetically engineered, protein-based optical biosensor of myosin II regulatory light chain phosphorylation. *J. Biol. Chem.*, **269**, 12880–12887.
10. Minta, A., Kao, J.P., and Tsien, R.Y. (1989) Fluorescent indicators for cytosolic calcium based on rhodamine and fluorescein chromophores. *J. Biol. Chem.*, **264**, 8171–8178.
11. Chen, C.-A., Yeh, R.-H., and Lawrence, D.S. (2002) Design and synthesis of a fluorescent reporter of protein kinase activity. *J. Am. Chem. Soc.*, **124**, 3840–3841.
12. Pearce, D.A., Jotterand, N., Carrico, I.S., and Imperiali, B. (2001) Derivatives of 8-hydroxy-2-methylquinoline are powerful prototypes for zinc sensors in biological systems. *J. Am. Chem. Soc.*, **123**, 5160–5161.
13. Shults, M.D., Pearce, D.A., and Imperiali, B. (2003) Modular and tunable chemosensor scaffold for divalent zinc. *J. Am. Chem. Soc.*, **125**, 10591–10597.
14. Walkup, G.K. and Imperiali, B. (1998) Stereoselective synthesis of fluorescent α-amino acids containing oxine (8-hydroxyquinoline) and their peptide incorporation in chemosensors for divalent zinc. *J. Org. Chem.*, **63**, 6727–6731.
15. Seitz, W.R. (1980) *CRC Critical Reviews in Analytical Chemistry: Fluorescence Derivatization*, Vol. 8, CRC Press, Boca Raton, FL.
16. Bronson, R.T., Montalti, M., Prodi, L., Zaccheroni, N., Lamb, R.D., Dalley, N.K., Izatt, R.M., Bradshaw, J.S., and Savage, P.B. (2004) Origins of "on-off" fluorescent behavior of 8-hydroxyquinoline containing chemosensors. *Tetrahedron*, **60**, 11139–11144.
17. Shults, M.D., Carrico-Moniz, D., and Imperiali, B. (2006) Optimal Sox-based fluorescent chemosensor design for serine/threonine protein kinases. *Anal. Biochem.*, **352**, 198–207.
18. Shults, M.D. and Imperiali, B. (2003) Versatile fluorescence probes of protein

kinase activity. *J. Am. Chem. Soc.*, **125**, 14248–14249.

19. Shults, M.D., Janes, K.A., Lauffenburger, D.A., and Imperiali, B. (2005) A multiplexed homogeneous fluorescence-based assay for protein kinase activity in cell lysates. *Nat. Methods*, **2**, 277–284.
20. Luković, E., González-Vera, J.A., and Imperiali, B. (2008) Recognition-domain focused chemosensors: versatile and efficient reporters of protein kinase activity. *J. Am. Chem. Soc.*, **130**, 12821–12827.
21. Luković, E., Vogel Taylor, E., and Imperiali, B. (2009) Monitoring protein kinases in cellular media with highly selective chimeric reporters. *Angew. Chem. Int. Ed.*, **121**, 6960–6963.
22. Rainey, M.A., Callaway, K., Barnes, R., Wilson, B., and Dalby, K.N. (2005) Proximity-induced catalysis by the protein kinase ERK2. *J. Am. Chem. Soc.*, **127**, 10494–10495.
23. Stains, C.I., Luković, E., and Imperiali, B. (2010) A p38α-Selective chemosensor for use in unfractionated cell lysates. *ACS Chem. Biol.*, **6**, 101–105.
24. Stains, C.I., Tedford, N.C., Walkup, T.C., Luković, E., Goguen, B.N., Griffith, L.G., Lauffenburger, D.A., and Imperiali, B. (2012) Interrogating signaling nodes involved in cellular transformations using kinase activity probes. *Chem. Biol.*, **19**, 210–217.

2 FLiK and FLiP: Direct Binding Assays for the Identification of Stabilizers of Inactive Kinase and Phosphatase Conformations

Daniel Rauh and Jeffrey R. Simard

2.1 Introduction – The Biological Problem

The catalytic transfer of the terminal phosphate of adenosine triphosphate (ATP) onto protein substrates is one of the most important post-translational modifications and is crucial for the regulation of many signaling pathways [1]. Kinases have complex functions in the regulation of every cellular process differentiation and proliferation. Kinase-catalyzed phosphorylation often activates or inactivates by changing the protein charge or conformation which, in turn, can influence the dynamics of multi-protein complexes or alter their subcellular localization. Intracellular kinase activity is counteracted by phosphatases, which dephosphorylate corresponding substrate proteins. Disregulation in the shuffling of these phosphate groups often manifested by genetic lesions such as mutations, amplifications, or deletions which are causative elements in many diseases, including cancer, diabetes mellitus, and Alzheimer's disease (Figure 2.1a), making kinases and phosphatases attractive targets for medicinal chemistry and chemical biology research [2].

2.1.1 Kinase Inhibitors – Stabilizing Inactive Enzyme Conformations

A major roadblock in protein kinase inhibitor research and development is the challenge of poor selectivity and the likelihood of unwanted off-target inhibition, which are largely a consequence of the highly conserved ATP binding site shared by all protein kinases. A way to work around this is to move away from classic ATP-competitive inhibitors and target alternative sites that, for example, become accessible when the kinase adopts an inactive conformation (Figure 2.1b) [3]. Additionally, it is becoming more evident that examining drug-target residence times will provide a more complete context for fully understanding kinase inhibitor selectivity *in vivo*. Ideally, inhibitors should have high rates of association (k_{on}) and slow rates of dissociation (k_{off}) to maximize residence time with the target enzyme [4]. In the case of kinases, an inhibitor which appears to be

Concepts and Case Studies in Chemical Biology, First Edition. Edited by Herbert Waldmann and Petra Janning.

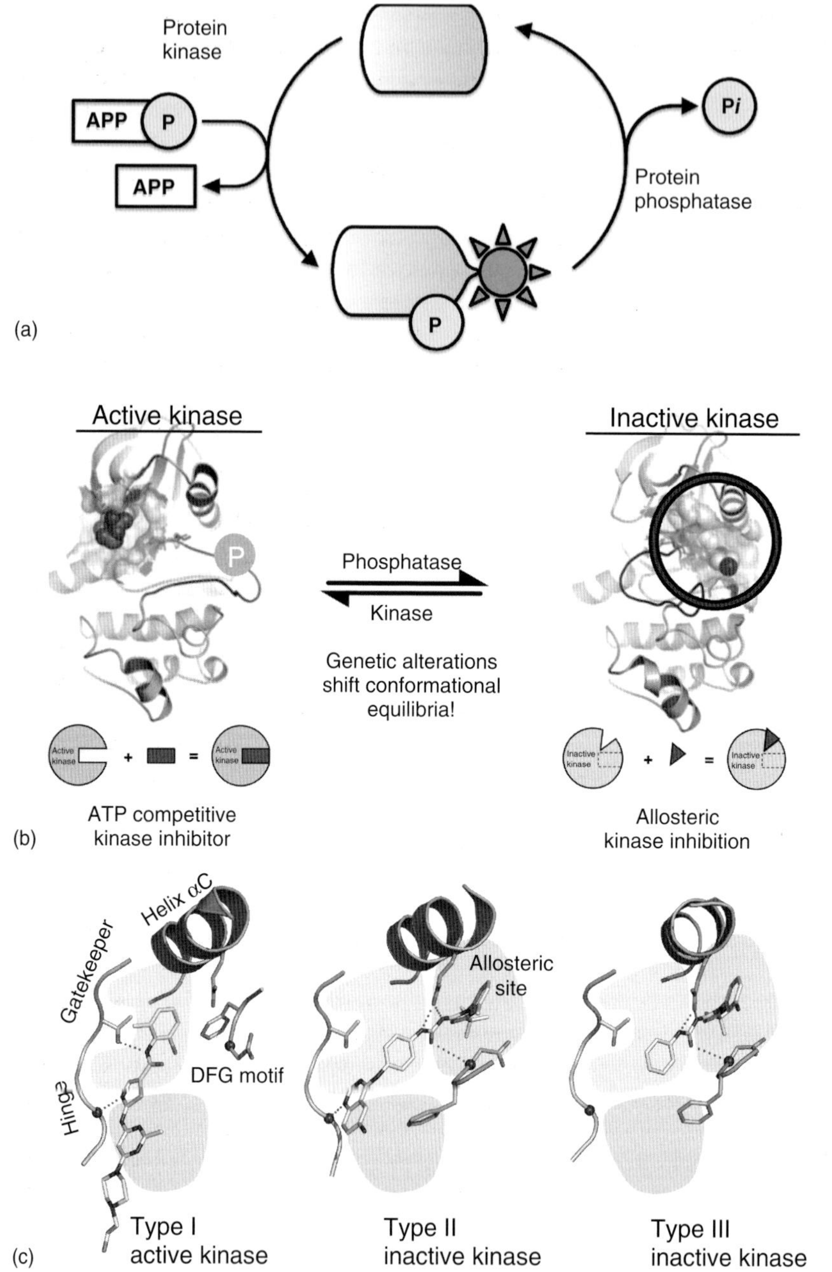

Figure 2.1 *Protein kinase regulation and inhibition*. Transfer of the γ-phosphate from ATP alters the activation state of substrate proteins (a). Protein kinases exist in active and inactive conformations. This dynamic interplay can be altered by mutations and is causative for many diseases such as cancer (b). Various types of kinase inhibitors (c). Type II and type III kinase inhibitors stabilize enzymatically inactive conformations.

relatively unselective *in vitro* can be rendered more selective if it dissociates most slowly from the kinase of interest. Thus, medicinal chemistry efforts should not judge selectivity based solely on inhibitor affinities (K_d) and potencies (inhibitor concentration 50, IC_{50}) obtained from *in vitro* profiling of compounds against the entire kinome. Although these types of studies undoubtedly contain valuable information and provide a solid groundwork for further inhibitor development, it is possible for a wide range of k_{on} and k_{off} values to result in the same overall affinity ($K_d = k_{off}/k_{on}$). Thus, in addition to affinity, lead optimization strategies should consider the kinetic components contributing to affinity. By identifying aspects of ligand structure which prolong k_{off} relative to k_{on}, medicinal chemistry efforts can facilitate the design of molecules with improved residence times for the desired kinase. This approach would likely minimize the unwanted consequences of high affinity off-target binding *in vivo*.

Emerging data suggest that the issue of kinase inhibitor selectivity can be addressed by moving away from classic ATP-competitive (Type I) inhibitors and targeting the DFG-out pocket with type II and type III inhibitors (Figure 2.1c). The DFG-out pocket is adjacent to the ATP binding site and is frequently referred to as an *allosteric pocket* or *kinase-switch pocket*. In comparison to the ATP binding site, the allosteric pocket is a more restrictive binding cavity and is only accessible upon a change in conformation. Therefore, this pocket tends to be less accessible to small molecules. However, the amino acids lining this pocket are much less conserved across the kinome, providing opportunities for additional H-bonding and hydrophobic interactions between the kinase and ligands which can bind within this pocket [5]. Once bound, these additional interactions tend to reduce the k_{off} of type II and type III inhibitors relative to k_{on}. Thus, a logical methodology for improving kinase inhibitor selectivity by prolonging drug-target residence times should focus on the identification and kinetic optimization of ligands which can bind preferentially to the DFG-out conformation.

The availability of the DFG-out pocket requires the kinase activation loop to adopt a catalytically deficient conformation in which the ATP binding site becomes partially occluded by the Phe side chain of the DFG motif. While the DFG-out conformation is more favorable in the unphosphorylated kinase, phosphorylation of the activation loop shifts conformational equilibria to the more active DFG-in conformation, increases kinase activity, and often reduces the affinity of type II and type III inhibitors [1]. Although the search for chemical scaffolds which have affinity for the DFG-out pocket is moving to the forefront of kinase inhibitor research, efforts have been constrained by the lack of high-throughput assay technologies which can identify and discriminate for ligands which bind to and stabilize enzymatically inactive kinase conformation.

2.1.2
Monitoring Conformational Changes upon Ligand Binding

We have developed FLiK as a widely-applicable assay system for both identifying and characterizing DFG-out binding ligands [6, 7]. Kinases are site-specifically labeled with an environmentally-sensitive fluorophore which reports on conformational changes induced by the binding of specific types of ligands. Changes in kinase conformation alter the charged microenvironment and solvation of the fluorophore, resulting in distinct and quantifiable changes in its emission spectrum which, in turn, provide a straightforward binding assay methodology for determining the K_d of the ligand. The assay also allows for follow-up characterization of identified compounds by permitting the determination of k_{on} and k_{off} to better understand the kinetic factors which contribute to the measured affinity. A key advantage of this approach is that enzyme activity is not required. Additionally, FLiK allows the use of the unphosphorylated inactive kinase, which enhances the sensitivity to detect ligands which bind preferentially to the DFG-out conformation [8]. This is in contrast to measuring IC_{50} values using traditional activity-based assays which rely on the use of phosphorylated active kinase. The FLiK approach has been used to successfully monitor conformational changes in the activation loop of both Ser/Thr and Tyr kinases associated with the slow binding of DFG-out inhibitors [6–9]. To date, we have also applied the FLiK approach to other kinase structural elements, including labeling of the P-loop (glycine-rich loop) to identify more selective type I ligands which engage the flexible P-loop in certain kinases [10]. Additionally, we have recently reported labeling strategies aimed at remote binding sites outside of the ATP binding cleft as a method for identifying more selective allosteric (Type IV) ligands, including assays for the mitogen activated protein (MAP) insert pocket of p38α, the myristate pocket of Abl kinase and the allosteric pocket of full length Akt [11, 12]. Likewise, we have successfully employed this approach to aid the discovery of novel phosphatase inhibitors which bind more favorably to inactive enzyme conformations [13]. Thus, similar to FLiK, the FLiP (Fluorescent Labels in Phosphatases) assay serves as conformation-specific binding assay for phosphatases.

In this chapter, we focus on the development and application of the FLiK and FLiP approaches, championed by our laboratories in Dortmund. Our group has employed these assays to address a number of kinases and phosphatases from different organisms and to facilitate the development of novel inhibitors and functional probes. We then utilize such probes and tool compounds to decipher phosphorylation and dephosphorylation events within complex disease states and to Foster drug development. Rather than detailing the findings of the different studies, this chapter will focus on the structure-based design of FLiK and FLiP protein constructs to enable these assays and their application in screening initiatives.

2.2
The Chemical Approach

The FLiK approach requires the removal of solvent exposed cysteines and the insertion of a cysteine into a desired position in the kinase of interest, which will serve as the attachment point for a thiol-reactive fluorophore. The kinase mutant is then expressed, purified, labeled, and characterized using standard biochemical and biophysical methods. Although this may seem somewhat straightforward, it is necessary to have a strategy in place when designing the protein construct. Identifying optimal fluorophore labeling positions is critical to enabling a high-throughput assay that reports specific ligand-induced conformational changes. Small fluorophores such as Acrylodan are commonly employed for generating fluorescent protein conjugates which report on conformational changes. Ideally, fluorophores should be highly sensitive to polarity and/or the charged microenvironment that is characteristic of nearby amino acid side chains in the protein. It is also advantageous to choose fluorophores which are thiol reactive. In contrast to amine reactive probes, labeling by thiol-reactive fluorophores is typically complete and more specific due to the lower abundance of free thiols in proteins. However, to ensure the site specific labeling for the FLiK assays, naturally occurring Cys residues which are solvent-exposed should be conservatively mutated and replaced by Ser or any amino acids which tends to be conserved at this position when compared to homologous proteins. For the FLiK assay, it is critical to insert a Cys residue into an amino acid position on the kinase that exhibits significant movement upon ligand binding and is somewhat solvent-exposed to enable the covalent attachment of the added fluorophore (Figure 2.2). To date, we have reported on several labeling strategies which we have successfully used to develop FLiK assays for various kinases [6–12, 14]. These assays enable rapid and specific detection of inhibitors and ligands with unique binding modes. Although the specific labeling site may vary for each kinase or binding site, a general approach can be applied in the design of kinase constructs compatible with FLiK. For any structural feature which is known to exhibit flexibility or undergo conformational changes upon ligand binding, a kinase construct which is compatible with the FLiK assay can be designed employing algorithms and software available in the public domain such as (i) basic local alignment search tool (BLAST) as a method of finding other kinases with the highest percent sequence identity to the kinase of interest (ii) PyMol and COOT to analyze protein crystal structures of the kinase of interest, if available. If no published structures are available, several online modeling tools such as (iii) Swiss Modeler or (iv) ESyPred3D can be used to generate 3D structural models based on available structural templates in the (v) Protein Data Bank (*http://www.rcsb.org*). The kinases identified in the BLAST search may serve as convenient starting points. Clustal W is useful when performing amino acid sequence alignments of the kinase of interest with a number of other highly homologous kinases. This method may help identify positions in the sequence which are tolerant of mutation and compatible with the FLiK approach. Sequence alignments may help identify highly conserved regions, common phosphorylation sites, or positions known to be involved in key structural interactions. Additionally, sequence alignments may reveal certain positions which have a naturally-occurring Cys. Such positions may tolerate

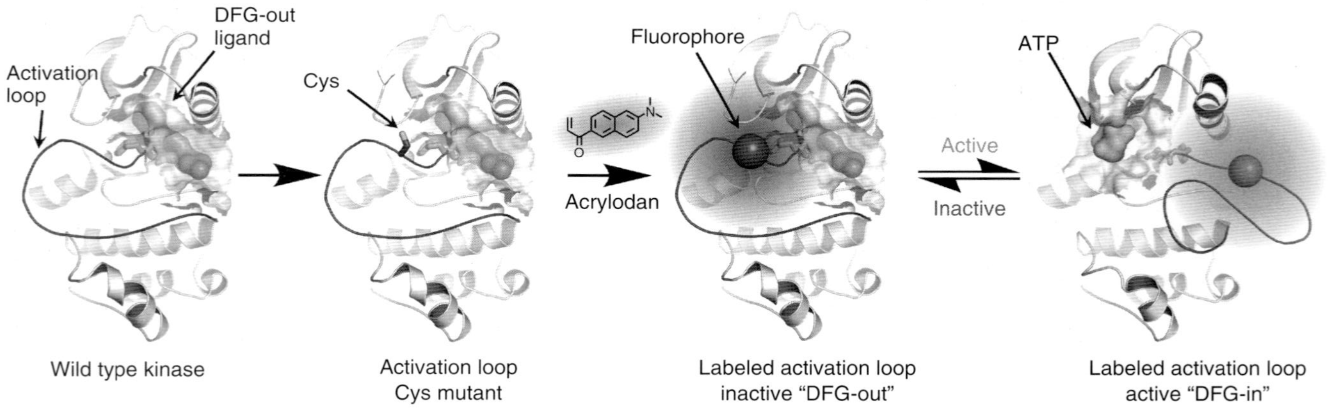

Figure 2.2 *FLiK constructs*. Structurally flexible are labeled with a thiol reactive fluorophore. Conformational changes triggered by ligand binding alter the emission properties of the fluorophores. Activation loop labeled protein kinases for the detection and discrimination of type II and type III inhibitors. (Adapted with permission from Macmillan Publishers Ltd: [6], copyright 2009.)

mutations in which a Cys is introduced for specific labeling with the desired fluorophore.

2.3 Chemical Biological Research/Evaluation

In a first set of experiments, we employed FLiK for the identification and development of kinase inhibitors to overcome the emerging problem of acquired drug resistance in targeted cancer therapies [1]. To achieve this, we introduced a point mutation into the activation loop of the kinase domain of cSrc to allow labeling with the thiol-reactive Acrylodan [6]. Acrylodan is a fluorophore sensitive to polarity changes in its environment. Conformational changes associated with the binding of allosteric inhibitors alter the environment of the fluorophore, thereby modifying its emission characteristics (Figure 2.3a). In a screening campaign, we identified pyrazoloureas as type III allosteric binders to cSrc with weak micromolar affinity. This chemical scaffold was not known to bind to cSrc previously. Crystal structures of the hit molecules in complex with cSrc and a drug resistant mutant variant confirmed the allosteric binding mode and stabilization of an enzymatically inactive conformation (Figure 2.3b). Furthermore, structural studies shed some light on the preference of cSrc for an N′-aryl moiety in the vicinity of the gatekeeper amino acid (Box 2.1) located in the hinge region of the kinase domain and inspired the design and synthesis of novel type II inhibitors. Fusion of the pyrazolourea scaffold (Type III, allosteric) with a quinazoline core (Type I, ATP-competitive) resulted in potent type II inhibitors with low nanomolar affinity that locked the kinase in the inactive DFG-out conformation. More importantly, these inhibitors were also active on drug resistant gatekeeper mutant forms of cSrc (cSrc-T338M), Abl (Abl-T315I), and KIT (KIT-T670I) [15, 16] (Figure 2.3c). KIT and Abl are essential targets in gastrointestinal stromal tumors (GISTs) and chronic myelogenous leukemia (CML). Drug resistance is a major problem in the long term treatment with targeted cancer therapies. Treatment of CML patients with imatinib often leads on average to a higher incidence of drug resistance mutations in the kinase domain of Bcr-Abl and intolerance to imatinib in 30% of examined patient populations in the first 5 years. Even more dramatically, in the case of patients suffering from solid GISTs, 14% of patients initially do not respond to imatinib treatment and 50% will develop resistance mutations in the kinase domain of the stem cell growth factor receptor KIT within the first 2 years.

In this first proof of concept study, labeling the activation of clinically relevant tyrosine kinases with fluorophore allowed for the sensitive and reliable identification of DFG-out stabilizers from compound libraries. The identified hits and subsequent compound optimization illustrates a generic alternative rationale to overcome drug resistance by generating type II inhibitors that have the intrinsic ability to adapt to the binding site distortions induced by these mutations while also locking the kinase in an inactive conformation.

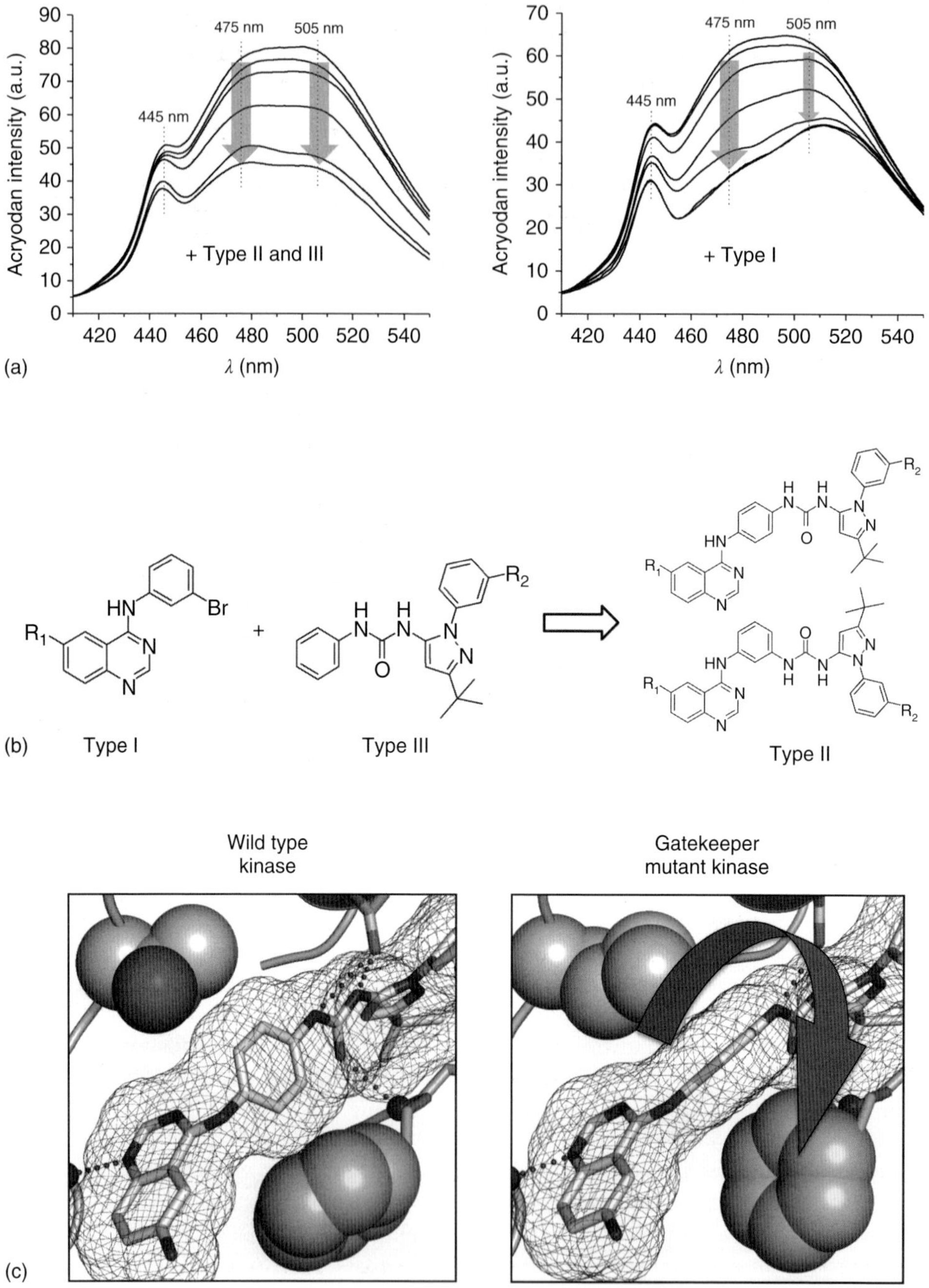

(a)
Acryodan intensity (a.u.)
λ (nm)
445 nm
475 nm
505 nm
+ Type II and III
+ Type I
(b)
HN
Br
R1
N
N
H
R2
O
Type I
Type III
Type II
(c)
Wild type kinase
Gatekeeper mutant kinase

Figure 2.3 *First HTS for the detection allosteric Src inhibitors.* In the absence of ligand, acrylodan-labeled cSrc shows two emission maxima at 475 and 505 nm. Type I ligands induce a robust loss of fluorescence intensity (arrows) at 475 nm, resulting in a red shift in the emission maxima to 510 nm (right panel). Type II and III inhibitors stabilize the inactive kinase conformation and elicit a different response in which the emissions at 475 and 505 nm are equally reduced. The emission signal at 445 nm is less sensitive to ligand binding and serves as an internal reference point, allowing for more stable ratiometric fluorescence measurements and K_d determinations (a). Rationally designed type II inhibitors based on the binding modes of type I 4-aminoquinazolines and type III pyrazoloureas bound to cSrc. (b) Fusion of the two fragments resulted in significantly higher binding affinities. In drug resistant cSrc, Abl, and central 1,4-substituted phenyl element of the type II hybrid compound can freely rotate to adapt to a larger gatekeeper residue (c). (Adapted with permission from Macmillan Publishers Ltd: [6], copyright 2009.)

Box 2.1 The Gatekeeper Residue

The gatekeeper residue is a conserved, residue in the ATP binding pocket of protein kinases and a critical determinant for selective inhibition within kinase families. In targeted cancer therapies, a single recurring mutation at the gatekeeper position results in drug resistance of several target kinases including BCR-Abl (CML), KIT (GIST), and epidermal growth factor receptor (EGFR) (non small cell lung cancer). The bulkier residue at the gatekeeper position sterically impedes binding of inhibitors in the active site of the kinase and often shifts the equilibrium inactive to active kinase conformations affinity to ATP. Interestingly, in chemical biology experiments, mutation of this naturally occurring bulky residue (larger than Ala) to smaller residues such as Ala or Gly generates a pocket not found in wild-type kinases. ATP-analog based competitive small molecules designed to complement the extended ATP binding pocket can be used to specifically target and inhibit the analog sensitive kinase. This elegant approach is also known as *bump and hole* and discussed elsewhere.

2.3.1
Finding the Unexpected

The majority of small molecules known to modulate kinase activity target the highly conserved ATP-pocket. Consequently, these inhibitors are less specific and can lead to the inhibition of multiple kinases. Selective modulation of kinase function remains a major hurdle in kinase inhibitor research. Therefore, ligands which bind to less conserved sites and target the non-catalytic functions of protein kinases provide new avenues to unique modes of inhibition. Several mitogen activated protein kinases (MAPKs), cyclin dependant kinases (CDKs), and glycogen synthase kinase 3 (GSK-3) contain a hydrophobic pocket at their C-terminus about 30 Å away from the ATP-pocket. This C-terminal insert regulates the intracellular localization of GSK-3, binds regulatory proteins in CDK2 and has been shown to bind substrates such as transcription factors and phosphatases in Erk2 (extracellular signal-regulated kinase). The same pocket exists in p38α MAPK and, although this pocket has no known biological function, potent ligands which specifically bind to this allosteric site may offer a valuable starting point for the development of chemical biology tool compounds for the investigation of its biological function.

To enable screening for compounds which bind to this unique binding site in p38α, we applied the FLiK approach to develop a fluorescent-labeled kinase assay system which takes advantage of ligand-induced conformational change of α-helices 1L14 and 2L14 of the p38α MAPK insert [17]. This HTS-amenable (high-throughput screen/screening) assay allowed the identification and characterization of 2-phenylquinazolines as ligands for this allosteric site. We were able to further develop higher affinity ligands and confirmed binding to this remote pocket using protein X-ray crystallography. We postulated that binding to this remote pocket might allosterically modulate kinase activity through some kind of intramolecular cross-talk mechanism with the active site of the kinase, but we were not able to observe any inhibition of kinase activity in activity-based assays for p38α. The compound was also profiled against 95 additional kinases in activity-based biochemical assays and did not appreciably inhibit the activity of any kinase.

Interestingly, there are naturally occurring isoforms of p38α which show altered signaling pathway preferences. Major differences between these isoforms and wild type p38α lie at the C-terminal end of the kinase, including regions within or proximal to the MAPK insert labeled in the FLiK assay. We propose that the MAPK insert may have a scaffolding function and somehow regulates p38α signaling independent of the activity of p38α itself. Future studies aimed at elucidating the biological function(s) of this site may rely on the use of tightly binding molecule identified with this unique FLiK assay. These molecules may perturb the conformation of the MAPK insert, disrupt, or enhance protein–protein interactions with this structural feature and provide alternative mechanisms for modulating the complex MAPK network.

2.3.2
Targeting Protein Interfaces – iFLiK

The advanced understanding in the orchestration of kinase function has clearly shown that kinase regulation extends beyond their capacity to phosphorylate other proteins [18]. In addition to the catalytically active kinase domain, most kinases feature highly dynamic regulatory domains that govern their activity. Protein kinases also serve as scaffolding proteins to form multi-enzyme complexes or by competing for and disrupting protein–protein interactions. Although many examples are known so far, even more functions remain unexplored as different kinase conformations allow for a variety of interactions with other binding partners [19]. The challenge for the next decade will be to understand these underlying scaffolding relationships in more detail. Investigating these interactions will not only provide novel insights into the complex issues of yet unexplored cellular signaling networks, it will also foster the development of innovative new drugs. From a chemical biology point of view, the generation of allosteric inhibitors that target interdomain interactions in protein kinases will provide a unique toolbox to probe and study protein dynamics in more complex biochemical and cellular settings.

2.3.3
Screening Akt

Although some small molecules are already known that modulate kinases via kinase-specific mechanisms (e.g., preventing the formation of activated complexes or mimicking by native activators) such discoveries are few in number, usually arise by chance and underline once more the urgent need for methods that can identify novel compounds with such modulating functions [19]. As these regulatory processes are associated with conformational changes in the complex architecture of intra- and intermolecular domain interactions, we developed iFLiK. iFLiK can detect more complex changes in protein kinase conformations such as intramolecular interdomain interactions. As a proof of concept study, we focused on the Ser/Thr kinase Akt [14]. Akt plays a key role in signaling pathways responsible for cell survival, proliferation, and apoptosis, and is a cancer drug target. The kinase domain of Akt is allosterically regulated by phosphatidylinositol lipids that bind to the adjoined PH domain of Akt (Figure 2.4a). In the inactive conformation ("PH-in" or "closed"), the PH domain moves in and forms tight interactions with the kinase domain via polar contacts and buries access to the substrate and ATP binding pockets. In 2005, researchers from Merck Sharpe and Dohme reported the serendipitous discovery of Akt inhibitors that only inhibited the full-length kinase but not the kinase domain alone. Structural investigations revealed that these inhibitors bind at the interface of the PH and kinase domain, locking Akt in an enzymatically inactive closed conformation. While these compounds were discovered by chance and their mode of action was initially unknown, these allosteric inhibitors proved to exhibit superior selectivity by addressing the unique activation mechanism of Akt and resulted in the development of the clinical candidate MK-2206, which is currently in phase II clinical trials for the treatment of various cancers. Given the great promise of allosteric Akt inhibitors, we developed iFLiK for the identification of molecules that target Akt and stabilize the enzyme in a catalytically inactive conformation. Following our structure-based design principles and analyzing the structural interactions between the PH- and kinase-domain, we proposed Glu49 on the surface of the PH-domain as a suitable site for the selective labeling of the thiol-reactive reporter fluorophore and introduced the mutation Glu49Cys (Figure 2.4b). In this case, we chose 1-(2-maleimidylethyl)-4-(5-(4-methoxyphenyl)oxazol-2-yl)pyridinium methanesulfonate (PyMPO) due to its excitation at longer wavelengths to avoid inhibitor auto-fluorescence. While classic ATP-competitive Akt inhibitors had no effect on PyMPO fluorescence, the allosteric inhibitors such as MK 2206 provoked a bathochromic shift of the emission spectrum, which allowed the determination of their K_d values.

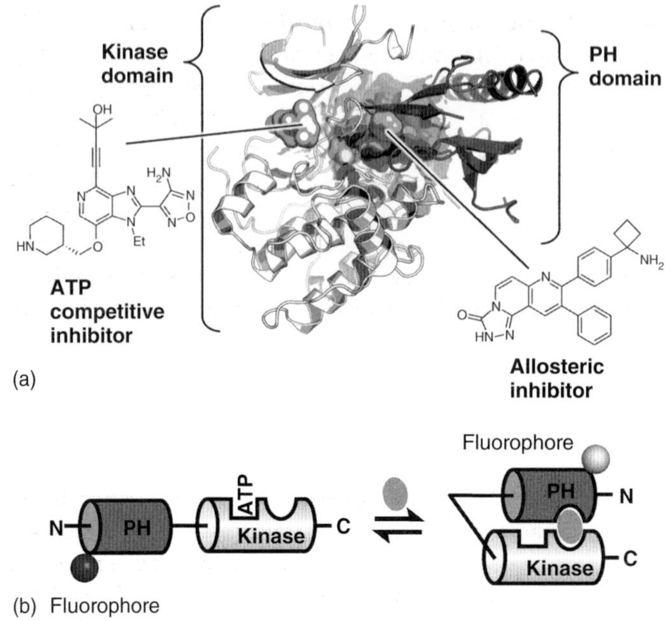

Figure 2.4 *Principle of iFLiK*. The PH domain of Akt is tightly associated with the kinase domain, forming a binding pocket for type IV allosteric inhibitors (spheres) at the domain interface, approximately 15 Å from the ATP binding pocket. MK-2206 is the first allosteric inter-domain inhibitor and currently in phase II clinical trials (a). Akt exists in an equilibrium between an inactive conformation, in which the PH domain binds to the kinase domain and obstructs access to the ATP binding pocket, and an active open conformation. Inter-domain allosteric inhibitors bind into a hydrophobic pocket formed by residues of the PH domain–kinase domain interface, locking Akt in the closed, enzymatically inactive, conformation. PH domain labeled covalently with reporter fluorophore *via* a synthetic cysteine (Glu49Cys). Transitions from open to closed conformations change the environment around the reporter fluorophore and therefore its fluorescence characteristics (b).

In a medium throughput screen, we utilized iFLiK to analyze a library of 10 000 compounds, consisting of both commercially purchased reagents as well as compounds synthesized in our laboratory. In this 1-point screen at 10 μM, the 24 compounds that displayed >25% binding relative to MK-2206 (as a positive control for the desired binding mode) were considered hits and selected for follow-up studies. Using dose-response measurements performed in triplicate, 13 out of 24 compounds were validated and produced a dose-dependent change in the emission spectrum of PyMPO. In an orthogonal activity-based assay, 12 of the 13 validated hits from the screen indeed inhibited Akt phosphorylation, including a series of pyrrolo[2,3-*d*]pyrimidines which were selective for the full-length kinase, as would be expected for allosteric inhibitors. To better understand the observed structure activity relationship (SAR), these hits were docked into published full-length crystal structures of Akt and revealed two main binding modes. To further understand the binding of these compounds, we obtained a focused library of 90 compounds based on the pyrrolopyrimidine scaffold from commercial sources and tested these to further explore the SAR. An additional seven responded in the iFLiK assay and exhibited selective inhibitory activity only on the full-length Akt. Most interestingly, upon treatment of an Akt-sensitive cancer cell line with these pyrrolopyrimidines, a dose-dependent decrease in cell viability was observed in the micromolar range. Western Blot analysis revealed a dose-dependent reduction in phosphorylation of the Akt specific substrate S6K, confirming Akt as the cellular target of these molecules. This complete story underlines the power of iFLiK for the selective detection of biologically active allosteric inter-domain stabilizers of Akt.

2.3.4
Targeting Phosphatases – FLiP

The importance of the fine-tuned and dynamic balance between phosphorylation and dephosphorylation is important for the homeostasis of a living cell and makes kinases as well as phosphatases highly attractive targets for chemical biology and medicinal chemistry research. Although phosphatases and their inhibitors have been heavily investigated for years, the development of clinically relevant phosphatase inhibitors faces major roadblocks and is often plagued by limited selectivity and unfavorable pharmacokinetics [20]. Until now no phosphatase inhibitor has been approved by the Food and Drug Administration (FDA) for the treatment of human diseases and calls for the development of next generation drugs and therapeutics that circumvent these major roadblocks in phosphatase inhibitor research. Reasons for these limitations arise at the molecular and structural level. Much like kinases, the active sites of phosphatases are structurally very well conserved, contributing to limited inhibitor selectivity. Even more importantly, the primary substrate pocket of a phosphatase is highly charged to complement the negative charge of the phosphorylated substrate. When screening for phosphatase inhibitors, this leads to the predominant discovery of negatively charged substrate mimetics that suffer from poor bioavailability and limited cell permeability, disqualifying them as lead candidates for further drug development. An approach to circumvent these limitations is the identification and exploitation of allosteric sites that are less conserved and, when addressed by small organic molecules, can lock the phosphatase in an inactive conformation. A prime example of a phosphatase drug target is the protein tyrosine phosphatase 1B (PTP1B). PTP1B activity is associated with type II diabetes and offers possibilities for the treatment of obese patients, since mice lacking the PTP1B gene showed resistance and decreased incidence of obesity and diabetes. The discovery of a druggable allosteric pocket distant from the catalytic site offers a promising new opportunity for the development of PTP1B modulators that lock the phosphatase in its inactive conformation and circumvent the problems of active site directed inhibitors (Figure 2.5a). However, methods which can identify specific allosteric PTP1B inhibitors fall short. To address this issue, we developed FLiP as a direct binding assay analogous to

FLiK, which enables detection of the binding of inhibitors independent of phosphatase activity [13]. To develop this assay, we labeled PTP1B on a C-terminal helix adjacent to a site known to accommodate allosteric modulators of phosphatase activity (Figure 2.5b). We identified Leu294 as suitable for replacement by a Cys residue and subsequent labeling with acrylodan to generate the allosteric site assay. The architecture of the allosteric site allows binding of ligands which trigger conformational changes that prevent the WPD-loop (Trp-Pro-Asp) from adopting

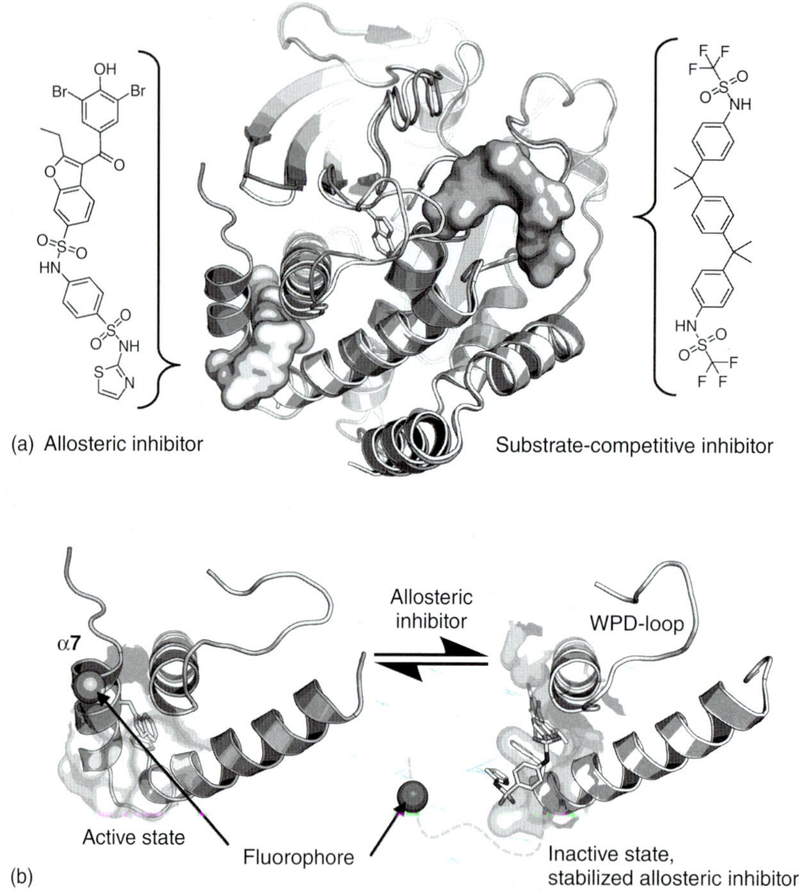

Figure 2.5 *Principle of the FLiP assay.* Overlay of the active and inactive forms of PTP1B (catalytic domain). Active site inhibitor and allosteric inhibitor are shown in a surface representation (a). In the active form, helix α7 docks onto helix α3. This interaction is stabilized by hydrophobic packing of W291 at the interface of helix α6 helix α3. The WPD loop closes over the substrate in the active site. Upon binding of allosteric inhibitors, a conformational rearrangement occurs, disabling substrate recognition. The disordering of helix α7 (dashed line) upon ligand binding is reported by the fluorophore (spheres) (b). (Adapted with permission from [13]. Copyright (2013) American Chemical Society.)

a catalytically competent conformation. The WPD-loop is located above the active site of the catalytic domain and regulates catalysis. Using an analogous strategy, we identified F182 in the highly-conserved WPD loop as suitable for replacement by a cysteine residue and subsequent labeling with acrylodan to allow detection of competitive active site inhibitors. In both assays, structural rearrangements of this site triggered by ligand binding can be monitored by detecting changes in acrylodan fluorescence. The change in fluorescence is clearly dependent on the inhibitor dose used and allows for K_d determinations. In addition, we showed that the system is robust enough to measure k_{on} and k_{off} of inhibitor binding. Overall, the FLiP assay is a strong tool in the search for novel selective and potent phosphatase modulators with drug-like chemical composition.

2.3.5 Lessons Learned from High-Throughput Screens

The FLiK assay technology can easily be adapted to high-throughput assay plates (96-, 384-, and 1536-well formats). Adaptation of the assay to small volume microtiter plates dramatically reduces the amount of kinase and compound required while increasing throughput to enable rapid screening of large compound collections. We have performed several successful HTS screens and identified valuable starting points for further compound development [6, 8, 14, 16, 17, 21–23]. A typical primary screen involves assay plates containing one compound per well at a maximum concentration of 10–20 μM. Typically, compounds which bind at least 50% compared to the positive control at 10–20 μM are chosen for follow-up screening. After a primary screen, hits are selected and rescreened in a dose-response format to confirm binding to the kinase and to determine the K_d value. As is the case with any fluorescence-based assay, intrinsic compound fluorescence can lead to difficult data analysis under certain excitation and emission conditions and result in the detection of false-positive and false-negative hits. The identification of false positives can be tricky and sometimes difficult. To assess background fluorescence of the compounds alone, each compound plate can be screened twice, once with only buffer and once with the FLiK kinase in the same buffer to enable background subtraction. This is especially the case where the compound fluorescence is more intense than the fluorophore at high compound concentrations (>10 μM), making simple background subtraction difficult. Additionally, the use of background plates is cost and time intensive and will not account for other types of fluorescence artifacts such as quenching or synergizing interactions between the fluorophore and certain compounds. The artifacts are more difficult to identify and may not necessarily be observed if the compounds are placed into suspension with free, unreacted fluorophore since the fluorescence characteristics and behavior of reactive fluorophores change upon conjugation to proteins. Since most organic molecules fluoresce intrinsically between 450 and 550 nm, we performed a systematic analysis and conjugated a large number of red-shifted fluorophores to different kinase constructs in hopes of developing FLiK assays which function

Table 2.1 Table with thiol-reactive fluorophores suitable for FLiK.

Brand name	ΔM^{a} (Da)	$\lambda_{exc,max}$ [b] (nm)	$\lambda_{em,max}$ [b] (nm)	Structure
Acrylodan	225	386	470	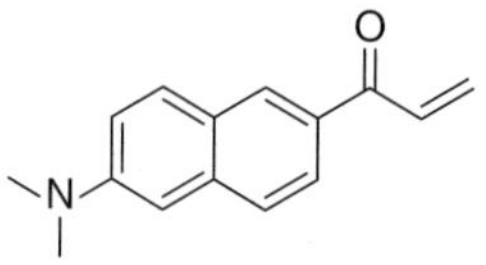
IAANS	355	326	462	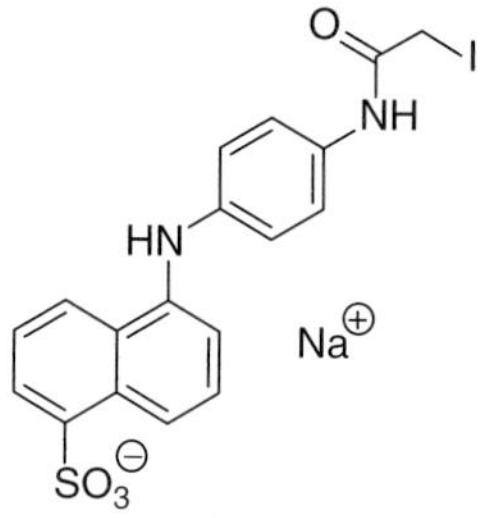
PyMPO	376	412	561	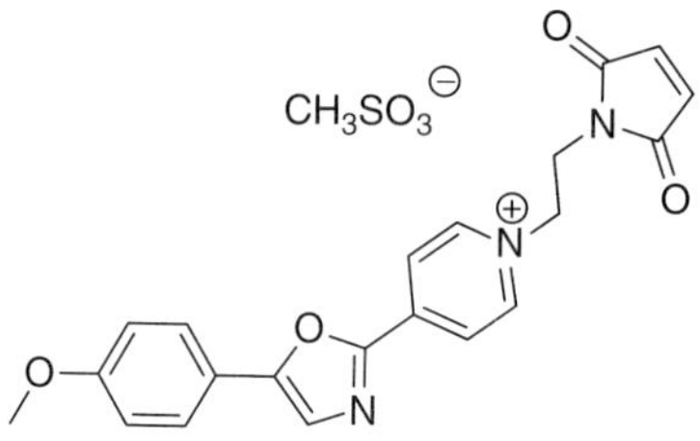
Texas Red C2	729	595	615	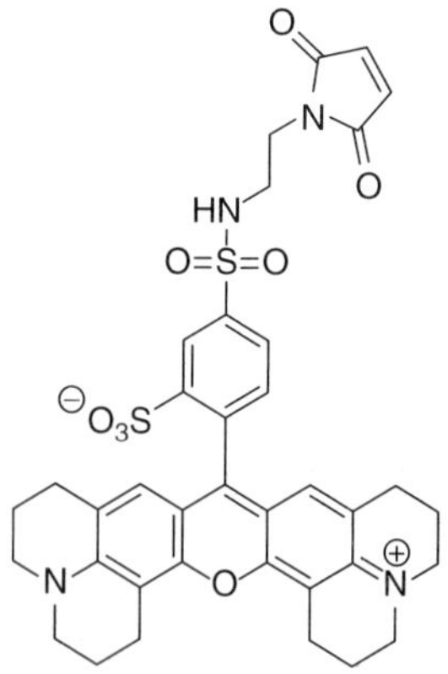

Table 2.1 (*Continued*)

Brand name	ΔM^{a} (Da)	$\lambda_{exc,max}{}^{b}$ (nm)	$\lambda_{em,max}{}^{b}$ (nm)	Structure
DY-647	764	653	672	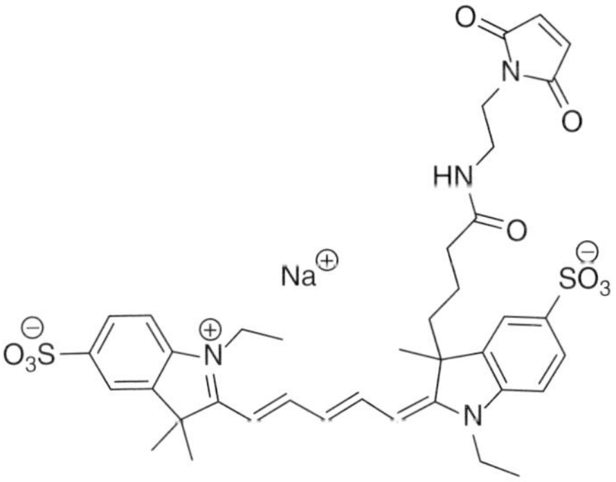
Atto 565	634	563	592	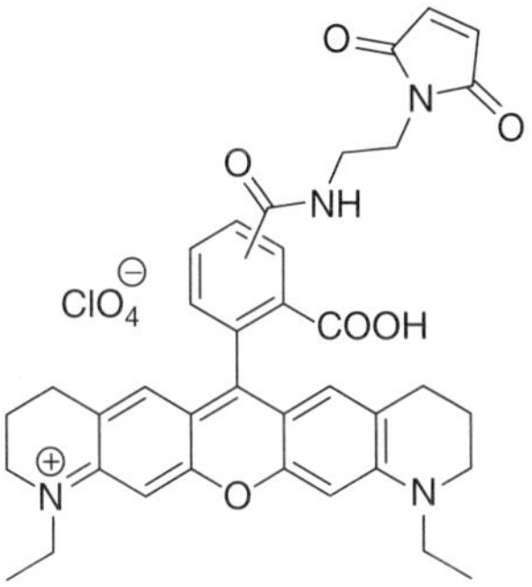
Atto 610	514	615	634	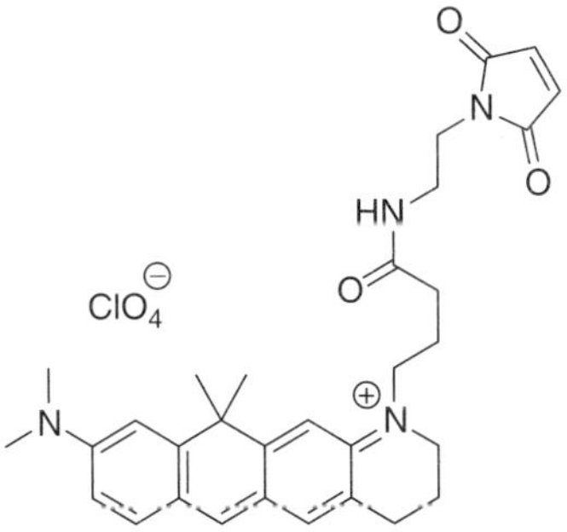

a) Excitation and emission maxima as provided by the supplier.
b) Neither structure nor exact molecular mass of Alexa Fluor 660 was provided by the supplier.

at emission ranges above 600 nm, conditions where neither intrinsic compound fluorescence nor light scattering occurred [24]. For all protein-fluorophore conjugates tested, alternative fluorophores were identified which can report this conformational change as reliably as Acrylodan, the original fluorophore used in the development of our various FLiK assays. With these red-shifted fluorophores, even highly rigid compounds with high intrinsic compound fluorescence (as recorded at lower wavelengths) could now be reliably measured.

In general, we found that the best performing red-emitting fluorophore exhibited a bathochromic shift upon ligand binding, allowing a ratio of intensities at two wavelengths to be calculated and used to determine K_d values (Table 2.1). By contrast, fluorophores exhibiting a simple intensity decrease of the emission maxima without a bathochromic shift translated into poor assay performance. The identification of red-shifted fluorophores suitable for FLiK represents a major improvement of our technologies that will enable a more straightforward discovery of complex enzyme inhibitors and reduce the number of fluorescence artifacts, false positives, and false negatives.

2.4 Conclusions

Protein phosphorylation and dephosphorylation are the most important ways cellular proteins are modified to regulate function and to transduce information between distinct cellular sites via signaling pathways. In spite of the enormous wealth of knowledge that has been accumulated to understand phosphate shuffling in association with disease states, the development of inhibitors is often hampered by limited selectivity, and lack of efficacy in complex biological systems. In this chapter, we have discussed various examples of protein engineering used to enable compound screening to address the bottlenecks in current kinase and phosphatase research, by moving away from classical ATP or substrate competitive inhibitors and targeting alternative conformations and allosteric binding sites. Profiling compounds against the entire kinome or phosphatome provides valuable information about the affinity and selectivity of compounds *in vitro*. However, optimizing compounds to lengthen drug-target residence times will ultimately provide a more complete context for fully understanding kinase inhibitor selectivity *in vivo*. By considering the kinetic components (k_{on} and k_{off}) of affinity, lead optimization strategies may improve *in vivo* selectivity and efficacy by directing medicinal chemistry efforts around improving the residence time of the ligand for the desired kinase.

For kinases, we have developed FLiK as a high-throughput screening technology which enables the rapid and robust identification of ligands which bind to and stabilize specific kinase conformations. The FLiK approach allows straightforward determination of not only ligand affinity (K_d values), but also kinetic characterization (k_{on} and k_{off}) to better understand the kinetic factors which contribute to the measured affinity. Moreover, FLiK does not require kinase activity or prior knowledge of the substrate, which may be advantageous when studying novel or less-characterized kinases. To date, the FLiK approach has been used to successfully monitor conformational changes in the activation loop of both Ser/Thr and Tyr kinases associated with the slow binding of DFG-out inhibitors. It has also been adapted to detect ATP-competitive inhibitors which engage the glycine-rich loop (P-loop) as well as type IV ligands which bind to remote allosteric binding sites outside of the ATP binding cleft. More recently, we have extended this approach to phosphatases and to full length kinases to monitor a variety of conformational changes which might be addressed through novel phosphatase inhibitor development.

References

1. Rabiller, M., Getlik, M., Klüter, S., Richters, A., Tückmantel, S., Simard, J.R., and Rauh, D. (2010) Proteus in the world of proteins: conformational changes in protein kinases. *Arch. Pharm. (Weinheim)*, **343** (4), 193–206.
2. Cohen, P. and Alessi, D.R. (2013) Kinase drug discovery--what's next in the field? *ACS Chem. Biol.*, **8** (1), 96–104.
3. Davis, M.I., Hunt, J.P., Herrgard, S., Ciceri, P., Wodicka, L.M., Pallares, G., Hocker, M., Treiber, D.K., and Zarrinkar, P.P. (2011) Comprehensive analysis of kinase inhibitor selectivity. *Nat. Biotechnol.*, **29** (11), 1046–1051.
4. Copeland, R.A. (2011) Conformational adaptation in drug-target interactions and residence time. *Future Med. Chem.*, **3** (12), 1491–1501.
5. Liu, Y. and Gray, N.S. (2006) Rational design of inhibitors that bind to inactive kinase conformations. *Nat. Chem. Biol.*, **2** (7), 358–364.
6. Simard, J.R., Klüter, S., Grütter, C., Getlik, M., Rabiller, M., Rode, H.B., and Rauh, D. (2009) A new screening assay for allosteric inhibitors of cSrc. *Nat. Chem. Biol.*, **5** (6), 394–396.
7. Simard, J.R., Getlik, M., Grütter, C., Pawar, V., Wulfert, S., Rabiller, M., and Rauh, D. (2009) Development of a fluorescent-tagged kinase assay system for the detection and characterization of allosteric kinase inhibitors. *J. Am. Chem. Soc.*, **131** (37), 13286–13296.
8. Simard, J.R., Grütter, C., Pawar, V., Aust, B., Wolf, A., Rabiller, M., Wulfert, S., Robubi, A., Klüter, S., Ottmann, C., and Rauh, D. (2009) High-throughput screening to identify inhibitors which stabilize inactive kinase conformations in p38alpha. *J. Am. Chem. Soc.*, **131** (51), 18478–18488.
9. Grütter, C., Simard, J.R., Mayer-Wrangowski, S.C., Schreier, P.H., Pérez-Martín, J., Richters, A., Getlik, M., Gutbrod, O., Braun, C.A., Beck, M.E., and Rauh, D. (2012) Targeting GSK3 from Ustilago maydis: type-II kinase inhibitors as potential antifungals. *ACS Chem.Biol.*, **7** (7), 1257–1267.
10. Simard, J.R., Grütter, C., Pawar, V., Aust, B., Wolf, A., Rabiller, M., Wulfert, S., Robubi, A., Klüter, S., Ottmann, C., and Rauh, D. (2010) Fluorophore labeling of the glycine-rich loop as a method of identifying inhibitors that bind to active and inactive kinase conformations. *J. Am. Chem. Soc.*, **132** (12), 4152–4160.
11. Schneider, R., Becker, C., Simard, J.R., Getlik, M., Bohlke, N., Janning, P., and Rauh, D. (2012) Direct binding assay for the detection of type IV allosteric inhibitors of Abl. *J. Am. Chem. Soc.*, **134** (22), 9138–9141.
12. Getlik, M., Simard, J.R., Termathe, M., Grütter, C., Rabiller, M., van Otterlo, W.A., and Rauh, D. (2012) Fluorophore labeled kinase detects ligands that bind within the MAPK insert of p38alpha kinase. *PLoS One*, **7** (7), e39713.
13. Schneider, R., Beumer, C., Simard, J.R., Grütter, C., and Rauh, D. (2013) Selective detection of allosteric phosphatase inhibitors. *J. Am. Chem. Soc.*, **135** (18), 6838–6841.
14. Fang, Z. *et al.* (2013) Discovery of inter-domain stabilizers – a novel assay system for allosteric akt inhibitors., submitted.
15. Richters, A., Ketzer, J., Getlik, M., Grütter, C., Schneider, R., Heuckmann, J.M., Heynck, S., Sos, M.L., Gupta, A., Unger, A., Schultz-Fademrecht, C., Thomas, R.K., Bauer, S., and Rauh, D. (2013) Targeting gain of function and resistance mutations in Abl and KIT by

hybrid compound design. *J. Med. Chem.*, **56** (14), 5757–5772.

16. Getlik, M., Grütter, C., Simard, J.R., Klüter, S., Rabiller, M., Rode, H.B., Robubi, A., and Rauh, D. (2009) Hybrid compound design to overcome the gatekeeper T338M mutation in cSrc. *J. Med. Chem.*, **52** (13), 3915–3926.
17. Getlik, M., Simard, J.R., Termathe, M., Grütter, C., Rabiller, M., van Otterlo, W.A., and Rauh, D. (2012) Fluorophore labeled kinase detects ligands that bind within the MAPK insert of p38alpha kinase. *PLoS One*, **7** (7), e39713.
18. Rauch, J., Volinsky, N., Romano, D., and Kolch, W. (2011) The secret life of kinases: functions beyond catalysis. *Cell Commun. Signaling*, **9**, 23.
19. Fang, Z., Grütter, C., and Rauh, D. (2013) Strategies for the selective regulation of kinases with allosteric modulators: exploiting exclusive structural features. *ACS Chem. Biol.*, **8** (1), 58–70.
20. Vintonyak, V.V., Waldmann, H., and Rauh, D. (2011) Using small molecules to target protein phosphatases. *Bioorg. Med. Chem.*, **19** (7), 2145–2155.
21. Getlik, M., Grütter, C., Simard, J.R., Nguyen, H.D., Robubi, A., Aust, B., van Otterlo, W.A., and Rauh, D. (2012) Structure-based design, synthesis and biological evaluation of *N*-pyrazole, *N*′-thiazole urea inhibitors of MAP kinase p38alpha. *Eur. J. Med. Chem.*, **48**, 1–15.
22. Grütter, C., Simard, J.R., Mayer-Wrangowski, S.C., Schreier, P.H., Pérez-Martín, J., Richters, A., Getlik, M., Gutbrod, O., Braun, C.A., Beck, M.E., and Rauh, D. (2012) Targeting GSK3 from Ustilago maydis: type-II kinase inhibitors as potential antifungals. *ACS Chem. Biol.*, **7** (7), 1257–1267.
23. Simard, J.R., Getlik, M., Grütter, C., Pawar, V., Wulfert, S., Rabiller, M., and Rauh, D. (2009) Development of a fluorescent-tagged kinase assay system for the detection and characterization of allosteric kinase inhibitors. *J. Am. Chem. Soc.*, **131** (37), 13286–13296.
24. Schneider, R., Gohla, A., Simard, J.R., Yadav, D.B., Fang, Z., van Otterlo, W.A., and Rauh, D. (2013) Overcoming compound fluorescence in the FLiK screening assay with red-shifted fluorophores. *J. Am. Chem. Soc.*, **135** (22), 8400–8408.

3
Strategies for Designing Specific Protein Tyrosine Phosphatase Inhibitors and Their Intracellular Activation

Birgit Hoeger and Maja Köhn

3.1
Introduction – The Biological Problem

3.1.1
Chemical Inhibition of Protein Tyrosine Phosphatase Activity

Protein phosphatases carry out dephosphorylation reactions on protein substrates that were previously phosphorylated by kinases, thereby enhancing or suppressing a flow of cellular information. The dephosphorylation leads to the modulation of activity or change of conformation of the protein substrates, or results in the loss of docking sites for downstream acting proteins. Classical protein tyrosine phosphatases (PTPs) are specialized in dephosphorylating phospho-tyrosine (pY) residues and constitute a class of related enzymes sharing active site features regarding sequence and structure (Box 3.1). The involvement of PTPs in numerous diseases such as diabetes, cancer, and immune disorders has led to a high demand for PTP inhibitors. However, the development of PTP active site inhibitors comes along with two major challenges that result from the intrinsic properties of the PTP active site. First, specificity is an issue resulting from the broad conservation of catalytic site characteristics and was long thought to not being achievable. Second, bioavailability is a major problem due to the substrate preference of tyrosine phosphatases consisting of negatively charged phosphate groups, which results in a high number of negatively charged compound hits from screening libraries – a characteristic that makes it difficult for molecules to pass cellular membranes [1, 2].

Concepts and Case Studies in Chemical Biology, First Edition. Edited by Herbert Waldmann and Petra Janning.

Box 3.1 Protein Tyrosine Phosphatases (PTPs) and Their Catalytic Mechanism

PTPs are classified on the basis of their structure and sequence [3]. They can be subgrouped into classical nonreceptor-type nontransmembrane PTPs, receptor-like membrane-localized PTPs, and the dual-specificity phosphatases (DUSPs) [4]. Classical PTPs mostly recognize phospho-tyrosine as their substrate, whereas DUSPs are also known to dephosphorylate proteins on serine or threonine, and even phospholipids or RNA can be substrates [3]. All PTPs contain a conserved CXXXXXR motive (single-letter amino acid code). The cysteine is the catalytically active moiety. Another residue that is involved in the catalytic mechanism (with some exceptions [5]) is an aspartic acid in the so-called WPD (Trp-Pro-Asp) loop, which is distinct from the active site [4] (Figure 1).

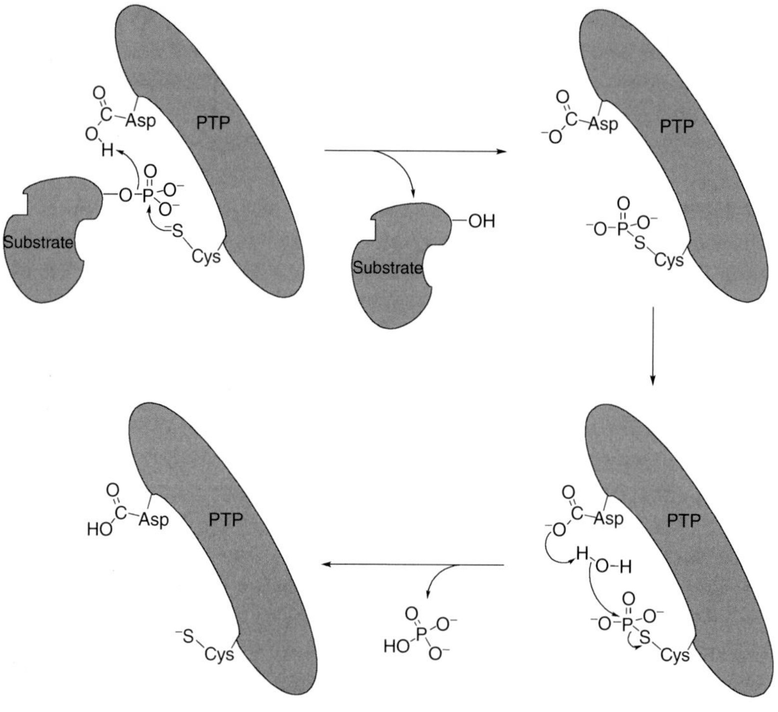

Figure 1 The catalytic Cys in a PTP's active site attacks a phosphate substrate, assisted by Asp as proton donor. Subsequent hydrolysis yields the free phosphate.

In this case study, the concept of bivalent ligands is described. This concept has successfully proved wrong the notion that PTPs cannot be targeted selectively [1], marking a strategy to achieve high specificity within a class of closely related enzymes. The question of how to design cell-membrane-permeable molecules is still struggling scientists working on related topics. To date, several methods have been developed to achieve cell permeability. Among them, reversible chemical

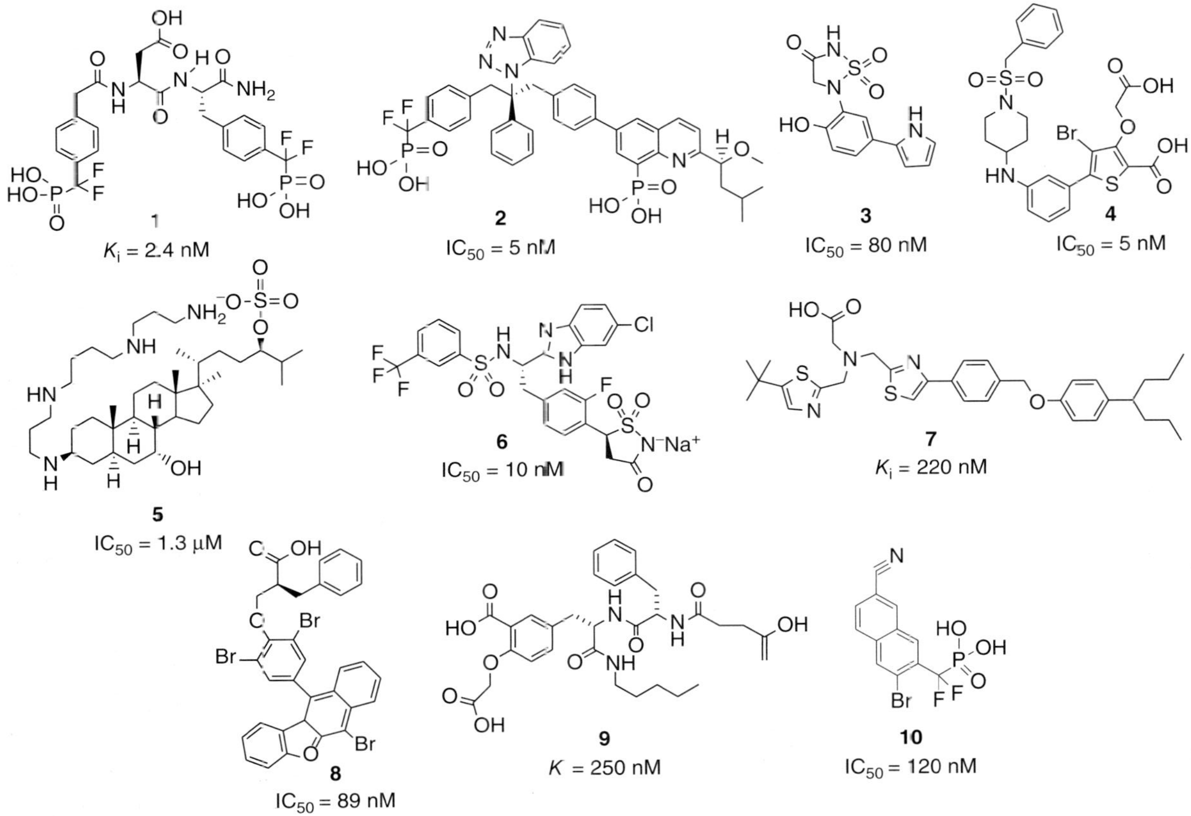

Figure 3.1 Chemical structures of some known PTP1B inhibitors [1, 14, 15].

masking of negatively charged groups or attachment of (removable) positively charged entities such as cell-penetrating peptide stretches are the most common [6, 7]. In this chapter we further discuss examples for both strategies.

3.1.2 PTP1B as Inhibitor Target

Protein tyrosine phosphatase 1B (PTP1B) is seen as the classic example of PTPs and represents the most studied member of the family. It is involved in several regulatory processes, with negative influence on insulin- and leptin-receptor-mediated signaling as its most famous roles (Box 3.2). Consequently, dysregulation of the enzyme contributes to type 2 diabetes and obesity [1, 2, 8, 9]. The enzyme was therefore identified as a highly attractive drug target. Furthermore, PTP1B has been shown to be involved in different types of cancer and several other disease mechanisms [10–12]. The search for PTP1B inhibitors has resulted in several molecules with inhibition potencies ranging from the high micromolar to the lower nanomolar range. In Figure 3.1, a diverse set of PTP1B modulators with their respective potencies is shown, including small molecules, peptides, and natural-product-like structures. Compound **1** thereby represents the most potent inhibitor developed to date. This compound is introduced in this chapter as an example of how to improve a first screening hit to gain cell-permeability and intracellular self-activation.

Box 3.2 PTP1B, TCPTP, and Their Signaling Pathways

PTP1B and T-cell protein tyrosine phosphatase (TCPTP) (existing in two variants sized 45 and 48 kDa) are close relatives [13]. The primary and tertiary structure similarity of their catalytic domains is very high (72% identity), and importantly, they also share the second phosphotyrosine-binding pocket, which makes it very difficult to find selective inhibitors of their active sites [1, 13]. Both phosphatases are involved in insulin and leptin signaling, although with nonredundant roles. The schematic represents the pathways in liver (insulin) and hypothalamus (leptin). Both phosphatases attenuate insulin signaling by dephosphorylating the insulin receptor (IR) either at the plasma membrane (TCPTP) or when it is internalized (PTP1B). They also regulate leptin signaling through PTP1B's action on phosphorylated Janus kinase 2 (JAK2) in the cytoplasm and TCPTP's dephosphorylation of signal transducer and activator of transcription 3 (STAT3) in the nucleus. Leptin is important for regulation of appetite, and insulin is crucial for glucose homeostasis [13] (Figure 2).

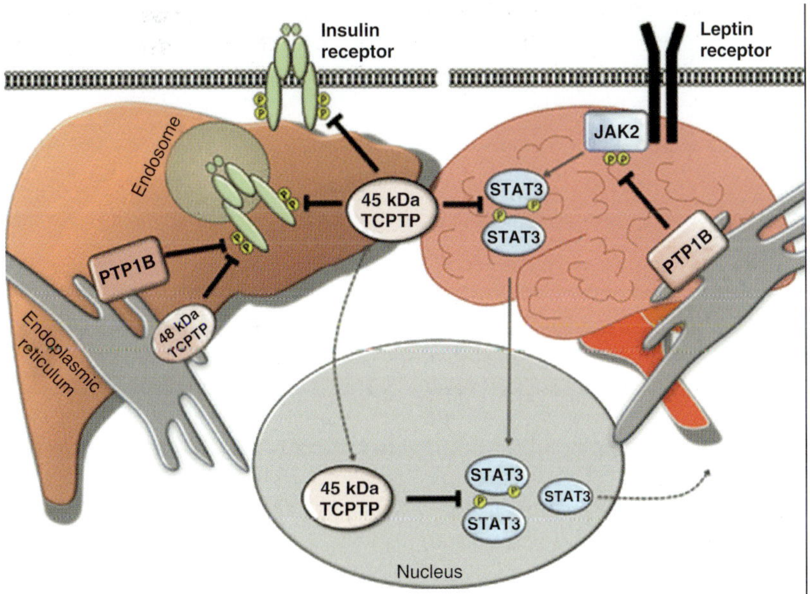

Figure 2 PTP1B and TCPTP regulate insulin signaling (liver) and leptin signaling (brain). Reprinted with permission from [13]. © 2012 The Author Journal compilation © 2012 FEBS.

3.2 The Chemical Approach

3.2.1 The Concept of Bivalent Ligands – Development of a Specific PTP1B Inhibitor

Subpockets bordering the active site of PTPs can substantially contribute to substrate recognition [10, 16, 17]. Thus, the phosphotyrosyl-binding pocket alone is not sufficient for substrate recognition. A crystal structure of PTP1B revealed a second, so-called *aryl phosphate-binding site* in close proximity to the active site pocket, which importantly is not conserved within the PTP family [18]. This second binding site was the clue to achieving selectivity within the PTP family by using bidentate ligands, which bind simultaneously to the conserved active site region and the nonconserved aryl-binding site in PTP1B [10].

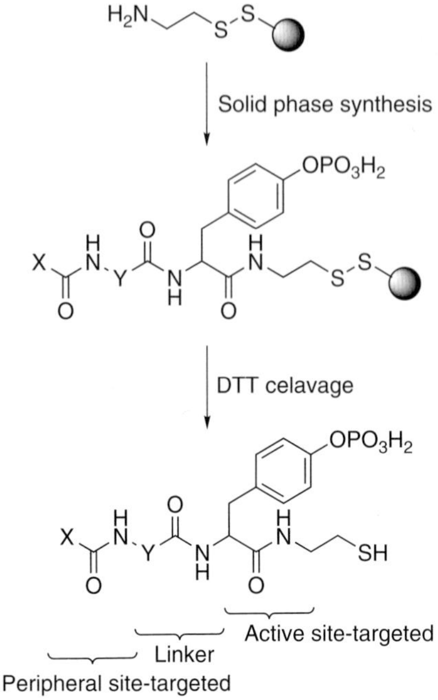

Figure 3.2 Parallel synthesis strategy for a small library of bivalent ligands targeting PTP1B, with Y representing various linker moieties and X representing various peripheral site-targeting entities [10].

Zhang and coworkers developed such a bivalent ligand that represents the most potent and selective PTP1B inhibitor to date (compound **1** in Figure 3.1) [10]. Their chemical strategy was based on a peptide containing nonhydrolyzable pY analogs both for targeting the active site cleft as well as the adjacent aryl-binding site. Nonhydrolyzable pY analogs cannot be cleaved by the phosphatase, and therefore the phosphatase binds them but cannot release them, leading to a blockage, and hence inhibition of the active site.

First, a small synthetic peptide-based library was prepared by parallel synthesis of active site-targeted building blocks attached to various arylic peripheral site-targeted building blocks, separated by various linkers. Thereby, the active site-binding moiety was attached to the so-called *TentaGels* via a cleavable disulfide bond, enabling rapid and easy solid-phase peptide synthesis. A scheme of this parallel synthesis approach is depicted in Figure 3.2. Hereby, the pY moieties were not yet replaced by their nonhydrolyzable counterparts because of synthetic ease and the experimental setup of the following affinity screen.

By screening the compounds of this library in an affinity-based enzyme-linked immunosorbent assay (ELISA) (discussed in Section 3.3), high-affinity binders for a catalytically inactive variant (not able to dephosphorylate the pY moieties) of PTP1B could be identified (Figure 3.3, with compound **11** as the best candidate found). The next step was to synthesize a nonhydrolyzable analog of

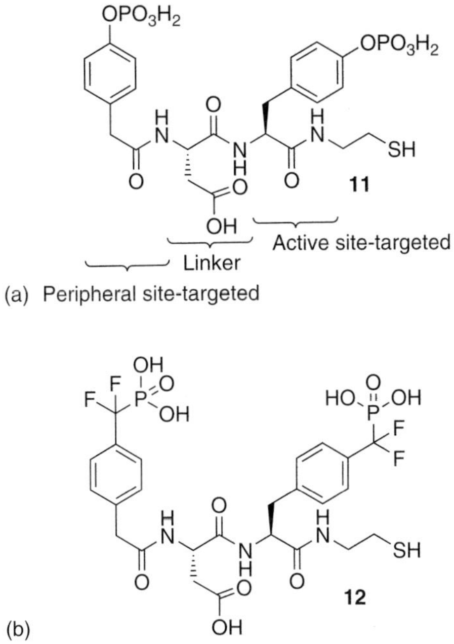

Figure 3.3 Chemical structure of (a) the best bivalent hit from a library screen and (b) its nonhydrolyzable analog [10].

the best candidate, containing difluoromethyl-phosphonate moieties instead of the hydrolyzable phosphates (compound **12**, Figure 3.3), in order to gain a high-affinity inhibitor of wild-type PTP1B. Chemical strategies exist for easy access to difluorophosphonate residues as pY mimetics for peptide synthesis [10, 19]. *In vitro* phosphatase assays (see Chapter 4, Box 3.2) are used to determine the activity of inhibitors toward phosphatases. *In vitro* activity determination, in this case using *p*-nitrophenyl phosphate (pNPP) (Box 4.2), of the final peptide analog (**12**) against wild type PTP1B and a panel of other PTPs including the closest structural homolog TCPTP (Box 3.2), revealed excellent to good selectivity of the compound [10]. Hence, it was demonstrated that it is possible to develop highly potent and especially selective inhibitors of a PTP by making use of a bivalent ligand approach.

3.2.2
Cell Permeability and Intracellular Activation of a Self-Silenced Inhibitor

Because phosphonate-containing peptidic inhibitors are generally not cell-permeable owing to their negative charges, Zhang and colleagues developed an elegant approach to overcome this problem [20]. By attaching a poly-arginine tail to the N-terminus of their non-cell-permeable inhibitor (**13** in Figure 3.4), the compound could be delivered over the plasma membrane through the

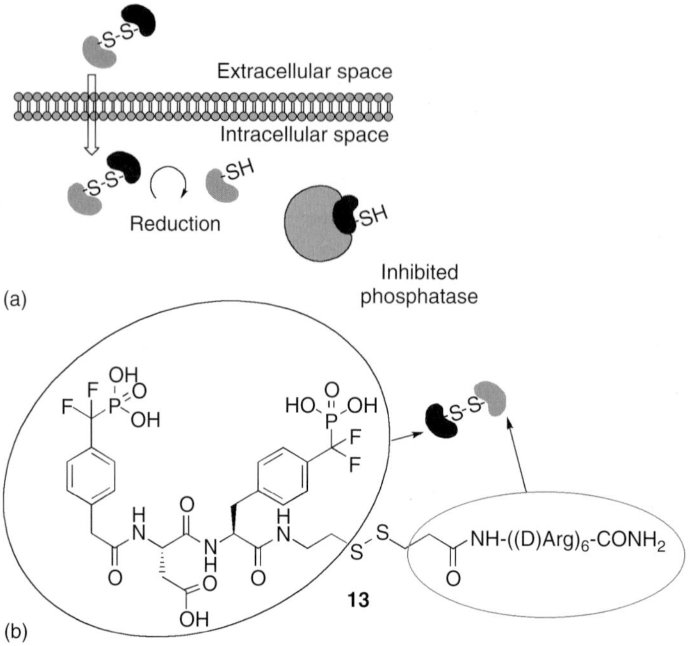

Figure 3.4 Intracellular activation of a self-silenced peptide inhibitor. (a) Schematic representation of intracellular reduction of a disulfide bond to release the activated phosphatase inhibitor (black) from the poly-arginine tail (gray). (b) Corresponding chemical structure of the self-silenced inhibitor [20].

positive charges of this arginine stretch that counteract the negative charges of the inhibitor. By using a disulfide bridge as linker between inhibitor and tag, the active peptide was released inside cells through reduction of the disulfide bond (Figure 3.4) [20]. Additionally, when the inhibitor was still attached to the poly-arginine tail, its activity was repressed through electrostatic interactions between the negatively charged inhibitor and the positively charged arginine residues. This way, an activated inhibitor is only achieved when reduction of the disulfide bond has occurred inside cells. Hence, the group named their construct an *intramolecularly self-silenced* probe. Hereby, they made use of a principle that is well established in nature by aminobenzamide (AB) protein toxins. These toxins are membrane permeable through a disulfide-linked transporter stretch that represses the activity of the toxin and is cleaved off once inside cells, releasing the active toxin [21].

3.2.3 A Prodrug Strategy to Gain Cell Permeability

The role of PTP1B in contributing to type 2 diabetes by dephosphorylating and hence negatively regulating insulin receptor β (IRβ) is well established. Consequently, the enzyme is a valuable drug target [1, 2, 8, 9].

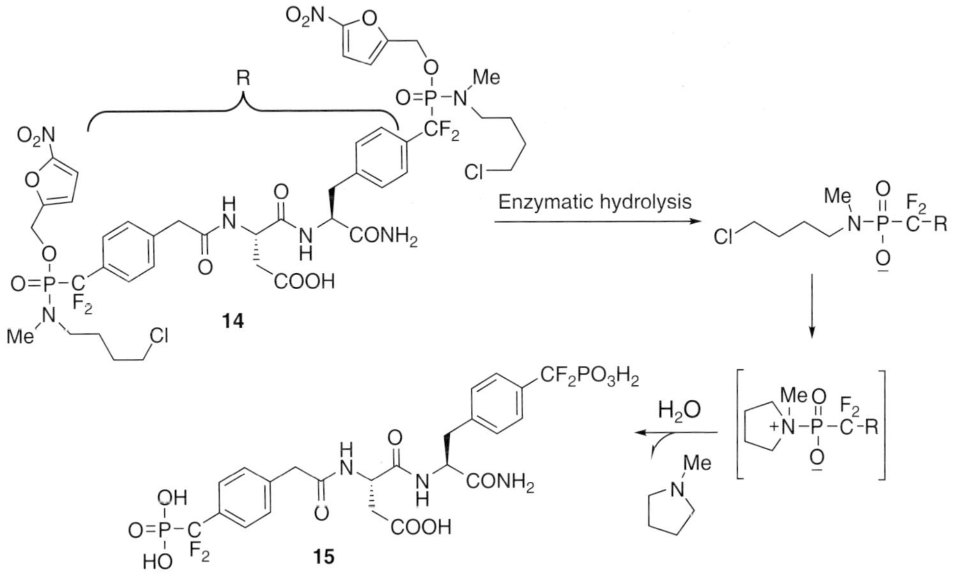

Figure 3.5 Intracellular activation of a phosphonate-based prodrug to its active counterpart [22].

The most potent and selective PTP1B inhibitor developed so far (**1** in Figure 3.1) faced the drawback of not being cell permeable owing to its negatively charged phosphonate moieties. The approach of Zhang and coworkers to introduce a polyarginine tail to deliver the peptide into cells is an excellent example of how to deliver chemical probes, but such an approach is not necessarily feasible for therapeutic agents owing to issues arising from the attachment of a polybasic polypeptide such as potential cytotoxicity [20, 22]. Therefore, Borch and coworkers developed a phosphonate-based prodrug strategy in order to provide a solution for the drug delivery problem [22]. By masking the two phosphonate moieties in inhibitor **1** with enzymatically cleavable, lipophilic protecting groups, the compound (**14**) can be delivered into cells and further be converted by enzymatic hydrolysis into its active form (**15**). Figure 3.5 shows the enzymatic activation of the nitrofurfuryl delivery group, its subsequent rearrangement and final hydrolysis to yield the activated drug **15**.

3.3 Chemical Biological Research/Evaluation

3.3.1 An Affinity-Based ELISA Assay to Identify Potent Binders

ELISAs are widely used biochemical assay techniques to identify binding partners. Generally, one assay component is immobilized on a solid support, while a soluble component – either an antibody or a probe–antibody conjugate – is probed for its ability to bind to the immobilized component. This interaction is read via an enzyme-linked second antibody, creating a photometrically detectable signal [23].

In order to probe their library of bivalent ligands (introduced in Section 3.2.1) for interaction with PTP1B, Zhang and coworkers [10] developed a competitive type of ELISA assay. In order to find high-affinity binding partners, a catalytically deficient GST-tagged (glutathione S-transferase) PTP1B variant (GST-PTP1B/C215S) that retains wild-type binding affinity but does not hydrolyze the ligand was used. In this competitive assay, a well-known biotinylated peptide substrate (biotinyl-caproic acid-DADEpYL-NH_2, single-letter amino acid code) with high affinity toward PTP1B was immobilized on a 96-well microtiter plate. Compounds from the library were individually incubated with GST-PTP1B/C215S and subsequently probed for competition with the immobilized peptide. After extensive washing, anti-GST antibodies were used as final readout, detecting enzyme bound to the immobilized peptide (Figure 3.6). This way, compounds that were displaced by the peptide substrate were regarded as nonbinders, whereas compounds preventing the enzyme from binding to the immobilized peptide were regarded as high-affinity PTP1B binders.

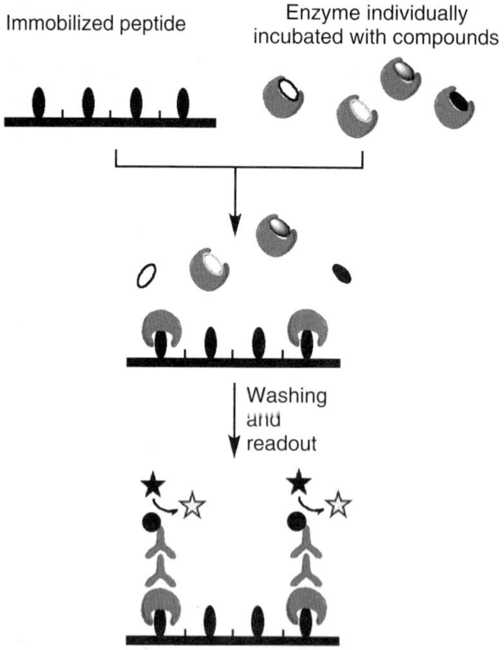

Figure 3.6 Principle of an affinity-based competitive ELISA assay. Only nondisplaced compounds are potent binders, represented by a very low signal of enzyme bound to the immobilized peptide.

3.3.2
Evaluation of Cell Permeability and Cellular Activity by Monitoring Insulin Receptor Signaling

In order to test synthetic peptides for their cell-penetrating ability and their inhibitory activity inside cells, evaluation of intracellular events such as decrease or increase in phosphorylation levels of downstream acting proteins of involved signaling networks can give an overview of the inhibitor's properties in living cells. Because PTP1B inhibitors act on a phosphatase directly involved in negative regulation of insulin signaling by dephosphorylating the IRβ [8, 9], monitoring of IR phosphorylation levels represents a direct readout of the inhibitor's performance.

For this purpose, mammalian cells were incubated with the self-silenced peptide analog (**13**) introduced in Section 3.2.2 as well as the prodrug (**15**) introduced in Section 3.2.3, and subsequently treated either with or without insulin. Cell lysates were resolved by SDS-PAGE (sodium docecylsulfate-polyacrylamide gel electrophoresis), transferred to nitrocellulose membranes, and probed with anti-phospho-Tyr or anti-phospho-IRβ antibodies [20, 22]. The prodrug was additionally probed with anti-phospho-ERK1/2 antibody (ERK: extracellular signal-regulated kinase) being a downstream target of the insulin receptor) [22]. Both compounds enhanced the pY levels of IRβ (with the prodrug also inducing ERK1/2 phosphorylation) [20, 22]. Hence, the compounds successfully inhibited PTP1B in living cells, consequentially blocking the PTP1B-mediated dephosphorylation of downstream effectors that therefore showed increased phosphorylation levels. It could thereby not only be proved that both peptides were able to penetrate cellular membranes but also that they are active inside cells and not only in the test tube.

3.4
Conclusions

In this chapter we introduced a concept of how to specifically target an enzyme within a structurally closely related family by making use of bivalent ligands. We discussed the development of a peptide-based inhibitor, both binding to the active site of phosphatase PTP1B as well as to an adjacent site, leading to a selective inhibitor with high potency toward the enzyme [10]. This approach can in theory be applied to any enzyme within a closely related family, provided a second ligand-binding site unique to this enzyme of interest exists. Notably, by targeting two distinct binding sites with a single ligand, a gain in potency can be achieved due to the additivity of free energy of binding [10]. Having potent and specific chemical modulators of enzyme activity in hand is a prerequisite for studying the enzyme's cellular functions in detail. Such tools often also represent good starting points for the development of novel therapeutic agents [1].

We further introduced ways to achieve cell membrane permeability using the example of the negatively charged bivalent PTP1B inhibitor [20, 22]. The question of how to gain cell permeability is a major issue in the field, as probes being only active *in vitro* but not being able to pass the cellular membrane barrier cannot be considered useful in most cases. By applying the developed specific PTP1B inhibitors to cell studies, more detailed insights into the biological mechanisms mediated by PTP1B action in health and disease can be achieved.

References

1. He, R., Zeng, L.-F., He, Y., Zhang, S., and Zhang, Z.-Y. (2013) Small molecule tools for functional interrogation of protein tyrosine phosphatases. *FEBS J.*, **280** (2), 731–750.
2. De Munter, S., Köhn, M., and Bollen, M. (2012) Challenges and opportunities in the development of protein phosphatase-directed therapeutics. *ACS Chem. Biol.*, **8** (1), 36–45.
3. Li, X., Wilmanns, M., Thornton, J., and Köhn, M. (2013) Elucidating human phosphatase-substrate networks. *Sci. Signal.*, **6** (275), rs10.
4. Tonks, N. (2006) Protein tyrosine phosphatases: from genes, to function, to disease. *Nat. Rev. Mol. Cell Biol.*, **7** (11), 833–846.
5. Tonks, N. (2013) Protein tyrosine phosphatases - from housekeeping enzymes to master regulators of signal transduction. *FEBS J.*, **280** (2), 346–378.
6. Garrido-Hernandez, H., Moon, K.D., Geahlen, R.L., and Borch, R.F. (2006) Design and synthesis of phosphotyrosine peptidomimetic prodrugs. *J. Med. Chem.*, **49** (11), 3368–3376.
7. Lundberg, P. and Langel, U. (2003) A brief introduction to cell-penetrating peptides. *J. Mol. Recogn.*, **16** (5), 227–233.
8. Elchelby, M., Payette, P., Michaliszyn, E., Cromlish, W., Collins, S., Lee Loy, A., Normandin, D., Cheng, A., Himms-Hagen, J., Chan, C.-C., Ramachandran, C., Gresser, M.J., Tremblay, M.L., and Kennedy, B.P. (1999) Increased insulin sensitivity and obesity resistance in mice lacking the protein tyrosine phosphatase-1B gene. *Science*, **283** (5407), 1544–1548.
9. Klaman, L.D., Boss, Q., Peroni, O.D., Kim, J.K., Martino, J.L., Zabolotny, J.M., Moghal, N., Lubkin, M., Kim, Y.B., Sharpe, A.H., Stricker-Krongrad, A., Shulman, G.I., Neel, B.G., and Kahn, B.B. (2000) Increased energy expenditure, decreased adiposity, and tissue-specific insulin sensitivity in protein tyrosine phosphatase-1B deficient mice. *Mol. Cell. Biol.*, **20** (15), 5479–5489.
10. Shen, K., Keng, Y.-F., Wu, L., Guo, X.-L., Lawrence, D.S., and Zhang, Z.-Y. (2001) Acquisition of a specific and potent PTP1B inhibitor from a novel combinatorial library and screening procedure. *J. Biol. Chem.*, **267** (50), 47311–47319.
11. Lessard, L., Stuible, M., and Tremblay, M.L. (2010) The two faces of PTP1B in cancer. *Biochim. Biophys. Acta*, **1804** (3), 613–619.
12. Dubé, N. and Tremblay, M.L. (2004) Beyond the metabolic function of PRP1B. *Cell Cycle*, **3** (5), 550–553.
13. Tiganis, T. (2013) PTP1B and TCPTP-nonredundant phosphatases in insulin signaling and glucose homeostasis. *FEBS J.*, **280** (2), 445–458.
14. Vintonyak, V.V., Waldmann, H., and Rauh, D. (2011) Using small molecules to target protein phosphatases. *Bioorg. Med. Chem.*, **19** (7), 2145–2155.
15. Zhang, Z.-Y. (2002) Protein tyrosine phosphatases: Structure and function, substrate specificity, and inhibitor development. *Annu. Rev. Pharmacol. Toxicol.*, **42**, 209–234.
16. Zhang, Z.-Y. (1997) Structure, mechanism, and specificity of protein-tyrosine phosphatases. *Curr. Top. Cell. Regul.*, **35**, 21–68.
17. Vetter, S.W., Keng, Y.F., Lawrence, D.S., and Zhang, Z.-Y. (2000) Assessment of protein tyrosine phosphatase 1B substrate specificity using "inverse alanine scanning". *J. Biol. Chem.*, **275** (4), 2265–2268.

18. Puius, Y.A., Zhao, Y., Sullivan, M., Lawrence, D.S., Almo, S.C., and Zhang, Z.-Y. (1997) Identification od a second aryl phosphate-binding site in protein tyrosine phosphatase 1B: a paradigm for inhibitor design. *Proc. Natl. Acad. Sci. U.S.A.*, **94** (25), 13420–13425.
19. Meyer, C. and Köhn, M. (2011) Efficient scaled-up synthesis of *N*-alpha-Fmoc-4-Phosphono(difluoromethyl)-L-phenylalanine and its incorporation into peptides. *Synthesis*, **20**, 3255–3260.
20. Lee, S.-Y., Liang, F., Guo, X.-L., Xie, L., Cahill, S.M., Blumenstein, M., Yang, H., Lawrence, D.S., and Zhang, Z.-Y. (2005) Design, construction and intracellular activation of an intramolecularly self-silenced signal transduction inhibitor. *Angew. Chem. Int. Ed.*, **44** (27), 4242–4244.
21. Falnes, P.O. and Sandvik, K. (2000) Penetration of protein toxins into cells. *Curr. Opin. Cell Biol.*, **12** (4), 407–413.
22. Boutselis, I.G., Yu, X., Zhang, Z.-Y., and Borch, R.F. (2007) Synthesis and cell-based activity of a potent and selective protein tyrosine phosphatase 1B inhibitor prodrug. *J. Med. Chem.*, **50** (4), 856–864.
23. Van Weemen, B.K. and Schuurs, A.H. (1971) Immunoassay using antigen-enzyme conjugates. *FEBS Lett.*, **15** (3), 232–236.

4
Design and Application of Chemical Probes for Protein Serine/Threonine Phosphatase Activation

Yansong Wang and Maja Köhn

4.1
Introduction

More than 30% of proteins are phosphorylated [1], and approximately 85% of all phosphorylation sites consist of a serine or a threonine [2]. Thus, protein serine/threonine (Ser/Thr) phosphorylation and dephosphorylation are ubiquitous and crucial mechanisms for the regulation of cellular signaling networks. The reversible phosphorylation and dephosphorylation are mediated by protein Ser/Thr kinases (PSTKs) and protein Ser/Thr phosphatases (PSTPs), respectively. Malfunctioning of either PSTKs or PSTPs contributes to the development of human diseases. However, compared to the extensive studies on PSTKs, the understanding of PSTP roles, regulation, pathways, and substrates is still limited on account of the fact that they were traditionally regarded as nonspecific housekeeping enzymes and because of their complex regulation (Box 4.1) [3]. Moreover, there are almost no selective chemical PSTP modulators available, a fact that complicates their investigation further [4]. This chapter describes the design and application of chemical probes for PSTP activation, focusing on protein phosphatase 1 (PP1).

Box 4.1 Protein Serine/Threonine Phosphatases (PSTPs) and Their Regulation

The classical superfamily of PSTPs can be divided into the family phosphoprotein phosphatases (PPPs) and the family of metal-dependent protein phosphatases (PPMs) [3, 5]. In addition, the phosphatases of the haloacid dehalogenase (HAD) superfamily also contains PSTPs [3, 5, 6]. The regulation of PSTPs, in particular PPPs, is highly complex. Protein serine/threonine kinases (>400 genes [3]) outnumber the PSTPs (70 [5]) by far. This signifies that the regulation of the PSTPs is very different from that of PSTKs, as PSTPs still need to counteract the PSTKs in a tightly controlled manner. To make it even more complex, kinases have evolved from one

Concepts and Case Studies in Chemical Biology, First Edition. Edited by Herbert Waldmann and Petra Janning.

common ancestor, while phosphatases have evolved from different ones, and are therefore structurally and mechanistically distinct [7]. The mechanism of the PPP and PPM families depends on the presence of metal ions, whereas the HADs and protein tyrosine phosphatases (PTP)s (see Box 3.1) follow other mechanisms [7].

Protein phosphatases 1–7 (PP1–7) of the PPP family share about 80% sequence identity across species and are thus highly conserved [3, 7]. Their regulation depends on regulatory proteins, which bind to the PPPs and form so-called *holoenzymes* with them [3]. The regulation of PP1 and PP2A is outlined below (Figure 1).

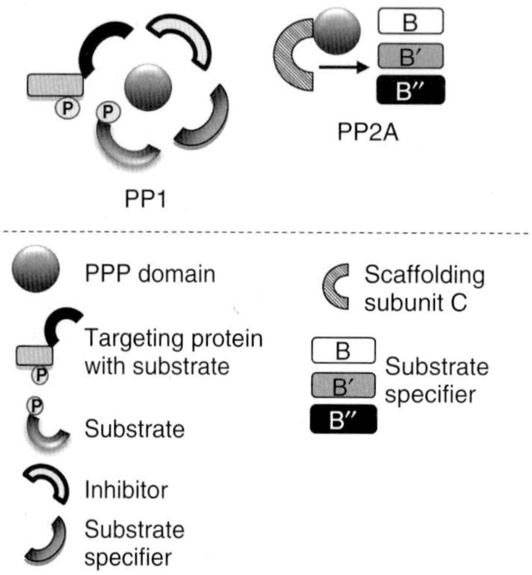

Figure 1 The regulation of PP1 and PP2A by their regulatory proteins via formation of holoenzymes. (Adapted from [8].)

4.2
The Biological Problem

PP1 and protein phosphatase 2A (PP2A) are the two most abundant PSTPs involved in basically all cellular processes ranging from mitosis [9] and Ca^{2+} signaling [10] to apoptosis [11]. PP1 and PP2A exist as holoenzymes composed of the catalytic subunit with structurally highly conserved catalytic cores and a distinct set of interacting proteins as regulatory subunits. These structurally and functionally different regulatory proteins control the subcellular localization, substrate specificity, and activity of PP1 and PP2A (Box 4.1) [12]. Four PP1 catalytic subunit isoforms (α, β, γ1, γ2), encoded by three genes, are expressed in mammalian cells with over 200 PP1 interacting proteins as regulators [12]. Many of these regulators bind to the PP1 catalytic subunit through a so-called *RVxF motif* (single letter amino acid code, x represents any amino acid except proline, the sequence can vary) that binds into a hydrophobic pocket on PP1 [13]. Other sequences flanking the RVxF motif on the regulatory proteins engage in other binding interactions with PP1 to control the affinity and specificity for different PP1 isoforms [14]. PP2A has two catalytic subunit isoforms that also interact with different regulators, but these regulators lack the RVxF motif. PP1 and PP2A are essential enzymes; however, the complex regulation, the high abundance, and the ubiquitous activity render their investigation extremely difficult. Therefore, the modulation of PP1 and PP2A holoenzyme activity with chemical probes would be a valuable approach to study their functions. Inhibitors or loss-of-function probes enable to perturb enzyme function reversibly, while activators or gain-of-function probes offer a way to address a particular cellular phenotype caused by an enzyme [15]. Thus, much effort was put into the identification and design of PP1 and PP2A inhibitors and, more recently, activators [4]. However, the issue of inhibitors being nonselective when targeting the active site of PP1 and PP2A as a result of their strong similarity is yet to be resolved [4]. Therefore, commonly used potent inhibitors such as calyculin A (Cal-A) or ocadaic acid (OA) show only marginal selectivity, rendering the probes not very useful for most applications [4, 16].

In contrast to the numerous inhibitors, thus far only few activators have been reported. The design of activators faces different challenges than the development of inhibitors. For activators, no general strategies are available, and binding sites that can be used for this purpose need to be identified individually. The immunosuppressant FTY720 (Figure 4.1), which is a structural analog of sphingosine and gets phosphorylated by sphingosine kinases in the cell [17], was serendipitously found to be a PP2A activator with no reactivity against PP1 [18]. It induces apoptosis in some types of cancer cells via activation of PP2A

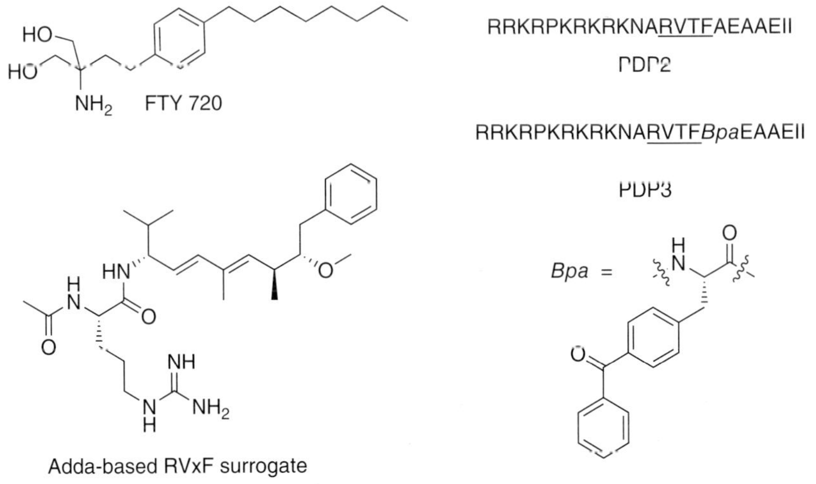

Figure 4.1 Structure of FTY720, a PP2A activator, as well as the Adda-based RVxF surrogate, PDP2, and PDP3. All three are PP1 activators. Adda = β-(2s,3s,8s,9s)-3-amino-9-methoxy-2,6,8-trimethyl-phenyldeca-4,6-dienoic acid. The RVxF motif in the PDPs is underlined.

to dephosphorylate Akt (also known as protein kinase B, PKB) pathway factors [18, 19]. However, aside from activating PP2A, FTY720 inhibits ceramide synthases [20] and phospholipase A_2 [21], and FTY720-phosphate acts as a potent sphingosine-1-phosphate (S1P) receptor agonist [17], demonstrating the complex cellular response to this chemical probe. The few reported PP1 activators target the RVxF-binding site on PP1 [16, 22]. A short peptide analog which mimics the RVxF motif using the aryl diene moiety "Adda" (β-(2s,3s,8s,9s)-3-amino-9-methoxy-2,6,8-trimethyl-phenyldeca-4,6-dienoic acid) as an "xF"-bioisostere (Figure 4.1) activates PP1 *in vitro* [22]. More recently, PP1-disrupting peptides (PDPs) (Figure 4.1) were introduced, which are able to activate PP1 inside cells and show selectivity over PP2A and PP2B, which is another PPP family member closely related to PP1 [16, 23]. The design and applications of these PDPs are described in this chapter.

4.3 The Chemical Approach

Regulatory proteins, such as inhibitor 2 (I2) and nuclear inhibitor of protein phosphatase 1 (NIPP1), bind to PP1 and inhibit its activity. Thus, the hypothesis behind the chemical approach to obtain a PP1 activator was that interruption of these interactions could lead to an unbound PP1 catalytic subunit, which could

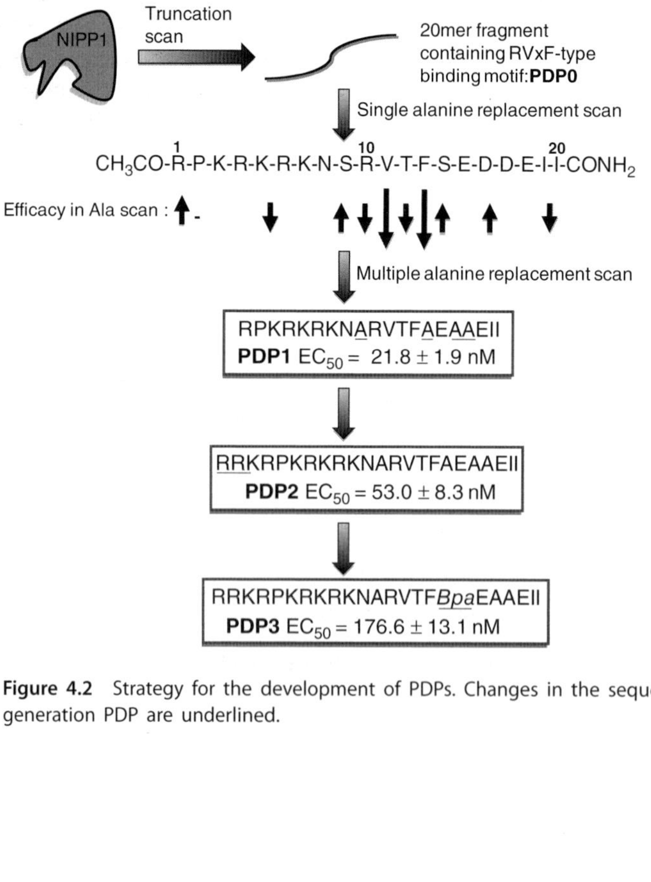

Figure 4.2 Strategy for the development of PDPs. Changes in the sequence of the next-generation PDP are underlined.

then dephosphorylate nearby substrates. Because most of the regulatory proteins, including I2 and NIPP1, bind to PP1 via the RVxF motif, the RVxF-binding site on PP1 appeared to be the ideal starting point to interrupt these interactions and to release the active PP1 catalytic subunit [16].

An RVxF motif containing 20mer peptide (PDP0), derived from NIPP1, was chosen as the starting point for developing a strong binder of the RVxF-binding site on PP1 (Figure 4.2). NIPP1 is a picomolar inhibitor of PP1 *in vitro* and was chosen because of its strong binding affinity to PP1 [24]. PDP0 showed a low nanomolar effective concentration 50 (EC_{50}) (Figure 4.2) in an *in vitro* phosphatase assay measuring the activity of PP1 toward its radioactively labeled substrate phosphorylase after deinhibition by the peptide (Box 4.2). After analysis of PDP0 applying a single alanine scan and a subsequent multiple alanine scan, it was found that the RVxF motif, the basic N-terminal stretch, and the two C-terminal isoleucines in PDP0 are important for the potent deinhibition of PP1. With this information, an optimized peptide (PDP1) with an EC_{50} of 21 nM was prepared (Figure 4.2) [16].

Box 4.2 Common Assays for Determining *in vitro* Phosphatase Activity

There are generally two ways to measure the *in vitro* activity of a phosphatase: (1) detection of the released phosphate and (2) detection of the dephosphorylated substrate.

1) These three methods are commonly used: the detection of radioactive phosphate containing ^{32}P [24], the colorimetric detection of a complex of phosphomolybdate and malachite green [25], and the colorimetric detection of a phosphorylated product obtained by a secondary enzymatic phosphate transfer using purine ribonucleoside phosphorylase (PNP) [26] (Figure 2).

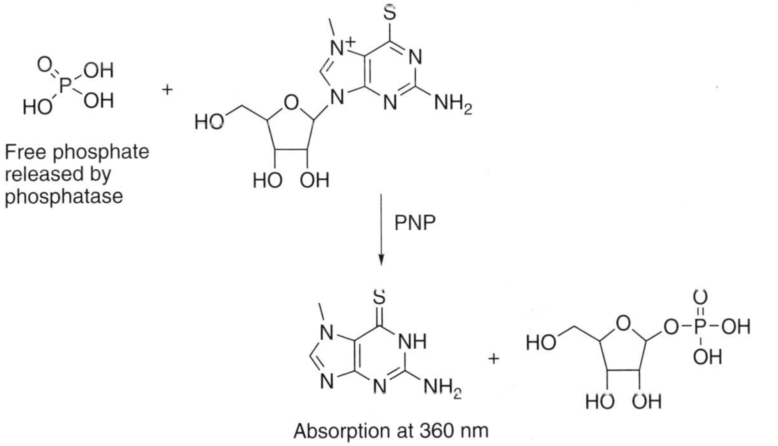

Figure 2 In vitro phosphatase assay using purine ribonucleoside phosphorylase for a colorimetric read-out.

2) Applied are fluorogenic compounds such as 6,8-difluoro-4-methylumbelliferyl phosphate (DiFMUP) or compounds changing the absorption spectra after dephosphorylation, such phospho-tyrosine, *para*-nitrophenol phosphate (pNPP) or 3-*O*-methylfluorescein phosphate (OMFP) [22, 26] (Figure 3). In a phosphatase inhibition assay, the phosphatase is first incubated with the inhibitor and then the phosphatase activity assay is carried out. Reduced activity due to inhibition is measured by one of the above-mentioned assay types, which are chosen depending on the phosphatase. A parameter often obtained in inhibition assays is the inhibitor concentration 50 (IC_{50}) (inhibition constant), which is the concentration where 50% of the enzyme is inhibited.

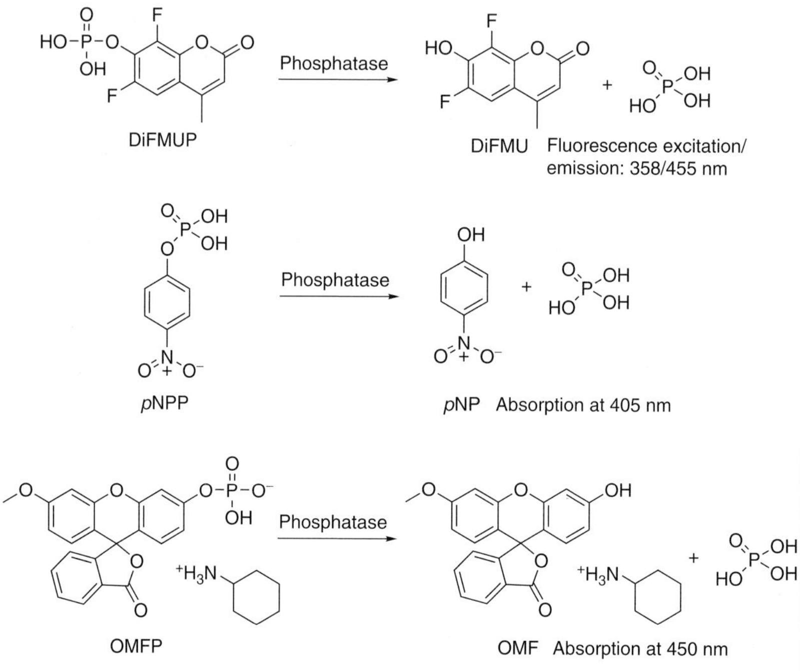

Figure 3 In vitro phosphatase assays using DiFMUP, pNPP and OMFP as substrates.

In a phosphatase activation assay, the phosphatase is first incubated with an inhibitor (e.g., PP1 with the protein I2 [16, 24]), which leads to reduced phosphatase activity. Then, the activator is added, followed by the measurement of the resulting increased activity; thus the activator deinhibits the phosphatase in such an assay. The parameter obtained is the EC_{50} (half-maximum effective concentration).

The use of peptides as chemical probes in cells is attached to two major challenges – the proteolytic stability and cellular uptake (see also Chapter 3). The cellular uptake of PDP1 was assessed by incubating live cells with 5-carboxyfluorescein (FAM)- labeled peptides and monitoring the uptake using confocal fluorescence microscopy. However, FAM-PDP1 showed very low cell permeability. To enhance the cell penetration without significantly compromising the PDP's efficacy, arginine and lysine residues were sequentially added to the N-terminus of PDP1 to afford PDP2 (see Chapter 3 for the strategy of adding positively charged amino acids). PDP2 penetrated the cells and was only slightly less efficacious in activating PP1 than the non-cell-permeable peptide PDP1 (Figure 4.2) [16].

PDP2, however, turned out to be not very stable against degradation inside cells. This can be measured by monitoring fluorescently labeled peptides (here FAM-PDP2) using fluorescence in-gel detection. Cells are incubated with the peptides for different amounts of time. The resulting fluorescence intensity of the samples at the expected peptide retention time in the gel (correlating with its mass) correlates with the amount of peptide left after a certain amount of incubation time. For the chemical stabilization of peptides, different methods are known, such as the incorporation of unnatural D-amino acids that are not recognized by peptide-digesting proteases. Also, in the case of PDP2, the incorporation of an unnatural amino acid (benzoyl phenylalanine, peptide PDP3) (Figures 4.1 and 4.2) led to enhanced cellular stability [16].

4.4 Chemical Biological Research/Evaluation

4.4.1 Selectivity of PDPs toward PP1 over PP2A and PP2B

The excellent selectivity of PDP2 and PDP3 toward PP1 was revealed by precipitating biotinylated PDP2 and PDP3 from cell lysates with streptavidin-coated Sepharose beads. PP1, but not PP2A, was detected in the following Western blot analysis, demonstrating that PDPs do not bind to PP2A [16]. Furthermore, it was shown that PP2B activity is not directly affected by PDP2. To this end, the FRET-based (Förster/fluorescence resonance energy transfer) PP2B phosphatase activity sensor calcineurin activity reporter 1 (CaNAR1) [27] was employed [23]. In CaNAR1, a PP2B activity-dependent molecular switch is used, leading to an increase in the FRET signal upon PP2B-mediated dephosphorylation of the probe (Box 4.3). Addition of PDP2 resulted in activation of PP2B; however, when the intracellular Ca^{2+} was chelated using BAPTA-AM (1,2-bis(*O*-aminophenoxy)ethane *N*,*N*,*N*′,*N*′ tetraacetic acid tetra(acetoxymethyl) ester), there was no detectable activation of PP2B by PDP2. PP2B requires Ca^{2+} in order to be active, and Ca^{2+} is released upon PDP-mediated PP1 activation (see Section 4.4.3). Therefore, this demonstrated that PDP2 *per se* did not, but the PDP2-induced Ca^{2+} release activated PP2B [23].

Box 4.3 Förster/Fluorescence Resonance Energy Transfer (FRET) and FRET Probes

FRET is a mechanism describing energy transfer between a donor chromophore, initially in its electronic excited state, to an acceptor chromophore through nonradiative dipole–dipole coupling [28]. The energy transfer depends on the distance between the two chromophores, which lies in the nanometer range. FRET probes measure enzyme activity inside cells by making use of a conformational switch or cleavage of the probe after having been modified by an enzyme. In case of CaNAR1, the probe consists of a phosphorylated substrate of PP2B sandwiched between the FRET chromophore pair. The probe undergoes a conformational change after dephosphorylation by PP2B, which leads to an increase in FRET [27]. The figure was adapted from [27]. λ_{ex} = wavelength of excitation of chromophore; λ_{em} = wavelength of emission of chromophore (Figure 4).

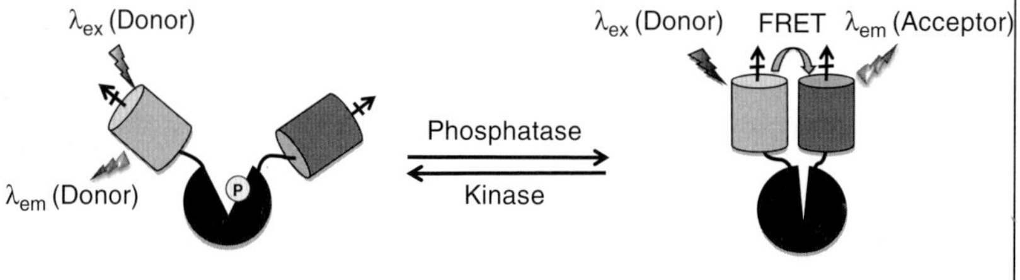

Figure 4 General design of a FRET-based phosphatase activity reporter.

4.4.2
Studying the Functions of PP1 in Mitosis with PDPs

In order to replicate, cells need to duplicate their chromosomes and then separate the identical sets before the cell divides into two. The process of separation of these identical sets of chromosomes is called *mitosis*. The so-called histones are proteins that package DNA and are part of chromosomes. At the end of mitosis, the holoenzyme complex of PP1 and the histone H3-PP1-targeting subunit Repo-Man dephosphorylates histone H3 on threonine 3 (H3T3), which is phosphorylated during mitosis [29]. The effects of PDP3-induced PP1 activation on the phosphorylation status of H3T3 and downstream effects were systematically studied [16].

PDP3 treatment of cells in mitotic arrest, where H3T3 is phosphorylated (H3T3ph), promoted histone H3T3ph dephosphorylation, overriding strong kinase activity that exists during mitosis. This demonstrated that PDP3 is efficacious in activating PP1 inside cells. The proposed mechanism, by which the PDPs disrupt PP1 regulatory protein complexes to release free, catalytically active

PP1, was corroborated with two further experiments. First, siRNA-mediated knockdown of Repo-Man resulted in a hyperphosphorylation of H3T3ph, which was expected because Repo-Man targets PP1 to H3T3ph and without it, PP1 does not act on H3T3ph in cells [29]. This phenotype was reversed by preincubation with PDP3 without any effect on the total level of PP1. This demonstrated that free PP1 catalytic subunit was released and it dephosphorylated the PP1 substrate H3T3ph in the absence of Repo-Man. Second, through pull-down of different regulatory proteins with and without PDP3 preincubation, and subsequent determination of the amount of coprecipitated PP1 using a phosphatase activity assay, it was shown that indeed there was less PP1 bound to the regulatory proteins after PDP3 treatment. These experiments substantiated that PDPs release free catalytically active PP1 subunit by disrupting PP1 holoenzymes [16].

The downstream effects of enhanced H3T3ph dephosphorylation by PDP3 treatment during mitosis were also studied. H3T3ph is known to serve as a docking site for Aurora B kinase, and mediates the targeting of this essential mitotic kinase to the centromeres (part of the chromosomes) during prometaphase (an early phase of mitosis). Using immunofluorescence, it was observed that PDP3-induced H3T3ph dephosphorylation caused the centromeric loss of Aurora B in mitotically arrested cells, confirming that the presence of H3T3ph (as opposed to H3T3) is required for the correct localization of Aurora B kinase during mitosis. Of note, Aurora B kinase is an oncogene and PDP-induced PP1 activation counteracted the function of this cancer-promoting protein, giving rise the to exciting hypothesis of using PP1 activation in cancer treatment [16].

4.4.3 Studying the Functions of PP1 in Ca^{2+} Signaling with PDPs

Ca^{2+} signaling is involved in multiple processes in an organism. Ca^{2+} ions regulate cellular live as second messengers in many aspects by altering electrostatic fields or binding to proteins to change their conformation [30]. Previous studies showed that PP1 inhibition by OA reduced Ca^{2+} levels [31] and suggested PP1 as the key player in the regulation of inositol-1,4,5-trisphosphate receptor (IP_3R)-dependent Ca^{2+} signaling [32]. In a subsequent study, the selective PDP probes were applied to test the effect of PP1 activity on the Ca^{2+} signal response [23].

PDP treatment induced rapid Ca^{2+} oscillations inside cells, which was detected by ratiometric Ca^{2+} live cell imaging using the fluorescent dye Fura-2 that binds Ca^{2+} ions [23]. The ratio of constant Fura 2 emission at 510 nm to the calcium-bound or -unbound dye emitting at 340 or 380 nm of light directly correlates with the amount of intracellular Ca^{2+}, independent of the dye concentration [33]. PDP2 acted reversibly, that is after washing the cells the Ca^{2+} oscillations were completely abolished, whereas PDP3-treated cells still showed sustained Ca^{2+} oscillations after washing. This indicated that PDPs trigger intracellular Ca^{2+} release in a permanent or reversible manner, and was explained with the different proteolytic stability of PDP2 and PDP3 [23]. In further experiments applying FRET probes combined with the PDPs, it was established that the PDP-induced

Ca^{2+} release originated from internal stores, in particular the endoplasmic reticulum, and involved IP_3Rs. Also, using different inhibitors such as OA and Cal-A and PDPs, it was shown that PP1 activity, in general, up-regulates the Ca^{2+} level inside the cell, and that PP2A and PP2B were not involved in the initial Ca^{2+} response [23].

4.5 Conclusion

Protein phosphorylation is one of the most important posttranslational modifications for cellular functions and signal transduction. As the phosphorylation state of a protein is controlled by the kinase and the phosphatase, modulators of kinases and phosphatases are valuable tools to study their functions. Compared to the large amount of kinase modulators, the development of phosphatase modulators is still limited and challenging. Over the years, protein phosphatases have been notoriously difficult to study.

In this chapter, we have presented examples of the design and applications of chemical probes for PSTP activation. With rational design, cell-permeable peptides that compete with endogenous RVxF motif-containing regulatory proteins for binding to PP1 in living cells were developed. They show high selectivity for PP1 over the closely related phosphatases PP2A and PP2B. The application of these peptides to study PP1 in mitosis and Ca^{2+} signaling are examples that these probes open up new routes to decipher PP1 functions in health and disease, and will assist in the design and development of innovative protein phosphatase directed therapeutics in the future.

References

1. Yang, C.Y., Chang, C.H., Yu, Y.L., Lin, T.C., Lee, S.A., Yen, C.C., Yang, J.M., Lai, J.M., Hong, Y.R., Tseng, T.L., Chao, K.M., and Huang, C.Y. (2008) PhosphoPOINT: a comprehensive human kinase interactome and phospho-protein database. *Bioinformatics*, **24** (16), i14–i20.
2. Roy, J. and Cyert, M.S. (2009) Cracking the phosphatase code: docking interactions determine substrate specificity. *Sci. Signal.*, **2** (100), re9.
3. Brautigan, D.L. (2013) Protein Ser/Thr phosphatases – the ugly ducklings of cell signaling. *FEBS J.*, **280** (2), 324–345.
4. Chatterjee, J. and Köhn, M. (2013) Targeting the untargetable: recent advances in the selective chemical modulation of protein phosphatase-1 activity. *Curr. Opin. Chem. Biol.*, **17** (3), 361–368.
5. Li, X., Wilmanns, M., Thornton, J., and Köhn, M. (2013) Elucidating human phosphatase-substrate networks. *Sci. Signal.*, **6** (275), rs10.
6. Seifried, A., Schultz, J., and Gohla, A. (2013) Human HAD phosphatases: structure, mechanism, and roles in health and disease. *FEBS J.*, **280** (2), 549–571.
7. Moorhead, G.B., De Wever, V., Templeton, G., and Kerk, D. (2007) Evolution of protein phosphatases in plants and animals. *Biochem. J.*, **8** (3), 401–409.
8. De Munter, S., Köhn, M., and Bollen, M. (2013) Challenges and opportunities in

the development of protein phosphatase-directed therapeutics. *ACS Chem. Biol.*, **8** (1), 36–45.

9. Wu, J.Q., Guo, J.Y., Tang, W., Yang, C.S., Freel, C.D., Chen, C., Nairn, A.C., and Kornbluth, S. (2009) PP1-mediated dephosphorylation of phosphoproteins at mitotic exit is controlled by inhibitor-1 and PP1 phosphorylation. *Nat. Cell Biol.*, **11** (5), 644–651.
10. Aoyama, H., Ikeda, Y., Miyazaki, Y., Yoshimura, K., Nishino, S., Yamamoto, T., Yano, M., Inui, M., Aoki, H., and Matsuzaki, M. (2011) Isoform-specific roles of protein phosphatase 1 catalytic subunits in sarcoplasmic reticulum-mediated Ca(2+) cycling. *Cardiovas. Res.*, **89** (1), 79–88.
11. Wang, R.H., Liu, C.W., Avramis, V.I., and Berndt, N. (2001) Protein phosphatase 1alpha-mediated stimulation of apoptosis is associated with dephosphorylation of the retinoblastoma protein. *Oncogene*, **20** (43), 6111–622.
12. Hendrickx, A., Beullens, M., Ceulemans, H., Den Abt, T., Van Eynde, A., Nicolaescu, E., Lesage, B., and Bollen, M. (2009) Docking motif-guided mapping of the interactome of protein phosphatase-1. *Chem. Biol.*, **16** (4), 365–371.
13. Egloff, M.P., Johnson, D.F., Moorhead, G., Cohen, P.T., Cohen, P., and Barford, D. (1997) Structural basis for the recognition of regulatory subunits by the catalytic subunit of protein phosphatase 1. *EMBO J.*, **16** (8), 1876–1887.
14. Terry-Lorenzo, R.T., Carmody, L.C., Voltz, J.W., Connor, J.H., Li, S., Smith, F.D., Milgram, S.L., Colbran, R.J., and Shenolikar, S. (2002) The neuronal actin-binding proteins, neurabin I and neurabin II, recruit specific isoforms of protein phosphatase-1 catalytic subunits. *J. Biol. Chem.*, **277** (31), 27716–27724.
15. Zorn, J.A. and Wells, J.A. (2010) Turning enzymes ON with small molecules. *Nat. Chem. Biol.*, **6** (3), 179–188.
16. Chatterjee, J., Beullens, M., Sukackaite, R., Qian, J., Lesage, B., Hart, D.J., Bollen, M., and Köhn, M. (2012) Development of a peptide that selectively activates protein phosphatase-1 in living cells. *Angew. Chem. Int. Ed.*, **51** (40), 10054–10059.
17. Mandala, S., Hajdu, R., Bergstrom, J., Quackenbush, E., Xie, J., Milligan, J., Thornton, R., Shei, G.J., Card, D., Keohane, C., Rosenbach, M., Hale, J., Lynch, C.L., Rupprecht, K., Parsons, W., and Rosen, H. (2002) Alteration of lymphocyte trafficking by sphingosine-1-phosphate receptor agonists. *Science*, **296** (5566), 346–349.
18. Matsuoka, Y., Nagahara, Y., Ikekita, M., and Shinomiya, T. (2003) A novel immunosuppressive agent FTY720 induced Akt dephosphorylation in leukemia cells. *Br. J. Pharmacol.*, **138** (7), 1303–1312.
19. Roberts, K.G., Smith, A.M., McDougall, F., Carpenter, H., Horan, M., Neviani, P., Powell, J.A., Thomas, D., Guthridge, M.A., Perrotti, D., Sim, A.T., Ashman, L.K., and Verrills, N.M. (2010) Essential requirement for PP2A inhibition by the oncogenic receptor c-KIT suggests PP2A reactivation as a strategy to treat c-KIT + cancers. *Cancer Res.*, **70** (13), 5438–5447.
20. Berdyshev, E.V., Gorshkova, I., Skobeleva, A., Bittman, R., Lu, X., Dudek, S.M., Mirzapoiazova, T., Garcia, J.G., and Natarajan, V. (2009) FTY720 inhibits ceramide synthases and up-regulates dihydrosphingosine 1-phosphate formation in human lung endothelial cells. *J. Biol. Chem.*, **284** (9), 5467–5477.
21. Payne, S.G., Oskeritzian, C.A., Griffiths, R., Subramanian, P., Barbour, S.E., Chalfant, C.E., Milstien, S., and Spiegel, S. (2007) The immunosuppressant drug FTY720 inhibits cytosolic phospholipase A2 independently of sphingosine-1-phosphate receptors. *Blood*, **109** (3), 1077–1085.
22. Tappan, E. and Chamberlin, A.R. (2008) Activation of protein phosphatase 1 by a small molecule designed to bind to the enzyme's regulatory site. *Chem. Biol.*, **15** (2), 167–174.
23. Reither, G., Chatterjee, J., Beullens, M., Bollen, M., Schultz, C., and Köhn, M. (2013) Chemical activators of protein phosphatase-1 induce calcium release inside intact cells. *Chem. Biol.*, **20** (9), 1179–1186.

24. Beullens, M., Van Eynde, A., Vulsteke, V., Connor, J., Shenolikar, S., Stalmans, W., and Bollen, M. (1999) Molecular determinants of nuclear protein phosphatase-1 regulation by NIPP1. *J. Biol. Chem.*, **274** (20), 14053–14061.
25. Feng, J., Chen, Y., Pu, J., Yang, X., Zhang, C., Zhu, S., Zhao, Y., Yuan, Y., Yuan, H., and Liao, F. (2011) An improved malachite green assay of phosphate: mechanism and application. *Anal. Biochem.*, **409** (1), 144–149.
26. McParland, V., Varsano, G., Li, X., Thornton, J., Baby, J., Aravind, A., Meyer, C., Pavic, K., Rios, P., and Köhn, M. (2011) The metastasis-promoting phosphatase PRL-3 shows activity toward phosphoinositides. *Biochemistry*, **50** (35), 7579–7590.
27. Newman, R.H. and Zhang, J. (2008) Visualization of phosphatase activity in living cells with a FRET-based calcineurin activity sensor. *Mol. BioSyst.*, **4** (6), 496–501.
28. Helms, V. (2008) *Principles of Computational Cell Biology*, Wiley-VCH Verlag GmbH, Weinheim, p. 202.
29. Qian, J., Lesage, B., Beullens, M., Van Eyde, A., and Bollen, M. (2011) PP1/Repo-man dephosphorylates mitotic histone H3 at T3 and regulates chromosomal aurora B targeting. *Curr. Biol.*, **21** (9), 766–773.
30. Clapham, D.E. (2007) Calcium signaling. *Cell*, **131** (6), 1047–1058.
31. Kuehnen, P., Laubner, K., Raile, K., Schöfl, C., Jakob, F., Pilz, I., Päth, G., and Seufert, J. (2011) Protein phosphatase 1 (PP-1)-dependent inhibition of insulin secretion by leptin in INS-1 pancreatic beta-cells and human pancreatic islets. *Endocrinology*, **152** (5), 1800–1808.
32. Devogelaere, B., Beullens, M., Sammels, E., Derua, R., Waelkens, E., van Lint, J., Parys, J.B., Missiaen, L., Bollen, M., and De Smedt, H. (2007) Protein phosphatase-1 is a novel regulator of the interaction between IRBIT and the inositol 1,4,5-trisphosphate receptor. *Biochem. J.*, **407** (2), 303–311.
33. Grynkiewicz, G., Poenie, M., and Tsien, R.Y. (1985) A new generation of Ca^{2+} indicators with greatly improved fluorescence properties. *J. Biol. Chem.*, **260** (6), 3440–3450.

5
Autophagy: Assays and Small-Molecule Modulators

Gemma Triola

5.1
Introduction

Autophagy (or self-eating) is a cellular pathway that regulates the degradation and recycling of obsolete organelles, long-lived proteins, protein aggregates, and pathogens. This process occurs under basal conditions and has a crucial role in cellular development, differentiation, survival, and homeostasis. Autophagy is complementary to the other major protein degradation pathway: the ubiquitin proteasome system (UPS). While UPS degrades short-lived nuclear or cytosolic proteins, autophagy is in charge of the clearance of long-lived cytoplasmic proteins, bigger protein complexes, and organelles. Three different pathways can be used for this purpose: the chaperon-mediated autophagy, microautophagy, and macroautophagy. In the chaperon-mediated autophagy, cytosolic proteins to be degraded are specifically selected by chaperons such as Hsc70 that recognize a pentapeptide in their sequence (Lys-Phe-Glu-Arg-Gln (KFERQ)-like) and by doing so target them to the lysosomal membrane where they will interact with proteins such as LAMP2 (lysosome-associated membrane protein type 2A). After getting unfolded, the proteins will get internalized in the lysosome and subsequently degraded by lysosomal hydrolases (Figure 5.1). While the other autophagy processes are activated 30 min after nutrient deprivation, the chaperon-mediated autophagy only starts after 10 h of starvation and can remain active for 3 days.

Macroautophagy, also known as *autophagy*, is the best characterized pathway. This mechanism is initiated with the formation of a phagophore, a cup-shaped double membrane that engulfs the cytoplasmic material to be degraded. The phagophore is then elongated and sealed to generate an autophagosome that will be fused with a lysosome, thereby delivering the cargo for degradation (Figure 5.1) [1]. Finally, the less-investigated microautophagy relies on the direct invagination of the lysosome to engulf cytosolic cargos.

A normal function of this process is crucial to maintain cell survival during normal cellular functioning, for example, under starving conditions (by enabling the recycling of cellular proteins, lipids, or carbohydrates to synthesize the required ones for ensuring survival), to prevent pathogen infection, or to eliminate

Concepts and Case Studies in Chemical Biology, First Edition. Edited by Herbert Waldmann and Petra Janning.

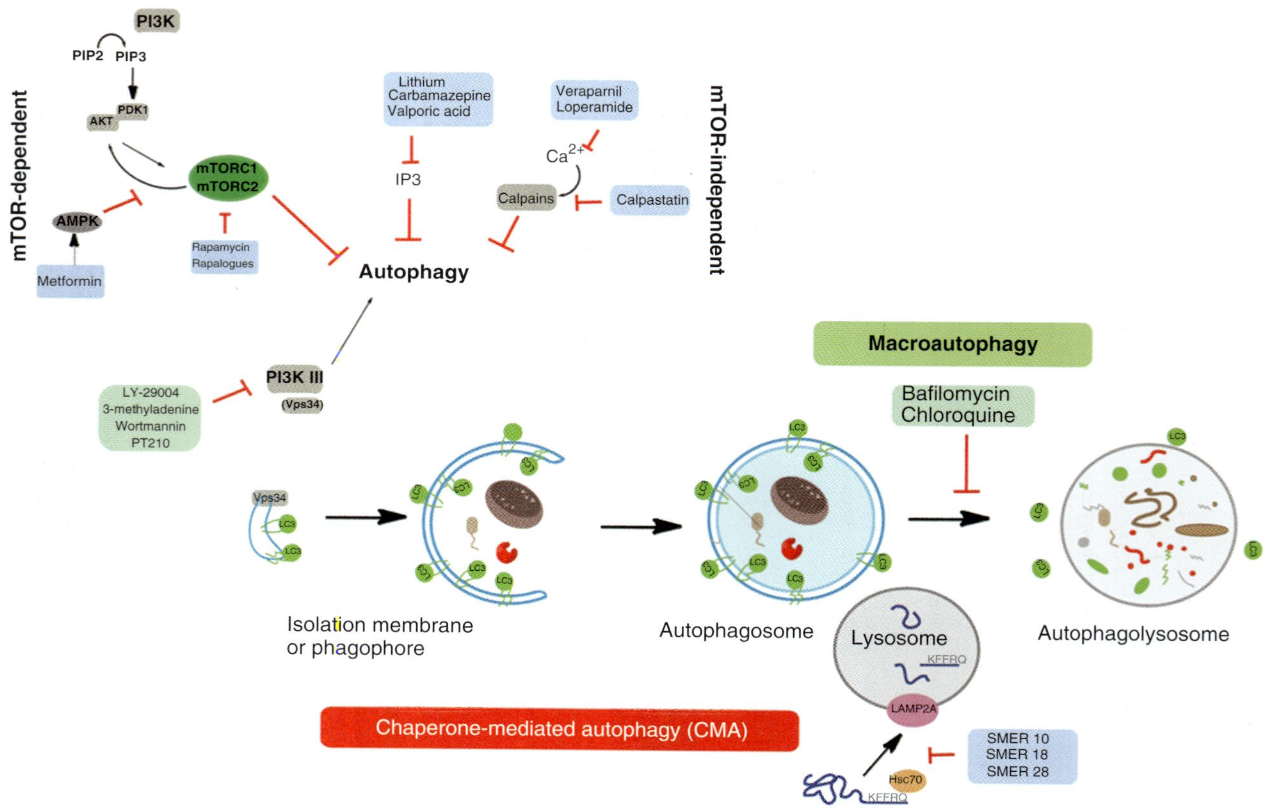

Figure 5.1 Schematic depiction of macroautophagy and chaperone-mediated autophagy, including some of the reported autophagy inhibitors (in blue) and inducers (in green).

protein aggregates. However, there is growing evidence that a malfunctioning of autophagy is also related to several pathologies and associated with aging [2]. Downregulation of autophagy seems to have a role in the progress of neurodegenerative diseases such as Huntington disease (HD), Parkinson's, or Alzheimer's, mainly due to the accumulation of aggregated proteins that cannot be cleared out. Dysfunction of the pathway has also been related to cancer, although in this case with controversial outcomes. Briefly, autophagy has been suggested to promote cell survival under the hypoxia conditions caused by chemotherapy and, contrarily, this process seems to function as an alternative apoptotic pathway in the early stages of tumor genesis [3]. Therefore, additional studies are required to determine the exact role of autophagy in cancer progression.

5.2 The Biological Problem

Autophagy is a complex and dynamic process regulated by different cellular pathways, that mainly act upstream of autophagosome formation. One of these pathways is the mammalian target of rapamycin (mTOR) signaling cascade, which has a key function in the regulation of autophagy. Hence, under nutrient-rich conditions, mTOR is activated resulting in autophagy suppression, whereas nutrient deprivation inhibits mTOR leading to activation of autophagy. mTOR signaling pathway can also be indirectly regulated by acting on the phosphoinositide-3-kinase (PI3K) signaling cascades. There are mainly two different classes of PI3K involved with different outputs. Class I PI3K, producing phospoinositide-3,4,5-triphosphate (PI(3,4,5)P3), leads to mTOR activation, and thus autophagy suppression, and class III PI3K/vps34 produces phosphatidylinositol-3-phosphate (PI3P) and positively regulates autophagy. Other key regulators of autophagy do not involve the mTOR network. Hence, modulation of the cellular calcium content or of the inositol 1,4,5-triphosphate (IP3) levels can also regulate autophagy in a mTOR-independent manner. Other indirect autophagy modulators act on lysosomal function by blocking lysosomal hydrolases or preventing its fusion with autophagosomes that cause autophagy inhibition (Figure 5.1).

In addition to these regulating pathways, a breakthrough in the study of autophagy was the identification of the so-called ATG (autophagy-related gene) proteins, essential for autophagosome generation, and its fusion with the lysosome [4]. One of these proteins is the yeast Atg8 and its mammalian orthologs LC3 (microtubule-associated protein 1 (MAP1) light chain 3), GATE-16 (golgi-associated ATPase enhancer of 16 kDa), and GABARAP (γ-aminobutyric acid type A receptor-associated protein). These proteins get associated with the autophagosome membrane after C-terminal conjugation to a phosphatidylethanolamine (PE) unit and are crucial for the autophagosome formation and closure and its fusion with lysosomes. Owing to this key role in autophagosome generation and maturation, these proteins have become *bona fide* markers of autophagy and have been widely used to monitor and study this process.

Although important advances have been seen in the past decade, still much work is needed to clarify the involvement of autophagy in health and disease and to elucidate the molecular mechanism controlling this process. As an example, the exact role of LC3 and other mammalian orthologs in autophagy progression is still unclear and more particularly the generation and maturation of autophagosomes. On the basis of this background, three main biological problems can currently be identified in the autophagy research field: the development of assays to identify and characterize autophagy, the identification of selective autophagy inhibitors or inducers, and the generation of chemical biology tools to elucidate the molecular mechanisms controlling this process.

5.2.1
Assays

An important field of research is the development of assays to detect and monitor autophagy, both at the cellular or protein level. Because autophagy is a dynamic and complex process involving many different cellular pathways at the same time (a change in the intracellular levels of cyclic AMP, the class I PI3Ks, the mTOR pathway, etc.), it is important to develop assays enabling a correct identification and monitoring of this process. The most employed methods for monitoring autophagy are largely based on the detection of autophagosome accumulation or on the identification of the LC3 protein (or the yeast Atg8), considered a specific autophagy marker. However, it is important to keep in mind that an accumulation of autophagosome or an increase in LC3 levels may be caused by autophagy induction but also can be the consequence of an inefficient fusion with the lysosomes or incorrect function of lysosomal hydrolases. Therefore, it is crucial to complement these studies with measurements of the overall autophagic flux (Box 5.1), in order to interpret the results correctly. A brief overview of the most commonly employed methods is summarized subsequently. It is also important to keep in mind that there are no absolute criteria to determine the autophagic status that can be systematically applied in all cellular and tissue systems. Therefore, the readers are encouraged to check out other complementary methods to determine cellular autophagy. An excellent review covering most of the existing assays has been recently published and may serve as a good basis to plan the required studies [5].

Box 5.1 Autophagic flux

Autophagic flux is the dynamic process including autophagosome synthesis, fusion with lysosomes, and the subsequent degradation of autophagic substrates.

5.2.2 Small-Molecule Modulators of Autophagy

The identification of novel, potent and selective autophagy inducers or inhibitors is required to further characterize this process and to be able to modulate it, either positively or negatively, both in basic research and with a final therapeutic aim. Some efforts have already been made to identify chemical modulators of autophagy. However, most of the detected compounds act simultaneously on different cellular pathways, which results in toxic or pleiotropic effects. Hence, there is still a clear need for more selective compounds that can block or induce autophagy without causing additional effects.

Rapamycin is, for example, one of the most used autophagy inducers. Rapamycin acts by forming a stable complex with FK506-binding protein 12 (FKB12) that stabilizes its association with the raptor mTOR and by doing this inhibits the kinase activity of mTOR. Although rapamycin is a highly selective inhibitor of mTOR, the high number of cellular processes, apart from autophagy, in which the mTOR pathway is involved, such as cell growth, protein synthesis, ribosome biogenesis, or nutrient metabolism, causes important side effects that are not compatible with a long-term use. Consequently, autophagy inducers that act independently of mTOR would be desired. Conversely, one of the most used autophagy inhibitors is 3-methyladenine. Its activity relies on the blockage of class I PI3K, but it also acts on class III PI3K and this should result in autophagy stimulation. However, the final outcome is usually autophagy inhibition because class III enzymes act downstream of the inhibited class I PI3K. Moreover, a dual role as autophagy activator and inhibitor depending on the nutrient conditions has been recently proposed for this compound [6] and therefore 3-methyladenine should be carefully used.

Autophagy inhibition has been also achieved upon general blocking of lysosomal pH acidification, which inhibits lysosomal function and prevents its fusion with autophagosomes. A significant example of this inhibitor class is bafilomycin, a macrolide that inhibits the vacuolar-type H^+ ATPase (V-ATPase) (ATPase, adenosine triphosphatase) responsible of the acid lysosomal pH required for its normal functioning. Another relevant example is chloroquine, an antimalarial drug that acts as a lysosomotropic agent (Box 5.2) and, as a result, causes the neutralization of the acid pH of these compartments.

Box 5.2 Lysosomotropic agents

Lysosomotropic agents are small molecules that get trapped in the lysosome as a result of protonation caused by the lysosomal acidic environment. As a consequence of this behavior, lysosomotropic agents accumulate preferentially in the lysosomes and ultimately cause the neutralization of the acid lysosomal pH.

In summary, there is still a clear need for novel chemical modulators that act selectively on this process, particularly if a final therapeutic aim is considered. Moreover, there is also a need for the identification of novel targets that enable a selective modulation of autophagy without affecting other cellular processes. Important attempts have already been done in this field and will be discussed subsequently. However, major advances are currently under way, which suggests that in the near future more selective inhibitors and novel targets will be identified.

5.3 The Chemical Approach

5.3.1 Assays

The most commonly employed methods for monitoring autophagy are largely based on the detection of the LC3 protein (or the yeast Atg8), that specifically localizes in autophagosome membranes and it is therefore considered a specific autophagy marker. LC3 detection is generally performed employing antibodies or using microscopy methods. Although most of the methods are widely applied and they have been very useful for the identification of autophagy processes, most of them present some limitations that should be taken into account. Therefore, complementary analysis using other approaches is generally recommended in all the cases.

As an example, a fluorescently labeled GFP-LC3 (green fluorescent protein) can be employed to monitor autophagy employing fluorescence microscopy. Autophagy induction is strongly correlated with the number of autophagosomes. As lipidated LC3-II is located in the autophagosomes, in cells expressing GFP-LC3 the number of positive GFP-LC3 vesicles, visible as fluorescent dots, is generally correlated with the number of autophagosomes, which enables its quantification. Although this method is one of the most employed and it has been very useful for determining autophagy, special care should be taken into account because of the large size of the GFP, which may cause artifacts in the localization, aggregation, or degradation of the protein (Figure 5.2a).

Another method to study autophagy relies on monitoring the direct increase in LC3-II levels as well as the ratio between the lipidated LC3-II and nonlipidated LC3-I on a Western blot employing antibodies against LC3 that recognize both proteins, the lipidated and the nonlipidated one. This is also a useful approach widely used in autophagy research, despite some limitations that have been encountered. Briefly, although a 10-fold increase in Atg8 levels is usually visible in yeast, this is not always the case in mammals, where autophagy induction may not always results in an increase in LC3-II levels due, for example, to a rapid turnover of the generated LC3-II that gets degraded in the lysosomes. Moreover, changes in LC3-II levels may be tissue and cell dependent, which complicates a systematic

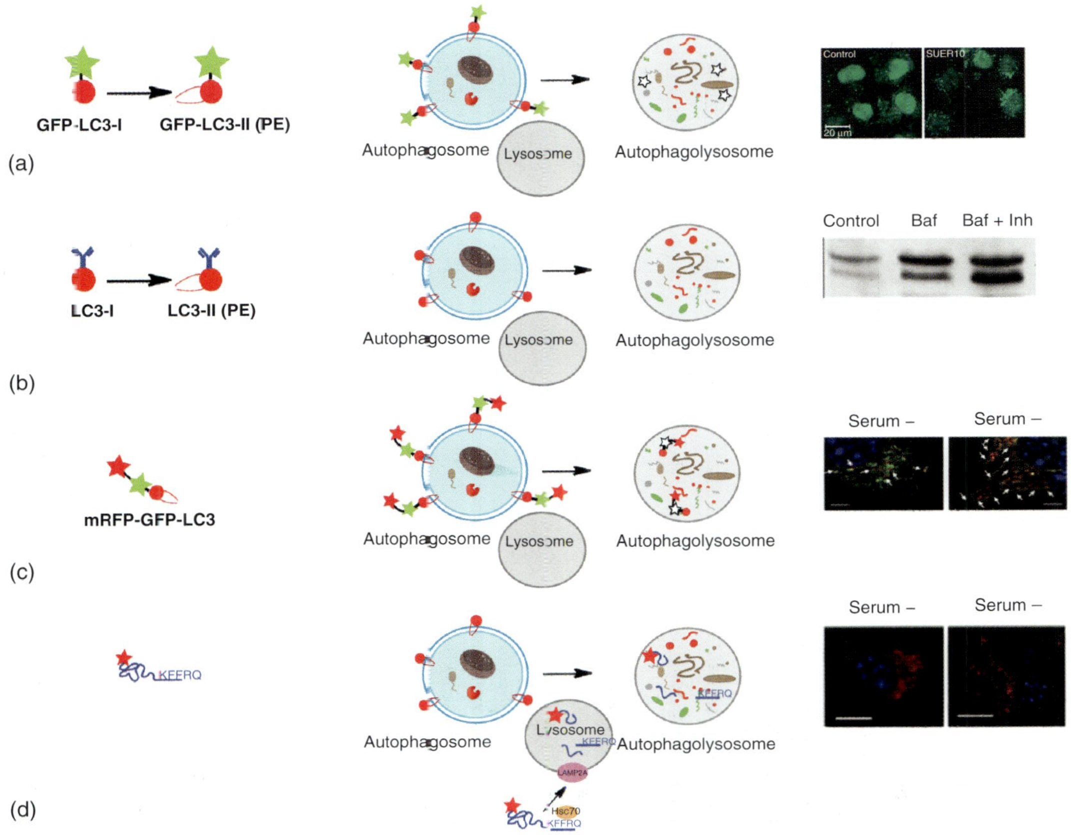

Figure 5.2 (a–d) Most commonly employed assays to monitor autophagy, autophagic flux, and chaperone-mediated autophagy (CMA). (Reprinted with permission from Macmillan Publishers Ltd: Nature Chemical Biology from [7], Copyright (2007) and [8] Copyright 2013).)

analysis using this approach. In addition, antibodies against LC3 may recognize differently the nonlipidated and the lipidated form. Generally, the antibodies available recognize an α-helix at the N-terminus of LC3. However, a probable conformational change of LC3 upon lipidation makes LC3-II recognition by the antibody better than the one of LC3-I, which may complicate the analysis.

An important fact that needs to be taken into account is that, as mentioned earlier, autophagosome or LC3-II accumulation may be produced as a result of an induction of autophagy but can also be produced for an inefficient fusion with lysosome or an inability of the lysosomes to degrade autophagosome content once fusion has taken place, caused, for example, by inhibition of lysosomal function after acidification of lysosomal pH. For these reasons, determination of LC3 levels should be always accompanied by measurements of autophagic flux (Box 5.1) in order to ensure a correct interpretation of the results.

Several different methods have been described to determine autophagic flux. Western blot analysis can be employed. In this case, it is necessary to measure LC3-I and LC3-II levels in cells treated with blockers of lysosomal degradation and compare them with the levels detected in nontreated cells. Lysosomal degradation can be blocked by treatment with protease inhibitors such as E-64d or by affecting lysosomal function, for example, upon treatment with inhibitors of the acidification of lysosomes such as bafilomycin or chloroquine. Further increase in LC3-II under these conditions implies that autophagic flux was previously working, while the detection of similar levels of LC3-II in treated and nontreated cells would mean that instead of autophagy induction there is a blocking in the lysosomal function (Figure 5.2b).

Separate detection of autophagosome and autolysosome numbers can help measure the autophagic flux. To this end, an increase in autophagosome and autolysosome numbers would indicate autophagy activation, while an increase in autophagosome with no changes in autolysosome numbers would suggest a block in maturation. To this end, a tandem monomeric LC3 protein (mRFP-GFP-LC3 or mCherry-GFP-LC3) that enables to detect selectively autophagosomes and lysosomes without requiring previous cell treatment with inhibitors has been extensively used. Because GFP fluorescence is sensitive to acid pH, autophagosomes will be visible in yellow as a result of the simultaneous fluorescence of GFP and monomeric red fluorescent protein (mRFP) or mCherry, while the acidic lysosomes will be visible in red (Figure 5.2c). This tandem fluorescent protein has been also employed to detect modulators of autophagy in high-throughput screen/screening (HTS) approaches.

The ultimate method to monitor autophagy and autophagic flux is the measurement of protein breakdown. Hence, an induction of autophagy should result in an increase in long-lived protein degradation (short-lived proteins are usually degraded by the proteasome system and therefore they should not be affected). The usual approach to investigate protein degradation relies on the labeling of protein upon incubation with a radioactively labeled amino acid for 24 h. This is followed by a nonlabeling step in order to enable the degradation of short-lived proteins. Then, the release of labeled amino acids in the medium coming from protein degradation is measured upon time using a scintillation counter. The same amino acid in a nonlabeled version is generally included in the medium to avoid the incorporation of the released labeled amino acid in the new protein synthesis.

In summary, there are several methods to study and monitor autophagy, mostly based on the presence of LC3-II. This protein is generated upon C-terminal lipidation of LC3-I with a unit of phosphatidylethanolamine and, once lipidated, gets preferentially associated with autophagosomes membranes and has therefore become a specific autophagy marker. However, it is always recommended to employ different complementary methods to detect and measure autophagy. Importantly, methods to independently detect autophagy and autophagy flux are required to discern if an increase in LC3-II is due to autophagy induction or to an inhibition in lysosome function.

5.4
Chemical Biological Evaluation

There have been some approaches to detect novel chemical modulators of autophagy. The initial key step is usually the establishment of a cellular screening assay enabling the detection of autophagy modulators. The activity of the identified small molecules on autophagy needs to be generally further investigated using other methods to confirm their effect.

Example 1 One of the first examples reported in this research area was a high-throughput, image-based screen developed to search for autophagy inducers [9]. The assay was based on the detection of molecules able to cause an increase in LC3-positive autophagosomes. To this end, a human glioblastome H4 cell line stably expressing a fluorescently labeled LC3 (GFP-LC3) was established and upon treatment with small molecules their effect on the number, size, and intensity of the fluorescence spots was investigated, detecting fluorescently labeled LC3 proteins/vesicles with a fluorescence microscopy. A compound library containing 480 small molecules with known bioactivity was screened and rapamycin and dimethyl sulfoxide (DMSO) were used as a positive and a negative control, respectively. As a result of this initial screening, 72 compounds were found active including known autophagy inducers such as tamoxifen, but there were also lysosomal function inhibitors such as bafilomycin, where the increase in LC3-positive autophagosome was not caused by increase of autophagy but due to a blockage in lysosomal function. Cellular toxicity of the active compounds was then investigated. Toxicity was assessed employing a nuclear staining (DAPI (4,6-diamidino-2-phenylindole), Box 5.3) and 23 compounds causing a 30% reduction in the cell numbers were discarded. The effect of the remaining 48 compounds was then investigated on an assay measuring long-lived protein degradation in the treated cells. To this end, cells were treated for 24 h with a medium containing a radioactive labeled amino acid, [3H] leucine, and after this time, the medium was replaced with a normal medium containing an excess of nonlabeled leucine and the cells were incubated for an additional 24 h to enable the degradation of labeled short-lived proteins. Small molecules were then added and the radioactivity present in the medium as a result of protein breakdown was measured at different time courses. After 24 h, the total radioactivity present in the cell lysate was measured and the degradation rate was calculated by dividing the average counts per minute (cpm) in medium with the total cpm (cpm medium + cpm in the cell lysate). As a result, eight compounds able to increase protein degradation were detected using this assay: fluspirilene, trifluoperazine, pimozide, niguldipine, nicardipine, amiodarone, loperamide, and penitrem A (Figure 5.3).

Box 5.3 DAPI Stain

DAPI is a dye employed for selective staining of DNA. DAPI acts by binding to the minor groove of double-stranded DNA, which results in a 20-fold increase in its fluorescence. It is one of the most commonly used methods for selective nuclear staining owing to its selectivity and high cell permeability.

Results were further confirmed by calculating the ratio between the lipidated LC3-II and LC3-I by Western blot. A ratio of 0.36 was detected in the DMSO control and 1.21 in the cells treated with rapamycin, whereas the cells treated with any of the other eight compounds had LC3-II/LC3-I ratios between 0.75 and 1.84, thus confirming its activity as autophagy inducers (Figure 5.4a,b).

As a final approach to investigate the effect of these small molecules on autophagy and therefore on protein degradation, a model protein containing polyglutamine (polyQ) repeats was chosen as an example. PolyQ repeat diseases are characterized by the presence of proteins containing polyQ tracts, generally longer than 35 amino acids and located at the N-terminus of proteins, and include, among others, the Hungtington disease (Box 5.4) or some spinocerebellar ataxies. It is believed that cells cannot correctly degrade proteins with extremely long polyQ sequences, which results in protein aggregation of the mutant proteins that can be seen inside neurons damaging these cells. To get an insight into the ability of these compounds to induce the clearance of polyQ-containing proteins, the cells were transfected with a fluorescently labeled polyQ sequence containing a hemagglutinin (HA) tag (GFP-Q79-HA). Detection of the protein using anti-HA blots indicated that except nicardipine the other autophagy inducers were able to reduce the accumulation of expanded polyQ in a dose-dependent manner, thus confirming its activity (Figure 5.4c,d).

Box 5.4 Huntington disease

HD, also called Huntington's chorea, causes a progressive degeneration of the nerve cells in the brain and it is caused by mutations in the Huntingtin gene. The resulting mutant protein (Huntingtin) contains a long stretch of polyQ at the N-terminus that may induce conformational changes and the formation of protein aggregates that cannot be cleared out by the cells.

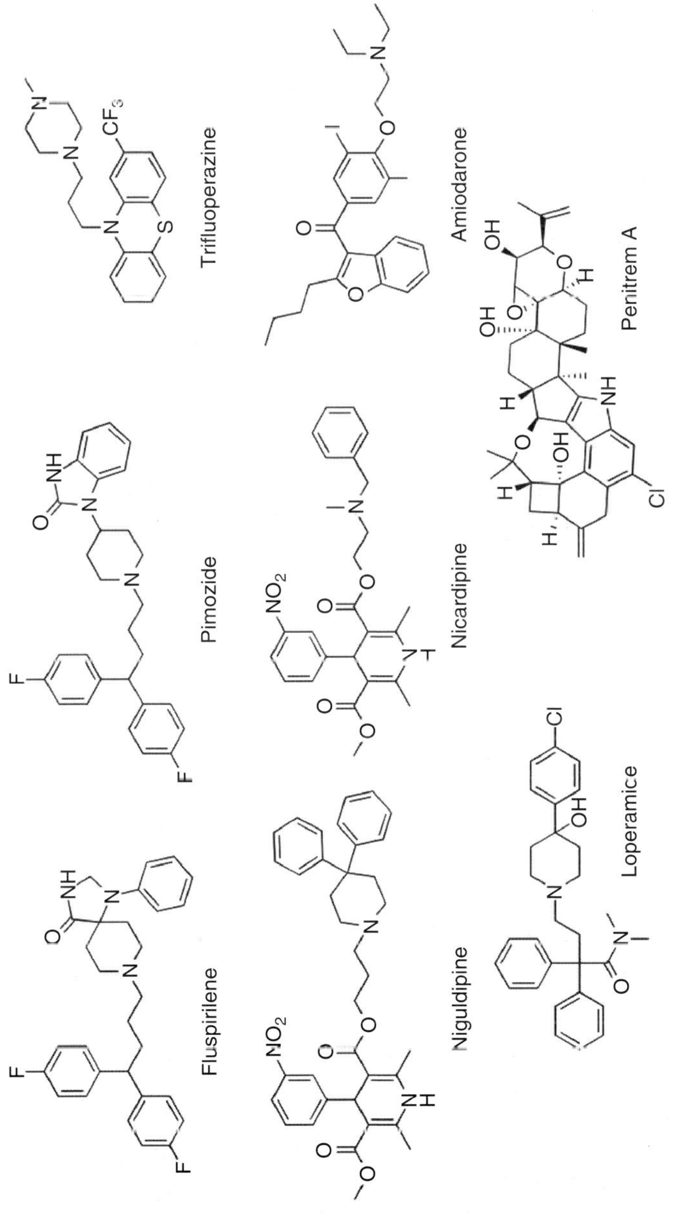

Figure 5.3 Small molecules identified as autophagy inducers.

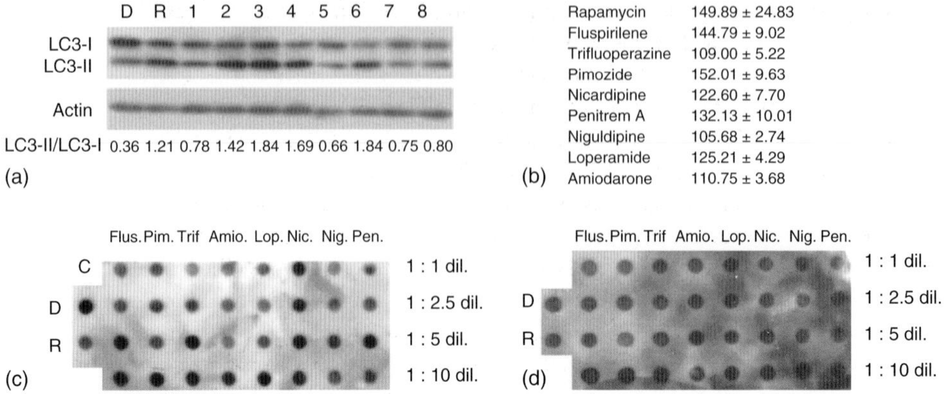

Figure 5.4 (a) Effect of the compounds on the ratio of LC3-I and LC3-II after 4 h of treatment with the indicated compounds. D, DMSO; R, rapamycin; 1, amiodarone; 2, niguldipine; 3, trifluoperazine; 4, loperamide; 5, penitrem A; 6, pimozide; 7, fluspirilene; 8, nicardipine. (b) Percentage of control long-lived protein degradation after 4 h of treatment. (c) Effect of the compounds on the accumulation of expanded polyglutamine (GFP-Q79-HA) as measured by dot blotting with anti-HA antibody. (d) Control experiment employing anti-actin antibody. (Reprinted with permission from [9], Copyright (2007) National Academy of Sciences, U.S.A.)

Although these identified drugs (fluspirilene, trifluoperazine, pimozide, niguldipine, nicardipine, amiodarone, loperamide, and penitrem A) are indicated for different diseases, they all share a similar mechanism of action. Fluspirilene, pimozide, and trifluoroperazine are antipsychotic drugs of the diphenylbutylpiperidine and the phenothiazine class, respectively. These drugs are used in different types of schizophrenia and act as dopamine receptor antagonists and calcium signaling blockers. Niguldipine, nicardipine, and amiodarone are drugs prescribed for the treatment of cardiovascular disorders. Niguldipine and nicardipine are mostly indicated for the treatment of hypertension, whereas amiodarone is an antiarrhythmic agent and they all act as calcium channel blockers. Finally, loperamide, employed against diarrhea, is an agonist of μ-opiod receptor and a nonselective calcium channel blocker, whereas penitrem A is a fungal toxin also related with calcium signaling. This common activity on calcium signaling is in agreement with other studies indicating that Calcium signaling may have a key role in autophagy regulation.

In summary, an HTS was established and employed to investigate the activity of a compound library containing 480 small molecules with known bioactivity. As a result, eight small molecules able to induce autophagy were identified and seven of them showed ability to increase protein degradation. More significantly, compounds were able to increase the clearance of proteins containing long polyQ tracts often related with neurodegenerative diseases such as the HD. Most of the detected compounds are calcium signaling modulators.

Example 2 In another recent approach to search for autophagy modulators, Rubinsztein and Schreiber used a yeast-based screen to search for small molecules inhibiting or enhancing the effect of rapamycin [7]. A library formed by 50 729 compounds was screened in this assay. Although the initial read-out was the suppression or the enhancement of the cytostatic effect of rapamycin in yeast, the effect of the identified active compounds on autophagy was then investigated in the absence of rapamycin in order to study their ability to induce clearance of relevant autophagy substrates. A53T α-synuclein, whose protein aggregates are involved in Parkinson's disease, was initially chosen as a working model. This assay revealed that from the initially identified active compounds, 13 caused an inhibition, and 4 activated the clearance of α-synuclein protein aggregates. The activity of the four enhancers was then further confirmed by investigating the clearance of another important autophagy substrate, Huntingtin. Mutant Huntingtin bears a long polyQ tract at the N-terminus, whic makes difficult its degradation and causes protein aggregates that are toxic to the cells (Box 5.4). Hence, as a model to study its clearance, COS-7 cells were transfected with an EGFP (enhanced green fluorescent protein)-labeled Huntingtin with 74 polyQ repeats (EGFP-HDQ74) and the levels of protein aggregates and cell death were measured upon treatment with the detected activators. One of the compounds was discarded because of toxic effects but the other three retained their activity in the enhancement of Huntingtin clearance (Figure 5.5).

Effect on autophagy was further confirmed in COS-7 and HeLa cells transfected with GFP-LC3, indicating that the three identified compounds were able to increase the number of autophagosomes per cell compared to control cells. Moreover, these compounds were also able to increase the levels of LC3-II in the presence of the lysosomal inhibitor bafilomycin A1 (Baf), as detected in Western blot, thus discarding an effect on lysosomal function or in the fusion with the autophagosomes (Figure 5.6).

To sum up, an HTS assay including 50 729 compounds has enabled the detection of three autophagy activators that may serve as a starting point for the synthesis of more potent and selective modulators of autophagy.

Example 3 Although most of the work in the search for chemical modulators of autophagy has been directed at targeting macroautophagy, recent efforts have been also made for the selective target of chaperone-mediated autophagy (CMA) [8]. CMA is involved in the progression of neurodegenerative diseases, and a decline in CMA activity with age seems to be strongly correlated with age-related disorders. CMA relies on the direct translocation of proteins to the

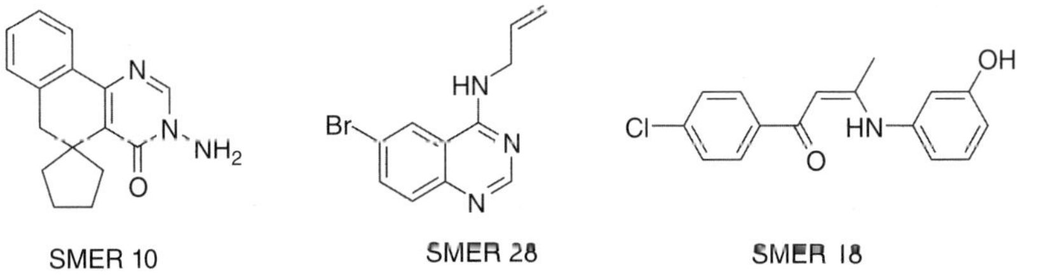

Figure 5.5 Small-molecule enhancers of rapamycin activity (SMER) detected in this study.

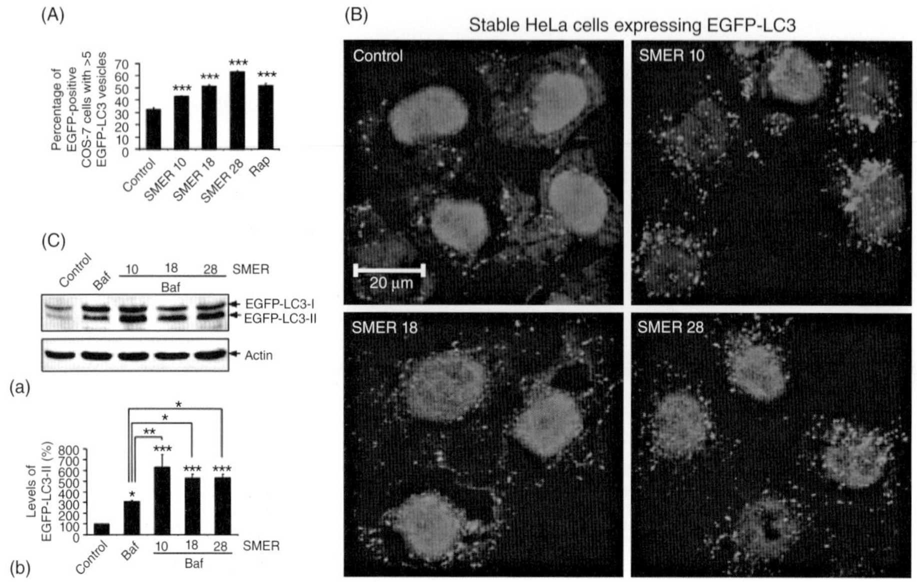

Figure 5.6 *SMERs 10, 18, and 28 induce autophagy in mammalian cells.* (A) COS-7 cells transfected with EGFP-LC3 construct for 4 h were treated with DMSO (control), 0.2 mM rapamycin (positive control), 47 μM SMER10, 43 μM SMER18, or 47 μM SMER28 for 16 h, and analyzed by fluorescence microscopy. (B) HeLa cells stably expressing EGFP-LC3 were treated with DMSO (control), 47 μM SMER10, 43 μM SMER18, or 47 μM SMER28 for 24 h. Confocal sections show cells containing EGFP-positive autophagic vesicles. Nuclei are stained with DAPI. Bar, 20 mM. (C) HeLa cells stably expressing EGFP-LC3 were treated for 4 h with DMSO (control) or 200 nM bafilomycin A1 (Baf), or with 200 nM bafilomycin A1 and 47 μM SMER10, 43 μM SMER18, or 47 μM SMER28. Cells were left untreated or pretreated with SMERs for 24 h before adding bafilomycin A1. (a) Levels of EGFP-LC3-II were determined by immunoblotting with antibody against EGFP and (b) densitometry analysis relative to actin. (Reprinted with permission from Macmillan Publishers Ltd: Nature Chemical Biology from [7], Copyright (2007).)

lysosomes mediated by chaperons and lysosomal proteins. There is currently a lack of selective modulators of CMA, mainly due to the fact that the cellular pathways controlling this pathway are nearly unknown [10]. A recent work has, however, shed light into this process by identifying a novel and selective target for CMA, the retinoic acid receptor alpha (RARα), which has enabled the synthesis of RARα antagonists that resulted in selective CMA induction without affecting macroautophagy.

The first hint of the key role of the RARα receptor in the regulation of CMA was revealed by studies showing that the knockdown of RARα using small hairpin ribonucleic acid (shRNA) resulted in a significant increase in the degradation rate of long-lived proteins (Figure 5.7A,B,C). The putative involvement of macroautophagy in this effect was first investigated. However, the reduced effect of 3-methyladenine, a well-characterized inhibitor of macroautophagy, in decreasing the rate of proteolysis (Figure 5.7D) and in the amount of LC3-II protein

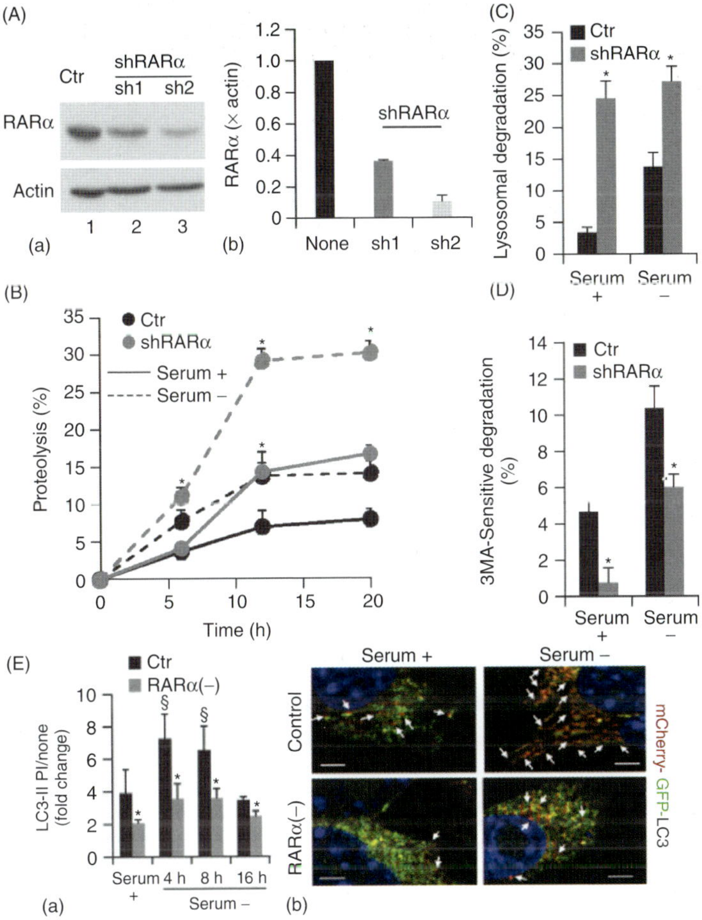

Figure 5.7 *Effect of knockdown of RARα on intracellular turnover of long lived proteins.* (A) Knockdown of RARα in NIH3T3 mouse fibroblasts was conducted using two different shRNAs, sh1, and sh2, compared to control (Ctr). (a) Representative immunoblot. Actin is shown as loading control and (b) amounts of RARα in control and knockdown cells determined by densitometric quantification of immunoblots represented by the one shown in (a). Values are normalized for actin and expressed as multiples of control (none) values; $n = 3$. (B) Rates of degradation of long-lived proteins in control and RAR(−) cells maintained in the presence or absence of serum for 12 h. Values are expressed as percentage of proteolysis; $n = 3$. (C, D) Percentage of degradation due to lysosomes (C) and macroautophagy (D) in cells assayed as in (B) but treated with inhibitors of lysosomal proteolysis (C) or with 3-methyladenine (3MA) to block macroautophagy (D). Values are expressed as percentage of total protein degradation sensitive to the lysosomal inhibitors; $n = 3$. In all panels, all values are mean s.e.m., and differences with control are significant for $*P < 0.05$. (E) Serum-deprived RAR(−) cells do not present higher amounts of macroautophagy as indicated by LC3-II levels (a) and the number of mCherry-GFP-LC3 containing lysosomes (b). (Reprinted with permission from Macmillan Publishers Ltd: Nature Chemical Biology from [8], Copyright (2013).)

detected suggested that the highest protein degradation rate was not caused by macroautophagy. This was further confirmed by a decrease in lysosome numbers when using a double-labeled mCherry-GFP-LC3, thus confirming a reduction in macroautophagy and indicating that the increase in protein degradation could not be caused by this process (Figure 5.7E).

To further investigate the involvement of CMA in this process, cells were transfected with a photoactivatable (PA) KFERQ-PA-Cherry. In CMA, proteins containing a KFERQ-like sequence are recognized by chaperons such as Hsc70 that directs them to the lysosome for their degradation. Hence, the KFERQ-Cherry is visible as a fluorescent diffused pattern but upon CMA activation it should change to fluorescence dots clearly visible in the lysosomes, owing to its internalization. As expected, an increase in fluorescent puncta was visible in the knockdown cells RARα(−) when compared to control cells, thus confirming the activation of CMA in RARα(−) cells (Figure 5.8). As an additional indication of the involvement of this process, cells were treated with all-*trans* retinoic acid (ATRA, Figure 5.9), the natural substrate for the RARα receptor and the activation of RARα upon binding of its substrate resulted in a reduced CMA activation in response to serum removal that was accompanied by a reduction of protein degradation rates.

Because the previous results demonstrated that ATRA does not affect macroautophagy and that the effect of ATRA on CMA was only dependent on RARα, it was envisaged that ATRA derivatives acting as RAR antagonists should result in selective activators of CMA without affecting other autophagy pathways.

To this end, a small library of 29 compounds based on retinoic acid derivatives was designed and synthesized. Modifications at the C4 position at the hydrophobic ring were incorporated to prevent its oxidation. Derivatives were grouped in four major families: aminonitrile retinoids (AmRs), boron-aminonitrile retinoids (BAmR), guanidine retinoids (GRs), and atypical retinoids (AR) (Figure 5.9).

Once the toxicity of the compounds on cells was discarded at a concentration of 50 μM, the compounds were tested on fibroblasts expressing the CMA reporter (KFERQ-Cherry) and their effect was investigated by measuring the increase in fluorescence puncta in cells and further confirmed upon investigation of the protein degradation rate using metabolic labeling methods. These studies enabled

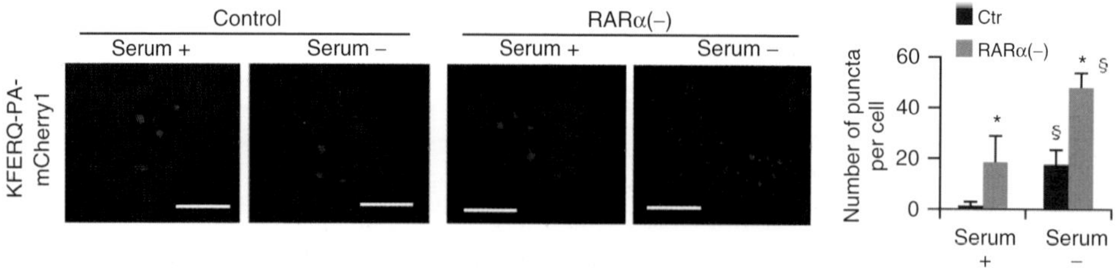

Figure 5.8 Effect of the identified compounds on the activation of chaperon-mediated autophagy. (Reprinted with permission from Macmillan Publishers Ltd: Nature Chemical Biology from [8], Copyright (2013).)

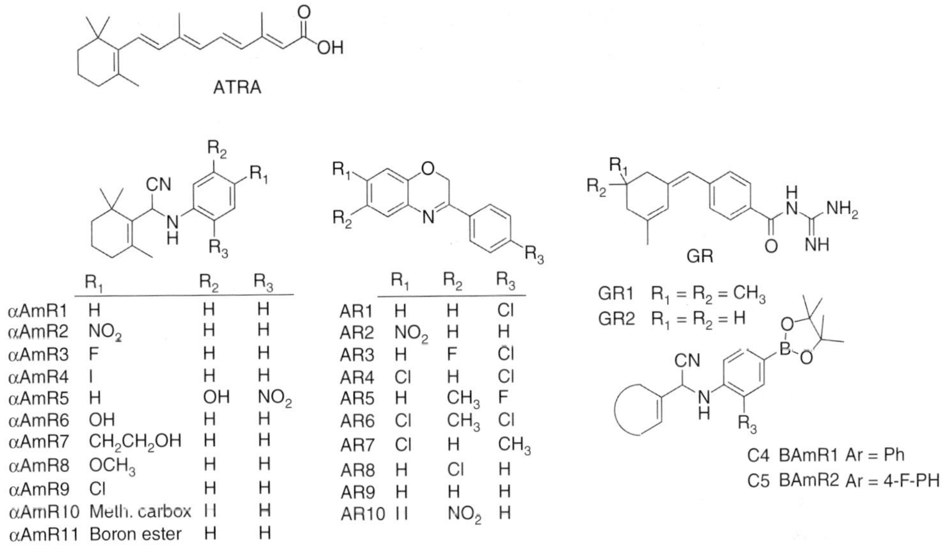

Figure 5.9 ATRA and ATRA derivatives tested for their activity as RARα receptor antagonists.

the identification of three compounds (AR7, GR1, and GR2) able to cause a 2.5-fold increase in the number of fluorescent puncta in serum-supplemented cells. The activity of these three compounds was further validated by investigating the protein degradation rates using metabolic labeling. Moreover, additional studies revealed that these three compounds were not able to produce an increase of CMA in cells knocked down for the lysosomal protein LAMP2A, crucial for the initiation and progression of CMA and considered a unique component of this type of autophagy. Conversely, compounds were still causing CMA activation in cells knocked down for LAMP2B, known to participate in other cellular functions such as macroautophagy, lysosomal biogenesis an cholesterol trafficking, thus further confirming their effect on the selective activation of CMA.

In summary, the selective activation of CMA achieved by the retinoic acid derivatives shows the potential selective targeting of autophagy processes using small molecules.

Although much work is still required for a better characterization of autophagic processes, the identification of small molecules capable of acting selectively on macroautophagy and CMA may have important therapeutic potentials. It is envisaged that in the near future more novel phenotypic screening assays will be established, enabling the identification of autophagy modulators hopefully with a better selectivity profile. Moreover, more specific assays focused on the identification and validation of novel targets will probably also contribute to the advancement in this research field.

5.5
Conclusion

Research in autophagy has experienced major advances in the past decade [11]. This is also illustrated by the growing interest in the scientific community that has been translated in a tremendous increase in the number of publications covering this research field.

A key step was the identification of autophagy-specific proteins LC3, GATE-16, and GABARAP that enabled the molecular characterization of this process, the identification of a stepwise mechanism and the establishment of multiple assays and screenings methods. Consequently, these previous works have strongly contributed to our current knowledge. However, there is still a need for chemical tools allowing a more in-depth characterization of the process. Analogously, although some important inhibitors and activators of autophagy have been described and widely used in basic research, their lack of selectivity usually results in pleiotropic and toxic effects. Some additional attempts have been made to identify novel and more selective inhibitors, although with limited success. Therefore, many challenges still lie ahead for the application of autophagy modulators to diseases. These limitations will be probably addressed in the near future with the development of novel HTS assays together with chemical biology techniques. One can well imagine that the application of chemical biology tools to autophagy will strongly contribute to advance our knowledge in this field by enabling the molecular characterization of this process, the identification of novel interesting targets, and the detection of potent and selective modulators for these targets.

References

1. Fleming, A., Noda, T., Yoshimori, T., and Rubinsztein, D.C. (2011) Chemical modulators of autophagy as biological probes and potential therapeutics. *Nat. Chem. Biol.*, 7, 9–17.
2. Rubinsztein, D.C., Marino, G., and Kroemer, G. (2011) Autophagy and aging. *Cell*, **146**, 682–695.
3. Rubinsztein, D.C., Codogno, P., and Levine, B. (2012) Autophagy modulation as a potential therapeutic target for diverse diseases. *Nat. Rev. Drug Discovery*, **11**, 709–730.
4. Shpilka, T., Mizushima, N., and Elazar, Z. (2012) Ubiquitin-like proteins and autophagy at a glance. *J. Cell Sci.*, **125**, 2343–2348.
5. Klionsky, D.J., Abdalla, F.C., Abeliovich, H., Abraham, R.T., Acevedo-Arozena, A., Adeli, K., Agholme, L., Agnello, M., Agostinis, P., Aguirre-Ghiso, J.A., Ahn, H.J., Ait-Mohamed, O., Ait-Si-Ali, S., Akematsu, T., Akira, S., Al-Younes, H.M., Al-Zeer, M.A., Albert, M.L., Albin, R.L., Alegre-Abarrategui, J., Aleo, M.F., Alirezaei, M., Almasan, A., Almonte-Becerril, M., Amano, A., Amaravadi, R., Amarnath, S., Amer, A.O., Andrieu-Abadie, N., Anantharam, V., Ann, D.K., Anoopkumar-Dukie, S., Aoki, H., Apostolova, N., Arancia, G., Aris, J.P., Asanuma, K., Asare, N.Y., Ashida, H., Askanas, V., Askew, D.S., Auberger, P., Baba, M., Backues, S.K., Baehrecke, E.H., Bahr, B.A., Bai, X.Y., Bailly, Y., Baiocchi, R., Baldini, G., Balduini, W., Ballabio, A., Bamber, B.A., Bampton, E.T., Banhegyi, G., Bartholomew, C.R., Bassham, D.C., Bast, R.C. Jr., Batoko, H., Bay, B.H., Beau, I., Bechet, D.M., Begley, T.J., Behl, C., Behrends, C., Bekri, S.,

Bellaire, B., Bendall, L.J., Benetti, L., Berliocchi, L., Bernardi, H., Bernassola, F., Besteiro, S., Bhatia-Kissova, I., Bi, X., Biard-Piechaczyk, M., Blum, J.S., Boise, L.H., Bonaldo, P., Boone, D.L., Bornhauser, B.C., Bortoluci, K.R., Bossis, I., Bost, F., Bourquin, J.P., Boya, P., Boyer-Guittaut, M., Bozhkov, P.V., Brady, N.R., Brancolini, C., Brech, A., Brenman, J.E., Brennand, A., Bresnick, E.H., Brest, P., Bridges, D., Bristol, M.L., Brookes, P.S., Brown, E.J., and Brumell, J.H. (2012) Guidelines for the use and interpretation of assays for monitoring autophagy. *Autophagy*, **8**, 445–544.

6. Wu, Y.T., Tan, H.L., Shui, G.H., Bauvy, C., Huang, Q., Wenk, M.R., Ong, C.N., Codogno, P., and Shen, H.M.J. (2010) Dual role of 3-methyladenine in modulation of autophagy *via* different temporal patterns of inhibition on class I and III phosphoinositide 3-kinase. *J. Biol. Chem.*, **285** (14), 10850–10861.
7. Sarkar, S., Perlstein, E.O., Imarisio, S., Pineau, S., Cordenier, A., Maglathlin, R.L., Webster, J.A., Lewis, T.A., O'Kane, C.J., Schreiber, S.L., and Rubinsztein, D.C. (2007) Small molecules enhance autophagy and reduce toxicity in Huntington's disease models. *Nat. Chem. Biol.*, **3**, 331–338.
8. Anguiano, J., Garner, T.P., Mahalingam, M., Das, B.C., Gavathiotis, E., and Cuervo, A.M. (2013) Chemical modulation of chaperone-mediated autophagy by retinoic acid derivatives. *Nat. Chem. Biol.*, **2013** (9), 374–382.
9. Zhang, L.H., Yu, J., Pan, H.L., Hu, P., Hao, Y., Cai, W.Q., Zhu, H., Yu, A.D., Xie, X., Ma, D.W., and Yuan, J.Y. (2007) Small molecule regulators of autophagy identified by an image-based high-throughput screen. *Proc. Natl. Acad. Sci. U.S.A.*, **104**, 19023–19028.
10. Finn, P.F., Mesires, N.T., Vine, M., and Dice, J.F. (2005) Effects of small molecules on chaperone-mediated autophagy. *Autophagy*, **1**, 141–145.
11. Klionsky, D.J. (2007) Autophagy: from phenomenology to molecular understanding in less than a decade. *Nat. Rev. Mol. Cell Biol.*, **8**, 931–937.

6
Elucidation of Protein Function by Chemical Modification

Yaowen Wu and Lei Zhao

6.1
Introduction

Studying protein function *in vitro* or in the context of live cells and organisms is of vital importance in biological research. The complexity is dramatically increased from the level of the genome to the proteome by posttranslational modifications of proteins (PTMs), such as phosphorylation, lipidation, glycosylation, ubiquitination, and so forth. PTMs play an important role in regulating structure and function of proteins. However, it is usually challenging to recombinantly produce posttranslationally modified proteins in terms of homogeneity and output. Moreover, it is not possible to manipulate the structure of proteins using traditional biochemical and molecular biology techniques. Chemical modification of proteins to generate posttranslationally modified proteins and/or to incorporate synthetic probes, such as fluorophores, affinity tags, and other functional labels is invaluable in investigating protein function *in vitro* and *in vivo* [1].

It is of great importance to visualize biological events in cells. Genetic tags such as fluorescent proteins (FPs) are widely used to detect and track proteins [2, 3]. Many organic dyes are superior to fluorescent proteins in terms of brightness, photostability, far red-emission, environmental sensitivity, and flexibility for modifications of their spectral and biochemical properties. Moreover, chemical modification is widely used for protein immobilization on microarrays [4]. Therefore, chemical modification of proteins has become an important strategy for the study of protein function. The emerging chemical labeling techniques have significantly expanded the range of manipulating protein structures.

Chemical modification can be introduced at the amino-acid side chains and/or the N/C-terminus of a protein. In this chapter, we show a few examples of chemical protein modification methods and application of these methods to solve biological questions.

Concepts and Case Studies in Chemical Biology, First Edition. Edited by Herbert Waldmann and Petra Janning.

6.2 The Biological Problem

6.2.1 Small GTPases

Guanosine triphosphate (GTP)-binding proteins (G-proteins, GTPases (guanosine triphosphatase)) play regulatory functions in many cellular processes including signal transduction, cell growth and differentiation, cytoskeletal rearrangement, and vesicular transport. Rab (rat sarcoma related in brain) GTPases function as key regulators of intracellular vesicular transport. As the other members in the rat sarcoma (Ras) superfamily, the switch between the inactive GDP-bound (guanosine diphosphate) form and the active GTP-bound form of Rab GTPases is highly regulated by GTPase-activating proteins (GAPs), which accelerate the intrinsic GTP hydrolysis of GTPases, and by guanine nucleotide exchange factors (GEFs), which facilitate exchange of GDP for GTP (Figure 6.1). The GTP-bound Rab proteins interact with their effectors involved in vesicular transport, including budding, delivery, docking, and fusion [5, 6].

Similar to Ras and Rho proteins, Rab proteins are posttranslationally modified at the C-terminus with prenyl (geranylgeranyl) groups that function as membrane anchors. Protein prenylation involves covalent attachment of the farnesyl (C-15 isoprenyl) or the geranylgeranyl (C-20 isoprenyl) moiety to one or two C-terminal cysteine residues of the protein substrate via a stable thioether linkage.

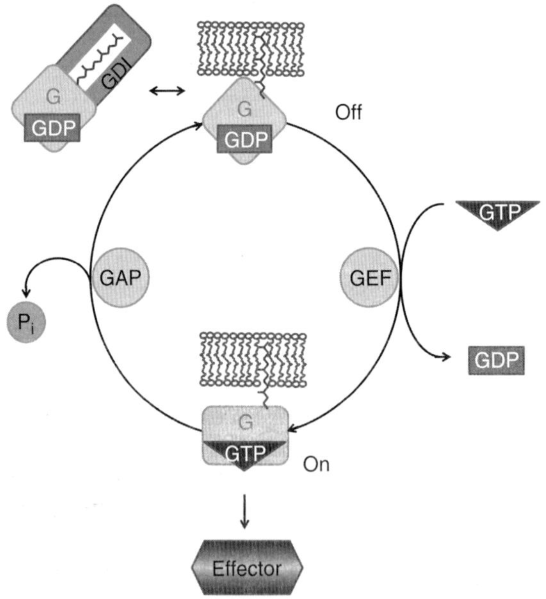

Figure 6.1 GTPase cycle of Ras-like proteins.

Prenylated small GTPases represent major hubs in most membrane-connected signaling networks. Rab prenylation is mediated by Rab geranylgeranyl transferase (RabGGTase), which works together with an adaptor protein, Rab escort protein (REP). Rab cycling between membranes and the cytosol is made possible by interaction with the GDP dissociation inhibitor (GDI). Both GDI and REP function as molecular chaperones, which can solubilize the prenylated Rab proteins in the cytosol.

Rab proteins compromise the largest subgroup of the Ras superfamily of small GTPases, with more than 60 members in humans. Each of the Rab GTPases in human cells regulates intracellular vesicular transport at a specific subcellular membrane. While the mechanistic basis for GDI-mediated extraction of Rab molecules from membrane is well understood [7], a thermodynamic model of GDI-mediated membrane delivery of Rab proteins remains to be established. As GDI is a generic regulator (only two isoforms in humans) for prenylated Rab proteins, it has been a perplexing question how individual Rabs are specifically targeted to their cognitive membrane compartments. Membrane-bound GDI displacement factors (GDFs) were proposed to disrupt the Rab:GDI complex, leading to insertion of the prenylated Rab into the membrane in the GDP form [8, 9]. Such molecules could play an important role in the Rab cycle as they would help determine localization of Rab proteins in the cell. However, the functional mechanism of such factors is unclear and only one GDF (Pra1 in humans and Yip3 in the yeast) with promiscuous activity on several different Rab proteins has been identified so far (Figure 6.2).

Study of the membrane trafficking process regulated by Rab requires prenylated Rab proteins. Previously, there were substantial difficulties in recombinant preparation of prenylated proteins and in obtaining prenylated Rab GTPases in defined nucleotide-bound states. Therefore, it was technically difficult to analyze the interaction between GDP/GTP-bound prenylated Rab and its regulators REP and GDI.

6.2.2
Autophagy

Autophagy is a catabolic process for the bulk degradation of intracellular materials, such as long-lived proteins, protein aggregates, and damaged organelles. During macroautophagy (autophagy hereafter), double-membrane vesicle structures, termed *autophagosomes*, sequester and engulf cytoplasmic components constitutively, or upon nutrient deprivation, or stress. The subsequent fusion of autophagosomes with lysosomes leads to the exposure of the sequestered materials to lysosomal hydrolases for degradation. Autophagy plays an important role in physiology, including cell development, and has been associated with diverse human diseases, including cancer, neurodegeneration, pathogen infection, and aging [10, 11].

In mammalian cells, autophagosomes initiate from isolation membranes (IMs). IMs expand, enfold cytosol, and finally close, forming autophagosomes. Microtubule-associated protein light chain 3 (LC3, the mammalian homolog

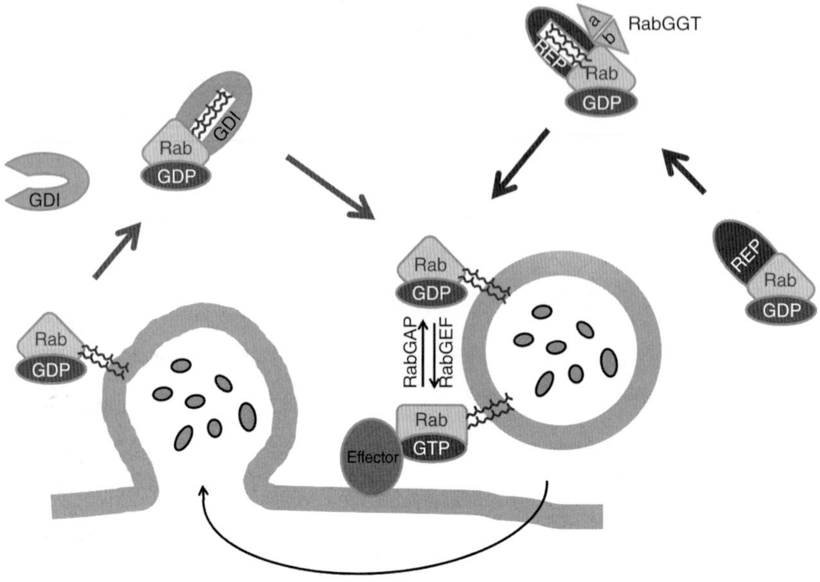

Figure 6.2 *Rab cycle through coordination between the GTPase cycle and the cycle of membrane attachment and detachment.* Rab proteins are intrinsically soluble and require a posttranslational modification for membrane association. They first associate with REP and form a complex that is the substrate for the subsequent dual prenylation of C-terminal cysteines by a heterodimetic RabGGTase (RabGGTα and ß). After lipid transfer, REP delivers the prenylated Rab to the membrane. The cycling of prenylated Rab proteins between the cytosol and membranes is facilitated by GDI. Both REP and GDI bind the GDP-bound inactive form of Rab. On the membrane, Rab proteins are activated by RabGEFs and deactivated by RabGAPs. In the active state, Rabs interact with structurally and functionally diverse effectors, including cargo sorting complexes on donor membranes, motor proteins involved in vesicular transport and tethering complexes that regulate vesicle fusion with acceptor membranes.

of yeast Atg8) plays a key role in the formation of autophagosomes and needs to be C-terminally modified with a phosphatidylethanolamine (PE) lipid for correct membrane localization and function. Lipidated LC3 (LC3-PE, also called *LC3-II*) has been used as a *bona fide* marker of the autophagosome and progression of autophagy (Figure 6.3). In cells, production of lipidated LC3 is controlled by two ubiquitin-like conjugation systems. Newly synthesized LC3 is processed by a protease, Atg4, to expose a C-terminal glycine. The resulting Atg8/LC3 serves as a substrate in a ubiquitin-like conjugation reaction mediated by Atg7 (E1) and Atg3 (E2) and is conjugated to PE, in a reaction controlled by the Atg12–Atg5:Atg16 complex (E3). The Atg12–Atg5:Atg16 complex is generated by another ubiquitin-like conjugation system. Atg12 is conjugated to the lysine side chain of Atg5 in sequential reactions catalyzed by Atg7 (E1) and Atg10 (E2). The E3-like enzyme is not required for Atg12–Atg5 conjugation. The Atg12–Atg5 conjugate further forms a complex with a multimeric protein, Atg16. The Atg12–5 conjugate promotes Atg8-PE formation,

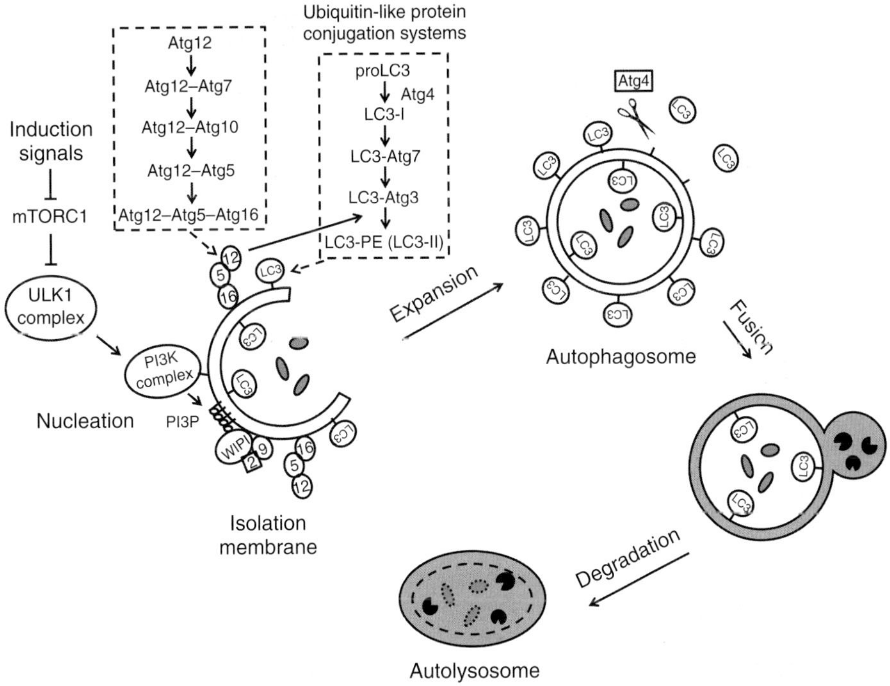

Figure 6.3 *The biogenesis of autophagosome in mammalian cells.* Upon activation of autophagy signaling pathways, autophagy proceeds *via* several mechanistically distinct steps. Cytosolic components are enclosed by an isolation membrane (also called *phagophore*), which elongates to form an autophagosome. The outer membrane of the autophagosome fuses with the lysosome to produce an autolysosome, in which the inner membranes of the autophagosome and the internal contents are degraded.

whereas Atg4 releases lipidated Atg8/LC3 from the outer surface of closed autophagosomes. LC3-PE locates at both sides of the double membrane of the pre-autophagosome.

It is still unclear how Atg8/LC3 regulates autophagosome biogenesis. Therefore, it is of great importance to be able to produce lipidated LC3 protein, in order to study the role of LC3 in autophagosome formation. However, it is challenging to generate lipidated LC3 protein by reconstituting the LC3-PE conjugation reaction *in vitro* with purified protein components, because of the difficulties in recombinant production of mammalian proteins involved in LC3-PE conjugation system.

Central to addressing those biological problems described earlier is the ability to produce posttranslationally modified (lipidated) proteins and to introduce reporter groups into these proteins as a read-out for protein activity. These issues are largely solved by using chemical methods as described in later sections.

6.3 The Chemical Approach

6.3.1 Expressed Protein Ligation and Click Ligation

In the early 1990s, Kent and coworkers introduced a breakthrough approach of native chemical ligation (NCL) [12], which is now a general method for chemical protein synthesis. In this method, two unprotected synthetic peptide fragments are coupled together under neutral aqueous conditions with the formation of a native peptide bond at the ligation site. The principle of NCL is shown in Figure 6.4a. This approach is based on a chemoselective reaction between a peptide containing an N-terminal cysteine and another peptide containing C-terminal thioester. Although the initial chemoselective reaction of transthioesterification is reversible, the subsequent S → N shift is irreversible and spontaneous. The scope of NCL application has been significantly widened upon introduction of the approach referred to as expressed protein ligation (EPL) from the Muir Laboratory [13]. In this method, both fragments containing C-terminal thioester and N-terminal cysteine can be produced recombinantly. EPL emerged as a result of the advances in self-cleavable affinity tags for recombinant protein purification using intein chemistry (Figure 6.4b). Inteins are protein insertion sequences flanked by host protein sequences (N- and C-exteins) and are eventually removed by a posttranslational process termed *protein splicing*. Inteins containing a C-terminal Asn to Ala substitution have been designed to keep their ability in the initial N → S acyl shift without further going through the later steps of protein splicing. Therefore, proteins fused to the N-termini of these engineered inteins can be cleaved by thiol reagents (such as 2-mercaptoethanesulfonate, MESNA) via an intermolecular transthioesterification reaction, releasing the α-thioester-tagged proteins. The recombinant polypeptide α-thioesters can then be ligated with a synthetic peptide or recombinant protein containing N-terminal cysteine.

Studies of the Rab protein functional cycle require methods that allow for generating preparative amounts of prenylated Rab proteins with new functionalities, such as fluorophores. Two approaches have been developed in our laboratory, including EPL and Cu(I)-catalyzed Huisgen 1,3-dipolar cycloaddition reaction (click ligation).

Here, we show an example for the semisynthesis of prenylated Rab7 proteins. C-terminal prenylated peptides were prepared using solid-phase peptide synthesis, while Rab protein thioesters with truncation of corresponding C-terminal amino-acid residues were generated recombinantly using intein chemistry. Chemical peptide synthesis allows for incorporation of fluorescent groups such as nitrobenzoxadiazole (NBD) and dansyl (dans) onto the lipid moiety or the lysine side chain. These fluorescent groups can serve as a reporter for protein–protein interactions, while they have minimal perturbation of protein function because of their small size. Ligation was performed in the presence of detergent cetyl trimethylammonium bromide (CTAB), which is able to facilitate

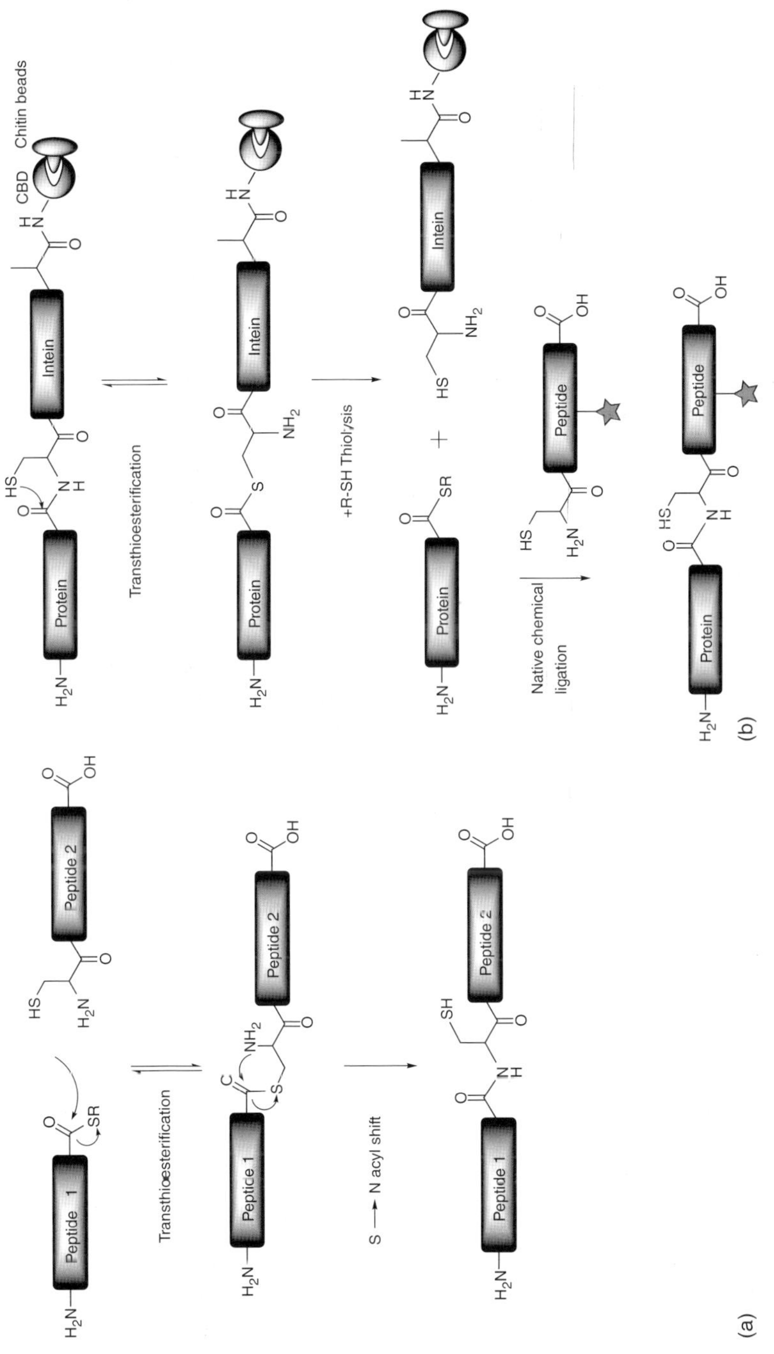

Figure 6.4 Principle of (a) native chemical ligation and (b) expressed protein ligation [14].

solubilization of prenylated peptides and efficiently promote the ligation. The resulting ligation products were washed extensively with organic solvents to remove excess peptides and detergents. Refolding of the denatured Rab proteins was achieved by a pulse-refolding approach in the presence of either GDP or guanosine 5′-*O*-[(β,γ)-imido]-triphosphate (GppNHp), a nonhydrolyzable analog of GTP (Figure 6.5). The denaturation and refolding approach not only facilitates purification of the protein but, in particular, also allows control of the nucleotide-binding state of the prenylated Rab proteins.

Prenylated Rab proteins can also be generated through a click ligation approach, which is much faster than NCL [16]. An azide-containing cysteine ($CysN_3$) was incorporated to the C-terminus of a recombinant Rab protein thioester using EPL. The resulting protein was subsequently ligated with prenylated peptides containing an alkyne group in the presence of Cu(I) and the ligand tris-(benzyltriazolylmethyl)amine (TBTA) (Figure 6.6). The ligation proceeded quantitatively within 30 min. As the ligation site is located at the unstructured C-terminal tail of Rab protein, it is unlikely that the triazole linker would affect protein folding. Indeed, the semisynthetic proteins are functional as determined by their reactivity in prenylation.

The synthesis of $CysN_3$ started from Fmoc-Cys(StBu)OH (Fmoc, fluorenylmethoxycarbonyl) and 2-azidoethanamine. The geranylgeranyled peptides with an alkyne group were synthesized on the basis of solid-phase lipidated peptide synthesis, as shown in Figure 6.6B.

These semisynthetic approaches resolve inherent problems in the analysis of posttranslationally modified proteins, where recombinant production of nonnative structures is either difficult or impossible. The constructed prenylated Rab protein probes were then further used to study the thermodynamics and kinetics of their interaction with regulatory proteins (see Section 6.4.1).

6.3.2 Site-Specific Modification of Proteins

Although NCL (EPL) provides a very useful approach to the C-terminal modification of proteins, a limitation of this method is that it usually leads to introduction of a cysteine residue at the ligation site of the target protein, regardless of whether this corresponds to the native structure or not. In addition, the NCL frequently proceeds at a relatively slow rate. Recent advancements in chemical methods have substantially expanded the tools for site-specific modification of proteins. In addition to NCL, nonnative chemical ligation methods represented by bioorthogonal chemistry have been widely used in chemical protein modification [14, 17, 18]. There is still a high demand for ligation reagents that display fast reaction rates under physiological conditions. Oxime-based reactions have found wide application in the conjugation of biomolecules, because of the absence of oxyamino groups and ketones in proteins and formation of stable oximes. Our laboratory has developed a facile method for C-terminal protein modification based on oxime ligation. A bisoxyamine molecule was first incorporated to the C-terminus of a

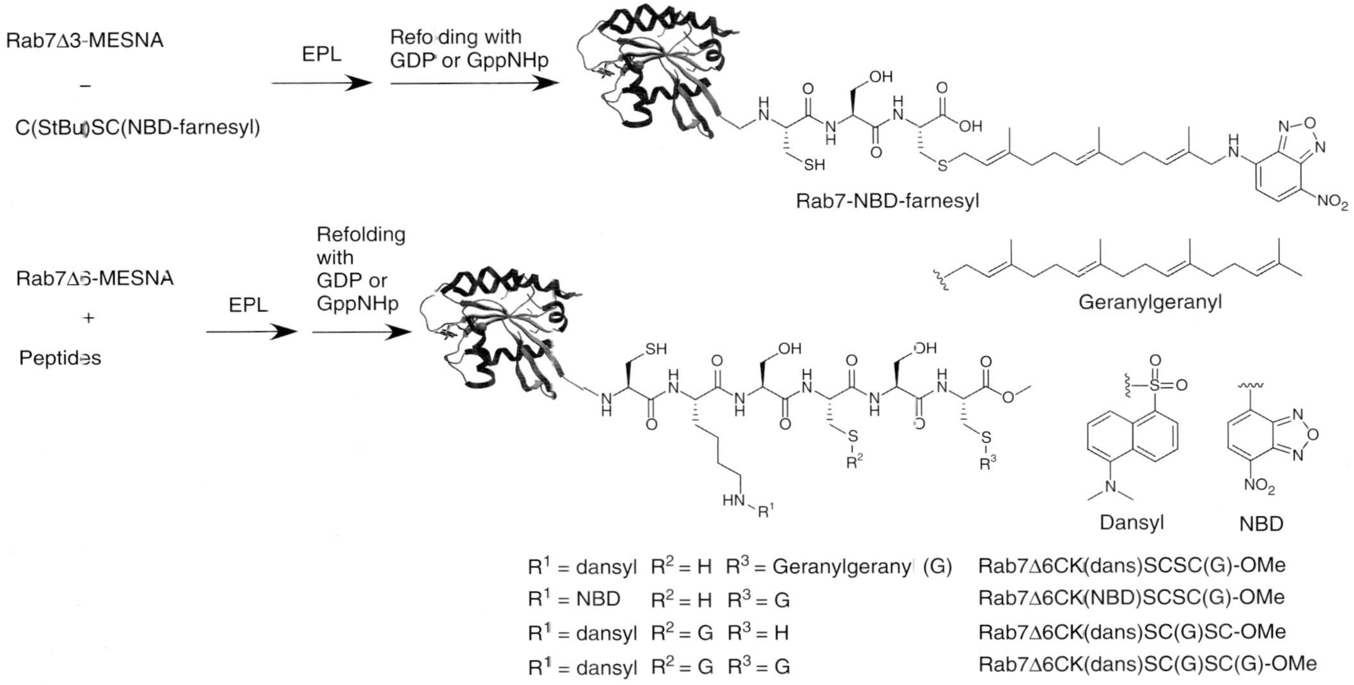

Figure 6.5 Construction of prenylated Rab7 proteins with a fluorophore on the isoprenyl moiety and mono- and digeranylgeranylated Rab7 proteins with a fluorophore at the peptide side chain. The NBD-farnesyl group is a fluorescent analog of a geranylgeranyl group [15].

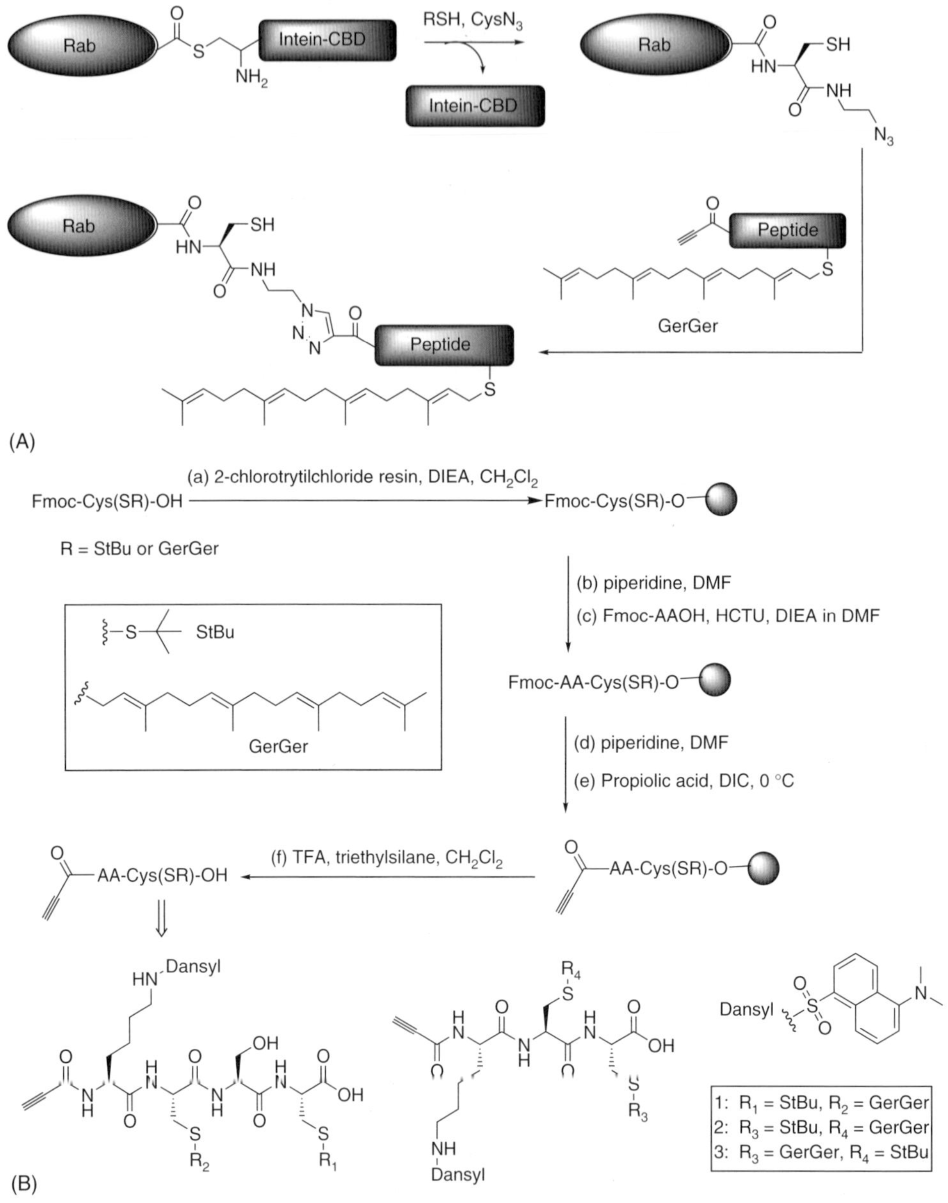

Figure 6.6 *Strategy for the semisynthesis of geranylgeranylated Rab proteins using click ligation.* (A) Reaction route. (B) Solid-phase peptide synthesis of the geranylgeranylated peptides with N-terminal alkyne moieties. The GerGer group was first incorporated into cysteine to produce the building block Fmoc-Cys(GerGer)OH. The building blocks were loaded and extended on the 2-chlorotrityl chloride (Trt) resin by standard Fmoc techniques. The geranylgeranylated fluorescently labeled peptides were obtained after cleavage from the Trt resin [16].

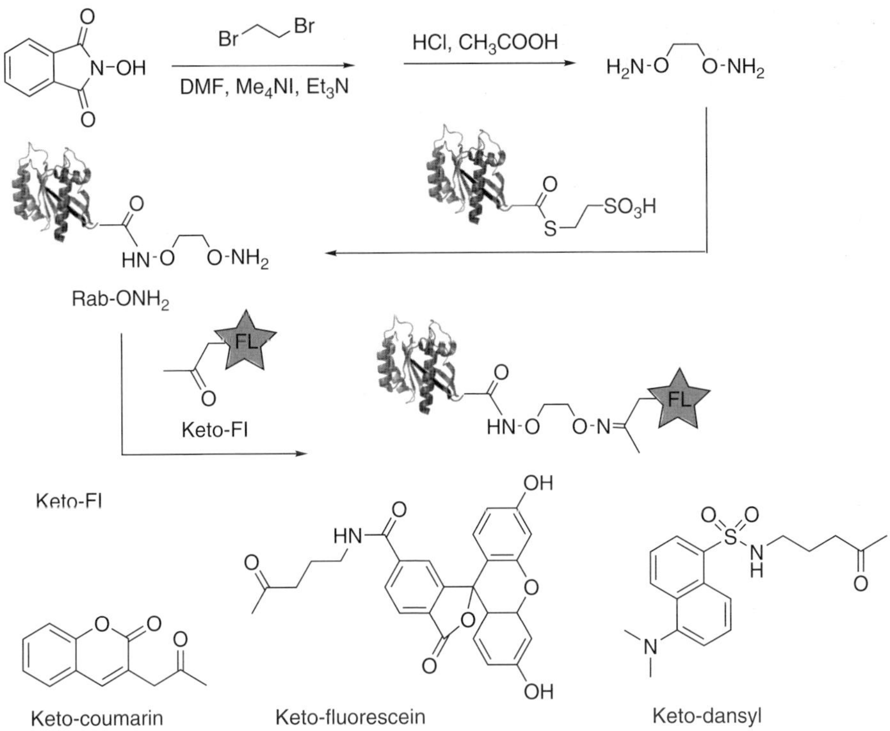

Figure 6.7 Site-specific labeling of proteins at the C-terminus [19].

protein through direct aminolysis of a protein thioester produced using intein chemistry (see Section 6.3.1). The resulting oxyamino-modified protein underwent efficient oxime ligation with fluorophores containing a ketone moiety in the presence of catalysis analine under physiological conditions (Figure 6.7) [19]. In contrast to NCL, oxime ligation does not require a cysteine residue for ligation and undergoes ligation fast ($t_{1/2} = 2.8$ h) and chemoselectively under mild conditions.

Multicolor labeling is an important technique for the characterization of proteins with respect to their structure, folding, and interactions at the single-molecule level and in cellular investigations. The key technique for such studies is based on fluorescence resonance energy transfer (FRET). FRET applications require the attachment of donor (D) and acceptor (A) molecules to specific sites of a given protein or interacting proteins. Such labeling is typically achieved through conjugation at cysteine residues or amino groups or by genetic fusion to different fluorescent proteins. However, site-specific incorporation of multiple fluorophores into a single protein remains a considerable challenge. Our laboratory has combined C-terminal oxime ligation and N-terminal NCL in one pot for dual labeling of a given protein [20].

As a proof-of-principle, this method has been used to generate a dual-labeled Rab7 GTPase. To generate Rab7 with an N-terminal cysteine, a tobacco etch virus

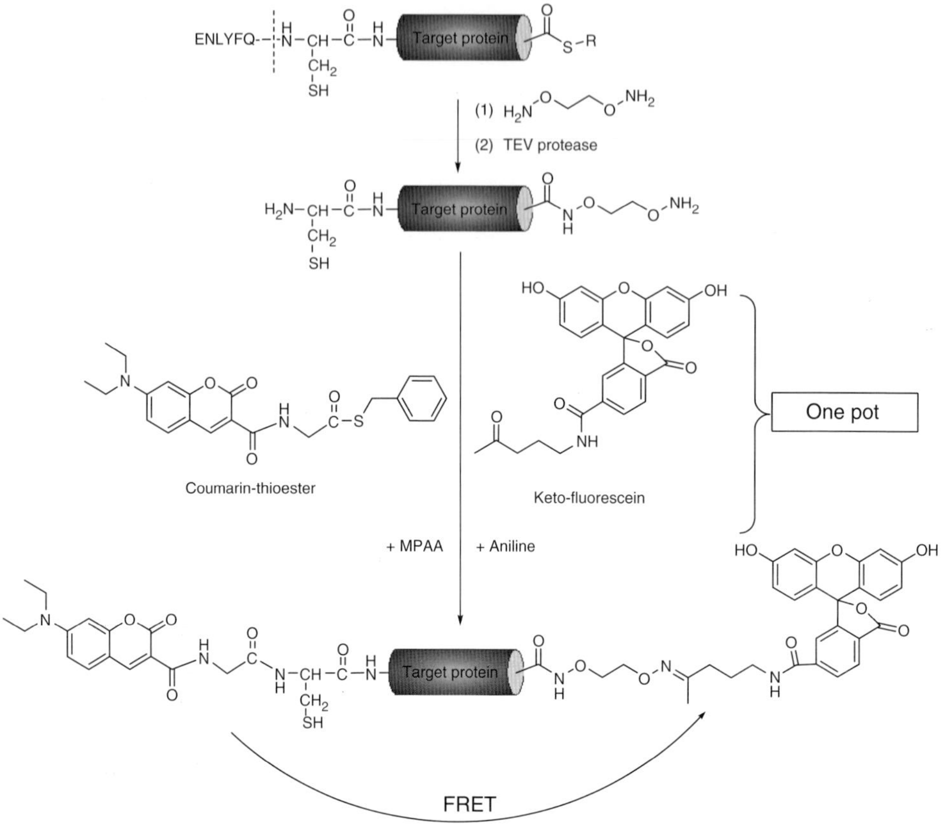

Figure 6.8 Strategy for the preparation of a two-color coumarin–fluorescein protein by one-pot chemoselective reactions [20].

(TEV) protease recognition sequence (ENLYFQ:C; the dashed line indicates the cleavage site) was fused to the N-terminus of Rab7. Intein-mediated incorporation of a bisoxyamine moiety into the Rab7 C-terminus was achieved as shown. The resulting oxyamine protein was treated with TEV protease to expose N-terminal cysteine. One-pot dual labeling could be achieved simply by incubation of both coumarin thioester (FRET donor) and keto-fluorescein (FRET acceptor) with the protein N-Cys-Rab7-ONH_2 on ice overnight in the presence of the catalysts (4-carboxylmethyl)thiophenol (MPAA) and aniline (Figure 6.8).

6.3.3 Semisynthesis of Lipidated LC3 Protein

Autophagy-associated lipidated LC3 protein was prepared using the EPL technique [21]. The choice of the ligation site was considered in order to meet some criteria. First, to perform the EPL under folding conditions, the ligation site should

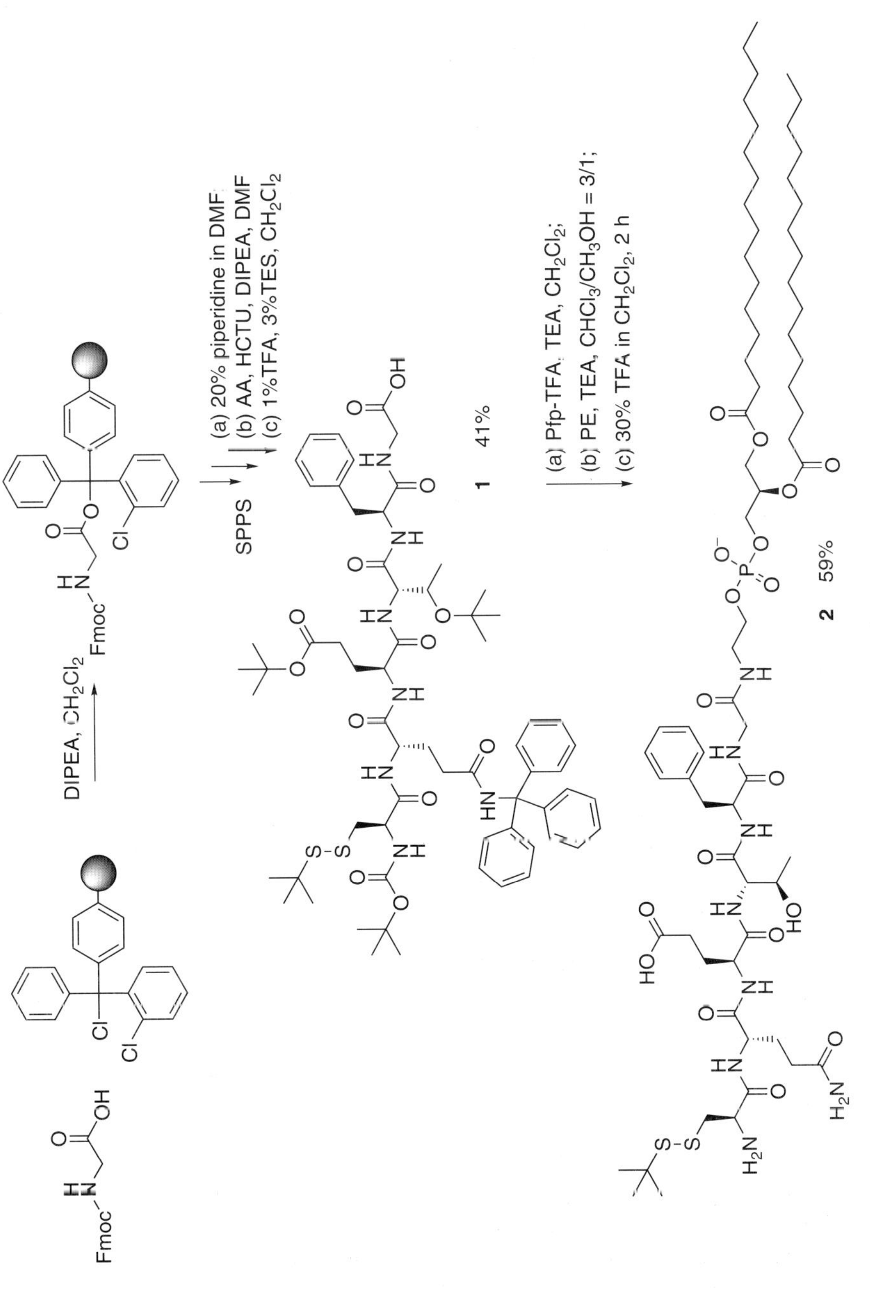

Figure 5.9 Synthesis of the C-terminal peptide of LC3.

be accessible at the protein surface. Moreover, the C-terminal truncation should not be detrimental for protein folding. Second, a short synthetic C-terminal peptide would be preferable in order to reduce the synthetic effort and the risk of perturbing protein folding. Third, introduction of a mutation to Cys at the ligation site should not influence LC3 function. After a number of tests, the ligation site was chosen at Ala114-Ser115.

The corresponding C-terminal lipidated peptide (residue 115–120) with Ser115 being replaced by Cys was synthesized by solid-phase peptide synthesis. The peptide chain was elongated using the Fmoc strategy. The resulting peptide **1** was subsequently activated as a pentafluorophenyl ester and coupled in solution to DPPE (1,2-dipalmitoyl-*sn*-glycero-3-phosphoethanolamine). After cleavage of all acid-sensitive side-chain protecting groups, the lipidated peptide **2** was obtained (Figure 6.9).

Unfortunately, $LC3^{1-114}$-thioester does not readily ligate with the lipidated peptide, albeit both the protein and the lipidated peptide being soluble in the presence of various detergents. But $LC3^{1-114}$-thioester undergoes NCL quantitatively with methylated cysteine, suggesting the protein thioester is competent for NCL. Previously, semisynthesis of lipidated proteins is frequently performed under denaturing conditions. A refolding step is usually required, without guarantee of success and protein quality. In addition, lipidated proteins are insoluble without detergents, which make them difficult to handle. Our laboratory introduced the strategy of using a facile TEV-cleavable maltose-binding protein (MBP) tag to facilitate lipidated protein ligation and solubilization of the lipidated protein. This method enables us to perform the ligation under folding conditions and to handle the lipidated protein in the absence of detergents.

The ligation reaction was performed by mixing the lipidated peptide **5** with MBP-$LC3^{1-114}$-thioester protein in the presence of MPAA as a catalyst. The

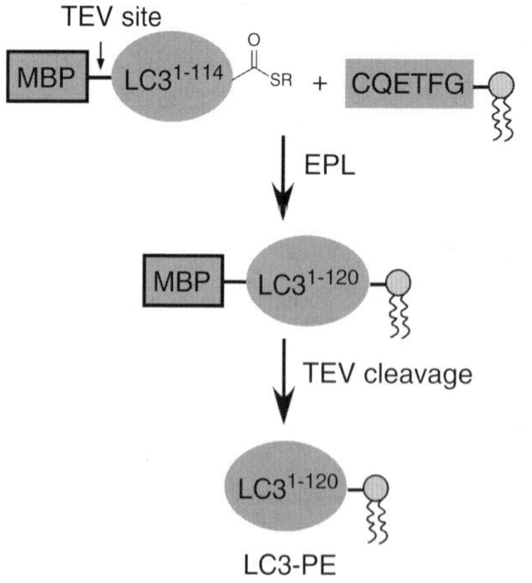

Figure 6.10 Synthesis of the lipidated LC3 protein.

ligation product was then purified by amylose affinity chromatography and size-exclusion chromatography to remove detergents and unreacted peptides. The resulting MBP-LC3-PE was soluble in buffer without detergents and could be used for later studies. LC3-PE was obtained by removal of the MBP tag using TEV protease cleavage (Figure 6.10).

6.4 Biological Research/Evaluation

6.4.1 Thermodynamic Basis of Rab Membrane Targeting

Elucidation of the thermodynamic basis of Rab membrane targeting requires analysis of interaction between prenylated Rab proteins (GDP/GTP-bound) and REP/GDI. Such analysis is made possible by generation of labeled prenylated Rab proteins. A set of Rab7-based protein probes with one or two isoprenyl moieties and fluorophores on the lipid moiety or the lysine side chain were prepared using the EPL technique as described in Section 6.3.1 (Figure 6.5). The semisynthetic method enables precise installation of GDP/GTP into Rab proteins to generate the "off" and "on" states, yielding for the first time homogeneous preparations of functionalized prenylated proteins in a well-defined nucleotide-bound state.

Thermodynamic and kinetic analyses of the interaction between prenylated Rab proteins and regulatory factors provide insights into the mechanism of Rab membrane targeting. For example, Rab7Δ6CK(NBD)SCSC(G)-OMe (Rab7NBD-G) displays a four- to fivefold fluorescence enhancement upon binding to REP-1 or GDI-1. This signal change was used to perform fluorescence titration experiments to determine K_d (dissociation constant) values of GDP:Rab7NBD-G complexes for REP-1 and GDI-1 (0.19 and 2.5 nM, respectively), and K_d values of GppNHp:Rab7NBD-G complexes for REP-1 and GDI-1 (228 and 1758 nM, respectively, Figure 6.11) [15]

In most cases, Rab proteins are doubly geranylgeranylated *in vivo*, with only a few Rabs being monogeranylgeranylated in mammalian cells. To obtain a digeranylgeranylated fluorescent Rab7 sensor protein, Rab7Δ6CK(dans)SC(G)SC(G)-OMe (Rab7dans-GG) was prepared. Fluorescence titration experiments showed that GDP:Rab7dans-GG binds to REP-1 and GDI-1 with K_d values of 1.4 and 4.5 nM, respectively, whereas no interaction could be detected between GppNHp:Rab7dans-GG and REP-1 or GDI-1 even at micromolar protein concentrations. These results indicated that replacement of GDP with GTP analog GppNHp leads to a reduction of the affinity of prenylated Rab proteins for their regulators REP-1 and GDI-1 by around three orders of magnitude. On the contrary, in the case of GTPase interaction with effector proteins, the affinity increases by several orders of magnitude on substitution of GDP by GTP. These reciprocal relationships are essential features of the Rab cycle, in which

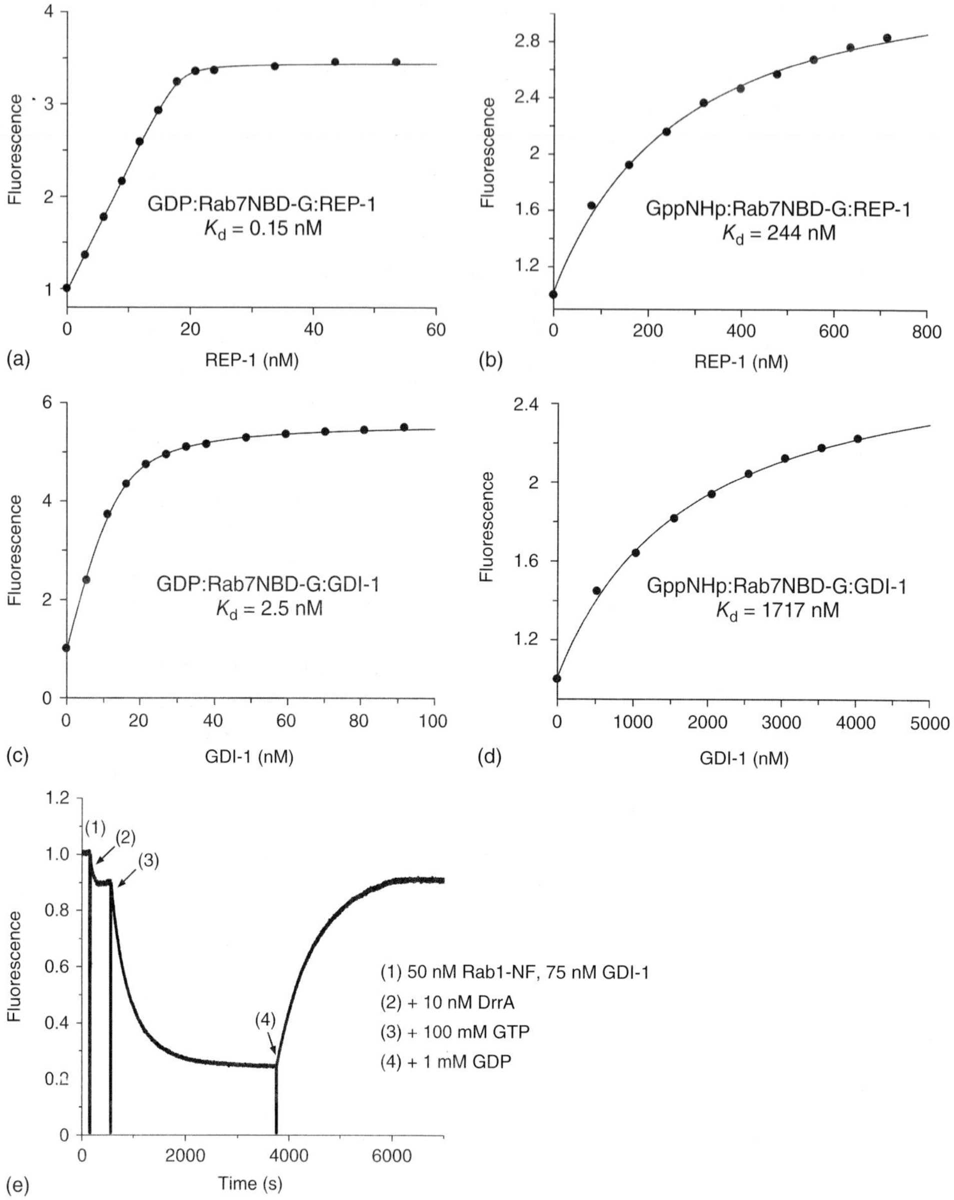

Figure 6.11 *Quantitative analysis of interaction of Rab7NBD-G with REP-1 and GDI-1.* (a,b) Titration of REP-1 to GDP/GppNHp-bound Rab7NBD-G. (c,d) Titration of GDI-1 to GDP/GppNHp-bound Rab7NBD-G. K_d values were obtained by fitting the data to quadratic equation. (e) DrrA-mediated displacement of GDI-1. Fifty nanometer Rab1-NF:GDI-1 complex was supplemented with 10 nM DrrA. Nucleotide exchange was triggered by adding 100 μM GTP. Fluorescence was recovered by adding an excess of GDP (1 mM GDP) [15].

nucleotide exchange coordinates membrane delivery, effector interactions, and retrieval of Rabs from membranes.

To further study the relationship between nucleotide exchange and Rab targeting to membranes, a RabGEF from *Legionella pneumophila* (DrrA, defects in Rab1 recruitment protein A) was used in investigating the effect of GEFs on the Rab:GDI complex. Kinetics of the interaction was monitored by a fluorescence change of Rab1-NBD-farnesyl (Rab1-NF). DrrA mediated the exchange for GTP or GDP and resulted in loss or recovery, respectively, of the Rab binding to GDI (Figure 6.11e). These measurements suggest GEF activity is sufficient to disrupt the Rab:GDI complex and could lead to membrane insertion.

GDFs were proposed to play a key role in the displacement of Rab:GDI complexes. However, a GDF model is problematic from the thermodynamic point of view, because displacement of GDI requires a tight association of GDF with prenylated Rabs. This would lead to the question as to how GDF would be replaced (Figure 6.12b, model I).

As shown in this study, GEF-mediated exchange of GDP for GTP dramatically reduces the affinity of Rabs to GDI and leads to an essentially irreversible dissociation of GDI. GEF-mediated nucleotide exchange plays a key role in providing the free energy to drive this process. The results obtained with DrrA suggest that GEF activity is necessary and sufficient to displace GDI, but the dissociation of the Rab:GDI complex is rate limiting in this process (Figure 6.12b, model III). Therefore, GTP/GDP exchange catalyzed by a membrane-specific GEF is the thermodynamic determinant for the delivery to and stabilization of Rab on a particular membrane or membrane domain (Figure 6.12).

6.4.2
Monitoring Protein Unfolding and Refolding Using a Dual-Labeled Protein

The dual-labeled *N*-coumarin-Rab7-fluorescein protein has been used for protein unfolding and folding studies. The intramolecular FRET signal provides a facile read-out for protein folding status, as the distance between N- and C-termini is significantly changed under folding and denaturing conditions. The FRET signal decreased in the presence of 8 M denaturant guanidine hydrochloride (GdnHCl). The unfolding kinetics could also be recorded using the FRET signal (Figure 6.13a,b). Recovery of the FRET signal was observed upon refolding by diluting the denatured protein into the refolding buffer in the presence of cofactor guanine nucleotides (Figure 6.13c). The refolding kinetics was monitored in real time with observed refolding rate constants of 0.16 and 0.24 min^{-1} in the presence of 50 mM GDP and GppNHp, respectively, in keeping with the fact that GTPases bind GDP and GTP in similar affinities (Figure 6.13d). In future applications, dual-labeled proteins with bright and photostabile dyes such as Cy3 and Cy5 could facilitate studies of protein folding and unfolding at a single-molecule level.

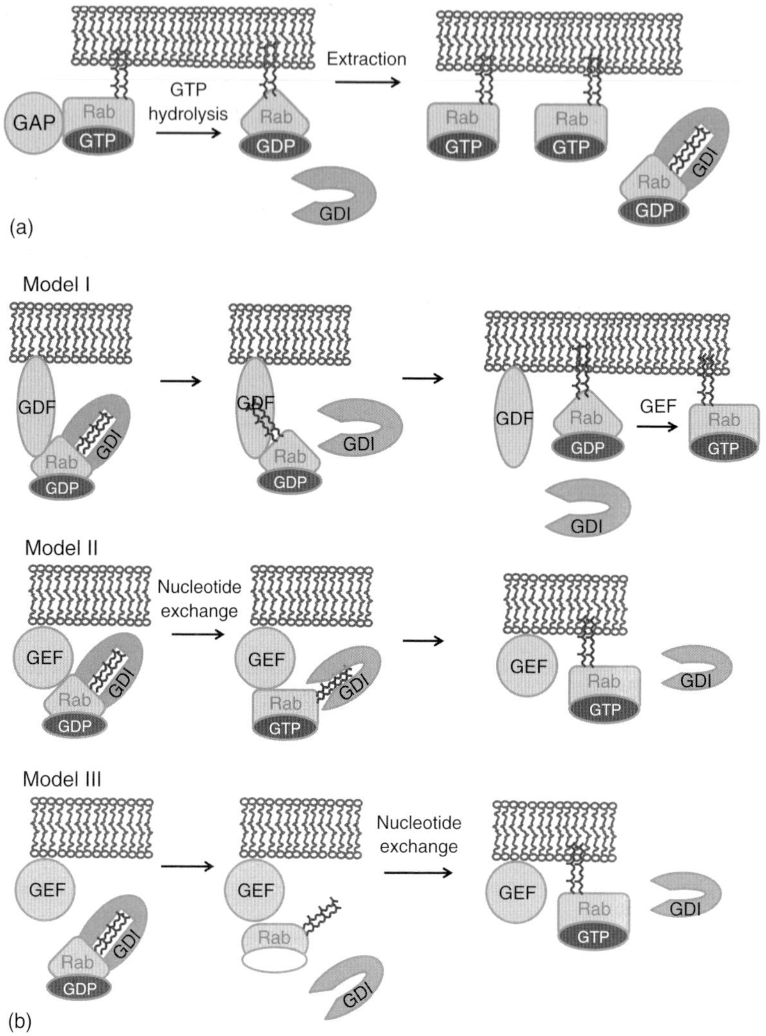

Figure 6.12 *Models of modulation of Rab recycling and targeting of Rabs to membranes by the state of bound nucleotide.* (a) The accepted minimal model of Rab retrieval from the membrane. (b) In model I, GDF-mediated displacement of GDI is followed by Rab membrane insertion and GEF-mediated nucleotide exchange. Models II and III depict GEF-mediated insertion of Rab into the membrane. In model II, a direct interaction of GEF with the Rab:GDI complex leads to nucleotide exchange and Rab dissociation. In model III, spontaneous dissociation of the Rab:GDI complex is rendered effectively irreversible by nucleotide exchange and membrane attachment. The work presented here supports model III [15].

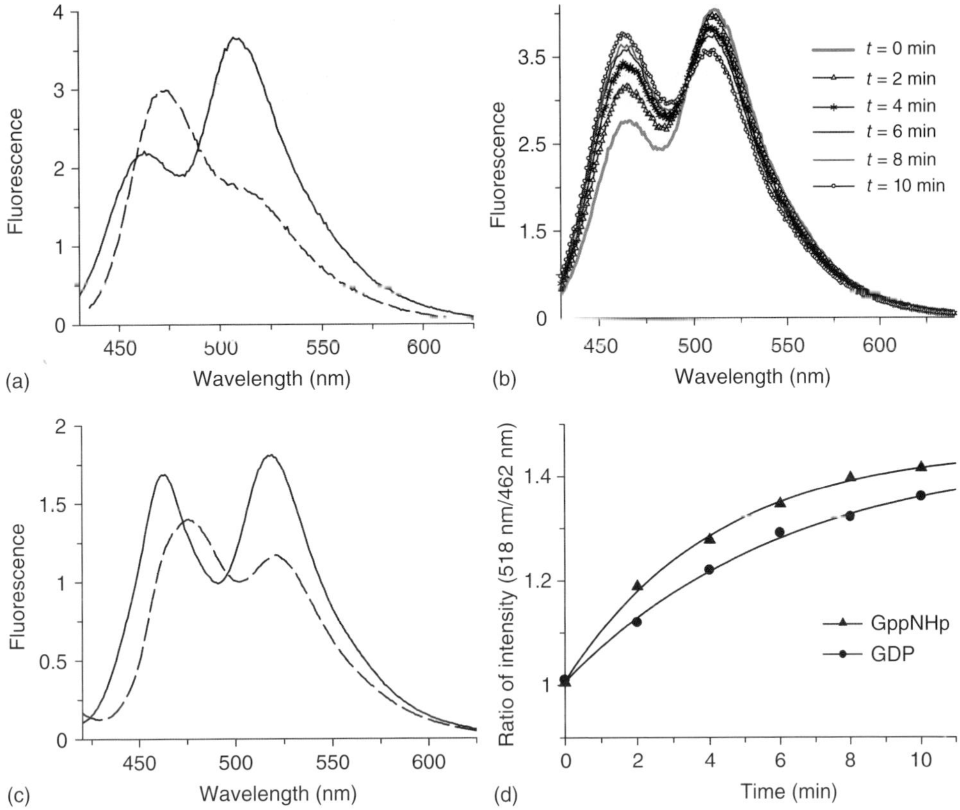

Figure 6.13 *Unfolding and refolding of the dual-labeled protein.* (a) Fluorescence spectra of *N*-coumarin-Rab7-fluorescein before (solid line) and after denaturation in 8 M GdnHCl for 30 min (dashed line). (b) Emission spectra of *N*-coumarin-Rab7-fluorescein in 2 M GdnHCl. Each spectrum was recorded at 2 min intervals. (c) Emission spectra of the denatured *N*-coumarin-Rab7Δ3-fluorescein in 8 M GdnHCl (dashed line) and after diluting into the refolding buffer (solid line). (d) The emission ratio of 518 nm/462 nm as function of time for refolding in the presence of GDP (circle) and GppNHp (triangle). Excitation was kept at 400 nm [20].

6.4.3 Semisynthetic Lipidated LC3 Protein Mediates Membrane Fusion

Atg8-PE/LC3-PE is required for the elongation of autophagosomal precursors. However, the function of Atg8-PE/LC3-PE in promoting membrane tethering and hemifusion remains controversial. Using *in vitro* reconstitution of Atg8 ubiquitin-like system, conjugation of yeast Atg8 to liposomes containing high concentrations (55%) of PE has been shown to promote the tethering and hemifusion of liposomes [22]. Cross-linking of LC3 to liposomes through maleimide-coupling strategy induces membrane tethering and fusion [23]. However, recent studies

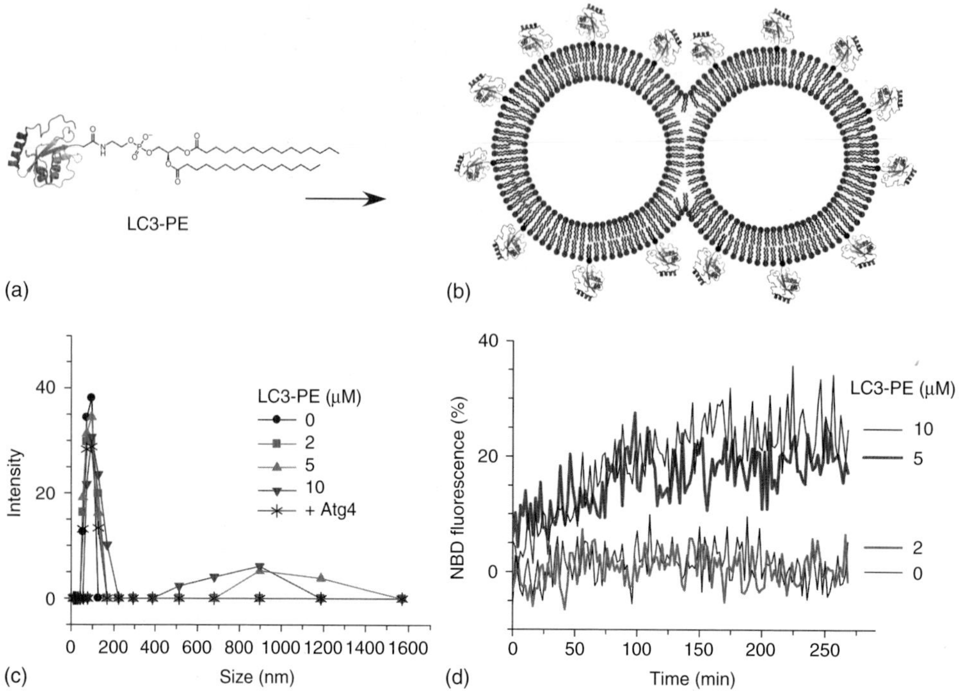

Figure 6.14 *Membrane tethering and fusion meditated by the semisynthetic MBP-LC3-PE protein in vitro.* (a) A schematic view of LC3-PE-mediated liposomal hemifusion. (b) LC3-PE induces membrane tethering. After incubation of liposomes with various amounts of LC3-PE for 4 h, the size distribution of the liposomes was examined by DLS. The measurement labeled "+Atg4" was performed with 10 μM LC3-PE in the presence of 25 nM Atg4B. (c) LC3-PE induces membrane fusion. A 4 : 1 mixture of the unlabeled and NBD + Rhod liposomes (0.35 mM lipids) was incubated with various concentrations of LC3-PE. Liposomes contain 30% PE. Experiments were performed in triplicate.

using both the reconstitution system and the maleimide-coupling strategy suggested that Atg8-PE/LC3-PE is not able to drive membrane fusion in the presence of physiological concentrations of PE (30%) [24]. LC3-PE was obtained in a multi-milligram scale using a semisynthetic approach (see Section 6.3.3). The semisynthetic LC3-PE allows for addressing the perplexing question on the membrane fusing activity of LC3-PE.

To this end, MBP-LC3-PE was used in liposomal assays, as it is soluble in aqueous solution without detergents (Figure 6.14a). The ability of MPB-LC3-PE to promote liposome tethering and fusion was determined by dynamic light scattering (DLS) and the lipid mixing assay, respectively [21]. Addition of MBP-LC3-PE to liposomes containing various concentrations of PE (30% and 55%) induced aggregate formation in a dose-dependent manner. In contrast, after treatment with catalytic amounts of Atg4B to cleave PE, MBP-LC3-PE had no effect on liposome size distribution, in line with the fact that lipidation of LC3 is essential for membrane association and function of LC3 (Figure 6.14b). Membrane fusion activity was measured by the lipid mixing assay, in which fluorescence energy transfer from NBD-labeled lipid to rhodamine B (Rhod)-labeled lipid is reduced when labeled liposomes fuse with unlabeled liposomes. A dose-dependent induction of membrane fusion by MBP-LC3-PE was observed in the presence of 30% PE (Figure 6.14c). These findings clearly demonstrate that LC3-PE mediates membrane tethering and fusion at physiological concentrations of PE. This is particularly important, as progress in these aspects has been impeded by the lack of methods for producing preparative amounts of native LC3-PE proteins.

6.5
Conclusion

Protein chemical modification approaches are a powerful tool for generating fluorescently labeled proteins and posttranslationally modified proteins. Here, we have introduced a set of site-specific protein modification methods by use of chemoselective reactions, including NCL, oxime ligation, and click chemistry. These approaches enabled us to make C- and N-terminal-labeled proteins, facilitating investigation of protein–protein interactions, protein unfolding, and refolding. The semisynthetic prenylated Rab protein probes facilitate the analysis of interactions between Rab and its regulators, enabling establishment of the thermodynamic basis for Rab membrane targeting. Moreover, we introduced a facile method for the synthesis of lipidated protein LC3-PE, which makes it possible to reveal its function in autophagosome biogenesis. This chapter shows examples of combining chemical and biophysical approaches to address complex biological problems.

References

1. Foley, T.L. and Burkart, M.D. (2007) Site-specific protein modification: advances and applications. *Curr. Opin. Chem. Biol.*, **11**, 12–19.
2. Zhang, J., Campbell, R.E., Ting, A.Y., and Tsien, R.Y. (2002) Creating new fluorescent probes for cell biology. *Nat. Rev. Mol. Cell Biol.*, **3**, 906–918.
3. Shaner, N.C., Steinbach, P.A., and Tsien, R.Y. (2005) A guide to choosing fluorescent proteins. *Nat. Methods*, **2**, 905–909.
4. Wong, L.S., Khan, F., and Micklefield, J. (2009) Selective covalent protein immobilization: strategies and applications. *Chem. Rev.*, **109**, 4025–4053.
5. Hutagalung, A.H. and Novick, P.J. (2011) Role of Rab GTPases in membrane traffic and cell physiology. *Physiol. Rev.*, **91**, 119–149.
6. Stenmark, H. (2009) Rab GTPases as coordinators of vesicle traffic. *Nat. Rev. Mol. Cell Biol.*, **10**, 513–525.
7. Wu, Y.W., Tan, K.T., Waldmann, H., Goody, R.S., and Alexandrov, K. (2007) Interaction analysis of prenylated Rab GTPase with Rab escort protein and GDP dissociation inhibitor explains the need for both regulators. *Proc. Natl. Acad. Sci. U.S.A.*, **104**, 12294–12299.
8. Sivars, U., Aivazian, D., and Pfeffer, S.R. (2003) Yip3 catalyses the dissociation of

endosomal Rab-GDI complexes. *Nature*, **425**, 856–859.

9. Dirac-Svejstrup, A.B., Sumizawa, T., and Pfeffer, S.R. (1997) Identification of a GDI displacement factor that releases endosomal Rab GTPases from Rab-GDI. *EMBO J.*, **16**, 465–472.
10. Mizushima, N., Levine, B., Cuervo, A.M., and Klionsky, D.J. (2008) Autophagy fights disease through cellular self-digestion. *Nature*, **451**, 1069–1075.
11. Mizushima, N. and Komatsu, M. (2011) Autophagy: renovation of cells and tissues. *Cell*, **147**, 728–741.
12. Dawson, P.E. and Kent, S.B. (2000) Synthesis of native proteins by chemical ligation. *Annu. Rev. Biochem.*, **69**, 923–960.
13. Muir, T.W., Sondhi, D., and Cole, P.A. (1998) Expressed protein ligation: a general method for protein engineering. *Proc. Natl. Acad. Sci. U.S.A.*, **95**, 6705–6710.
14. Wu, Y.W. and Goody, R.S. (2010) Probing protein function by chemical modification. *J. Pept. Sci.*, **16**, 514–523.
15. Wu, Y.W., Oesterlin, L.K., Tan, K.T., Waldmann, H., Alexandrov, K., and Goody, R.S. (2010) Membrane targeting mechanism of Rab GTPases elucidated by semisynthetic protein probes. *Nat. Chem. Biol.*, **6**, 534–540.
16. Yi, L., Abootorabi, M., and Wu, Y.W. (2011) Semisynthesis of prenylated Rab GTPases by click ligation. *ChemBioChem*, **12**, 2413–2417.
17. Hackenberger, C.P. and Schwarzer, D. (2008) Chemoselective ligation and modification strategies for peptides and proteins. *Angew. Chem. Int. Ed.*, **47**, 10030–10074.
18. Sletten, E.M. and Bertozzi, C.R. (2009) Bioorthogonal chemistry: fishing for selectivity in a sea of functionality. *Angew. Chem. Int. Ed.*, **48**, 6974–6998.
19. Yi, L., Sun, H., Wu, Y.W., Triola, G., Waldmann, H., and Goody, R.S. (2010) A highly efficient strategy for modification of proteins at the C terminus. *Angew. Chem. Int. Ed.*, **49**, 9417–9421.
20. Yi, L., Sun, H., Itzen, A., Triola, G., Waldmann, H., Goody, R.S., and Wu, Y.W. (2011) One-pot dual-labeling of a protein by two chemoselective reactions. *Angew. Chem. Int. Ed.*, **50**, 8287–8290.
21. Yang, A., Li, Y., Pantoom, S., Triola, G., and Wu, Y.W. (2013) Semisynthetic lipidated LC3 protein mediates membrane fusion. *ChemBioChem*, **14**, 1296–1300.
22. Nakatogawa, H., Ichimura, Y., and Ohsumi, Y. (2007) Atg8, a ubiquitin-like protein required for autophagosome formation, mediates membrane tethering and hemifusion. *Cell*, **130**, 165–178.
23. Weidberg, H., Shpilka, T., Shvets, E., Abada, A., Shimron, F., and Elazar, Z. (2011) LC3 and GATE-16 N termini mediate membrane fusion processes required for autophagosome biogenesis. *Dev. Cell*, **20**, 444–454.
24. Nair, U., Jotwani, A., Geng, J., Gammoh, N., Richerson, D., Yen, W.L., Griffith, J., Nag, S., Wang, K., Moss, T., Baba, M., McNew, J.A., Jiang, X., Reggiori, F., Melia, T.J., and Klionsky, D.J. (2011) SNARE proteins are required for macroautophagy. *Cell*, **146**, 290–302.

7
Inhibition of Oncogenic K-Ras Signaling by Targeting K-Ras–PDEδ Interaction

Gemma Triola

7.1
Introduction

Ras (rat sarcoma) proteins are involved in many cellular processes regulating cell proliferation and differentiation. As a result, mutated Ras proteins leading to uncontrolled cell growth can be found in around 30% of all cancers. For correct localization and function, Ras proteins require proper membrane association and precise localization. However, the molecular mechanisms regulating the cellular distribution of these proteins, that is, the involvement of membrane diffusion processes or transport proteins, are largely unexplored. In this chapter, a chemical biological approach is described directed at investigating delta subunit of phosphodiesterase (PDEδ) as a prenyl-binding protein and characterizing its role in sustaining Ras protein function and localization in cells. To this end, a fully modified K-Ras4B protein was obtained by means of a novel semisynthetic approach and the resulting lipidated protein was employed in biochemical and biophysical studies, providing detailed insights into the structural requirements of Ras–PDEδ interaction. Additional cellular studies confirmed the key role of PDEδ in sustaining Ras function and correct transport between cellular endomembranes. This information has been crucial for the characterization of PDEδ as a new target and resulted in the identification of small-molecule inhibitors of the Ras–PDEδ interaction that are able to block oncogenic Ras signaling, thus opening novel therapeutic opportunities for the treatment of cancers bearing mutations in Ras oncogenes.

7.2
The Biological Problem

G-proteins or guanosine-nucleotide-binding proteins are proteins involved in a diverse range of cellular processes such as signal transduction, vesicular transport, proliferation, differentiation, cell cycle, or nuclear import. G-proteins can be divided in two main classes: the heterotrimeric G-proteins (formed by three

Concepts and Case Studies in Chemical Biology, First Edition. Edited by Herbert Waldmann and Petra Janning.

subunits, i.e., α, β, and γ) and the monomeric G-proteins, for which the Ras family of small guanosine triphosphatases (GTPases) is one of the most remarkable examples.

G-proteins function as molecular switches and can exist in two different conformations. Proteins bound to guanosine-5′-triphosphate (GTP) are in a so-called active or on-state, whereas the binding to guanosine-5′-diphosphate (GDP) promotes the formation of an inactive or off-state. Proteins in the active state interact with effectors and by doing so initiate and regulate many signaling processes. The hydrolysis of the GTP to GDP, with the help of GTPase-activating proteins (GAPs), results in the deactivation of the protein.

Owing to the importance of these molecular switches, their misregulation often results in proteins that are permanently arrested in the GTP-bound state, and thus they are continuously sending proliferation signals that result in uncontrolled cell growth and ultimately in cancer. This is the case of Ras proteins, important oncogens found mutated in around 30% of all cancers and up to 90% in pancreatic, lung, or colorectal cancer.

Four closely related Ras isoforms that share almost complete sequence identity have been described: H-Ras, N-Ras, and the two splice variants K-Ras4A and K-Ras4B. These proteins mainly differ in the C-terminally located hypervariable region (HVR) that targets them to different cellular compartments and is responsible for membrane association and for their precisely regulated cellular localization, both key for a correct function of Ras proteins (Figure 7.1) [1, 2]. The four isoforms contain important and distinct posttranslational modifications. They all possess a carboxymethylated and farnesylated cysteine at the C-terminus. Although this prenylated cysteine is essential for membrane attachment, additional motifs are required to ensure a stable membrane association. In the case of N- and H-Ras, this is achieved via palmitoylation by protein acyl transferases (PATs) of one or two cysteine residues, respectively, whereas for K-Ras this is mediated by a polybasic stretch formed by six consecutive lysines located near the prenylated cysteine. These different lipidation patterns not only mediate membrane attachment but are also responsible for the different compartmental localization of Ras proteins, which determines their interaction with different effectors and their spatiotemporal activity. Palmitoylthioesters are reversible modifications that can be cleaved by the action of acyl protein thioesterases (APTs). This reversibility results in a palmitoylation/depalmitoylation cycle that regulates the localization and function of N- and H-Ras. Briefly, proteins are

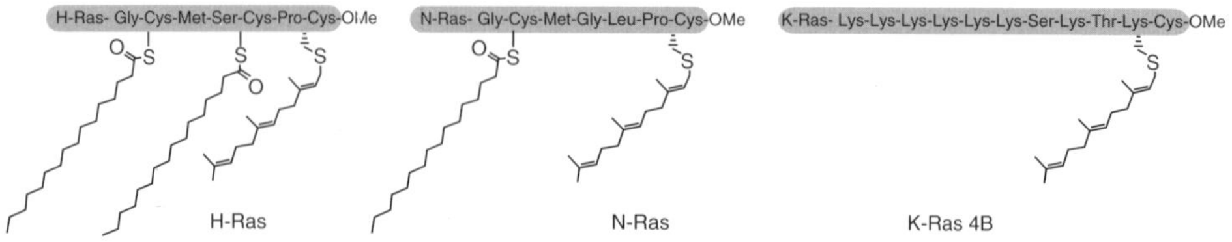

Figure 7.1 C-terminal lipidation motifs for H-, N-, and K-Ras4B.

palmitoylated at the Golgi and then transported to the plasma membrane by the secretory pathway. Subsequent depalmitoylation occurs by action of APTs and the resulting monofarnesylated proteins are then redirected to the Golgi, where they are kinetically trapped, to start a new palmitoylation round. As a result of this dynamic cycle, monopalmitoylated N-Ras is mainly localized to the Golgi, while the doubly palmitoylated H-Ras, owing to the presence of two palmitoyl residues, has a longer residence time at the plasma membrane where it preferably localizes. In the case of K-Ras4B, the C-terminal polybasic stretch determines its location by specifically interacting with acidic phospholipids such as phosphatidylserine or phosphatidylinositol that are preferentially located in the internal leaflet of the plasma membrane.

Because Ras proteins are farnesylated in the cytosol, then targeted to the endoplasmic reticulum for subsequent processing, and, if required, directed to the Golgi for palmitoylation before their ultimate signaling destination, the question arises how the specific distribution of these proteins is maintained. Traffic by vesicles or diffusion processes were proposed for N- and H-Ras proteins, and the involvement of transport proteins has been suggested for K-Ras. However, the factors regulating the transport or distribution of Ras proteins over the different cellular membranes are still poorly understood, mainly due to the lack of appropriate tools enabling this study.

Despite the high importance of the posttranslational modifications for their correct localization, function, and interaction with effectors, most of the studies with these proteins were typically performed on the soluble part of the protein, that is, the protein without the C-terminus. The main reason for that is that the generation of the required preparative amounts of pure fully lipidated proteins by means of molecular biology techniques is challenging, time consuming, and in most of the cases not practical or applicable. Moreover, biological approaches do not enable the preparation of proteins bearing nonnatural modification or tags, such small fluorophores that facilitate their study by biochemical or biophysical methods. A chemical biology approach was therefore developed to facilitate the production of such modified proteins by means of a combination of expression techniques and organic synthesis. This strategy combines the expression of a truncated protein and the synthesis of the C-terminal peptide using solid-phase approaches. The ligation of the protein with the lipidated peptide should permit the generation of semisynthetic proteins containing natural and nonnatural modifications. This approach has been already successfully employed for N-Ras and the resulting semisynthetic proteins have become invaluable tools in many biochemical, biophysical, nuclear magnetic resonance (NMR), or cellular studies giving important insights into their function, interaction with effectors and membranes or distribution in membrane microdomains [3–5].

The monofarnesylated K-Ras4B does not contain palmitoylatable cysteines and thus there is no palmitoylation/depalmitoylation cycle controlling its cellular distribution. In this case, the involvement of transport proteins, such as the PDEδ 6, has been hypothesized. However, a detailed study to elucidate the factors regulating Ras/PDEδ interaction, pivotal for the true understanding of the biological role of PDEδ in the distribution of prenylated Ras proteins, was lacking. In this chapter, a chemical biological approach is presented that gives access to fully lipidated K-Ras4B proteins, thereby enabling the study of its interaction with PDEδ and the characterization of the role of PDEδ in maintaining K-Ras distribution and function in the cell. Moreover, the gained information could be used for the establishment of a screening assay aimed to identify small-molecule inhibitors able to block this interaction and, in consequence, oncogenic Ras signaling.

7.3
The Chemical Approach

The generation of semisynthetic proteins relies on two different technologies that have seen a major evolution in the past decade: the synthesis of lipidated peptides, using solid-phase techniques, and the development of protein ligation methods enabling the chemical synthesis of proteins [6]. The chemistry and the development of both techniques, as pursued for the generation of semisynthetic K-Ras4B, are described here.

7.3.1
Chemical Synthesis of Proteins

In the past decade, numerous advances have been reported on the synthesis of proteins by ligation methods, thereby providing the necessary platform for combining large recombinant protein cores and peptides generated by organic synthesis. As a result, the reported ligation methods have become an attractive alternative to biological methods of protein production facilitating the generation of tailor-made proteins bearing natural or nonnatural modifications in preparative amounts.

One of the most relevant examples of chemical ligation techniques is the native chemical ligation (NCL) developed by Kent and coworkers [7]. In NCL, a peptide with a C-terminal thioester reacts with another peptide or protein bearing an N-terminal cysteine to yield a synthetic protein bound via a native amide bond. Later, Muir and coworkers [8] established the expressed protein ligation (EPL) version where the protein C-terminal thioester is generated by expression techniques employing appropriate vectors. Briefly, after expression and purification of an intein fusion protein, a thiolysis step yields the desired protein thioester that can react with a peptide containing an N-terminal cysteine residue to yield a semisynthetic protein (Figure 7.2). Hence, a combination of EPL and lipopetide synthesis was chosen as an efficient strategy for the synthesis of a fully modified K-Ras4B protein. For the efficient synthesis of K-Ras4B by EPL, the Gly174-Lys175 bond in the C-terminal flexible region was chosen as the ligation site. Thus, a farnesylated K-Ras4B peptide bearing an additional cysteine and a truncated protein thioester were required.

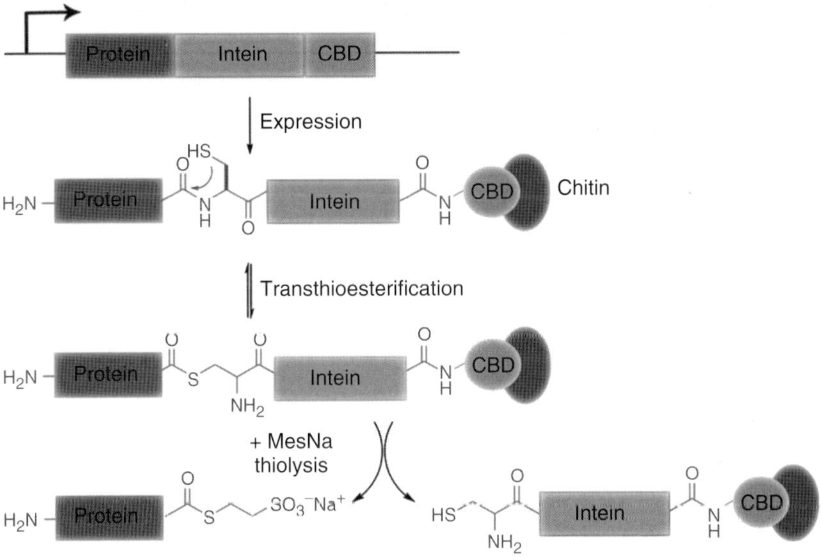

Figure 7.2 *Expressed protein ligation (EPL)* requires a protein C-terminal thioester that can be obtained by employing pTWIN vectors bearing an intein domain and a chitin-binding domain (CBD). The fusion protein is purified using chitin beads and a thiolysis step using sodium methanethiolate (MesNa) affords the desired protein thioester required for ligation with a peptide bearing an N-terminal cysteine.

7.3.2
Synthesis of Lipidated Ras Peptides

Strategies for the synthesis of lipidated peptides have been developed in the past decade both using solid-support and solution approaches. However, lipidated peptide synthesis using solution-phase protocols is time consuming and low yielding; therefore, solid-phase approaches are usually preferred. The synthesis of C-terminally lipidated peptides of Ras proteins has some requirements that pose limitations to the synthesis. Briefly, owing to the presence of the acid-sensitive farnesyl group, the C-terminal ester, and the labile thioester in the palmitoylated H- and N-Ras or the polybasic sequence in K-Ras, the synthesis of these peptides requires the choice of appropriate linkers and protecting groups and synthetic strategies that are compatible with the strict limitations of these peptides. That is, for example: (i) the chemical strategy should enable the presence of a C-terminal methyl ester. This can be achieved either with the choice of an appropriate linker, and usually included at the final stage of the synthesis or by using alternative linkers and adapting the chemical strategy. (ii) Strong acid-labile protecting groups and linkers need to be avoided because of the presence of the acid-sensitive farnesyl group; (iii) the same holds true for protecting groups that have to be removed under hydrogenolytic conditions; (iv) for palmitoylated peptides, the labile thioester can be cleaved by strong nucleophiles or undergo an *S,N*-acyl shift during peptide synthesis to an N-terminally unmasked cysteine; and (v) the synthetic approach should enable the incorporation of tags such as fluorophores, and nonhydrolyzable analogs. For N- and H-Ras, different strategies have been already developed. A linker that has given good results is the hydrazide linker that can be cleaved by oxidation to an acyldiazene, which is then attacked by a suitable nucleophile such as methanol, thus releasing the desired peptide as C-terminal methyl ester [9]. Another linker that has given good results for the synthesis of such peptides is the sulfonamide or Ellman linker, in which the peptide is attached to the linker as a stable acylsulfonamide that becomes sensitive to nucleophilic attack by methanol upon *N*-selective methylation with iodoacetonitrile [10].

For synthesis of the C-terminal K-Ras4B peptide, an alternative solid-phase strategy was developed. In this case, the acid-sensitive trityl resin was chosen because of its orthogonality to the farnesyl group, enabling the release of the peptide from the solid support using low concentrations of acid that do not affect the acid-labile prenyl moiety. To ensure the introduction of the C-terminal cysteine methyl ester, the peptide was anchored to the resin by the side-chain amino group of the allyl-protected lysine 184. After selective removal of the allyl ester, the S-farnesylated cysteine methyl ester was coupled. The peptide was then elongated using appropriate protecting groups, for lysines (Alloc), that enable their orthogonal palladium (0)-catalyzed deprotection, and Thr (Trt), Ser (Trt) that will be deprotected at the same time that the peptide is released from the solid support. Finally, the Cys required for ligation was added at the N-terminus protected as a *tert*-butyl disulfide. After deprotection of the Alloc groups, the resin was treated with a low concentration of acid, resulting in the release of the peptide from the solid support and the concomitant deprotection of the trityl groups. The resulting peptide was then purified with preparative high-performance liquid chromatography (HPLC) and obtained in a 15% yield (Figure 7.3) [11].

7.3.3 Synthesis of K-Ras4B Protein

For the synthesis of K-Ras4B protein by EPL, a truncated K-Ras protein thioester needs to be generated via expression in an appropriate vector. Protein thioesters are usually generated as intein fusion proteins. In this case, the IMPACT-TWIN (intein mediated purification with an affinity chitin-binding tag-two intein) system was chosen for the expression of the required K-Ras 1–174 thioester. The IMPACT system contains an additional chitin-binding tag that facilitates the purification of the fusion protein (Figure 7.2). Briefly, the gene coding for the truncated K-Ras4B thioester (1–174) was cloned into the IMPACT vector pTWIN1. The resulting plasmid was transformed into *Escherichia coli* BL21 (D3 cells) and the recombinant fusion protein was then purified by chitin-affinity column chromatography. The IMPACT system relies on the inducible self-cleavage activity of inteins to separate the target protein from the affinity tag as a C-terminal thioester upon treatment with a thiol. Thus, the desired thioester was

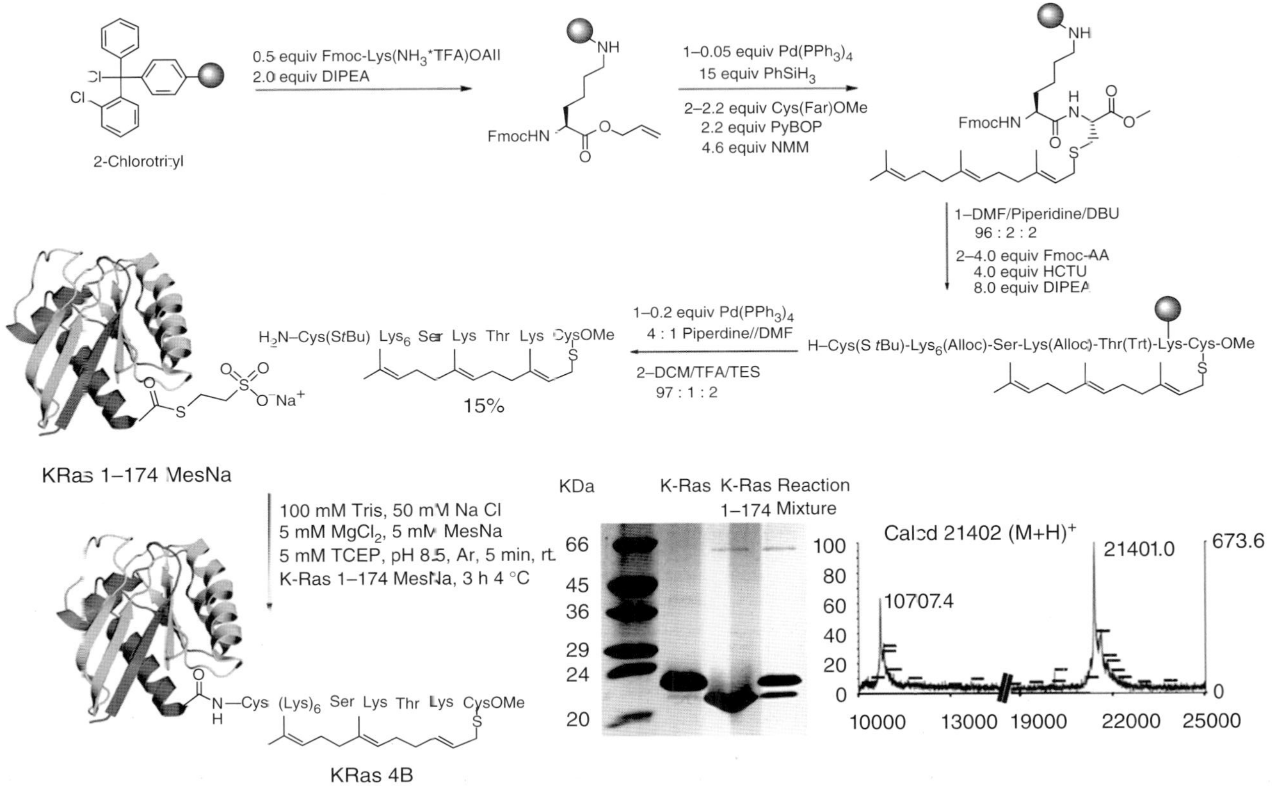

Figure 7.3 *K-Ras4B C-terminal peptide* was synthesized employing the trityl linker and after purification was reacted with a K-Ras thioester (K-Ras 1–174 MesNa) via EPL, yielding the semisynthetic K-Ras4B. Purity and identity of the purified protein were characterized by SDS-PAGE and MALDI-TOF. SDS-PAGE: sodium dodecyl sulfate/polyacrylamide gel electrophoresis, MALDI-TOF: matrix-assisted laser desorption ionization/time of flight.

released from the column by treatment with sodium methanethiolate (MesNa), yielding the corresponding K-Ras4B thioester that was purified by size-exclusion chromatography and characterized by electrospray mass spectrometry (ESI-MS) (Figure 7.3). The same strategy could be employed to prepare an oncogenic G12V K-Ras4B thioester bearing a mutation in residue 12, a valine instead of a glycine, that renders a permanently active protein arrested in the active GTP-bound state.

The synthetic farnesylated peptide corresponding to the C-terminus of K-Ras4B was then ligated to the protein thioester in the presence of tris(2-carboxyethyl)phosphine (TCEP), a phosphine that reduces the disulfide present at the N-terminal cysteine, thereby enabling the initiation of the chemical ligation. After 4 h of incubation, an 80% conversion of the ligation product could be detected. The ligated protein was then isolated employing a cation-exchange chromatography. This technique exploits the different physical properties of the ligated K-Ras4B protein, which has an isoelectric point pI (Box 7.1) of 8.2, and the truncated K-Ras thioester (pI 5.3) yielding the desired ligated protein in a 50–70% yield (Figure 7.3). A fluorescently labeled version of the ligated K-Ras4B could also be prepared by this strategy with the prior labeling of the protein core thioester with an NHS-activated (*N*-hydroxysuccinimide) fluorophore that reacts selectively with the amino groups of lysine side-chains followed by subsequent ligation of the fluorescently labeled K-Ras thioester with the lipidated peptide [11].

Box 7.1 Concepts and Techniques

Isoelectric point or pI is the pH at which in one molecule the negative and positive charges are equal.

Fluorescence polarization. Fluorescence polarization or anisotropy has been widely used for the study of protein–ligand interactions. This technique relies on the excitation of a fluorophore with plane-polarized light causing a polarized fluorescence. If the dye is attached to a small, rapidly rotating molecule, its random orientation results in low-fluorescence polarization because the planes into which the light will be emitted can be different from the plane used for initial excitation. Conversely, the binding of the fluorophore to large, slow rotating molecules (i.e., proteins) will result in high polarization.

siRNA – small interfering ribonucleic acid are double-stranded RNA, usually about 20–25 nucleotides long, that interfere with the expression of particular proteins by specifically degrading certain messenger ribonucleic acid sequences (mRNA).

Fluorescence lifetime imaging microscopy (FLIM)-based quantitative fluorescence resonance energy transfer (FRET). Direct detection of biomolecular interactions is possible with FRET measurements, where a donor fluorophore transfers the energy to an acceptor fluorophore in case they are close in space. The combination of this technique together with FLIM, based on the decrease in donor fluorophore lifetime that is induced by FRET, has recently enabled the quantitative assessment of the protein-interacting fractions [12].

FLAP (fluorescence loss after photoactivation) is a useful technique to investigate protein dynamics. The protein of interest can be fused to a photoactivatable green fluorescent protein (GFP) that upon irradiation increases fluorescence intensity around 100 times. Hence, after photoactivation of a certain cellular region, the protein diffusion rate can be determined by specifically tracking the photoactivated proteins [13].

FRAP (fluorescence recovery after photobleaching) is a complementary technique to FLAP. In FRAP, a specific cellular region is bleached after irradiation of a beam of laser and protein diffusion of a fluorescently labeled protein can be determined by detecting fluorescence recovery in the bleached area.

Alpha assays require two bead types, a donor bead that in this case will interact with the biotinylated peptide and an acceptor bead that recognizes the His-tagged protein. The donor beads contain a phthalocyanine that upon irradiation at 680 nm converts ambient oxygen to singlet oxygen. This singlet oxygen will then diffuse to the acceptor bead and react with a thioxene derivative, which results in the emission of a highly amplified chemiluminescence signal that can be monitored. As the singlet oxygen species can only diffuse 200 nm in solution, if the interaction is blocked by the presence of a small molecule it will not be able to reach the donor bead, thus causing the loss of the signal.

K-Ras-dependent cells are cancer cell lines that require sustained K-Ras function for viability. These cells were identified from a large panel of human cancer cell lines harboring mutant K-Ras after treatment with specific shRNAs directed to K-Ras. In K-Ras-dependent cells, a decrease in K-Ras protein expression results in a marked growth suppression, whereas in K-Ras-independent cells, a decrease in protein expression is not correlated with growth inhibition [14].

Surface plasmon resonance (SPR) allows monitoring label-free interactions between biomolecules in real time. In SPR measurements, one of the biomolecules is immobilized on a sensor surface and the other one is transported in solution across the surface. Binding of the second biomolecule will cause an increase in mass and a proportional increase in refractive index that can be monitored. Gold surfaces are the most commonly used ones and they can be modified or functionalized for the specific binding of particular molecules. For example, biotinylated surfaces enable the immobilization of streptavidin-labeled proteins, and surfaces modified with lipophilic groups allow the immobilization of vesicles creating membrane-like structures to study protein association and dissociation with membranes.

7.4
Chemical Biological Evaluation

The functionality and molecular integrity of the semisynthetic proteins was explored with the oncogenic G12V K-Ras4B by investigating its ability to induce differentiation of the rat pheochromocytoma PC12 cell line. Under normal growth conditions, this cell line has a chromaffin-cell-like morphology, but after transfection of oncogenic Ras genes or microinjection of oncogenic Ras proteins these cells develop neurite-like outgrowths that can be employed to prove the functionality of the semisynthetic proteins. Hence, both a 100 μM solution of the semisynthetic G12V K-Ras4B protein and a 200 μM solution of a full-length recombinant oncogenic G12V K-Ras4B (that will be processed in the cell generating the C-terminal farnesylated and carboxymethylated cysteine) were microinjected. As a result, the cells differentiated in all the cases to neurite-like cells, while no response could be seen upon microinjection of a 200 μM solution of the truncated protein thioester lacking the C-terminal region essential for membrane association and cellular activity. These results proved the functionality of the obtained semisynthetic proteins.

The generated proteins with a correct folding and functionality and containing all the required posttranslational modifications are key tools to investigate the function of proteins in biological processes. Therefore, the semisynthetic K-Ras4B was employed to investigate the suggested role of PDEδ in the transport of prenylated proteins. Previous and preliminary results have indicated the function of PDEδ as a prenyl-binding protein; however, the structural parameters dictating this protein–protein interaction and the exact role of PDEδ in regulating Ras signaling are yet to be defined [15]. With the aim of characterizing the Ras/PDEδ interaction and to determine the structural requirements of the binding, the affinity of the synthesized protein as well as some truncated model peptides were first measured using a fluorescence polarization assay (Box 7.1). For that, the semisynthetic K-Ras4B protein was loaded with a fluorescently labeled nonhydrolyzable analog of GTP (mant-Gpp-NHp, guanosine-5′-*O*-[(β,γ)-imido]-triphosphate) that binds to the GDP/GTP binding domain of Ras proteins. In addition, a (Gly-Cys) dipeptide and a peptide corresponding to the C-terminus of K-Ras4B bearing a dansyl fluorophore at the N-terminus were also synthesized. Similar measurements were performed with a semisynthetic Rheb (Ras homology enriched in brain, another member of the Ras family of small GTPases) and a peptide corresponding to the C-terminus of Rheb both obtained following a parallel strategy. Rheb contains also a farnesylated cysteine methyl ester but lacks the polybasic sequence characteristic for K-Ras4B. The results of the fluorescence polarization measurements indicated that PDEδ binds with low affinity to the solely farnesylated cysteine and this affinity increases up to 200 nM when longer peptides corresponding to the C-terminus of both proteins are employed. However, although the residues next to the cysteine seem to be important to increase the affinity, the presence of the polybasic sequence in K-Ras4B has no effect on improving or decreasing affinity when compared to the neutral sequence of Rheb. Moreover, similar affinities were obtained with the generated semisynthetic proteins, thus indicating that although the residues next to the cysteine are important, no binding with the rest of the protein core seems to take place. Similar results were obtained with Rheb when bound to GDP or to Gpp-NHp as a nonhydrolyzable analog of GTP, indicating that interaction of PDEδ with farnesylated Ras proteins occurred regardless of the nucleotide-bound state of the protein (Figure 7.4b).

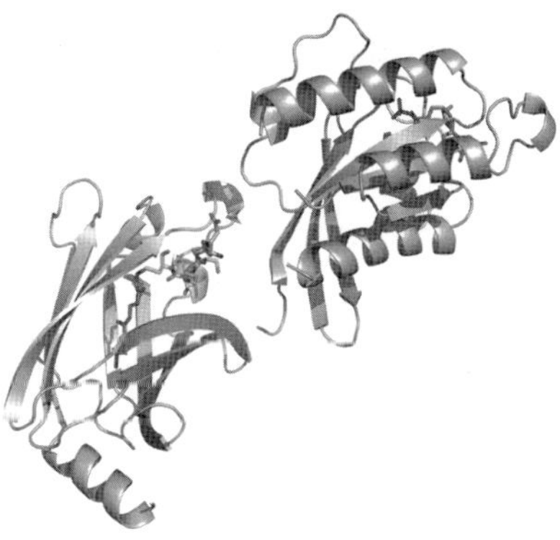

(a)

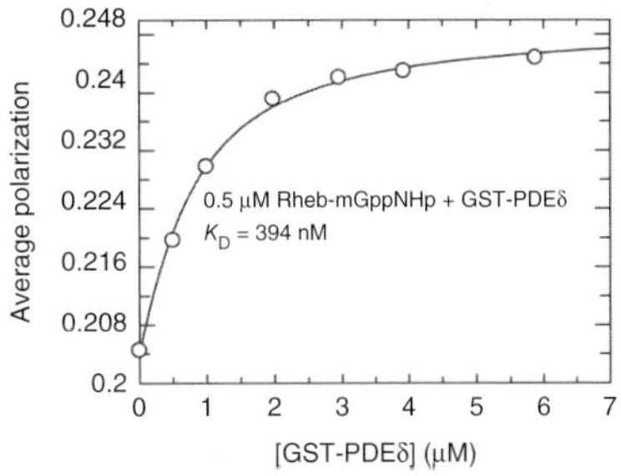

(b)

Vs PDEδ	K_D
Rheb mant-GppNHp	394 nM
K-Ras4B mant-GppNHp	302 nM
Dansyl-GKSSC(Far)-OMe (from Rheb)	102 nM
H-Cys(S*t*Bu)K(Dan)K_5SKTKC(Far)-OMe (from K-Ras 4B)	227 nM
Dansyl-GC(Far)-OMe	616 nM

Figure 7.4 (a, b) Crystal structure of the Rheb–PDEδ complex and affinities of PDEδ for fluorescently labeled proteins and C-terminal peptides from Rheb and K-Ras as measured by a fluorescence polarization assay.

The interaction between PDEδ and the farnesylated Rheb protein was further investigated by X-ray crystallography. The 1.7 Å structure of PDEδ in complex with the farnesylated Rheb protein showed that the protein recognizes mainly the prenyl group that is deeply buried in a hydrophobic pocket while no major interactions with the rest of the protein occur, thus confirming the results obtained in the fluorescence polarization studies (Figure 7.4a) [16].

The biological significance of this interaction was next investigated in cells [17]. First, the effect of PDEδ on Ras protein distribution in cellular membranes was investigated by knockdown of PDEδ by specific siRNA (Box 7.1), which resulted in a shift from plasma membrane and Golgi-specific H/N-Ras localization to a randomized endomembrane distribution, thus demonstrating the physiological key role of PDEδ in maintaining the spatial organization of Ras proteins. Then, madine-darby canine kidney (MDCK) cells were transiently cotransfected with various Ras proteins fused to the monomeric yellow fluorescent protein Citrine and PDEδ fused to the monomeric red fluorescent protein mCherry and the cytoplasmatic distribution of Ras proteins was investigated after PDEδ expression by fluorescence microscopy. Indeed, PDEδ showed a solubilizing effect for N-Ras that lost membrane localization and was redistributed to the cytoplasm (Figure 7.5a).

Then, the direct interaction between PDEδ and farnesylated Ras and Rheb proteins was investigated in cells by carrying out FLIM-based quantitative FRET (Box 7.1) measurements to quantify the protein-interacting fractions. N-Ras labeled with mCitrine was employed as the donor, whereas dCherry fused to the C-terminus of PDEδ was used as the acceptor. The results obtained in the FLIM-FRET measurements proved that a high interacting fraction of N-Ras and PDEδ was found in the cytosol (Figure 7.5b).

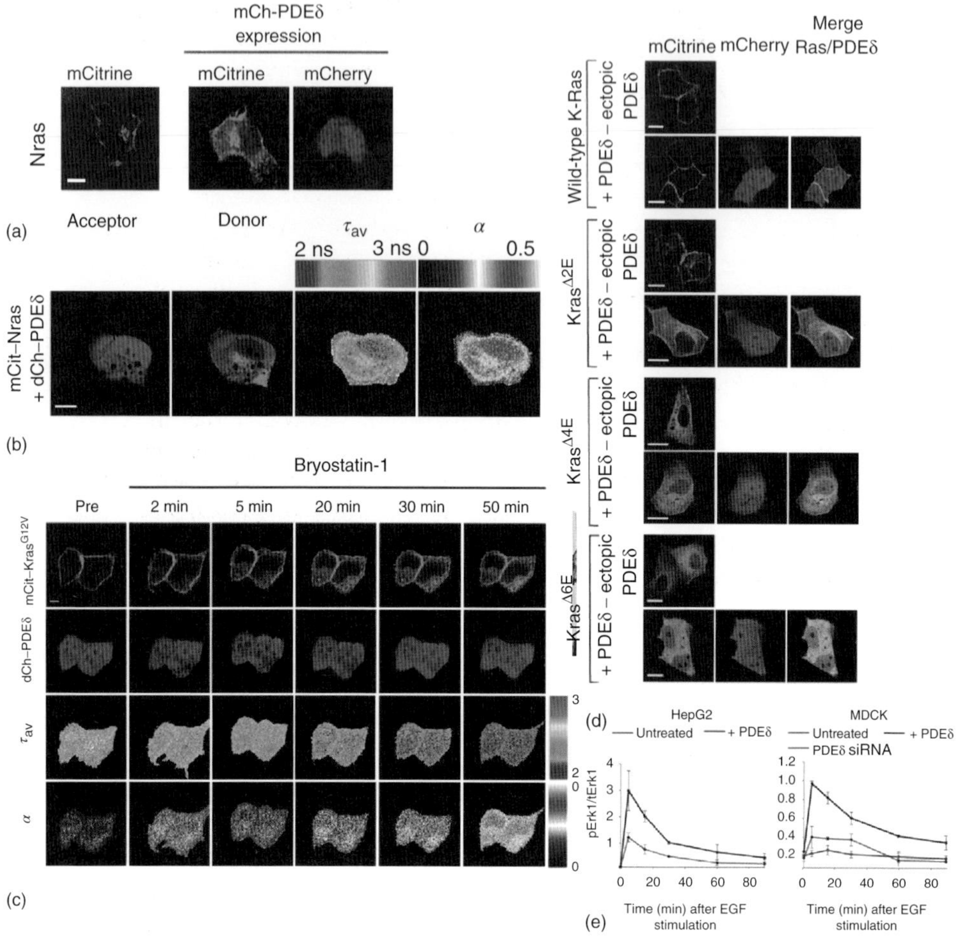

Figure 7.5 (a) PDEδ has a solubilizing effect on citrine-labeled N-Ras that lost membrane localization and got redistributed to the cytoplasm after PDEδ expression. (b) FLIM-FRET enables measuring the protein-interacting fraction between N-Ras and PDEδ. (c) Treatment of cells with the PKC agonist bryostatin-1 causes the dissociation of K-Ras4B from the plasma membrane and an increase in the interacting K-Ras-PDEδ fraction. (d) Reduction of basic charge in K-Ras4B results in membrane dissociation and in an increase in interacting K-Ras-PDEδ fraction. (e) Expression of PDEδ in HepG2 and MDCK cells results in an enhancement of pErk1 levels. On the contrary, knock-down of PDEδ by RNA interference reduced pErk1 levels. (Reprinted with permission from Macmillan Publishers Ltd: Nature Cell Biology from [17], Copyright (2012).)

Next, the role of PDEδ on the mobility of Ras proteins between cellular membranes was investigated by FLAP (Box 7.1). To this end, the effective diffusion of a photoactivatable Ras was measured after ectopic expression or downregulation of PDEδ by siRNA. In the first case, an increased diffusion of Ras could be observed, whereas the downregulation of PDEδ resulted in a reducing effect on the protein mobility. This was further supported by FRAP (Box 7.1) experiments employing mCit–N-Ras in mCh–PDEδ-expressing cells, showing an increased level of fluorescence recovery at the Golgi after photobleaching. This indicates that PDEδ indeed enhances the effective diffusion of N-Ras, thereby increasing the probability of encounter with the Golgi and thus the rate at which Ras gets trapped by repalmitoylation.

Next, the role of PDEδ as a solubilizing factor was investigated for the polycationic K-Ras4B. Direct interaction between K-Ras4B and PDEδ was first investigated in cells, however without much success. One of the reasons for this lack of detectable interaction in the cellular environments may be the tight affinity of K-Ras4B for the plasma membrane. To explore this possibility in more detail, the charge in the HVR of K-Ras4B was modified by gradual replacement of the basic lysines by acidic glutamates or neutral glutamine residues. As expected, this replacement resulted in a progressive release of K-Ras from the plasma membrane and in an increased interaction with PDEδ as observed by FLIM-FRET measurements, thus confirming that the high affinity of K-Ras for the plasma membrane prevents its interaction with PDEδ at steady-state conditions (Figure 7.5d).

The interaction of the polybasic K-Ras with the plasma membrane has been suggested to be regulated by the action of protein kinase C (PKC), which phosphorylates a serine residue located in the vicinity of the polylysine stretch, thus diminishing the electrostatic interaction and resulting in the detachment of K-Ras from the plasma membrane. To investigate in more detail the K-Ras4B/PDEδ interaction, similar studies were performed after previous treatment of cells with the PKC agonist, bryostatin-1. This treatment caused the dissociation of K-Ras from the plasma membrane and its redistribution to the endomembranes together with an increase of the interacting K-Ras PDEδ fraction as measured by FLIM-FRET, indicating that although PDEδ is not able to extract K-Ras from the plasma membrane owing to the high affinity of K-Ras for this acidic membrane, it interacts tightly with K-Ras when it is localized at any other endomembrane (Figure 7.5c).

Next, PDEδ involvement in Ras signaling and cell proliferation was investigated by measuring the levels of phosphorylated extracellular regulated kinase (pErk)1/2 (pErk) after epidermal growth factor (EGF) stimulation. First, HepG2 cells, which lack PDEδ, were taken as a model. Ectopic expression of PDEδ in HepG2 resulted in a several-fold increase in pErk1/2 levels, compared to untransfected HepG2 cells. Equally, MDCK cells also showed further enhancement of pErk1/2 levels after stimulation with EGF on ectopic expression of mCh–PDEδ. Conversely, knockdown of PDEδ by RNA interference reduced pErk1/2 levels (Figure 7.5e). These results show that the PDEδ-mediated enrichment of Ras at the plasma membrane enhances Ras activation and its downstream signaling. Analogously, a similar effect on the modulation of Ras signaling was observed in cancer cells, which was accompanied by a decrease in cell proliferation.

These cellular studies shed light on the role of PDEδ on Ras distribution and function [17]. Hence, the main role of PDEδ is to allow Ras proteins to shuttle between cellular membranes by solubilizing them in the cytosol and to facilitate the intracellular Ras diffusion and enhance its trapping at the right compartment. Briefly, PDEδ tightly interacts with monofarnesylated N- and H-Ras proteins and, as a consequence, increases their solubility in the cytosol and enhances their diffusion to the Golgi, where they will get repalmitoylated. K-Ras4B, once farnesylated, will interact with the most abundant endomembrane system of the endoplasmic reticulum, and will slowly get trapped to the highly negatively charged plasma membrane by electrostatic interactions. This slow process is accelerated by PDEδ by competing with endomembranes for binding with K-Ras4B and enhancing the diffusion of the resulting solubilized protein to the target plasma membrane. Once in the plasma membrane, the tight affinity of K-Ras4B for this particular membrane prevents the interaction with PDEδ, which is not able to extract it from the plasma membrane. In this case, the phosphorylation of a serine residue near the polybasic stretch is probably required to decrease the electrostatic interaction and promote K-Ras distribution to endomembranes where the cycle will start again.

Evidence at the molecular level for such a mechanism could also be obtained by a combination of biophysical techniques such as SPR (Box 7.1), frequency-domain fluorescence anisotropy, atomic force microscopy (AFM), or infrared reflection absorption (IRRA). For example, SPR measurements using immobilized membranes on chips also indicated that PDEδ seems to function as a solubilizing factor for K-Ras4B in the cytosol, and effectively delivers farnesylated K-Ras4B to the plasma membrane. Briefly, when immobilized membranes were treated with K-Ras4B followed by PDEδ, PDEδ was not able to extract immobilized K-Ras. However, when both proteins were incubated at the same time with the membranes, K-Ras association with the membrane was clearly accelerated compared to the addition of K-Ras4B only (Figure 7.6) [18].

On the basis of this information, which indicates the crucial role of PDEδ in sustaining Ras localization and signaling activity, a screening assay to detect small-molecule inhibitors of the Ras–PDEδ interaction was developed with the final aim of validating PDEδ as a novel target for the treatment of cancers characterized by oncogenic Ras signaling. To this end, a biotin-labeled peptide corresponding to the C-terminus of K-Ras4B and a His-tagged PDEδ were employed to establish an assay based on the Alpha-Screen technology (Box 7.1). The assay was then employed to screen a 150000 compound library, yielding several hits such as benzimidazoles ($K_D = 165 \pm 23$ nM). The identified hits were further characterized by an alternative fluorescence polarization assay and by crystal structure analysis of the benzimidazole hit in complex with PDEδ, which further confirmed its inhibitory activity and suggested the presence of two benzimidazole units in the PDEδ-binding pocket (Figure 7.7a). Next, a second-generation inhibitor class bearing two linked benzimidazoles was designed on the basis of the information obtained from the crystal structure. A focused library of this compound class containing different linkers between the two benzimidazoles (ester, ether) was synthesized, resulting in inhibitors with significantly increased affinity (such as Deltarasin, $K_D = 38 \pm 16$ nM, Figure 7.7a,b) whose binding was further confirmed

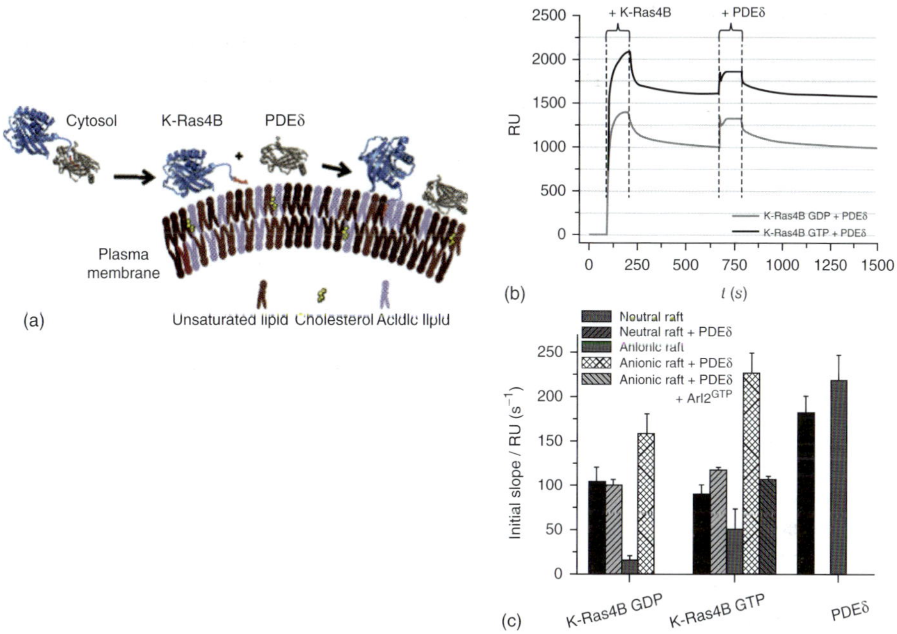

Figure 7.6 (a) Surface plasmon resonance measurements indicate that PDEδ is not able to extract K-Ras4B once associated with membranes (b upper level, sensogram). The addition to immobilized membranes of K-Ras4B together with PDEδ resulted in an increased membrane association of K-Ras4B (b, lower level). (Reprinted (adapted) with permission from [18], Copyright (2012) American Chemical Society.)

by X-ray analysis and direct titration using a 5-carboxytetramethyl rhodamine (TAMRA) labeled inhibitor [19].

The effect of deltarasin on oncogenic K-Ras signaling was then investigated in different human pancreatic ductal adenocarcinoma (PDAC) cell models: Panc-Tu-I and Capan-1 cells that depend on oncogenic K-Ras signaling for survival, the K-Ras mutated but independent PANC-1, and BxPC-3 that express wild-type K-Ras. The treatment of these cells with 5 μM of deltarasin caused a reduced proliferation and cell death of K-Ras-dependent cells (Panc-Tu-I and Capan-1, Box 7.1) in a dose-dependent manner, while almost no effect could be seen in the other cell lines (Figure 7.7c). Moreover, the antitumor effect of deltarasin was also investigated in nude mice bearing subcutaneous human Panc-Tu-I tumor cell xenografts. To this end, deltarasin was injected intraperitoneally once (QD) or twice (BID) per day ($10\,\mathrm{mg\,kg^{-1}}$ QD, $15\,\mathrm{mg\,kg^{-1}}$ QD, and $10\,\mathrm{mg\,kg^{-1}}$ BID) and the effect of the treatment on tumor size was then measured, revealing a clear dose-dependent reduction or even blockage of Panc-Tu-I tumor growth rate in treated mice with respect to the vehicle-injected controls (Figure 7.7d) [19].

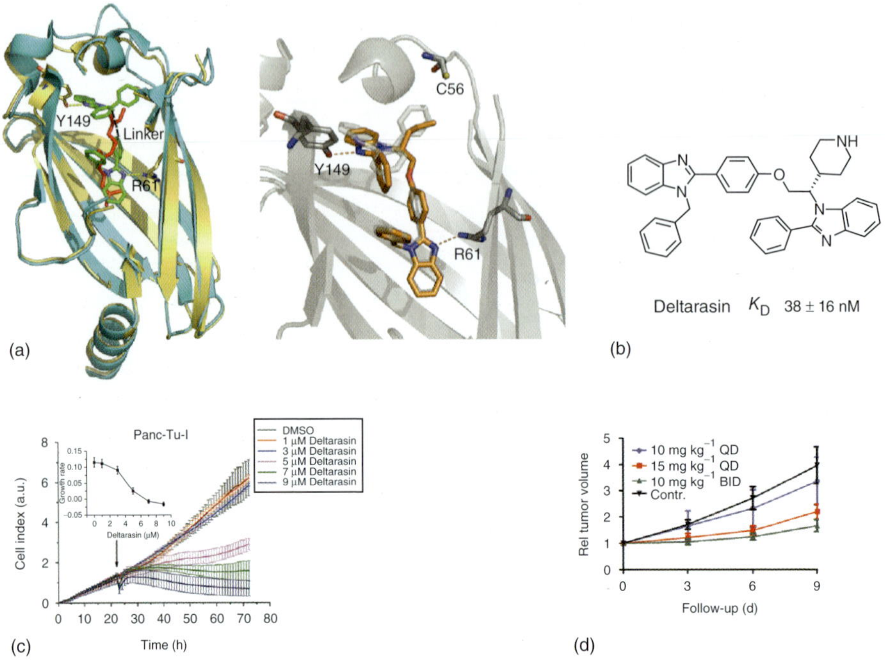

Figure 7.7 (a) Ribbon diagram of PDEδ structure in complex with two benzimidazoles hits obtained in the HTS screening (yellow), and overlay with the previously obtained crystal structure of farnesylated Rheb peptide with PDEδ (cyan). Structure of two linked benzimidazoles (orange sticks) in complex with PDEδ. Overlaid is the structure of two molecules of benzimidazoles (faint gray sticks) in complex with PDEδ. (b) Structure of deltarasin and binding affinity for PDEδ. (c) Real-time cell analysis of deltarasin dose PDAC cell proliferation response (Panc-Tu-I cells). The inset shows deltarasin dose versus growth-rate response. (d) Tumor volume measurements of Panc-Tu-I xenografts treated with deltarasin at the dosages indicated. HTS: high-throughput screen/screening. (Reprinted with permission from Macmillan Publishers Ltd: Nature from [19], Copyright (2013).)

7.5 Conclusions

In summary, a chemical biology approach was established to characterize the role of the prenyl-binding protein PDEδ in maintaining Ras localization and signaling. First, a fully lipidated K-Ras4B protein could be obtained by means of a combination of protein expression and lipidated peptide synthesis. The resulting semisynthetic protein was essential for determining the structural requirements of PDEδ for the recognition of prenylated proteins, showing that PDEδ is a general binder of farnesylated proteins that recognizes the farnesyl group as well as some amino acids located near the C-terminus. There is no further binding with the protein core, which could be further confirmed by X-ray crystallography. The suggested role of PDEδ in the transport and signaling of Ras proteins was then investigated in cellular and biophysical studies, indicating that PDEδ has a crucial role in sustaining Ras cellular distribution and activity. The knockdown of PDEδ by genetic methods resulted in Ras redistribution to endomembranes and a decrease in signaling and cell proliferation, thus indicating that the Ras–PDEδ interaction may be a novel and interesting target for the treatment of cancers depending on oncogenic Ras signaling. On the basis of these results, a screening assay aimed to detect inhibitors of the Ras–PDEδ interaction was established and employed to screen a 150000 compound library. After validation of the identified hits, a structure-based design was followed to obtain inhibitors with increased affinities such as deltarasin. Cellular studies with adenopancreatic cancer cells depending on K-Ras for survival showed that deltarasin blocks Ras signaling, reduces proliferation, and causes cell death in K-Ras-dependent cells, while almost no effect can be detected in K-Ras-independent cells. Moreover, the administration of deltarasin to mice bearing subcutaneous human Panc-Tu-I tumor cell xenografts caused a dose-dependent reduction of tumor growth. These results validate PDEδ as a novel target and open new avenues for the treatment of cancers characterized by oncogenic Ras signaling.

References

1. Konstantinopoulos, P.A., Karamouzis, M.V., and Papavassiliou, A.G. (2007) Post-translational modifications and regulation of the RAS superfamily of GTPases as anticancer targets. *Nat. Rev. Drug Discovery*, **6**, 540–555.
2. Triola, G., Waldmann, H., and Hedberg, C. (2012) Chemical biology of lipidated proteins. *ACS Chem. Biol.*, **7**, 87–99.
3. Vogel, A., Reuther, G., Weise, K., Triola, G., Nikolaus, J., Tan, K.T., Nowak, C., Herrmann, A., Waldmann, H., Winter, R., and Huster, D. (2009) The lipid modifications of Ras that sense membrane environments and induce local enrichment. *Angew. Chem. Int. Ed.*, **48**, 8784–8787.
4. Weise, K., Triola, G., Brunsveld, L., Waldmann, H., and Winter, R. (2009) Influence of the lipidation motif on the partitioning and association of N-Ras in model membrane subdomains. *J. Am. Chem. Soc.*, **131**, 1557–1564.
5. Rocks, O., Gerauer, M., Vartak, N., Koch, S., Huang, Z.P., Pechlivanis, M., Kuhlmann, J., Brunsveld, L., Chandra, A., Ellinger, B., Waldmann, H., and Bastiaens, P.I.H. (2010) The palmitoylation machinery is a spatially organizing system for peripheral membrane proteins. *Cell*, **141**, 458–471.
6. Brunsveld, L., Kuhlmann, K., Alexandrov, K., Wittinghofer, A., Goody, R.G., and Waldmann, H. (2006) Lipidated ras and rab peptides and proteins – synthesis, structure, and function. *Angew. Chem. Int. Ed.*, **45**, 6622–6646.
7. Dawson, P.E., Muir, T.W., Clarklewis, I., and Kent, S.B.H. (1994) Synthesis of proteins by native chemical ligation. *Science*, **266**, 776–779.
8. Muir, T.W., Sondhi, D., and Cole, P.A. (1998) Expressed protein ligation: a general method for protein engineering. *Proc. Natl. Acad. Sci. U.S.A.*, **95**, 6705–6710.
9. Lumbierres, M., Palomo, J., Kragol, G., Roehrs, S., Müller, O., and Waldmann, H. (2005) Solid phase synthesis of lipidated peptides. *Chem. Eur. J.*, **11**, 7405–7415.
10. Triola, G., Gerauer, M., Görmer, K., Brunsveld, L., and Waldmann, H. (2010) Solid-phase synthesis of lipidated ras peptides employing the ellman sulfonamide linker. *Chem. Eur. J.*, **16**, 9585–9591.

11. Chen, Y.X., Koch, S., Uhlenbrock, K., Weise, K., Das, D., Gremer, L., Brunsveld, L., Wittinghofer, A., Winter, R., Triola, G., and Waldmann, H. (2010) Synthesis of the Rheb and K-Ras4B GTPases. *Angew. Chem. Int. Ed.*, **49**, 6090–6095.
12. Walther, K.A., Papke, B., Sinn, M.B., Michel, K., and Kinkhabwala, A. (2011) Precise measurement of protein interacting fractions with fluorescence lifetime imaging microscopy. *Mol. BioSyst.*, **7**, 322–336.
13. Patterson, G.H. and Lippincott-Schwartz, J. (2002) A photoactivatable GFP for selective photolabeling of proteins and cells. *Science*, **297**, 1873–1877.
14. Singh, A., Greninger, P., Rhodes, D., Koopman, L., Violette, S., Bardeesy, N., and Settleman, J. (2009) A gene expression signature associated with "K-Ras Addiction" reveals regulators of EMT and tumor cell survival. *Cancer Cell*, **15**, 489–500.
15. Zhang, H.B., Liu, X.H., Zhang, K., Chen, C.K., Frederick, J.M., Prestwich, G.D., and Baehr, W. (2004) Photoreceptor cGMP phosphodiesterase delta subunit (PDE delta) functions as a prenyl-binding protein. *J. Biol. Chem.*, **279**, 407–413.
16. Ismail, S.A., Chen, Y.X., Rusinova, A., Chandra, A., Bierbaum, M., Gremer, L., Triola, G., Waldmann, H., Bastiaens, P.I.H., and Wittinghofer, A. (2011) Arl2-GTP and Arl3-GTP regulate a GDI-like transport system for farnesylated cargo. *Nat. Chem. Biol.*, **7**, 942–949.
17. Chandra, A., Grecco, H.E., Pisupati, V., Perera, D., Cassidy, L., Skoulidis, F., Ismail, S.A., Hedberg, C., Hanzal-Bayer, M., Venkitaraman, A.R., Wittinghofer, A., and Bastiaens, P.I.H. (2012) The GDI-like solubilizing factor PDE delta sustains the spatial organization and signalling of Ras family proteins. *Nat. Cell Biol.*, **14**, 148–158.
18. Weise, K., Kapoor, S., Werkmuller, A., Mobitz, S., Zimmermann, G., Triola, G., Waldmann, H., and Winter, R. (2012) Dissociation of the K-Ras4B/PDEdelta complex upon contact with lipid membranes: membrane delivery instead of extraction. *J. Am. Chem. Soc.*, **134**, 11503–11510.
19. Zimmermann, G., Papke, B., Ismail, S., Vartak, N., Chandra, A., Hoffmann, M., Hahn, S.A., Triola, G., Wittinghofer, A., Bastiaens, P.I.P., and Waldmann, H. (2013) Small molecule inhibition of the KRAS-PDEdelta interaction impairs oncogenic KRAS signalling. *Nature*, **497**, 638–642.

8
Development of Acyl Protein Thioesterase 1 (APT1) Inhibitor Palmostatin B That Revert Unregulated H/N-Ras Signaling

Frank J. Dekker, Nachiket Vartak, and Christian Hedberg

8.1
Introduction

This chapter describes the combination of bioinformatics, organic synthesis, *in vitro* inhibition studies, and live-cell imaging to elucidate the function of acyl protein thioesterase 1 (APT1) in regulation of protein palmitoylation. APT1 critically influences the localization and function of several palmitoylated peripheral membrane proteins of the rat sarcoma (Ras) and Rous sarcoma oncogene cellular homolog (Src) family, which themselves have pivotal roles in cancer signaling. Protein structure similarity clustering (PSSC) was employed to identify proteins that are structurally similar to APT1. Gastric lipase was identified among a variety of proteins that have similar 3D structure to APT1, and thus the natural lipase inhibitor lipstatin and its derivatives served as guiding structures for the creation of an analogous APT1 inhibitor. A compound library created on this basis was then screened in biochemical and phenotypic assays to identify and generate potent and cell-permeable inhibitors of APT1. These inhibitors were then utilized to study the effects of APT1 inhibition on Ras localization and signaling in live cells using advanced imaging approaches. The resultant inhibitors were demonstrated to perturb oncogenic Ras localization and signaling, eventually leading to phenotypic reversion of oncogenic cells to a normal phenotype.

8.2
The Biological Problem – The Role of APT1 in Ras Signaling

The Ras guanosine triphosphatases (GTPases) exist in three isoforms in human cells – H/N/K-Ras. All three Ras isoforms are localized to the plasma membrane and play a crucial role in growth-factor-derived signals that lead to cell proliferation (Box 8.1). Oncogenic mutations in Ras genes, which result in the formation of a constitutively active Ras protein, are often found in several types of cancers. Such constitutively active Ras transmits a continuous proliferative signal from the

Concepts and Case Studies in Chemical Biology, First Edition. Edited by Herbert Waldmann and Petra Janning.

plasma membrane even in the absence of growth factors, which leads to unregulated tumor growth [1]. Perturbation of Ras localization to the plasma membrane is therefore expected to disrupt this oncogenic proliferative signal, making it a viable strategy for cancer therapeutics [2]. The plasma membrane localization of the three isoforms is distinct, and conferred by differences in posttranslational modifications and physicochemical features of the C-terminus of these proteins. Firstly, all three Ras isoforms undergo irreversible farnesylation at the CAAX box, a sequence feature present at their C-terminal. The farnesyl moiety is hydrophobic, and this imparts some a specific affinity toward membranes. Further, H/N-Ras undergo reversible S-palmitoylation at cysteine residues in the C-terminus, which further increases the hydrophobicity and membrane affinity of these proteins [3].

The reversible protein S-palmitoylation forms the basis of a reaction-diffusion mechanism termed the *acylation cycle* (Figure 8.1), responsible for generating an enrichment of Ras on the plasma membrane and preventing its localization on other cellular membranes [4]. S-palmitoylation of H/N-Ras is catalyzed by protein-acyl transferases (PATs) present on the Golgi apparatus. The resultant increase in membrane affinity of Ras molecules traps them on Golgi membranes. These Ras molecules are then transferred to the plasma membrane via vesicular transport of the Golgi–PM secretory pathway [5]. Thus, a nonequilibrium enrichment of Ras is generated at the plasma membrane. However, several processes lead to the eventual "leakage" of Ras proteins from the plasma membrane. These entropic processes include membrane dynamics such as endocytosis and membrane mixing. These leaked Ras proteins will "mislocalize," especially to all

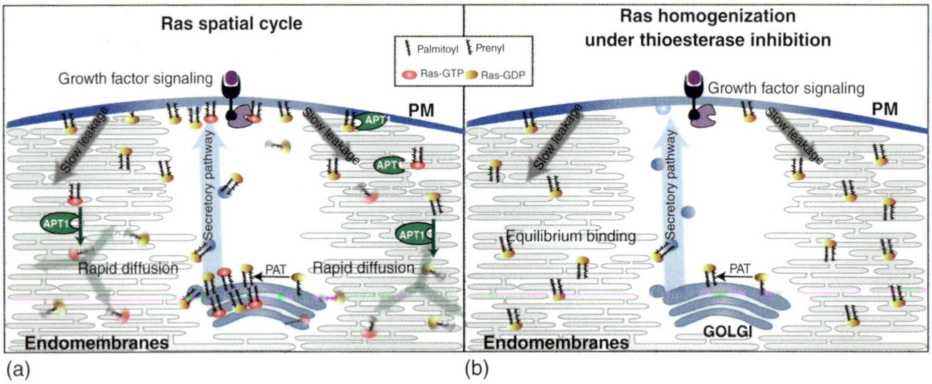

Figure 8.1 (a) The maintenance of the specific subcellular localization of Ras GTPases by a dynamic cycle of ubiquitous depalmitoylation and palmitoylation on the Golgi system. Palmitoylation on the Golgi redirects Ras GTPases to the plasma membrane through the secretory pathway. This plasma membrane enrichment allows Ras to function as a signal transducer in growth factor signaling. The slow "leakage" of this enrichment to endomembranes is corrected by APT1-mediated depalmitoylation, so that Ras proteins may diffuse to the Golgi and be repalmitoylated. (b) The interruption of the acylation cycle by inhibition of APT1, which leads to mislocalization of Ras on endomembranes.

endomembranes at equilibrium, because of exchange processes. To correct for mislocalization, thioesterases such as APT1 [6] depalmitoylate Ras molecules ubiquitously on endomembranes. The resulting reduction in membrane affinity of Ras molecules allows them to diffuse rapidly throughout the cellular interior. These diffusing Ras molecules thus have a high probability of encountering the Golgi apparatus, where they are repalmitoylated and transferred once again to the plasma membrane, completing the acylation cycle. The function of APT1 is therefore to rectify mislocalization of Ras proteins and allow the acylation cycle to reinstate their physiological plasma membrane localization. If APT1 is inhibited, this correction can no longer occur and Ras remains distributed on endomembranes. The absence of Ras from the plasma membrane leads to a corresponding attenuation of Ras-mediated signaling. In the context of cancer therapeutics, APT1 inhibition leads to the downregulation of oncogenic Ras signaling, with beneficial effects against neoplastic transformation, metastasis, and tumor growth.

Box 8.1 Ras GTPases

The family of Ras GTPases serves as molecular switches in different cellular signaling events. Ras GTPases cycle between the active guanosine triphosphate (GTP)-bound form and the inactive guanosine diphosphate (GDP)-bound form. Extracellular stimulation of transmembrane receptors activates intracellular guanine nucleotide-exchange factors (GEFs) and GTPase-activating proteins (GAPs), which triggers the conversion of Ras GTPases from their inactive GDP-bound form to their active GTP-bound form. Activation of Ras GTPases triggers downstream signaling events. Protein–membrane and protein–protein interactions play an important role in spatial and temporal activation of Ras GTPases.

Small-molecule inhibitors of APT1 were developed using a knowledge-based approach [7]. These inhibitors were used for chemical knockdown of APT1 in functional studies of Ras palmitoylation in cell-based assays. These studies revealed that long-term inhibition of cellular thioesterase activity by APT1 inhibitors leads to an entropy-driven loss of the precise localization of palmitoylated Ras proteins. As a consequence, oncogenic Ras signaling is downregulated [7].

8.3 The Chemical Approach

8.3.1 The Challenge to Make Small-Molecule Modulators of Protein Function

Selective and potent small-molecule inhibitors for the protein of interest are required for reverse chemical genetics studies. The emergence of high-throughput screening and combinatorial synthesis of compound collections enables screening of many compounds on many targets. However, initial expectations that screening of large compound libraries on many targets will result in the discovery of many new hit and lead structures for reverse chemical genetics studies and drug development were not met. PSSC is a knowledge-based approach in which the ligand-sensing core of proteins are clustered and knowledge about known ligands for members of such a cluster can be employed to guide compound library development for other members of the cluster (Box 8.2). Proteins with a high structural similarity and a low sequence similarity are the most interesting cases for PSSC, because they represent distantly related targets that have a good chance to bind to molecules with similar core structures [8, 9]. This chapter provides a case study in which PSSC was applied to find small-molecule inhibitors of the enzyme APT1 (Figure 8.2a). The inhibitors were applied in reverse chemical genetics investigations of Ras localization and signaling in cell-based studies [7].

Box 8.2 Protein Structure Similarity Clustering (PSSC)

PSSC is a knowledge-based approach to develop small-molecule inhibitors of protein function. Protein domains or cores with similar three-dimensional structures are clustered in so-called PSSCs. Knowledge about known ligands for members of such a cluster can be employed to guide the design of focused compound collections for screening on the other cluster members. Increased hit rates can be expected using this approach.

8.3.2
Bioinformatics – Target Clustering

In order to develop a small-molecule APT1 inhibitor, a PSSC was constructed on the basis of the structural similarity to the ligand-sensing core of APT1 (protein data base, PDB code 1FJ2) [10]. A high structural similarity between the enzyme's gastric lipase (PDB code 1K8Q) [11], and APT1 was discovered (Figure 8.2b). Gastric and pancreatic lipases are inhibited by the natural product lipstatin and its derivative tetrahydrolipstatin 1 [12, 13]. Tetrahydrolipstatin was chosen as the biologically prevalidated starting point for compound library design. A compound collection was designed on the basis of the β-lactone core motif that is present in lipstatin and tetrahydrolipstatin (Figure 8.2a) [7].

8.3.3
Compound Collection Synthesis

A focused compound collection of 99 compounds with a β-lactone core structure was synthesized (Scheme 8.1) [7]. Thirty compounds with a *syn*-configuration around the β-lactone core were synthesized by *syn* selective aldol reactions of ethyl esters and aliphatic aldehydes [14]. The *syn* aldol products were saponified

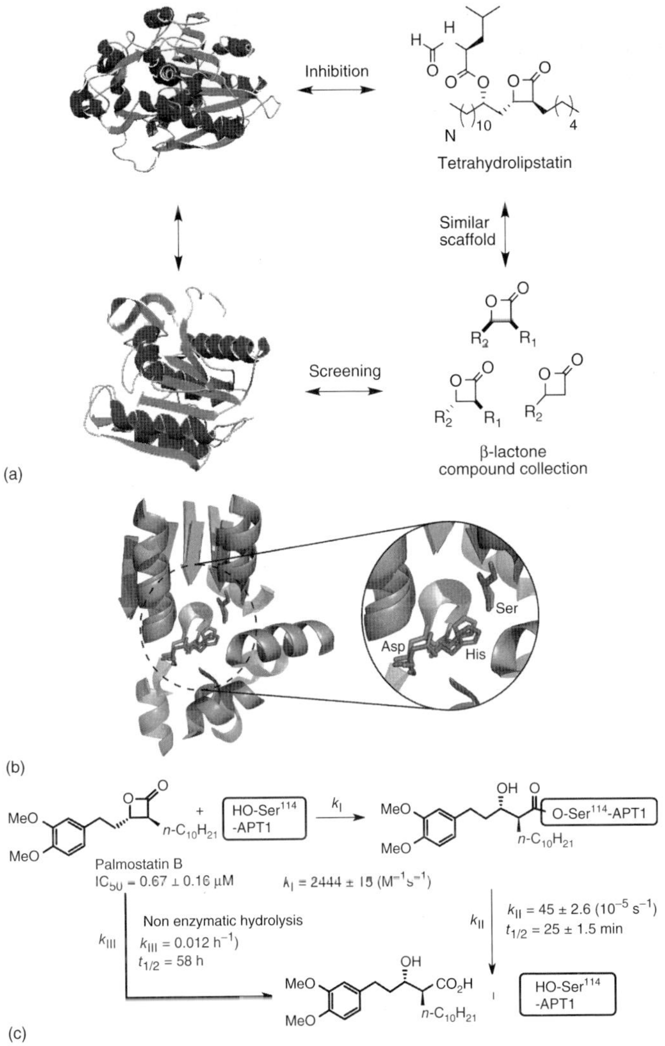

Figure 8.2 (a) PSSC is a strategy for hit finding based on a known inhibitor–enzyme combination, in this case gastric lipase and tetrahydrolipstatin. The protein structure of the enzyme is used for selection of a target cluster and the chemical structure of the inhibitor is used as inspiration source for synthesis of a compound collection. The ligand-sensing cores of the enzymes APT1 and gastric lipase show a high structural similarity despite a low sequence similarity and fits to the guidelines formulated for PSSC. The β-lactone core structure of the lipase inhibitor tetrahydrolipstatin was chosen as a starting point for compound library design. (b) Overlay of the ligand-sensing cores of gastric lipase (dark) and APT1 (light). The catalytic triad (Asp, His, Ser) of each enzyme is shown as sticks (also shown in the enlarged image). (c) The palmostatins inhibit the enzyme APT1 as slowly converted substrates that inactivate the enzyme quickly (k_I) followed by slow regeneration of the active enzyme (k_{II}). Nonenzymatic hydrolysis of the β lactone (k_{III}) also occurs, characterizing its stability in aqueous solution. (Reprinted with permission from [7].)

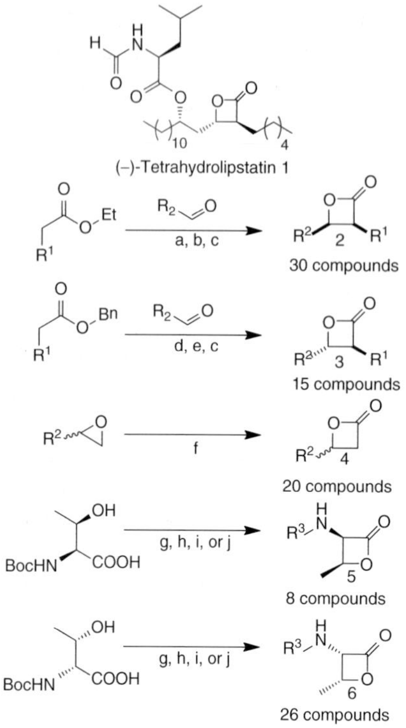

Scheme 8.1 *Synthesis of a focused compound collection with a β-lactone as core structure.* (a) $(nBu)_2BOTf$, DIPEA, CH_2Cl_2; (b) LiOH, dioxane, H_2O; (c) $PhSO_2Cl$, pyridine; (d) $(Chx)_2BOTf$, Et_3N, CH_2Cl_2; (e) Pd/C, H_2, MeOH; (f) carbonylation using $[(salph)Cr(THF)_2][Co(CO)_4]$ and CO (salph = *N,N*-*o*-bis (3,5-di-*tert*-butylsalicylidene)-1,2-phenylenediamine; THF = tetrahydrofuran); (g) PyBOP, Et_3N, CH_2Cl_2; (h) trifluoro acetic acid (TFA), *p*-toluenesulfonic acid (PTSA); (i) R_3COOH, $ClCO_2Et$, Et_3N, pyridine, CH_2Cl_2; and (j) R_3COOH, PyBOP, hydroxybenzotriazole (HOBT), DIPEA, CH_2Cl_2.

and converted to the β-lactones. The enantiomerically enriched *syn* isomers were synthesized by aldol reactions of one chiral starting material using titaniumtetrachloride and sparteine or *N*,*N*-diisopropylethylamine (DIPEA) as a base for enolization [15]. By this method, both series of *syn* aldol products can be obtained by changing the base. The chiral auxiliaries were removed and the resulting β-hydroxyacids were converted to the β-lactones. Compounds with general structure **3** were synthesized by antiselective aldol reactions [14]. The *anti*-aldol products were subjected to hydrogenolysis followed by lactonization. The synthesis of the enantiomerically enriched *anti*-isomers of the obtained hits was performed by aldol reactions using dicyclohexylboron triflate on carboxylates esterified to chiral (+) or (−) norephedrin-derived chiral auxiliaries [14]. The chiral auxiliaries were removed and the resulting β-hydroxyacids were converted to the β-lactones. Compounds with general structure **4** were synthesized by metal-catalyzed carbonylation of epoxides using procedures described by

Schmidt and coworkers [16]. Compounds with general structure **5** and **6** were synthesized starting from Boc-protected L- or D-threonine that were lactonized with benzotriazole-1-yl-oxy-tris-pyrrolidino-phosphonium hexafluorophosphate (PyBOP) and Et_3N. Subsequently, the Boc group was cleaved and carboxylates were coupled to provide the products.

8.3.4
In vitro Enzyme Inhibition Studies

The compound collection was investigated for APT1 and lipase inhibition in a 96-well formatted assay monitoring the release of 4-nitrophenol from the corresponding octanoate by absorbance at 405 nm [7]. The most potent APT1 inhibitor was denoted palmostatin and the four isomeric forms were subjected to IC_{50} determination and mechanistic analysis (Table 8.1). Palmostatins B and C provided the lowest IC_{50} values, whereas the IC_{50} for palmostatin A was threefold higher and for palmostatin D 10-fold higher. The corresponding β-hydroxyacids of palmostatins A and B showed an IC_{50} value higher than 50 μM for APT1 inhibition and the β-hydroxyacids of palmostatins C and D showed less than 50% inhibition of APT1 at 1 mM. This shows that the β-lactone core has a major contribution to binding.

8.3.5
Mechanistic Investigation on APT1 Inhibition

A detailed study of the mechanism of APT1 inhibition by the palmostatins was performed [7]. Lineweaver–Burke analysis revealed that with varying inhibitor concentration K_m is increased and V_{max} remains constant, which indicates competitive inhibition. It was proposed that the palmostatins behave as slowly converted substrates that inhibit the enzyme APT1 according to the model shown in Figure 8.2c. The enzyme-active site serine attacks the palmostatins by nucleophilic opening of the β-lactone characterized by a rate constant k_I, followed by regeneration of the active enzyme by hydrolysis of the active site ester characterized by a rate constant k_{II}. Nonenzymatic hydrolysis of the β-lactone characterized by a rate constant k_{III} can also occur (Figure 8.2c). k_{III} for palmostatin B (*anti*-isomer) in aqueous solution (pH 7.0) is $0.012 \pm 0.001\ h^{-1}$ corresponds to a half-life of 58 h and for palmostatin C (*syn* isomer) $0.018 \pm 0.001\ h^{-1}$ corresponds to a half-life of 38 h. A presteady-state kinetic analysis for palmostatins A–C revealed that a quick initial interaction k_I was followed by a slow reactivation of the enzyme k_{II} (Table 8.1, Figure 8.2c), whereas for palmostatin D binding and hydrolysis were comparably fast and no separated rate constant could be determined. The β-lactones behave as slowly converted substrates in comparison to 4-nitrophenol octanoate, which was used as a substrate in the assay. The octanoylated enzyme species is hydrolyzed with a k_{cat} of $0.54\ s^{-1}$ corresponding to a half-life of 1.3 s. The β-hydroxyacylated enzymes are hydrolyzed with a rate that is 600–8000 times slower than the octanoylated enzyme. The slower hydrolysis rate can most likely be attributed to displacement of a conserved water

Table 8.1 IC_{50} values and kinetic parameters for inactivation and reactivation of APT1 by palmostatins A–D according to the model shown in Figure 8.2c.

Compound	IC_{50} (μM) APT1	k_I ($M^{-1}\,s^{-1}$) inactivation	k_{II} ($10^{-5}\,s^{-1}$) reactivation
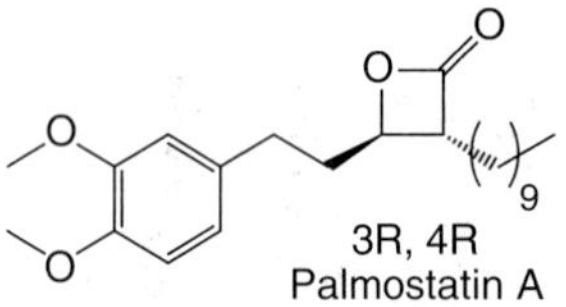 3R, 4R Palmostatin A	2.2 ± 0.2	964 ± 16	7.3 ± 1.5 $t_{1/2} = 158 \pm 33$ min
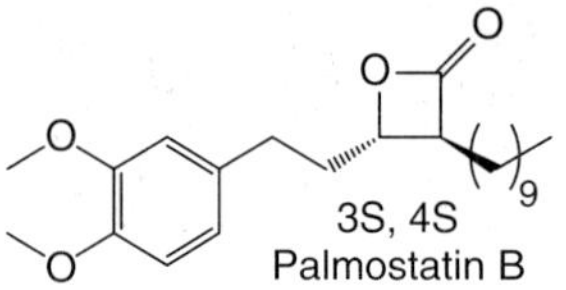 3S, 4S Palmostatin B	0.67 ± 0.02	2444 ± 15	45 ± 2.6 $t_{1/2} = 25 \pm 1.5$ min
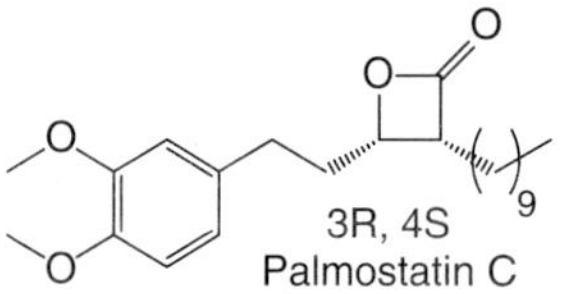 3R, 4S Palmostatin C	1.3 ± 0.1	>4000	85 ± 2.5 $t_{1/2} = 14 \pm 0.4$ min
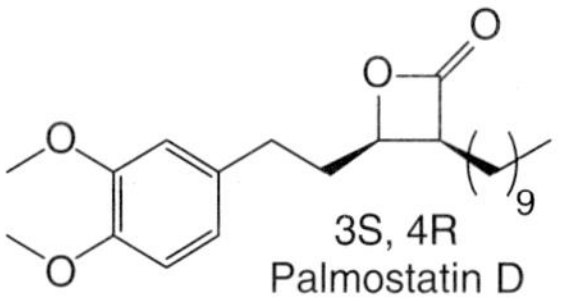 3S, 4R Palmostatin D	27 ± 3	Inactivation and reactivation comparably fast	

Palmostatin A *ee* = 80%, palmostatin B *ee* = 80%, palmostatin C > 99%, palmostatin D *ee* > 99%, n = 3, standard deviations from three determinations are reported.

molecule in the binding site that is positioned to hydrolyze the acylated enzyme. Furthermore, the substitution in the α- and β-position and the stereochemistry around these centers most likely further delay the hydrolysis and would explain the 10-fold difference between the stereoisomers. These results demonstrate that palmostatin B fulfills the requirements to be used as a small-molecule tool in a reverse chemical genetics study on the poorly defined role of APT1 in Ras signaling in cellular assays.

8.4
Chemical Biological Research/Evaluation

8.4.1
In vivo Enzyme Inhibition Studies

In order to determine the efficacy of palmostatin B to inhibit APT1 in cells, fluorescence lifetime imaging microscopy (FLIM) was performed with

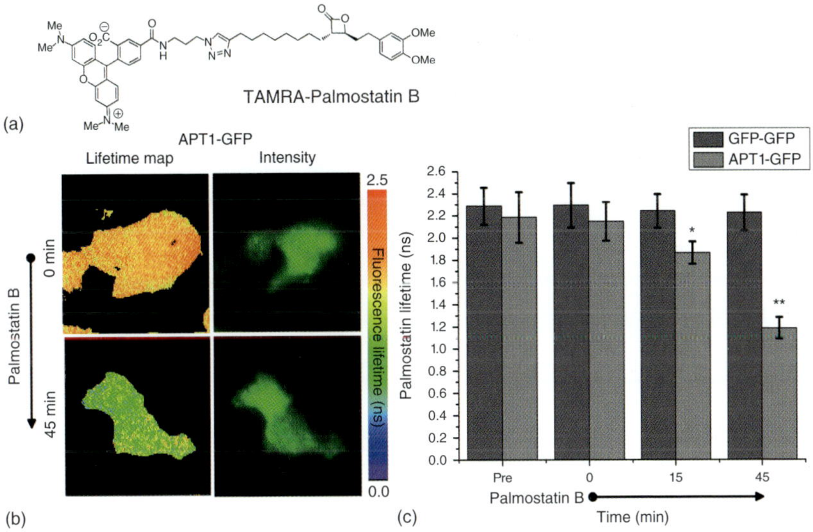

Figure 8.3 (a) The structure of TAMRA-labeled palmostatin B. (b) Widefield frequency-domain fluorescence lifetime imaging of live cells. The specific binding of APT1-GFP (donor) to TAMRA-labeled palmostatin B (acceptor) is measured as a reduction in fluorescence lifetime of APT1-GFP throughout the cell of up to 1.26 ns. The corresponding intensity images of the cells are shown for comparison. (c) Graph shows reduction in APT1-GFP lifetime as compared to GFP–GFP (negative control) lifetime upon incubation with TAMRA-palmostatin B. Lower limit for APT1/palmostatin-B-bound fraction is estimated to be 40%. Number of "*" indicates a significance in multiples of SE. Error bars indicate mean + SEM ($n = 20$). (Reprinted with permission from [7].)

TAMRA-labeled (*N,N,N′,N′*-tetramethyl-6-carboxyrhodamine) derivative of palmostatin B in cells expressing APT1-GFP (green fluorescent protein). A specific time-dependent reduction in the fluorescence lifetime of APT1-GFP was detected upon incubation with 1 μm TAMRA-palmostatin B, indicating that palmostatin B was binding APT1 effectively in cells. Incubation of TAMRA-palmostatin B with a GFP–GFP construct did not show a reduction in the fluorescence lifetime of GFP, indicating that a measured drop in the lifetime of APT1-GFP was caused by specific binding of palmostatin B to the active site of APT1. The maximum fluorescence lifetime reduction occurred at 60 min, corresponding to an APT1-palmostatin-B-bound fraction of at least 40% (Figure 8.3, Box 8.3).

Box 8.3 Fluorescence Lifetime Imaging Microscopy

In FLIM, the nanosecond decay kinetics of the electronic excited state of fluorophores such as mCitrine, characterized by the fluorescence lifetime "τ," are mapped spatially using a microscope equipped with a detector capable of high-frequency modulation. The fluorescence lifetime is sensitive to excited-state reactions such as fluorescence resonance energy transfer (FRET), and changes in τ can be used to detect macromolecular associations within living cells. The advantage of measuring the fluorescence lifetime of fluorophores is that this parameter is directly dependent upon excited-state interactions but independent of parameters such as fluorophores concentration and optical-path length, which are difficult to control inside a cell. Fluorescence lifetime image acquisition is also rapid enough to make measurements in live cells feasible.

8.4.2
Palmostatins Inhibit Depalmitoylation of Ras GTPases

The effect of palmostatin B on cellular palmitoylation and depalmitoylation of Ras GTPases was studied using time-resolved confocal fluorescence microscopy (Box 8.4) on semisynthetic Ras proteins that are microinjected (Box 8.5) into the cell. The localization of N-Ras in these cells was visualized through ectopic expression of mCitrine-NRas and the Golgi apparatus was marked through ectopic expression of GalT-mCFP. The semisynthetic Ras construct (Box 8.6) denoted as CysFar is a non-palmitoylated, solely farnesylated Ras GTPase, and is a probe for membrane anchoring by palmitoylation (Figure 8.4a). The construct denoted PalFar is palmitoylated and farnesylated and is a probe for depalmitoylation following membrane anchoring by palmitoylation (Figure 8.4a). Microinjection of CysFar into madine-darby canine kidney (MDCK) cells resulted in rapid accumulation of the probe on the Golgi in both untreated and palmostatin-B-treated cells (Figure 8.4b). This shows that *palmitoylation* of Ras GTPases was not blocked in the presence of palmostatin B. A behavior similar to CysFar was observed when PalFar was microinjected in untreated cells. However, specific Golgi accumulation was abolished upon microinjection of

Figure 8.4 *Palmostatin B specifically inhibits depalmitoylation.* (a) Schematic illustration of the Cy3-labeled semisynthetic lipoproteins (Cy3 is a fluorescent label). The two differentially lipid-modified carboxyterminal N-Ras heptapeptides were coupled via a maleimidocaproyl linker to the carboxyterminal cysteine of recombinant expressed N-Ras (1–181). GalT-CFP is a marker for the Golgi system. Citrine-N-Ras is a fluorescently labeled N-Ras construct. (b) Confocal time-lapse images of MDCK cells expressing the Golgi marker GalT-CFP and Citrine-N-Ras before and after microinjection of CysFar (i) or PalFar (ii). Cells were incubated for 80 min with 1 μM palmostatin B before the experiment. (c) Quantitative ratiometric analysis of CysFar accumulation at the Golgi in the presence of palmostatin B using Citrine-N-Ras as reference (i). Plateau values of Cy3 fluorescence at the Golgi were normalized to one ($n = 5$). The N-Ras accumulation on the Golgi by palmitoylation in palmostatin-B-treated cells was equally quick as in untreated cells. Contrast of PalFar fluorescence at the Golgi over adjacent areas (membranes/cytosol, $n = 5$) compared to CysFar fluorescence contrast ($n = 5$) in the presence of 1 μM palmostatin B (ii). The palmostatin B blocks the depalmitoylation of N-Ras, thereby preventing N-Ras attachment to the Golgi by repalmitoylation. (Reprinted with permission from [7].)

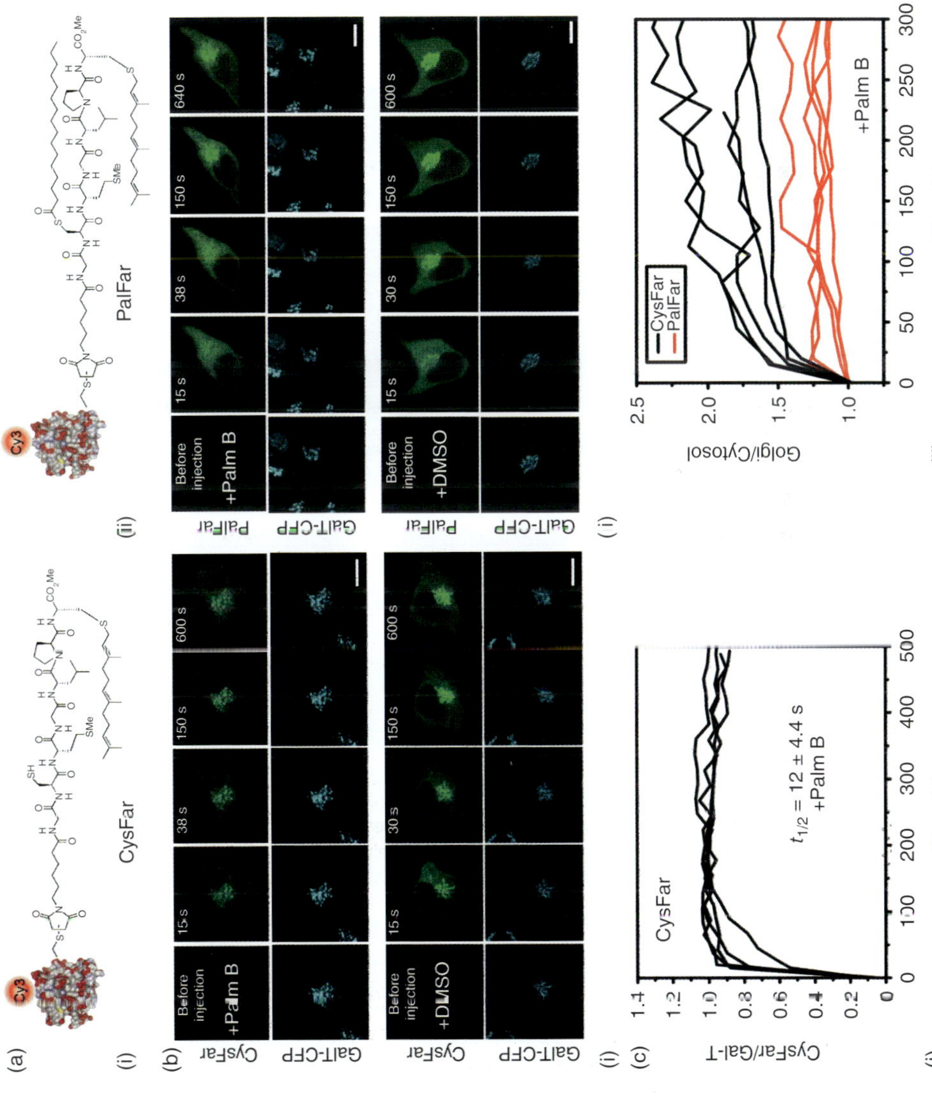
(a)
(i)
(ii)
Cy3
CysFar
PalFar
(b)
CysFar
GalT-CFP
PalFar
Before injection
+Palm B
+DMSO
15 s
38 s
30 s
150 s
600 s
640 s
(c)
CysFar/Gal-T
CysFar
$t_{1/2}$ = 12 ± 4.4 s
+Palm B
Time (s)
Golgi/Cytosol
—CysFar
—PalFar
+Palm B
Time (s)

PalFar into palmostatin-B-treated cells (Figure 8.4c). These results demonstrate that palmostatin B selectively inhibits cellular depalmitoylation of Ras GTPases without affecting their palmitoylation.

Box 8.4 Time-Resolved Fluorescence Microscopy

Microscopy on single cells using proteins that are fluorescently labeled. Fluorescent protein labels that are often used are yellow fluorescent protein (YFP), GFP, or Citrine. Fluorescent chemical labels are Cy3 and Cy5. This technology can be used for steady-state studies or for time-resolved studies. For time-resolved studies, proteins are microinjected or photoactivated/bleached and fluorescence changes are monitored over time.

Box 8.5 Microinjection

Microinjection is a process in which very fine needles are used to inject substances into a single living cell under the microscope. Small needles of roughly 0.5–5 μm in diameter are used to inject the desired contents into the desired subcellular compartment.

Box 8.6 Semisynthetic Proteins

Semisynthetic proteins are proteins that are partly made by organic synthesis and partly by biotechnology. One part of a protein, which is truncated at the N- or C-terminus, is made by biotechnology. The truncated part is made by organic synthesis including the desired chemical modifications. Both parts are fused using ligation technologies such as, for example, native chemical ligation or coupling via a maleimide functionality to a cysteine thiol. This method provides the original proteins including the desired chemical modifications, which is useful for functional studies.

8.4.3
Palmostatins Disturb the Localization of Ras GTPases

It was investigated how palmostatin B affects the steady-state localization of palmitoylated Ras GTPases [7]. MDCK cells expressing mCitrine-labeled N-Ras and a mCherry-labeled unpalmitoylatable mutant of H-Ras (HRasC181/184) were treated with palmostatin B (Figure 8.5). After 40 min and 5 h, the local N-Ras concentrations had changed significantly, indicating that palmostatin B influences the palmitoylation/depalmitoylation cycle. At 90 min, mCitrine-NRas had reached the same equilibrium distribution over all membranes, which was indicated by the unpalmitoylatable mCherry-HRasC181S, C184S. APT1 inhibition therefore had interrupted the acylation cycle, and led to a random

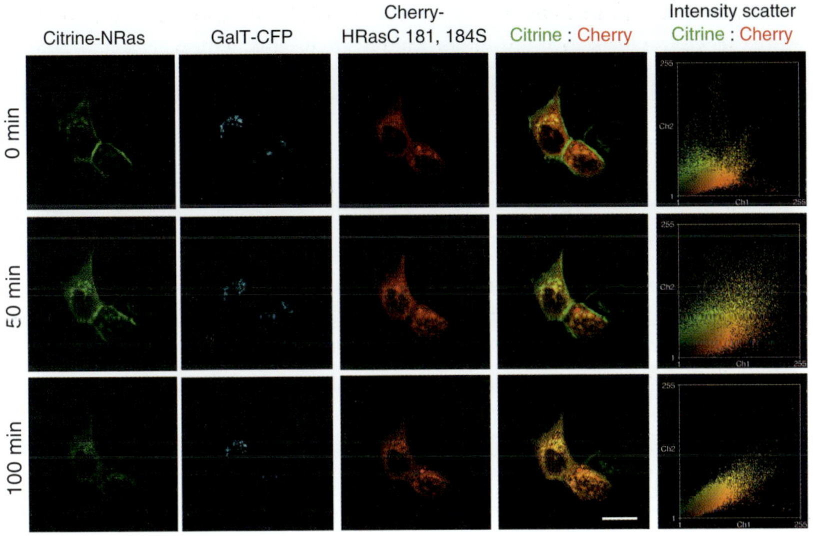

Figure 8.5 *Changes in steady-state localization of mCitrine-labeled N-Ras upon treatment with palmostatin B.* N-Ras is localized on the Golgi and on the plasma membrane in untreated cells. Upon treatment with palmostatin B, the specific subcellular localization of the N-Ras is lost as it redistributes over all endomembranes. The endomembrane distribution is indistinguishable from the localization of HRasC181S,C184S – an unpalmitoylatable Ras mutant that distributes randomly over all membranes. The intensity scatter plots show the complete colocalization between these proteins upon palmostatin B treatment. Scale bars represent 10 μm. (Reprinted with permission from [7].)

distribution on N-Ras on endomembranes, instead of its plasma membrane enrichment. Furthermore, after overnight incubation it was observed that the effect of palmostatin B was reversible, which indicates a similar mechanism of inhibition in live cells as observed *in vitro* for APT1. A similar palmostatin-B-induced redistribution was observed in cells expressing Citrine-H-Ras. However, this Ras isoform exhibited more persistent plasma membrane localization after palmostatin B treatment.

8.4.4
Palmostatins Inhibit Downstream Signaling of Ras GTPases

Next it was studied if the observed redistribution of H-Ras to endomembranes leads to decreased signaling of constitutively active oncogenic H-RasG12V [7]. Ras transformed MDCK-F3 cells show a long and spindle-like phenotype and grow in multiple layers with a reduced number of cell–cell contacts. These cells show a loss of contact inhibition, which is accompanied by the loss of E-cadherin expression at the cell surface. Palmostatin B caused reversal to the round phenotype that is characteristic of untransformed MDCK cells, which is comparable to the reversal induced by the specific MAPK/Erk kinase (MEK) (MAPK, mitogen activated protein kinase) inhibitor U0126. The capacity of oncogenic H-RasG12V to couple into the MEK-Erk1/2 (extracellular signal-regulated kinase) pathway was studied in order to determine whether the palmostatin-B-induced backtransformation of MDCK-F3 cells is due to a reduced signaling output downstream of oncogenic H-RasG12V [7]. Palmostatin B treatment of MDCK-F3 cells led to a reduction in Erk1/Erk2 phosphorylation. Overexpression of oncogenic H-RasG12V in these cells increased the phosphorylation of both Erk1 and Erk2 and doubled the time for palmostatin-B-induced loss of phosphorylation. This shows that palmostatin B uncouples oncogenic H-RasG12V signaling via Erk1/2 by the redistribution of H-RasG12V to endomembranes (Figure 8.6).

8.5 Conclusions

This study shows that a combination of bioinformatics, organic synthesis, enzyme inhibition studies, and cell biology is a very powerful approach in reverse chemical genetics studies to elucidate the function of genes and gene products – in this case, APT1. Signaling of H- and N-Ras is dependent on their proper and dynamic localization, which is maintained by S-palmitoylation and abundant S-depalmitoylation. S-depalmitoylation counteracts entropy-driven unspecific distribution of Ras among cellular membranes. APT1 was identified as a relevant thioesterase in S-depalmitoylation and APT1 inhibitors induced unspecific distribution among cellular membranes and consequently impaired Ras signaling. These results demonstrate that APT inhibitors are powerful tools to downregulate oncogenic Ras signaling and have thus potential for the development of cancer therapeutics. Further work in the field has led to even more potent inhibitors of acyl protein thioesterases [17, 18].

Figure 8.6 (a) Palmostatin-B-induced phenotypic reversion of HRasG12V-transformed MDCK-F3 cells. Untransformed MDCK cells treated with palmostatin B or dimethyl sulfoxide (DMSO) (vehicle control) are shown as controls. Right panel: cell circularity distribution ($n > 400$ cells for each case) of HRasG12V-transformed MDCK-F3 cells approaches that of untransformed MDCK cells upon treatment with palmostatin B. In all cases, 20 μM U0126-treated MDCK-F3 cells serve as positive control for phenotypic reversion, while DMSO-treated MDCK-F3 cells serve as negative control. Scale bars represent 20 μm. (b) Western blots and densitometric quantification of fraction of phospho-Erk1/2 over total Erk1/2 from gel-shift and specific phosphoErk specific antibody. MDCK-F3 cells as well as MDCK-F3 cells overexpressing Citrine-HRasG12V (upper and lower panels, respectively) show significant reduction in Erk1/2 phosphorylation after palmostatin B treatment. Error bars indicate mean + SEM ($n = 3$ gels). (Reprinted with permission from [7].)

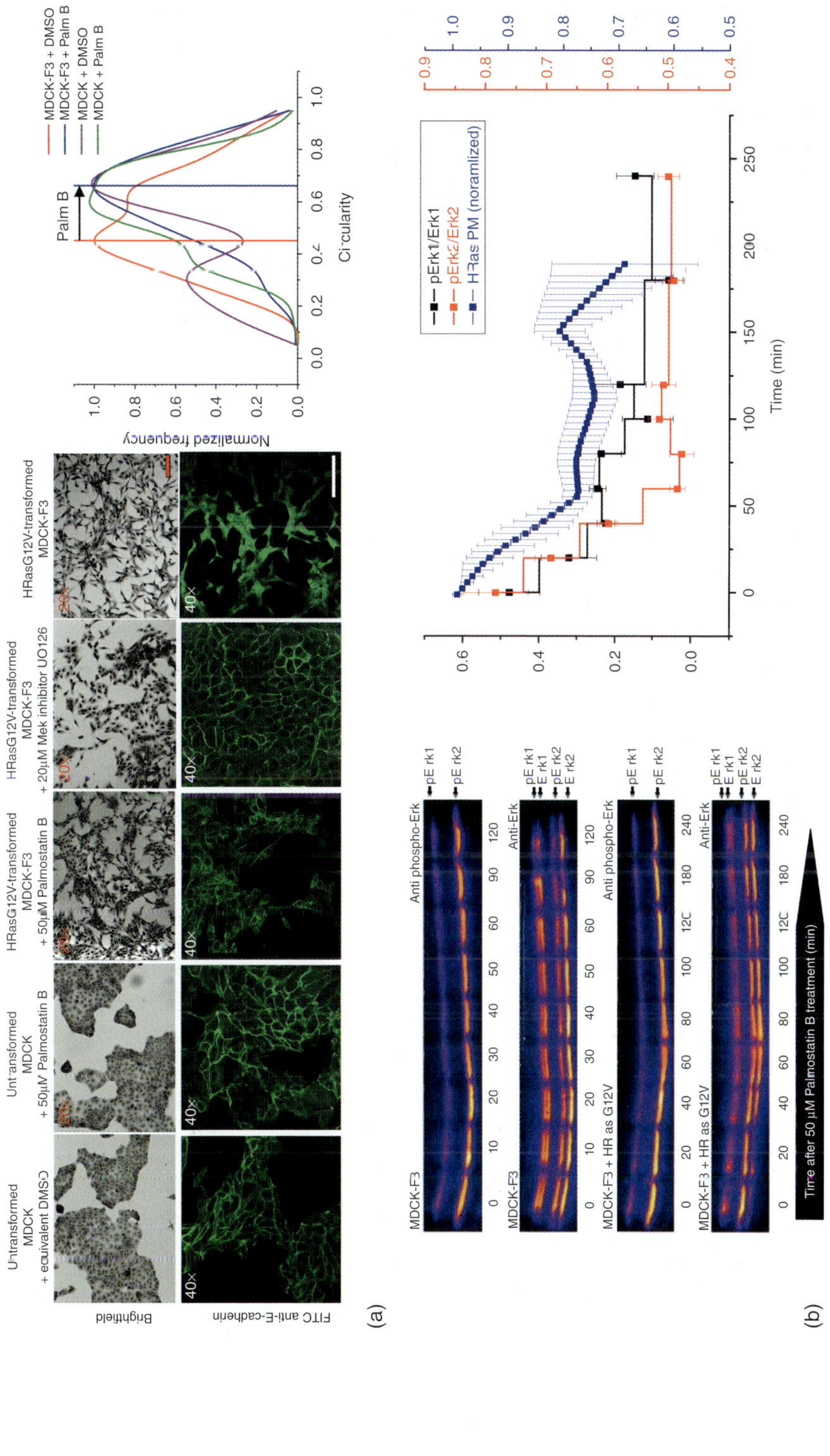
(a)
(b)
Brightfield
FITC anti-E-cadherin
Untransformed MDCK + equivalent DMSO
Untransformed MDCK + 50µM Palmostatin B
HRasG12V-transformed MDCK-F3 + 50µM Palmostatin B
HRasG12V-transformed MDCK-F3 + 20µM Mek inhibitor UO126
HRasG12V-transformed MDCK-F3
Normalized frequency
Circularity
Palm B
MDCK-F3 + DMSO
MDCK-F3 + Palm B
MDCK + DMSO
MDCK + Palm B
Anti phospho-Erk
Anti-Erk
MDCK-F3
MDCK-F3 + HRas G12V
Time after 50 µM Palmostatin B treatment (min)
pErk1/Erk1
pErk2/Erk2
HRas PM (noramlized)
Time (min)

References

1. Schubbert, S., Shannon, K., and Bollag, G. (2007) Hyperactive Ras in developmental disorders and cancer. *Nat. Rev. Cancer*, 7, 295–308.
2. Konstantinopoulos, P.A., Karamouzis, M.V., and Papavassiliou, A.G. (2007) Post-translational modifications and regulation of the RAS superfamily of GTPases as anticancer targets. *Nat. Rev. Drug Discovery*, **6**, 541–555.
3. Triola, G., Waldmann, H., and Hedberg, C. (2012) Chemical biology of lipidated proteins. *ACS Chem. Biol.*, **7** (1), 87–99.
4. Rocks, O., Peyker, A., Kahms, M., Verveer, P.J., Koerner, C., Lumbierres, M., Kuhlmann, J., Waldmann, H., Wittinghofer, A., and Bastiaens, P.I.H. (2005) An acylation cycle regulates localization and activity of palmitoylated Ras isoforms. *Science*, **307**, 1746–1752.
5. Rocks, O., Gerauer, M., Vartak, N., Koch, S., Huang, Z.-P., Pechlivanis, M., Kuhlmann, J., Brunsveld, L., Chandra, A., Ellinger, B., Waldmann, H., and Bastiaens, P.I.H. (2010) The palmitoylation machinery is a spatially organizing system for peripheral membrane proteins. *Cell*, **3**, 458–471.
6. Duncan, J.A. and Gilman, A.G. (1998) A cytoplasmic acyl-protein thioesterase that removes palmitate from G protein alpha subunits and p21(RAS). *J. Biol. Chem.*, **273**, 15830–15837.
7. Dekker, F., Rocks, O., Vartak, N., Menninger, S., Hedberg, C., Balamurugan, R., Wetzel, S., Renner, S., Gerauer, M., Schölermann, B., Rusch, M., Kramer, J.W., Rauh, D., Coates, G.W., Brunsveld, L., Bastiaens, P.I.H., and Waldmann, H. (2010) Small-molecule inhibition of APT1 affects Ras localization and signaling. *Nat. Chem. Biol.*, **6**, 449–456.
8. Koch, M.A., Wittenberg, L., Basu, S., Jeyaraj, D.A., Gourzoulidou, E., Reinecke, K., Odermatt, A., and Waldmann, H. (2004) Compound library development guided by protein structure similarity clustering and natural product structure. *Proc. Natl. Acad. Sci. U.S.A.*, **101**, 16721–16726.
9. Dekker, F.J., Koch, M.A., and Waldmann, H. (2005) Protein structure similarity clustering (PSSC) and natural product structure as inspiration sources for drug development and chemical genomics. *Currr. Opin. Chem. Biol.*, **9**, 232–239.
10. Devedjiev, Y., Dauter, Z., Kuznetsov, S.R., Jones, T.L., and Derewenda, Z.S. (2000) Crystal structure of the human acyl protein thioesterase I from a single X-ray data set to 1.5 A. *Structure*, **8**, 1137–1146.
11. Roussel, A., Miled, N., Berti-Dupuis, L., Riverère, M., Spinelli, S., Berna, P., Gruber, V., Verger, R., and Cambillau, C.J. (2002) Crystal structure of the open form of dog gastric lipase in complex with a phosphonate inhibitor. *J. Biol. Chem.*, **277**, 2266–2274.
12. Hadváry, P., Lengsfeld, H., and Wolfer, H.J. (1988) Inhibition of pancreatic lipase in vitro by the covalent inhibitor tetrahydrolipstatin. *Biochem. J.*, **256**, 357–361.
13. Hadváry, P., Sidler, W., Meister, W., Vetter, W., and Wolfer, H.J. (1991) The lipase inhibitor tetrahydrolipstatin binds covalently to the putative active site serine of pancreatic lipase. *J. Biol. Chem.*, **266**, 2021–2027.
14. Inoue, T., Liu, J., Buske, D.C., and Abiko, A.J. (2002) Boron-mediated aldol reaction of carboxylic esters: complementary *anti-* and *syn*-selective asymmetric aldol reactions. *J. Org. Chem.*, **67** (15), 5250–5256.
15. Crimmins, M.T., King, B.W., Tabet, E.A., and Chaudhary, K.J. (2001) Asymmetric aldol additions: use of titanium tetrachloride and (-)-sparteine for the soft enolization of N-acyl oxazolidinones, oxazolidinethiones, and thiazolidinethiones. *J. Org. Chem.*, **66** (3), 894–902.
16. Schmidt, J.A., Lobkovsky, E.B., and Coates, G.W. (2005) Chromium(III) octaethylporphyrinato tetracarbonylcobaltate: a highly active, selective, and versatile catalyst for epoxide carbonylation. *J. Am. Chem. Soc.*, **127**, 11426–11435.

17. Hedberg, C., Dekker, F.J., Rusch, M., Renner, S., Wetzel, S., Vartak, N., Gerding-Reimers, C., Bon, R.S., Bastiaens, P.I.H., and Waldmann, H. (2011) Development of highly potent inhibitors of the Ras-targeting human acyl protein thioesterases based on substrate similarity design. *Angew. Chem. Int. Ed.*, **50**, 9832–9837.
18. Rusch, M., Zimmermann, T.J., Buerger, M., Dekker, F.J., Goermer, K., Triola, G., Brockmeyer, A., Janning, P., Boettcher, T., Sieber, S.A., Vetter, I.R., Hedberg, C., and Waldmann, H. (2011) Identification of acyl protein thioesterases 1 and 2 as the cellular targets of the Ras-signaling modulators palmostatin B and M. *Angew. Chem. Int. Ed.*, **50**, 9838–9842.

9
Functional Analysis of Host–Pathogen Posttranslational Modification Crosstalk of Rab Proteins

Christian Hedberg, Roger S. Goody, and Aymelt Itzen

9.1
Introduction

Posttranslational modification (PTM) with functional groups is a universal mechanism for diversifying the activities of proteins. PTMs can affect many properties of proteins, such as localization, activity status, interaction networks, solubility, folding, turnover, or stability. It is therefore of vital importance to accurately determine the identities of modified proteins, the modified amino acid residues, and the covalently attached group. This chapter describes the process of PTM identification using the adenylylation (i.e., the covalent transfer of an adenosine monophosphate (AMP)) of rat sarcoma related in brain (Rab) proteins by *Legionella pneumophila* enzymes as an example. It also deals with the development of PTM-specific antibodies from synthetic peptides. This account underlines the importance of chemical biology in the elucidation of PTMs.

9.2
The Biological Problem

9.2.1
Posttranslational Modifications

PTMs (e.g., phosphorylation) can massively expand the functions and activities of proteins beyond the chemistry that is dictated by the biogenic amino acids. The consequences of PTMs can be manifold and a thorough analysis of their consequences is vital to understanding the biological implications. Therefore, we need to study the site(s) of modifications in a given protein and/or the scope of modified proteins in the proteome. It is also necessary to determine the specific biochemical effects on the target protein(s) with respect to activity, dynamic turnover, and function. In this respect, the analysis of PTMs exerted upon mammalian host proteins by enzymes from bacterial pathogens is especially interesting and can reveal previously unknown or unrecognized modifications. Recently, it was discovered

Concepts and Case Studies in Chemical Biology, First Edition. Edited by Herbert Waldmann and Petra Janning.

that the human pathogen *L. pneumophila* can modify proteins of the Rab subfamily by adenylylation and phosphocholination. Little or nothing was previously known about these PTMs. In this chapter we describe chemical and biochemical approaches to the preparative introduction of adenylylation into target proteins, to the analysis of the consequences of this PTM, and to the generation of reagents that allow global detection of adenylylated proteins.

9.2.2
Adenylylation of Small GTPases

Ras (rat sarcoma)-like small guanosine triphosphatases (GTPases) are essential regulators of diverse intracellular signaling processes. The Rab proteins constitute the largest subfamily of small GTPases and are involved in controlling intracellular vesicular trafficking. One important process regulated partially by Rabs is the uptake and elimination of bacterial pathogens. Some bacteria have evolved strategies to interfere with Rab function in order to ensure survival and replication. Rab proteins act by functioning as molecular on/off switches, with the activation state being determined by the phosphorylation state of a bound guanosine nucleotide. Thus, Rabs are "active" in the guanosine triphosphate (GTP)-bound form and "inactive" when complexed to guanosine diphosphate (GDP). Guanosine nucleotide exchange factors (GEFs) and GTPase activating protein (GAPs) control the activation and deactivation of Rabs, respectively. In the active form, Rabs interact specifically with effector molecules and thereby promote intracellular signaling. Rabs cycle between membrane-associated and cytosolic forms. The attachment to a membrane is mediated by one or two posttranslationally attached geranylgeranyl lipids at the C-terminus of Rabs. GDP dissociation inhibitors (GDIs) can solubilize inactive (i.e., GDP-bound) Rabs from the membrane and thereby recycle Rabs from a membrane at the end of their activity cycle.

Bacterial pathogens that survive intracellularly frequently interfere with the activity cycle of small GTPases in a specific manner in order to evade destruction by the host cell. For example, *Legionella* bacteria release a variety of proteins that specifically target the Rab1-family of small G-proteins and affect its nucleotide state (e.g., defects in Rab1 recruitment protein A (DrrA) has Rab1-GEF activity, LepB is a Rab1-GAP) [1]. In addition, many bacterial enzymes can alter the activity of small GTPases by introducing PTMs [1, 2]. The activity states of small GTPases are communicated to binding partners mainly via two highly conserved loops (switch I and switch II), and therefore these switch regions are frequently targeted with PTMs by bacterial enzymes. Using a structure-guided approach it was recently discovered that the N-terminal domain of the *Legionella* protein DrrA contains adenylyltransferase (ATase) activity [3]. The ATase of DrrA utilizes adenosine triphosphate (ATP) to adenylylate Rab1, that is, to modify the GTPase by addition of an AMP moiety (Figure 9.1). Adenylylation of mammalian proteins was previously unknown (note: adenylylation of Rho proteins has been reported recently [4, 5]) and thus several questions arise from this observation: is adenylylation a general mammalian PTM that also occurs in the absence of

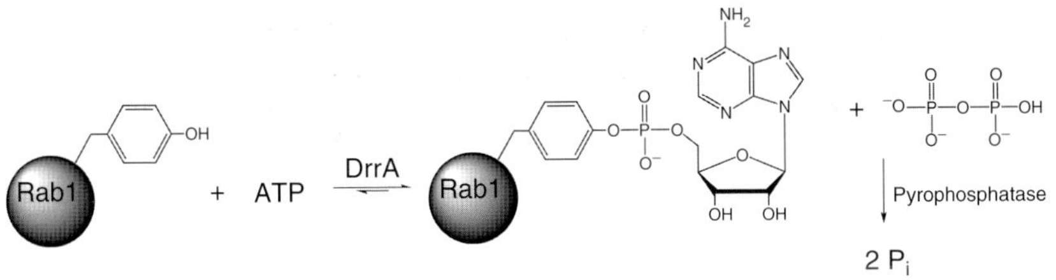

(a)

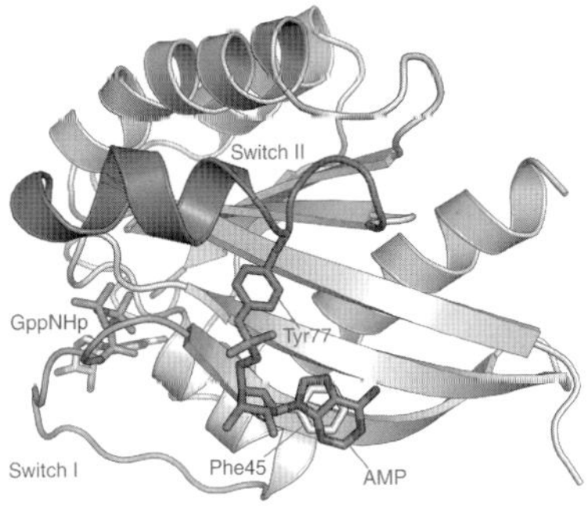

(b)

Figure 9.1 *Adenylylation as posttranslational modification.* (a) Reaction scheme of enzyme-catalyzed adenylylation (e.g., by DrrA). Here, an AMP is covalently attached from an ATP precursor to a tyrosine residue. The emerging pyrophosphate (PP_i) is subsequently hydrolyzed into phosphate by the enzyme pyrophosphatase, thereby shifting the reaction to the product side. (b) Cartoon representation of the X-ray crystal structure of adenylylated Rab1b:GppNHp. The adenylylated tyrosine 77 (Tyr77) is located in the important switch II region (dark grey) and makes a stacking interaction with the conserved Phe45.

bacterial infections (such as *Legionella*)? What are the functional consequences of adenylylation of Rab proteins? Where is the site of adenylylation of Rab1?

Addressing these basic questions requires analytical approaches that combine enzyme biochemistry, biophysics, and chemical synthesis. In this chapter we present one possible route to analyze new PTMs exemplified by the adenylylation of Rab1.

9.3 The Chemical Approach

The analysis of adenylylation requires preparatively adenylylated proteins in order to determine the site of the PTM, to determine the structure of the modified proteins, and to establish the functional consequences for Rab1 activity. Also, the systematic investigation of adenylylation in live cells and their distribution among different prokaryotic and eukaryotic proteomes requires specific tools such as antibodies in order to track and enrich modified proteins/peptides. The generation of adenylylation-specific antibodies depends on a stable molecular probe, that is, an adenylylated peptide that can be used in immunization procedures.

In this subchapter we therefore describe the preparation of adenylylated proteins and peptides for *in vitro* studies.

9.3.1 Preparative Adenylylation of Rab1

Preparative amounts of homogeneously adenylylated proteins facilitate the identification of the site of modification by mass spectrometry (MS) or X-ray crystallography. For these purposes, profound knowledge of the enzymatic properties, specificities, and kinetics is essential to reproducibly generate fully modified substrates. DrrA is a multidomain enzyme in which the ATase domain can only be generated by heterologous expression in *Escherichia coli* as either the full length protein or fused to the central Rab1-GEF domain. Because the presence of the central Rab1-GEF domain may affect the nucleotide state of the substrate and can therefore interfere with the kinetic characterization of the ATase domain, the GEF activity needs to be eliminated using targeted amino acid substitutions [6]. The initial preparative adenylylation experiments on Rab1:GDP using DrrA indicated that the reaction stalls at about 80% yield of Rab1:GDP-AMP (depending on the exact conditions), indicating that free energy change of the modification is not sufficient to drive the modification to completion. This is at least partially due to the buildup of pyrophosphate. The addition of pyrophosphatase that hydrolytically cleaves pyrophosphate into free inorganic phosphate eliminated this problem and allowed the production of homogeneously adenylylated Rab1:GDP [7].

Further analysis of the substrate properties of DrrA revealed that the enzyme prefers active Rab1 over the inactive protein by a quite large factor. The adenylylation of Rab1 can be monitored by the incorporation of radioactive AMP using ATP [α-^{32}P] as a substrate [3]. Although a very sensitive method, the use of a radioactive probe necessitates quenching the reaction after defined times and immobilization of the modified proteins on a support such as nitrocellulose filters for subsequent quantification of radioactivity, and is therefore an inconvenient method. A more convenient technique allowing direct monitoring of the reaction with time would be a spectroscopic method. Advantageously, the modification of Rab1 with AMP can be followed by the change in the intrinsic tryptophan fluorescence signal and therefore permits convenient determination of the enzyme kinetics of DrrA [7, 8]. The k_{cat}/K_m-value of DrrA was about 300 times higher for Rab1:GTP than for Rab1:GDP *in vitro* [3, 7], suggesting that Rab1 is activated (i.e., GDP is replaced by GTP) *in vivo* before adenylylation.

In principle, not only ATP but also GTP, cytidine triphosphate (CTP), and uridine triphosphate (UTP) may serve as substrates of DrrA and could therefore lead to heterogeneous nucleotidylylation of Rab1 *in vivo*. Using the assays described it was confirmed that DrrA can nucleotidylylate Rab1 in various ways, but that ATP is the preferred substrate [7].

The detailed characterization of the enzyme kinetics of DrrA clearly established that Rab1:GTP and ATP are the preferred enzyme substrates. The enzymatic parameters were used to conveniently estimate the reaction time for multi-milligram amounts of homogeneously adenylylated Rab1:GTP or the nonhydrolyzable derivative of GTP, Rab1:GppNHp (guanosine-5′-*O*-[(β,γ)-imido]-triphosphate).

9.3.2
Identification of the Site of Adenylylation

The identification of the site of adenylylation is important to develop hypotheses about the functional consequences of Rab1-adenylylation. Also, the modified amino acid sequence needs to be known in order to produce the correct adenylylated peptide that can be applied to generate antibodies (see Sections 9.3.3 and 9.3.4). The determination of the crystal structure of adenylylated Rab1:GppNHp revealed that DrrA targets the switch II region of Rab1 and covalently attaches an AMP moiety via a phosphodiester linkage to a specific tyrosine 77 (Tyr77) (Figure 9.1b). Although the crystal structure unambiguously confirms the adenylylation of Rab1 in an important regulatory region of the GTPase, the possibility cannot be excluded that adenylylation occurs at several sites but that only a subspecies out of the complete Rab1-AMP ensemble (i.e., Rab1-(Tyr77)-AMP) has been successfully crystallized.

An orthogonal technique to analyze the pattern of adenylylation of Rab1 is MS in combination with proteolytic digestion. For this purpose, a sample of homogeneously adenylylated and unmodified Rab1 is digested completely using the protease trypsin. To facilitate the total hydrolysis of the rather stable GTPase domain, a specific detergent can be added to the sample before the addition of the protease. The Rapigest^{TM} detergent is added to the protein solution, heated to 60 °C for 30 min, and allowed to cool to ambient temperature. After complete digestion with trypsin, the solution is acidified with trifluoroacetic acid, resulting in decomposition of the detergent. The hydrophobic component of the detergent precipitates and is removed by low-speed centrifugation. In contrast, the hydrophilic head group is not retained on the reversed column used for subsequent liquid chromatography coupled with tandem mass spectrometry (LC MS/MS) peptide analysis and therefore does not interfere with the peptide separation.

This procedure will unambiguously identify all adenylylated peptides. The combination of the first MS with a second collision-induced peptide fragmentation and MS allows the identification of the modified amino acid in a given peptide. In the case of adenylylation of Rab1 by DrrA, only one tryptic peptide fragment was found to be modified specifically on Tyr77 (^{72}TTTSSY^{77}-(AMP)YR^{79}), thereby confirming the results obtained by X-ray crystallography (covered in more detail in Section 9.3.5).

9.3.3 Synthesis of Site-Specifically Adenylylated Peptides

The investigation of protein adenylylation using synthetic reference material for MS method development, as well as for generation of specific AMP antibodies, has been hampered by the lack of efficient synthesis methods, thereby not allowing for the preparation of pure milligram quantities of peptides bearing adenylylated amino acids. Post synthesis functionalization strategies have used on-resin adenylylation via the *H*-phosphonate-method after completed peptide synthesis (Figure 9.2a), which carries a number of inherent problems, including poor compatibility with other functionalities, as well as not being applicable to tyrosine owing to the low reactivity of the phenolic oxygen. Previous attempts to prepare adenylylated peptides by Fmoc (fluorenylmethoxycarbonyl) amino acid building block approaches were largely unsuccessful, mainly leading to loss of adenine from the adenosine moiety upon acid-induced cleavage of the peptide from the solid support. The reason for the low yield in the case of adenylylation might be that the protective group strategy leads to extensive depurination when subjected to acidolytic release from the solid support. The depurination reaction of the commonly employed 2′,3′-bis-ester-protected adenosine is induced by participation of the 2′-ester group, acting as an internal nucleophile, expelling the adenine moiety (Figure 9.2b). Recently, we hypothesized that a protective group change could avoid the depurination reaction of the adenosine (Figure 9.2c). Here, 2′,3′-bis-ester protective groups of adenosine were replaced by an 2′,3′-isopropylidene acetal, as well as introduction of the deactivating N^6-bis-boc protection at the adenosine nitrogen to further stabilize the system. As a temporary protective group for the phosphodiester linkage, an *O*-cyanoethyl (CNE) was employed, allowing for instant deprotection during the first Fmoc-removal, thus stabilizing the phosphodiester linkage in the mono-anionic form. The tyrosine building block was successfully employed for the standard Fmoc-SPPS of a number of peptides (Figure 9.2c), all relying on global acidolytic cleavage, including tryptic fragments of the switch II region of adenylylated human Rab1b, which were isolated in 40–60% yield. To further extend the Fmoc-building block approach to peptides adenylylated on serine and threonine, we developed a strategy relying on the corresponding unprotected phosphodiesters (Figure 9.2d). Here, the allyl protective group of the phosphodiester linkage was removed in the last synthesis step, thus making the phosphodiester mono-anionic before activation and peptide coupling, thus avoiding β-elimination.

9.3.4 Generation and Application of α-AMP-Tyr/Ser/Thr-Antibodies

Antibodies represent a powerful tool for specific detection and enrichment of posttranslationally modified proteins from different sources, including cell lysates. Only recently, the production and possible applications of antibodies

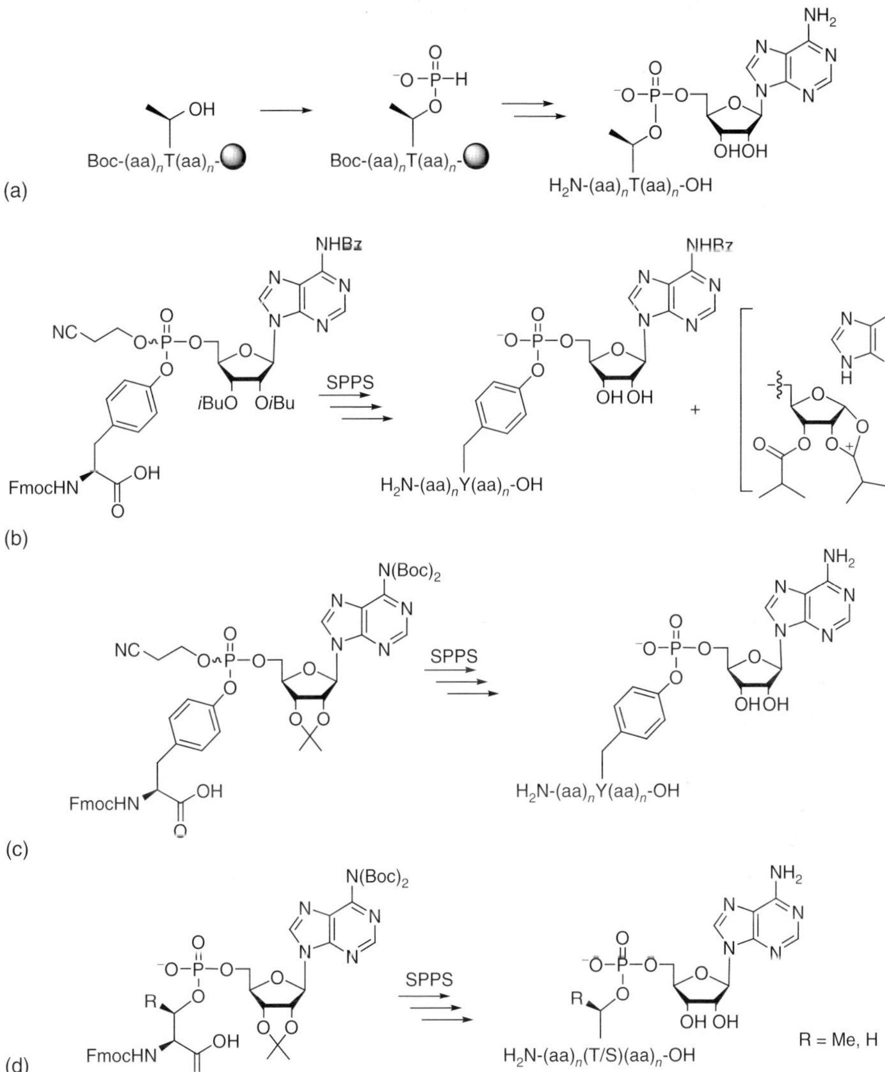

Figure 9.2 *Synthesis of adenylylated amino acids.* (a) On resin approach via *H*-phosphonation of the completed peptide. Limited compatibility with amino acids in the backbone. (b) The building block approach for SPPS with ester-protected adenosine gives rise to depurination under acidic conditions. (c) Switching the protective group strategy to 2′,3′-isopropylidene- and N6,N6-bis-Boc-protected adenosine allows efficient synthesis of peptides adenylylated on tyrosine. (d) For the threonine and serine building block approach, the mono-anionic form of the phosphodiester prevents β-elimination of phosphoadenosine under the basic conditions of SPPS.

specific for adenylylated proteins have been described using homogeneous adenylylated peptides for immunization [9, 10]. Nucleotidylylation, in general, is expected to be well enrichable by antibodies, given the polar character, hydrophobicity, and size of the modification. In order to decrease the influence of the proteins bearing the PTM, antibodies are typically produced using peptides containing the corresponding PTM. The availability of amino acid building blocks adenylylated on Ser, Thr, and Tyr residues in peptides will greatly simplify the production of specific antibodies against these PTMs. In this respect, experiments for enrichment of PTM-modified proteins from cell lysates could be performed after general proteolytic digestion of the cell lysates to both increase the accessibility of the PTMs and decrease nonspecific binding of the antibodies to random proteins in the lysate. Recently, we demonstrated the production of specific antibodies against adenylylated proteins and peptides [9, 10]. For Tyr-AMP antibodies, immunization in rabbits with KLH-conjugates of adenylylated Rab1-derived sequences resulted in a strong immune response as measured by immunosorbent assays (ELISA, enzyme linked immunosorbent assay). Isolation of total IgG and affinity purification against the adenylylated antigen peptide, followed by depletion against the unmodified peptide backbone resulted in milligram amounts of monoselective polyclonal antibodies, suitable for further biological investigations (Figure 9.3).

We performed a Western blot with decreasing amounts of Rab1b-AMP and unmodified Rab1b to investigate affinity and specificity of the derived antibody. As a result, Rab1b-AMP could be clearly discriminated from wild-type Rab1b (Figure 9.3a). The level of α-Tyr-AMP antibody binding to Rab1b-AMP is approximately 20-fold higher than for unmodified Rab1b. This suggests a major contribution of the AMP moiety to the antibody binding and only a weak recognition of the peptide backbone that was part of the immunization. Under these experimental conditions, down to 10 ng Rab1b-AMP could be detected with the α-Tyr-AMP antibody. In order to investigate the specificity of the derived α-Tyr-AMP antibody, the experiments were repeated with different adenylylated proteins. In addition to Rab1b-AMP, the BSA-AMP-peptide, and adenylylated Cdc42 (Cdc42-AMP) that had been preparatively adenylylated using VopS, were chosen. VopS has been reported to adenylylate Cdc42 specifically on Thr35 and can serve as a control of whether the α-Tyr-AMP antibody can discriminate between adenylylated threonines and tyrosines (Figure 9.3b). The α-Tyr-AMP antibody recognizes specifically adenylylated BSA-AMP-peptide and Rab1b-AMP over the nonmodified proteins. However, the antibody also recognizes Cdc42-AMP, indicating that it can also detect adenylylated threonines. This again suggests a major contribution of the AMP group to antibody binding with only a small influence by the amino acid side chain modified and the peptide backbone. Next, we performed nucleotide competition experiments. The binding of the α-Tyr-AMP to BSA-AMP-peptide (with BSA as control), AMP-Rab1b (with Rab1b as control), and AMP-Cdc42 (with Cdc42 as control), was performed in the presence and absence of either guanosine monophosphate (GMP) or AMP (Figure 9.3c). In the presence of GMP, no impairment of α-Tyr-AMP

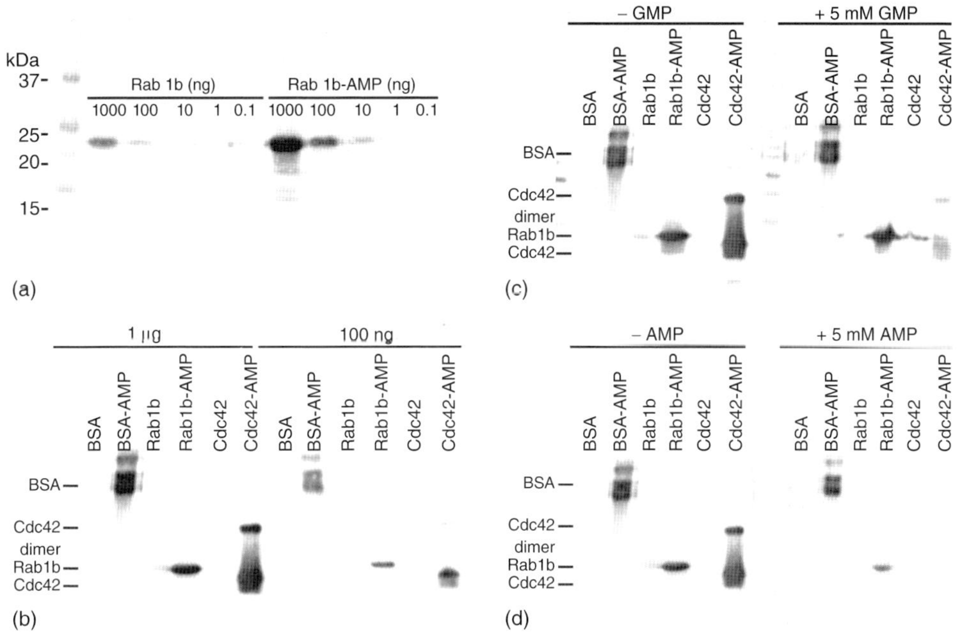

Figure 9.3 *Affinity and specificity of the derived α-Tyr-AMP-antibody.* (a) Western blot with indicated amounts of highly purified Rab1b and Rab1b-AMP using the α-Tyr-AMP-antibody (1 : 100 dilutions), demonstrating the specific recognition of the Tyr-AMP modification. (b) Western blot analysis of highly purified adenylylated and unmodified forms of BSA, Rab1b, and Cdc42 (1 or 0.1 μg, respectively, protein sample per lane). The α-Tyr-AMP-antibody (1 : 100 dilution) strongly binds to all tested adenylylated proteins, indicating additional binding activity for Thr-AMP (Cdc42). (c) Competition of α-Tyr-AMP-antibody in the presence of GMP or AMP. Samples (0.1 μg each) have been prepared as indicated in (b). The Western blots have been incubated with the α-Tyr-AMP antibody in the presence and absence of 5 mM GMP or AMP. AMP, but not GMP, competes moderately with antibody binding to Rab1b-AMP and BSA-AMP. Both AMP and GMP compete with antibody for Cdc42-AMP detection. (In all Western blots, IRDye800-conjugated donkey anti-rabbit IgG was used as secondary antibody.) (Reprinted with permission from [9]. Copyright © 2011 Wiley-VCH Verlag GmbH & Co. KGaA, Weinheim.)

antibody binding to BSA-AMP-peptide or Rab1b-AMP could be detected. However, when incubating with AMP, the signal of α-Tyr-AMP antibody binding decreased significantly, thus demonstrating the relevance of the adenine base of the AMP moiety to interaction with the antibody. Intriguingly, both GMP and AMP competed with α-Tyr-AMP antibody binding to Cdc42-AMP, although the competition with AMP appeared to be slightly more effective than with GMP. This observation could possibly hint at a strong recognition of the furanoside residue and the phosphate group of GMP/AMP, leading to a substantial degree of competition of AMP and GMP in the absence of the Rab1b peptide sequence.

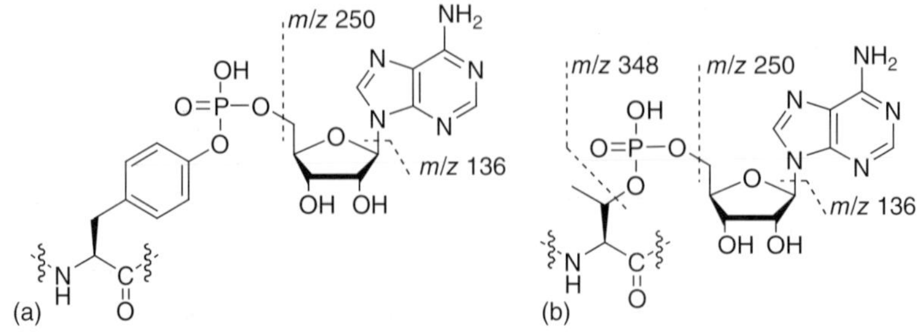

Figure 9.4 Mass spectrometric fragmentation patterns of the AMP group at tyrosine (a) and threonine (b) in MS/MS mode.

9.3.5
Detection of Adenylylation by MS Techniques

A number of MS investigations report putative adenylylated peptides, with the majority only conducted at the MS level, with the exception of a few, which were conducted at the tandem mass spectrometry (MS/MS) level. The distinct mass shift upon adenylylation of amino acid residues (serine, threonine, or tyrosine) renders it a good target for MS detection and identification. However, the fragmentation of adenylylated tryptic peptides derived from adenylylated proteins has only recently been systematically investigated. We demonstrated that adenylylated peptides show loss of parts of the AMP upon different fragmentation techniques (Figure 9.4).

As expected, electron transfer dissociation (ETD) yields less complicated spectra, with minimum fragmentation of the AMP itself. In contrast, CID (collision-induced dissociation) and high-energy collision dissociation (HCD) fragmentation caused AMP to fragment, generating characteristic ions suitable for identification of adenylylated peptides. The characteristic ions and losses upon CID and higher energy collision fragmentation from the AMP group turned out to be highly dependent on which amino acid was adenylylated, with different reporter ions for adenylylated threonine and tyrosine. The results showed that upon CID as well as HCD fragmentation, the whole AMP group is prone to leave if attached to a threonine, creating a loss of 347 for the majority of the fragments containing this moiety. If the adenylylation is positioned on a tyrosine, the predominant losses are either adenine (−135) or adenosine (−249). Upon ETD fragmentation, the modification is stable, and thereby the fragmentation spectra are easier to interpret manually as well as by search engines, such as Mascot (Figure 9.4) [11].

9.4
Chemical Biological Research/Evaluation

The preparatively adenylylated Rab1-protein and the generated antibodies can serve as excellent tools to further analyze the consequences of Rab1 adenylylation and the distribution of adenylylation as PTM in various proteomes.

9.4.1 Functional Consequences of Adenylylation

The production of preparative amounts of homogeneously adenylylated Rab1 allows detailed binding studies of the modified protein with regulatory factors (that are GEF, GAP, GDI) and downstream effectors. As the site of adenylylation is situated in the highly important switch II region, this modification would hypothetically interfere with the binding of most proteins. The facile production of the modified protein ensures that even interactions of moderate-to-weak affinity can be quantified.

As stated earlier, DrrA preferentially adenylylates active Rab1:GTP [3, 7]. We have therefore analyzed the binding of interaction partners that also show a preference for active Rab1:GTP over Rab1:GDP (that are GAPs and effectors) with respect to Rab1-adenylylation. The binding of the effectors *Legionella* LidA and human microtubule-associated monooxygenase, calponin, and LIM domain containing 3 (MICAL3) to Rab1:GTP and Rab1:GTP-AMP have been determined by analytical size exclusion chromatography. Clearly, MICAL3 but not LidA is unable to bind to adenylylated Rab1. This finding makes sense because the adenylylation of Rab1 by DrrA maintains the binding to the *Legionella* protein LidA but abrogates the interaction with the human factor MICAL3. Also, we monitored the GAP-catalyzed GTP hydrolysis of Rab1:GTP-AMP in comparison to the wild-type Rab1:GTP. Here we have used a reversed-phase chromatography approach in which the transition of GTP into GDP could be directly quantified [3]. In addition, we have used fluorescent techniques using intrinsic fluorophores (that are Rab tryptophanes) and extrinsic probes (i.e., mantGTP) with which the conversion of adenylylated and nonadenylylated Rab1:GTP or Rab1:mantGTP could be followed spectroscopically. All of these experimental approaches clearly established that GTP hydrolysis of active Rab1 AMP by the human GAP TBC1D20 and the *Legionella* GAP LepB was profoundly impaired. Therefore, adenylylation of Rab1 locks the GTPase in the GTP-state and causes discriminatory binding of effector proteins. Also, binding the recycling protein GDI is severely impaired by Rab-adenylylation and therefore promotes membrane binding.

The molecular basis for the observed changes in Rab1-binding patterns can be analyzed by X-ray crystallography of the adenylylated Rab1-protein. X-ray crystallography usually requires multi-milligram quantities of protein and therefore the establishment of preparative Rab1-adenylylation by DrrA is a prerequisite for using this technique. Structure determination of Rab1:GppNHp-AMP revealed that the AMP attached to Tyr77 of switch II in Rab1 acts as a bulky group. The AMP is positioned in the main interaction site of most binding partners (GAPs, effectors, GDI) and therefore sterically interferes with their binding. A stacking interaction between the nucleobase of AMP with Phe45 of Rab1 appears to stabilize the orientation of the adenylylated Tyr77 and therefore appears to inhibit a rotation that could uncover the blocked switch II region (Figure 9.1b).

Thus, we can conclude that a profound biochemical analysis of modified proteins can give invaluable insights into changes in protein-binding patterns. These

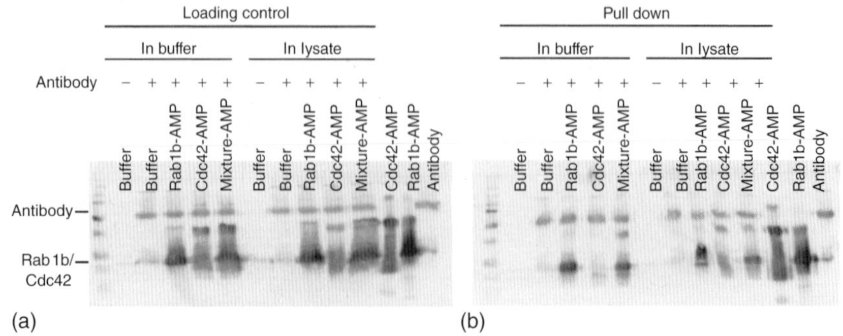

Figure 9.5 *Detection of adenylylated proteins in mammalian cell lysates.* Preparative adenylylated Rab1 and Cdc42 (0.1 μg each) have been exogenously added to buffer or 100 μg of Cos7 cell lysate. A biotinylated α-Tyr-Rab1-AMP antibody has been immobilized on magnetic streptavidin beads and used to pull down the Rab1-AMP and Cdc42-AMP samples. (a) Loading control containing sample mixtures before pull down. (b) Pull-down experiment of (a). (In all Western blots, IRDye800-conjugated donkey anti-rabbit IgG was used as secondary antibody.) (Reprinted with permission from [9]. Copyright © 2011 Wiley-VCH Verlag GmbH & Co. KGaA, Weinheim.)

results allow formulating hypotheses for the role of the PTM *in vivo* that can be tested further.

9.4.2
Detection of Adenylylated Proteins in Mammalian Cell Lysates

To further investigate the applicability of the Rab1 antibody, we performed pull-down experiments of adenylylated proteins from cell lysates. For this purpose, we added adenylylated Rab1 and Cdc42 to exogenously mammalian (simian) Cos7 cell lysates. We were able to pull down adenylylated Rab1 preferably, indicating the specificity of the α-Tyr-AMP antibody for Rab1-AMP binding even in the presence of the competitive environment of the cell lysate (Figure 9.5).

9.5
Conclusions

The example of adenylylation of Rab1 described here demonstrates the importance of a combined structural biological/biochemical/chemical approach to identifying and unraveling biological mechanisms, in this case a PTM. It also shows how this can lead to development of techniques for wider scale application to address the question of the prevalence of the particular modification. The basic principle here is that a thorough biochemical characterization has led to the ability to design and synthesize probe molecules and analytical techniques for this purpose.

References

1. Goody, R.S. and Itzen, A. (2013) Modulation of small GTPases by *Legionella*. *Curr. Top. Microbiol. Immunol.*, **376**, 117–133.
2. Aktories, K. (2011) Bacterial protein toxins that modify host regulatory GTPases. *Nat. Rev. Microbiol.*, **9** (7), 487–498.
3. Müller, M.P., Peters, H., Blümer, J., Blankenfeldt, W., Goody, R.S., and Itzen, A. (2010) The *Legionella* effector protein DrrA AMPylates the membrane traffic regulator Rab1b. *Science*, **329** (5994), 946–949.
4. Yarbrough, M.L., Li, Y., Kinch, L.N., Grishin, N.V., Ball, H.L., and Orth, K. (2009) AMPylation of Rho GTPases by *Vibrio* VopS disrupts effector binding and downstream signaling. *Science*, **323** (5911), 269–272.
5. Worby, C.A., Mattoo, S., Kruger, R.P., Corbeil, L.B., Koller, A., Mendez, J.C., Zekarias, B., Lazar, C., and Dixon, J.E. (2009) The fic domain: regulation of cell signaling by adenylylation. *Mol. Cell*, **34** (1), 93–103.
6. Schoebel, S., Oesterlin, L.K., Blankenfeldt, W., Goody, R.S., and Itzen, A. (2009) RabGDI displacement by DrrA from *Legionella* is a consequence of its guanine nucleotide exchange activity. *Mol. Cell*, **36** (6), 1060–1072.
7. Müller, M.P., Shkumatov, A.V., Oesterlin, L.K., Schoebel, S., Goody, P.R., Goody, R.S., and Itzen, A. (2012) Characterization of Enzymes from *Legionella pneumophila* involved in reversible adenylylation of rab1 protein. *J. Biol. Chem.*, **287** (42), 35036–35046.
8. Goody, P.R., Heller, K., Oesterlin, L.K., Muller, M.P., Itzen, A., and Goody, R.S. (2012) Reversible phosphocholination of Rab proteins by *Legionella pneumophila* effector proteins. *EMBO J.*, **31** (7), 1774–1784.
9. Smit, C., Blümer, J., Eerland, M.F., Albers, M.F., Muller, M.P., Goody, R.S., Itzen, A., and Hedberg, C. (2011) Efficient synthesis and applications of peptides containing adenylylated tyrosine residues. *Angew. Chem. Int. Ed.*, **50** (39), 9200–9204.
10. Albers, M.F., van Vliet, B., and Hedberg, C. (2011) Amino acid building blocks for efficient fmoc solid-phase synthesis of peptides adenylylated at serine or threonine. *Org. Lett.*, **13** (22), 6014–6017.
11. Hansen, T., Albers, M., Hedberg, C., and Sickmann, A. (2013) Adenylylation, MS, and proteomics-Introducing a "new" modification to bottom-up proteomics. *Proteomics*, **13** (6), 955–963.

10
Chemical Biology Approach to Suppression of Statin-Induced Muscle Toxicity

Bridget K. Wagner

10.1
Introduction

Millions of people worldwide take 3-hydroxy-3-methylglutaryl-coenzyme A (HMG-CoA) reductase inhibitors, or statins. A major dose-limiting side effect of statin use is myopathy. However, the mechanism of this toxicity is not fully clear, and the ability to suppress muscle toxicity with a small molecule could transform clinical treatment. We took approaches involving chemical profiling, chemical epistasis analysis, and high-throughput suppressor screening to better understand the muscle toxicity induced by statins. The results suggested that a Rab (rat sarcoma related in brain) prenylation event plays an important role in statin-induced muscle toxicity. Further, we identified a protein kinase C (PKC) inhibitor, Gö6976, that suppressed the toxicity of simvastatin in a mouse muscle cell line. This strategy illustrates the power and potential of chemical biology to have an ultimate impact on clinical treatment and human health.

10.2
The Biological Problem

Millions of people in the world suffer from cardiovascular disease, and it is a leading cause of death in both men and women. Elevation in plasma low-density lipoprotein (LDL) cholesterol levels is a major risk factor for myocardial infarction (heart attack) in these patients. Drugs to reduce dyslipidemia have included niacin and the fibrate class, but each of these has clinical limitations, such as low efficacy or toxic side effects. The development of HMG-CoA reductase inhibitors, or statins, has had an enormous clinical impact on the treatment of heart disease and prevention of heart attack, and these are taken by tens of millions of patients worldwide [1]. One of the first such drugs, lovastatin, was discovered in the 1970s as a fungal natural product [2] and lowered lipid levels in animals and healthy volunteers. Problems with the development of another early statin, compactin, halted advancement of lovastatin to regular clinical use until the late 1980s. Since then,

Concepts and Case Studies in Chemical Biology, First Edition. Edited by Herbert Waldmann and Petra Janning.

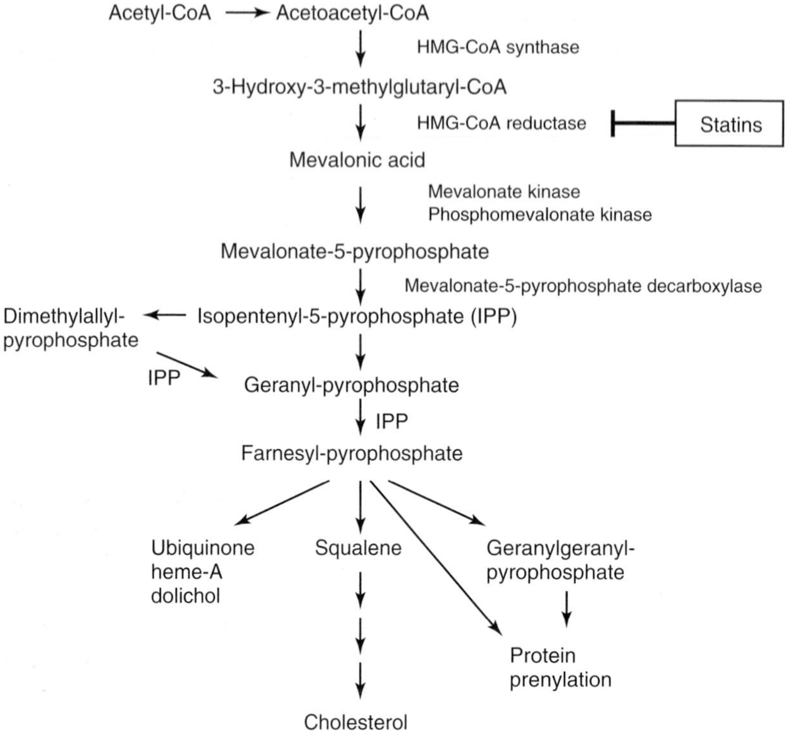

Figure 10.1 *Cholesterol biosynthesis pathway*. Statins inhibit the rate-limiting step, HMG-CoA reductase.

the statin class has been among the best-selling drugs in the world, and clinicians have even considered preventative administration to healthy patients [3].

HMG-CoA reductase is the rate-limiting step in cholesterol biosynthesis, resulting in the generation of mevalonate (Figure 10.1). This pathway is responsible for the synthesis of not only cholesterol but also dolichol and isoprenoid units such as farnesyl pyrophosphate (FPP) and geranylgeranyl pyrophosphate (GGPP), which are used to modify proteins (e.g., rat sarcoma (Ras)) and small molecules (e.g., CoQ10). Statins thus inhibit the synthesis of all steps downstream of mevalonate. Statins exert their clinical effect by causing a compensatory upregulation of the LDL receptor in the liver, enabling greater clearance of LDL from the bloodstream.

A major dose-limiting side effect associated with statin use is muscle toxicity, which can be exacerbated by vigorous exercise [4]. Although muscle toxicity can be difficult to quantify clinically, it ranges from muscle weakness and cramps reported by the patient, to severe myopathy, to rhabdomyolysis, which is rare (0.1–0.5%) but can be life threatening [5]. Muscle-related symptoms limit the statin dose achievable clinically, and for some patients, a change in, or even halting of, statin treatment is required [6]. Thus, the optimal lowering of LDL levels cannot be realized. Although not fully understood, this side effect is thought to be on-mechanism. In other words, statin effects on HMG-CoA reductase may directly cause myopathy, rather than an effect on another cellular target. The several branches of this pathway downstream of HMG-CoA reductase thus provide candidate mechanisms for causing myopathy. Overall, we rationalized that identifying the cellular basis of statin-induced muscle toxicity and targeting it chemically may allow the medical community to more fully harness the therapeutic potential of these drugs.

10.3 The Chemical Approach

10.3.1 Generation of a Compendium of Mitochondrial Activity

Our initial efforts focused on chemical profiling of mitochondrial activity in the mouse muscle cell line C2C12. We developed five phenotypic assays (calcein viability, adenosine triphosphate (ATP) levels, 3-(4,5-dimethylthiazol-2-yl)-2,5-diphenyltetrazolium (MTT) activity, mitochondrial membrane potential, and reactive oxygen species (ROS) levels) and one gene-expression-based assay to measure the mitochondrial effects of nearly 2500 known biologically active, or bioactive, compounds [7]. The analysis of this compendium enabled the generation of some interesting hypotheses. In particular, six statins (Figure 10.2) were present in the collection, and we found that four of these compounds induced a profile reminiscent of mitochondrial toxicity in C2C12 myotubes. Upon closer examination, the other two statins, although not showing this profile at the screened concentrations, induced similar cellular effects at higher concentrations. This result led us to hypothesize that statins induce a mitochondrial toxicity in muscle cells. We then determined the nearest-neighbor compounds that resulted in a similar signature of activity. Of the top 10 clinically used drugs with a similar signature, 5 were reported in the literature at least once to induce myopathy in a clinical setting. Further investigation of one of these drugs, propranolol, confirmed that it actually did induce myopathy in patients with greater frequency than related atenolol or metoprolol [8]. Thus, we were able to generate and test a clinical hypothesis as a result of chemical profiling.

In order to identify novel suppressors of statin-induced muscle toxicity, we describe a chemical screening approach involving the testing of known bioactives. Additional screening efforts are focusing on the collection of compounds derived from diversity-oriented synthesis (DOS) at the Broad Institute [9], but these are not discussed here. The use of bioactives enables the development of new hypotheses regarding the mechanisms of action of effective compounds, and is exceptionally useful in developing phenotypic assays, for which target identification is highly challenging. We also used chemical tools to dissect the various branches of the mevalonate pathway. This chemical–genetic epistasis analysis was very useful in generating the hypothesis that the inhibition of transfer of

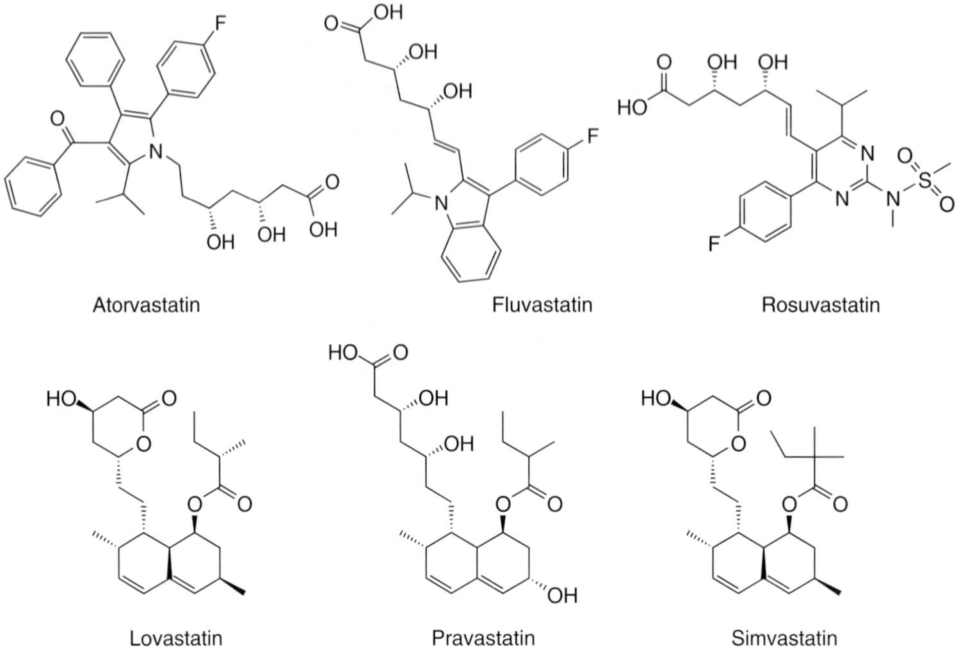

Figure 10.2 Chemical structures of clinically used statins.

geranylgeranyl groups to the Rab family of proteins by statins appears to be the relevant biological activity causing muscle toxicity.

10.4 Chemical Biology Research/Evaluation

10.4.1 Chemical Epistasis Analysis

First, we sought to develop an effective assay that could enable rapid measurement of statin-induced toxicity and its suppression by small molecules [10]. Our previous work to generate a mitochondrial compendium in muscle cells provided us the preliminary data needed for this assay. We seeded C2C12 myoblasts in optical 384-well plates, and when the cells reached confluence, induced differentiation to myotubes by incubating with 2% horse serum for 4–6 days. Differentiation was confirmed visually by observing cell fusion and syncytia formation characteristic of myotubes. We then treated differentiated myotubes for 48 h with 10 μM simvastatin, and measured cellular ATP levels with a commercially available kit (CellTiter-Glo, Promega). As expected, we observed a twofold reduction in ATP levels compared to dimethyl sulfoxide (DMSO) controls.

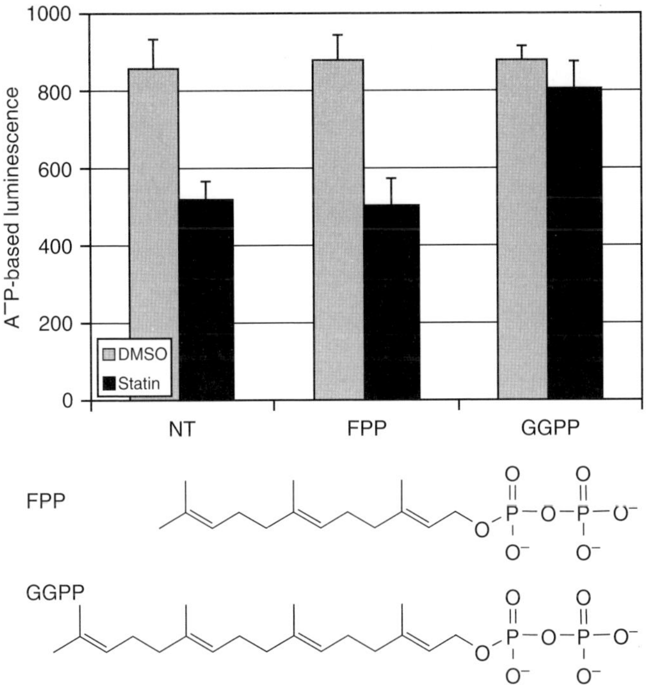

Figure 10.3 *Geranylgeranyl pyrophosphate (GGPP) suppresses statin-induced muscle toxicity*. C2C12 myotubes were treated for 48 h with simvastatin in the absence or presence of FPP or GGPP. Cellular ATP levels were used as a surrogate for viability. The structures of FPP and GGPP are shown.

In order to prepare for chemical screening, we sought to determine whether a positive control could inform us about the biological mechanism of muscle toxicity. Thus, we treated C2C12 myotubes with simvastatin in the absence or presence of various intermediates in the cholesterol biosynthesis pathway. Only co-treatment of muscle cells with GGPP was able to suppress the loss of ATP levels (Figure 10.3). FPP, cholesterol, or coenzyme Q had no effect on the effects of statin treatment on ATP levels. These results indicated that muscle toxicity induced by statin treatment may be dependent on GGPP-regulated events in the cell.

GGPP is used by the cell for prenylation steps, in which these isoprenoid units (Figure 10.3) are added to proteins at the C-terminus. Two enzymes are responsible for the transfer of GGPP to proteins (Figure 10.4). Geranylgeranyltransferase-I (GGTase-I) transfers GGPP to Rac and Rho proteins, and can recognize the CaaX sequence (where "a" is an aliphatic amino acid and "X" is usually leucine). Alternatively, geranylgeranyltransferase-II (GGTase-II), or Rab GGTase, transfers GGPP to the Rab family of proteins by recognizing two cysteines at the C-terminus, without the need for a CaaX sequence. Chemical inhibitors of these transferase enzymes enable the dissection of protein prenylation events responsible for a particular phenotype. Because GGTase-I has been an attractive target for cancer

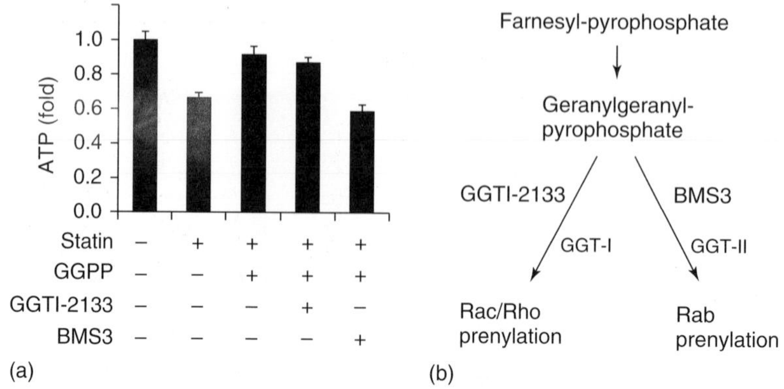

Figure 10.4 *Rab prenylation events are involved in statin-induced muscle toxicity.* (a) Co-treatment with statin and GGPP restores ATP levels. However, addition of BMS3, an inhibitor of geranylgeranyltransferase-II, results in a loss of suppression. (b) The steps involved in transfer of geranylgeranyl groups to proteins.

chemotherapy, there are a number of commercially available inhibitors such as GGTI-2133 for this enzyme. Less attention has been paid to GGTase-II, but one compound, BMS3 [11], is a selective inhibitor of this enzyme.

We thus reasoned that if we suppressed the effects of statins on muscle toxicity by co-treatment with GGPP, we could then determine whether Rac/Rho prenylation or Rab prenylation was the important factor by additional treatment with GGTase inhibitors. This experiment revealed that treatment with simvastatin, GGPP, and GGTI-2133 did not decrease ATP levels, but treatment with simvastatin, GGPP, and BMS3 did result in a decrease in cellular ATP levels (Figure 10.4). These results showed that a Rab prenylation event was responsible for statin-induced muscle toxicity.

10.4.2
High-Throughput Screening

While supplementation with GGPP completely restores myotube viability in the presence of simvastatin, we wished to identify small-molecule suppressors of statin toxicity that could be developed as potential clinical leads. We started by screening a collection of 2240 bioactive small molecules. We differentiated C2C12 myotubes in clear-bottomed 384-well plates, treated with 10 μM simvastatin and pin-transferred the compound collection, for 48 h co-treatment at an average screening concentration of 10 μM. We identified several compounds that suppressed the loss of ATP levels, but decided to focus on the most potent compound, Gö6976 (Figure 10.5). This compound is annotated as a PKC α and β inhibitor. Co-treatment with simvastatin and Gö6976 resulted in increased ATP levels in C2C12 myotubes, but not to as great an extent as GGPP (Figure 10.5). After identification of Gö6976, we performed a similar epistasis analysis as we

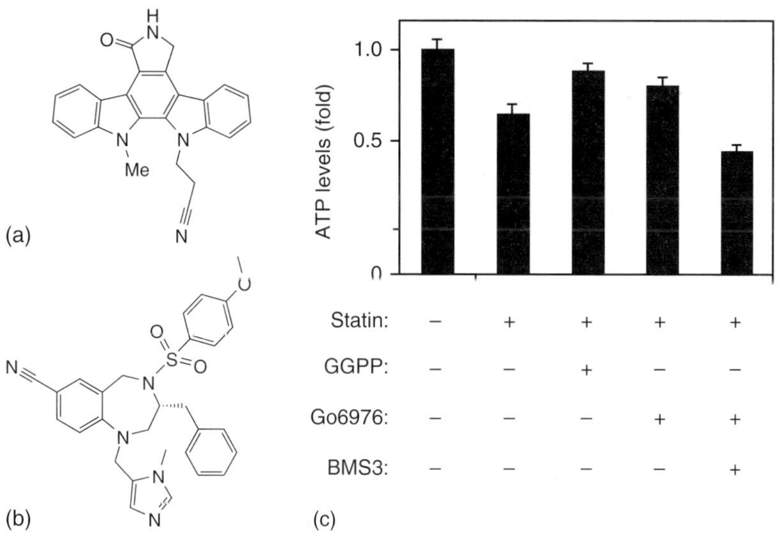

Figure 10.5 *High-throughput screening identified Gö6976* (a) as a suppressor of statin-induced muscle toxicity. (b) Chemical structure of BMS3. (c) The suppressive effects of Gö6976 are, as in the case of GGPP, lost by further addition of BMS3.

did with GGPP. GGTI-2133 had no effect on Gö6976 suppression, but BMS3, the GGTase-II inhibitor, resulted in the loss of ATP levels (Figure 10.5). This result suggests that Gö6976 may have an effect on restoring GGPP levels, such that inhibiting the transferase is downstream of the suppressive phenotype and negates the beneficial effect of Gö6976. Importantly, we also noted that Gö6976 had no effect on the gene-expression levels of the low-density lipoprotein receptor (LDLR) in either muscle or liver cells. This result is important because it reveals that Gö6976 is not likely to have a direct effect on cholesterol biosynthesis inhibition by statins. Further, we validated that Gö6976 suppresses statin-induced muscle toxicity in zebrafish.

10.5 Conclusion

This study provides evidence that Rab prenylation is important to statin-induced muscle toxicity, and that it is possible to identify suppressive small molecules that should not inhibit the beneficial effects of statins on blood cholesterol levels. The use of small molecules to dissect biosynthetic pathways is certainly not new, but affords a precise and rapid understanding of the phenotypic consequences of cellular perturbations. For the future, modern chemical biology techniques, including affinity labeling of isoprenoids [11], provide an attractive opportunity to identify the specific Rabs responsible for statin-induced muscle toxicity. As a common laboratory tool compound, Gö6976 is unlikely to become a clinical candidate.

Further screening has the power to identify chemical series with improved properties and more selective inhibition of statin-induced muscle toxicity.

References

1. Wenner Moyer, M. (2010) The search beyond statins. *Nat. Med.*, **16**, 150–153.
2. Tobert, J.A. (2003) Lovastatin and beyond: the history of HMG-CoA reductase inhibitors. *Nat. Rev. Drug Discovery*, **2**, 517–526.
3. Wald, D.S., Morris, J.K., and Wald, N.J. (2012) Randomized Polypill crossover trial in people aged 50 and over. *PLoS One*, **7**, e41297.
4. Thompson, P.D., Clarkson, P., and Karas, R.H. (2003) Statin associated myopathy. *J. Am. Chem. Soc.*, **289**, 1681–1690.
5. Graham, D.J., Staffa, J.A., Shatin, D., Andrade, S.E., Schech, S.D., La Grenade, L., Gurwitz, J.H., Chan, K.A., Goodman, M.J., and Platt, R. (2004) Incidence of hospitalized rhabdomyolysis in patients treated with lipid-lowering drugs. *J. Am. Chem. Soc.*, **292**, 2585–2590.
6. Siddiqi, S.A. and Thompson, P.D. (2009) How do you treat patients with myalgia who take statins? *Curr. Atherosclerosis Rep.*, **11**, 9–14.
7. Wagner, B.K., Kitami, T., Gilbert, T.J., Peck, D., Ramanathan, A., Schreiber, S.L., Golub, T.R., and Mootha, V.K. (2008) Large-scale chemical dissection of mitochondrial function. *Nat. Biotechnol.*, **26**, 343–351.
8. Setoguchi, S., Higgins, J.M., Mogun, H., Mootha, V.K., and Avorn, J. (2010) Propranolol and the risk of hospitalized myopathy: translating chemical genomics findings into population-level hypotheses. *Am. Heart J.*, **159**, 428–433.
9. Nielsen, T.E. and Schreiber, S.L. (2008) Towards the optimal screening collection: a synthesis strategy. *Angew. Chem. Int. Ed.*, **47**, 48–56.
10. Wagner, B.K., Gilbert, T.J., Hanai, J.I., Imamura, S., Bodycombe, N.E., Bon, R., Waldmann, H., Clemons, P.A., Sukhatme, V.P., and Mootha, V.K. (2011) A small-molecule screening strategy to identify suppressors of statin myopathy. *ACS Chem. Biol.*, **6**, 900–904.
11. Nguyen, U.T., Guo, Z., Delon, C., Wu, Y., Deraeve, C., Franzel, B., Bon, R.S., Blankenfeldt, W., Goody, R.S., Waldmann, H., Wolters, D., and Alexandrov, K. (2009) Analysis of the eukaryotic prenylome by isoprenoid affinity tagging. *Nat. Chem. Biol.*, **5**, 227–235.

11
A Target Identification System Based on MorphoBase, ChemProteoBase, and Photo-Cross-Linking Beads

Hiroyuki Osada, Makoto Muroi, Yasumitsu Kondoh, and Yushi Futamura

11.1
Introduction

In this chapter, we describe a target identification system in mammalian cells that enables identification of the molecular targets of bioactive compounds. This system is based on three approaches for target identification that are classified as either phenotypic profiling (indirect) or affinity beads-based pull down (direct). Successful target identification should utilize a combination of these methodologies. Here, newly developed methodologies for target identification that are MorphoBase and ChemProteoBase as indirect approaches and photo-affinity beads as a direct approach, are described. MorphoBase identifies molecular targets of bioactive compounds based on morphological changes of cancer cell lines induced by the compounds, while ChemProteoBase identifies molecular targets based on proteomic changes induced by the compounds. Photo-cross-linking beads are used for searching bioactive compounds-interacting target based on direct interaction of target proteins and bioactive compounds. We describe the case studies of these methodologies for target identification of the bioactive compounds NPD6689/NPD8617/NPD8969, BNS-22, methyl-gerferin (M-GFN), and xanthofumol.

11.2
The Biological Problem

Novel bioactive compounds derived from natural resources have been frequently used as probes to enhance our understanding of complex biological systems. Certain chemical probes, such as FK506 and trichostatin, have been at the forefront of new fields in biological sciences [1–4]. To discover such compounds, considerable efforts have been made on the basis of broth library screening, biological-activity-guided isolation, and structural elucidation. The use of recent technological advances, such as high-throughput screening (HTS) and chemical libraries, has accelerated the identification of new desirable compounds. Thus, for

Concepts and Case Studies in Chemical Biology, First Edition. Edited by Herbert Waldmann and Petra Janning.

academic research, large chemical libraries have been constructed, and HTS programs have been implemented, including the Molecular Library Program at the NIH and the Compound Management and Screening Center in the Max Planck Institute of Molecular Physiology. RIKEN has founded the Natural Products Depository (NPDepo, *http://npd.riken.jp/*) as a public chemical bank in Japan and has provided a chemical library that primarily focuses on natural products and their derivatives [5]. In NPDepo, a fraction library of natural products, which consists of semipurified natural products systemically collected from microbial fermentation broth, was also constructed to overcome the incompatibility of crude extracts with HTS [6, 7]. We have screened this library in an attempt to discover novel biological compounds.

Recent trends in screening methods have shifted to target-based approaches, in which direct interaction or direct inhibitory activity are usually assessed *in vitro;* these screening methods are the most rational and powerful approaches to identify compounds interacting with target proteins. However, this approach is limited to the well-established targets, and undesired off-target effects of the compound cannot be predicted. Traditionally, phenotypic screens have been the primary method used to discover bioactive compounds, and many novel compounds have been identified using such methods. For example, FK506 was originally isolated as an immunosuppressant [8], and the antibiotic trichostatin was rediscovered as an inducer of erythroleukemia differentiation before the identification of its specific histone deacetylase inhibitory activity [9]. Unbiased cell-based phenotypic screens mostly promise to identify small molecules with biologically relevant properties and the ability to modulate complex cellular processes. However, target identification of compounds obtained by cell-based screening is often a difficult and time-consuming process. Indeed, other reviews have described target identification for small bioactive molecules as being similar to "finding the needle in the haystack" [10].

There are two fundamentally different approaches for identification of molecular targets: direct and indirect. Direct approaches are based on the analysis of direct interactions of target proteins and small molecules. In this approach, using bait such as chemical beads or fluorescent probes, we can catch the target molecule from the cell lysate directly; however, nonspecific binding may occur. In addition, target proteins with low expression in cell lysates or targets exhibiting only weak interactions with the identified small molecule may not be able to be detected. In contrast, indirect approaches facilitate target identification through validation procedures using well-characterized compounds or genome-wide mutations. Construction of a comprehensive phenotypic database as a reference and linkage of external databases allow us to rapidly search for pathways associated with target proteins. However, when the reference or the phenotypic assay system is not adequate for the target, predictions from the search results are not always focused and accurate.

In recent decades, much progress has been made in establishing methods for both direct and indirect approaches to the target identification for small molecules [11]. In direct approaches, compared to the pioneering work of Schreiber and coworkers [1], who used an affinity matrix conjugated with FK506, improvement of technology required for such approaches has been made, including matrices, linkers, and conjugation of compounds. Protein identification by mass spectrometry (MS) has allowed for the rapid identification of a large number of proteins from small amounts of sample. Other techniques and *in vitro* protein expression systems, such as phage display, have also been used to overcome low amounts of target proteins in samples. Two-dimensional electrophoresis for detection of target proteins bound to fluorescent probes, isothermal titration calorimetry (ITC), and surface plasmon resonance (SPR) have been used for screening and validation of ligand protein interactions. Recently, a method to identify targets using protease sensitivity on SDS-PAGE (sodium docecylsulfate-polyacrylamide gel electrophoresis) was reported [12]. In addition, for indirect approaches, we have been able to obtain large amounts of data through the development of suitable analytic equipment and by progress in omics research. Mutant libraries for chemical genomics have been prepared. Using such datasets, many profiling systems have been developed, including those based on chemical-genomics-, transcriptomics-, proteomics-, metabolomics-, and cell imaging-based profiling.

As described, various methods are available for the identification of targets of compounds. However, a general methodology applied successfully for the majority of the cases, that is, a "gold standard," has not yet been established. Therefore, it is important to analyze a combination of adequate methods. Here, we introduce some originally developed methodologies for target identification at RIKEN and case studies of analysis of novel compounds in combination with these methodologies.

11.3 Chemical Approaches

11.3.1 MorphoBase

Tumor cells often dynamically and specifically change shape depending on the mode of action of a drug, and experienced cell biologists can judge the presumed molecular target of a test compound by simple observation of typical morphological changes. This prompted us to accumulate information on morphological changes induced by various compounds with known mechanisms and construct a database linking morphology to drug function, termed *MorphoBase*, which may be helpful in the target identification of a drug and the discovery of unique bioactive compounds.

In general, morphological traits judged by visual inspection are subjective, often making it difficult to obtain quantitative and reproducible results. To circumvent errors inherent to subjective measurements, we constructed MorphoBase replete with high-content imaging and statistical analyses of multidimensional morphological parameters (Figure 11.1) [13]. Specifically, we designed a high-content imaging method to segment the cells and quantify the effects of various authentic compounds on morphological changes in two cancer cell lines, *src*^{ts}-NRK and HeLa cells. Because bright-field cell images do not produce a clearly defined cell outline, nuclear staining was introduced. In addition, the unique textures induced by a drug on the cell surface or inside the cell, such as granules and vacuoles, were defined by the descriptor "granular" to subdivide these components. Our method provides 12 morphological parameters each from properly segmented "cell," "nuclear," and "granular" fragments of two cancer cell lines, resulting in a total of 71 descriptors to characterize the vast variety of cell-shape changes induced by mechanistically distinct compounds.

Next, we developed a data analysis program that incorporated multivariate statistical tools to automatically analyze, visualize, and rank multiparametric phenotype datasets; principal component analysis (PCA) was applied to visualize phenotypic responses, and the similarities in morphological changes were defined by two statistical computations: (i) "similarity ranking" determined by Euclidean distance metrics and (ii) our original index, termed *probability scores.* A test set of 207 authentic compounds, supplied by the RIKEN NPDepo and Screening Committee of Anticancer Drugs (SCADS) inhibitor kits, was created, and their multiparametric phenotypic responses were analyzed. As a result, inhibitors with a common mode of action formed clusters in PC1-PC2 scatter plots and were mutually ranked in their top 20 nearest neighbors, suggesting that MorphoBase can be used to successfully profile phenotypes by drug function. To expand the function of MorphoBase profiling, we further developed a training algorithm and applied it to a well-validated drug set comprising 54 drugs with 14 molecular targets encompassing a typical antitumor mechanism of action. This training algorithm could be used to determine which of the 14 target classes in the training set was a plausible candidate for a test compound by "probability scores," the mean z-scores for a test compound relative to the median point of each target class.

In summary, we developed a high-content imaging method and a phenotype profiling system using similarity search software, based on statistical analyses of multiparametric phenotype responses, to identify the molecular targets of compounds of interest with an "unbiased eye." We discuss a case study using the MorphoBase system later (see Section 11.4.1).

11.3.2 ChemProteoBase

In addition to cell shape, expression levels and modifications of proteins are changed by drug treatment depending on the mechanism of action of the drug. Recently, we developed a proteomic profiling system for target analysis of compounds based on proteome analysis by two-dimensional difference gel electrophoresis (2D-DIGE) [14]. This profiling system was termed *ChemProteoBase* (Figure 11.2).

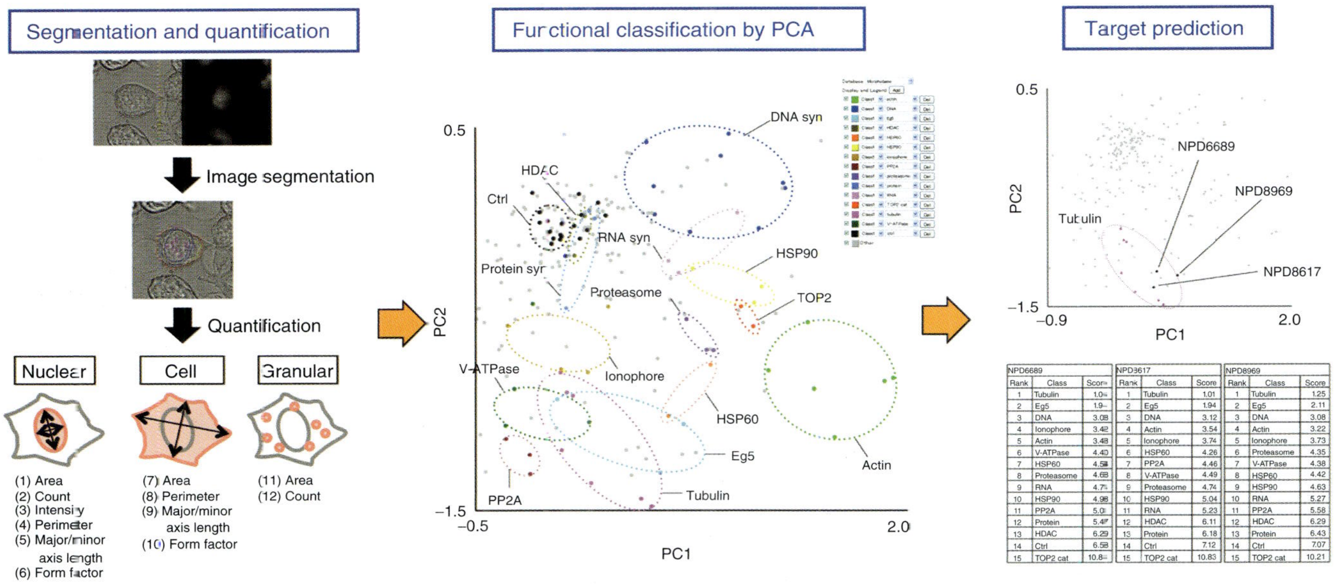

NPD6689		
Rank	Class	Score
1	Tubulin	1.0[illegible]
2	Eg5	1.9[illegible]
3	DNA	3.08
4	Ionophore	3.42
5	Actin	3.48
6	V-ATPase	4.40
7	HSP60	4.5[illegible]
8	Proteasome	4.68
9	RNA	4.7[illegible]
10	HSP90	4.98
11	PP2A	5.0[illegible]
12	Protein	5.47
13	HDAC	6.29
14	Ctrl	6.58
15	TOP2 cat	10.8[illegible]

NPD8617		
Rank	Class	Score
1	Tubulin	1.01
2	Eg5	1.94
3	DNA	3.12
4	Actin	3.54
5	Ionophore	3.74
6	HSP60	4.26
7	PP2A	4.46
8	V-ATPase	4.49
9	Proteasome	4.74
10	HSP90	5.04
11	RNA	5.23
12	HDAC	6.11
13	Protein	6.18
14	Ctrl	7.12
15	TOP2 cat	10.83

NPD8969		
Rank	Class	Score
1	Tubulin	1.25
2	Eg5	2.11
3	DNA	3.08
4	Actin	3.22
5	Ionophore	3.73
6	Proteasome	4.35
7	V-ATPase	4.38
8	HSP60	4.42
9	HSP90	4.63
10	RNA	5.27
11	PP2A	5.58
12	HDAC	6.29
13	Protein	6.43
14	Ctrl	7.07
15	TOP2 cat	10.21

Figure 11.1 *Overview of MorphoBase profiling.* The input images (bright-field and the same Hoechst33342-stained cells) are used for "nuclear," "cell," and "granular" segmentation. After cells are segmented, 12 morphological parameters are quantified for each cell. Phenotypic descriptors of 207 test compounds profiled by multivariate statistical analysis. MorphoBase profiling results are visualized by PCA. The target protein of a test compound is predicted on the basis of the similarities in morphological changes between a compound of interest and reference compounds defined by statistical computations.

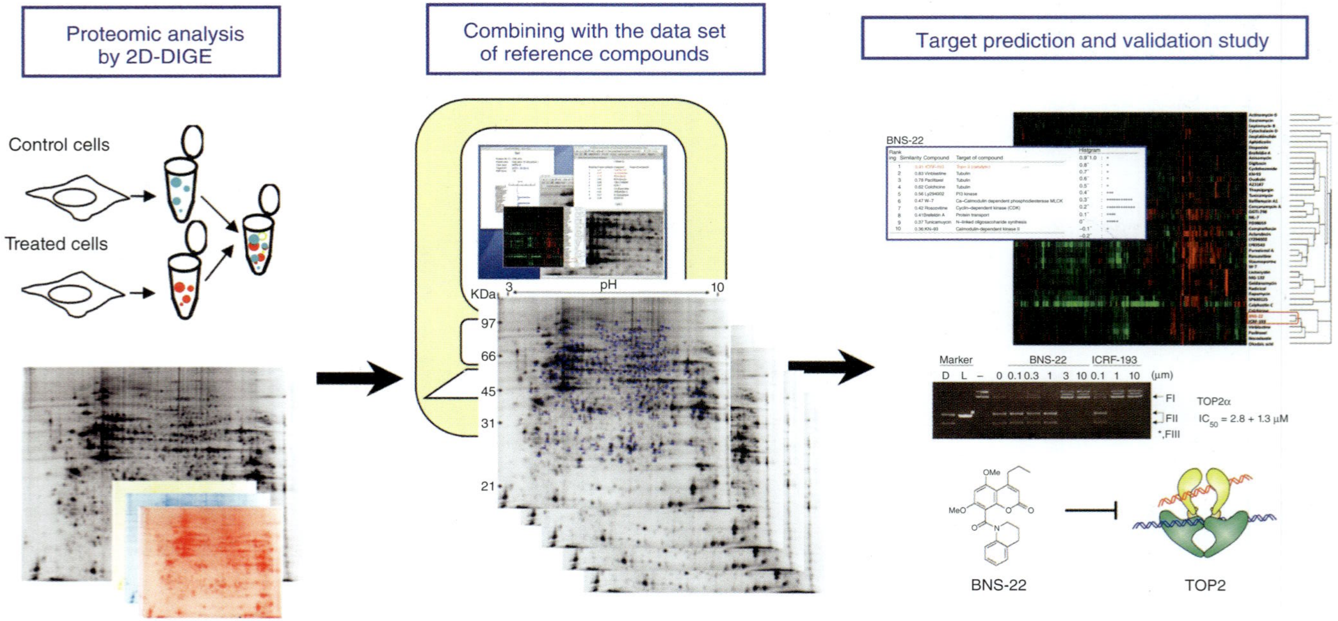

Figure 11.2 *Overview of ChemProteoBase profiling.* Proteomic analysis is performed by 2D-DIGE, and expression data of around 300 spots are acquired. Compared with data sets of well-characterized compounds in ChemProteoBase, the plausible target is predicted by finding the most similar compounds in proteomic profile. On the basis of the prediction, validation studies are performed.

The expression levels of proteins in HeLa cells treated with 19 well-known inhibitors were successfully classified by cluster analysis according to the inhibitors' mechanisms of action. Thus, we have since constructed a database from which expression data from separate experiments can be compared. Proteomic analyses of HeLa cells are performed after 18 h of exposure to test compounds at a concentration that inhibits cell growth by 80% or more. The expression data obtained from the around 300 spots, reproducibly detected in 2D-electrophoresis images of HeLa cell lysates, were used to calculate the ratio of each spot in treated and control cells and compare these ratios to the dataset in ChemProteoBase.

MS-based proteome analysis has been significantly improved in the past decade and it allows for the detection of more types of proteins and proteins with low abundances compared to 2D-DIGE-based proteome analysis. However, in our system, it is sufficient to compare a proteomic profile of the around 300 spots and it is not necessary to identify each spot.

Compared with gene expression profiling, which can simultaneously measure the expression of more than 20 000 genes, proteome analysis by 2D-DIGE can trace at most 1000 proteins. However, proteomic profiling results would be expected to provide different information and may be informative for target identification in some cases because it has the advantage of analyzing protein expression directly and may include information on protein modification. We discuss a case study using the ChemProteoBase system later (see Sections 11.4.1 and 11.4.2).

11.3.3
Photo-Cross-Linking Beads

Affinity chromatography is a traditional direct approach of target identification that uses immobilized small molecules to pull down its binding proteins from a complex protein mixture, such as cell lysates (Figure 11.3). For immobilization of small molecules, it is necessary that functional groups dispensable for bioactivity are revealed by structure–activity relationship (SAR) studies. Small molecules are attached to an affinity tag (e.g., biotin) or solid matrix (e.g., agarose beads) using these nonessential site(s). Then, the small-molecule-immobilized beads are incubated with cell lysates, followed by extensive washing to eliminate nonspecifically bound proteins. Tightly binding proteins are then eluted under highly denaturing conditions and analyzed by SDS-PAGE. The protein bands are identified by MS.

The major limitation of affinity chromatography is the need to derivatize the small molecules of interest in SAR studies. SAR studies are time consuming, and laborious, and require medicinal chemistry knowledge. Moreover, many small molecules cannot be modified without loss of bioactivity. The lack of a suitable functional group may interrupt immobilization in a functional group-dependent manner. We have developed an easy-to-use, nonselective universal coupling method that enables the attachment of a variety of small molecules to agarose beads using a photo-cross-linking reaction without requiring SAR studies and

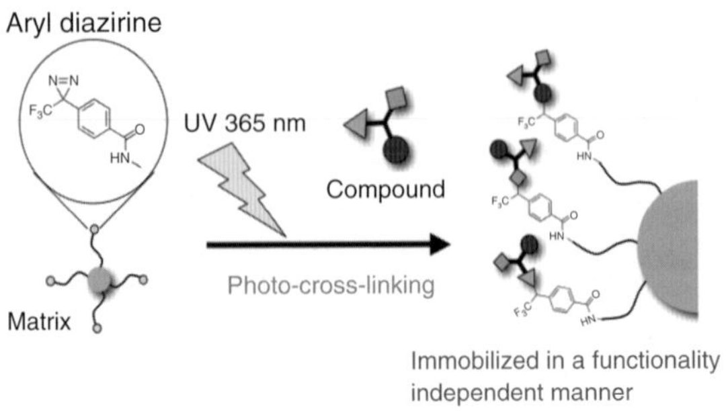

Aryl diazirine
UV 365 nm
Compound
Photo-cross-linking
Matrix
Immobilized in a functionality independent manner

(a)

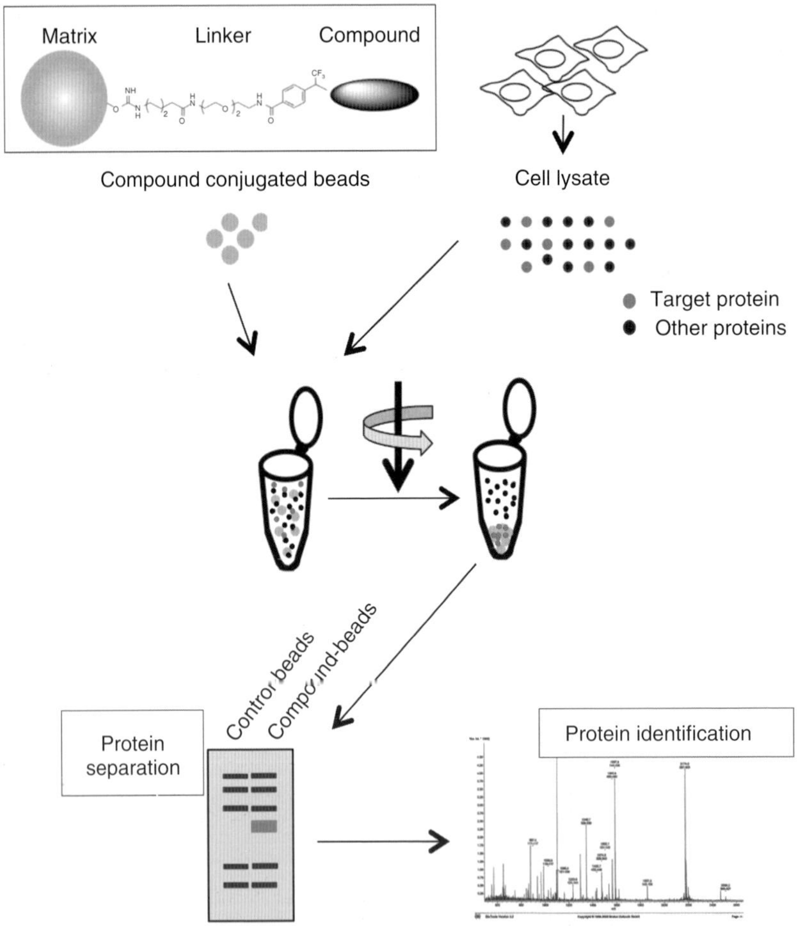

Matrix
Linker
Compound
Compound conjugated beads
Cell lysate
Target protein
Other proteins
Control beads
Compound-beads
Protein separation
Protein identification

(b)

derivatization of small molecules (Figure 11.3a) [15]. This method was originally developed for chemical arrays to introduce a variety of small molecules onto glass slides in a functional group-independent manner [16]. Upon UV irradiation, aryl-diazirine groups covalently introduced onto agarose beads are transformed into highly reactive carbenes, which in turn bind to or insert irreversibly into a proximal small molecule in a manner that is independent of the functional groups. This method is useful for immobilization of small molecules, especially natural products that mostly have complex structures and lack functional groups available for modification without loss of bioactivity, to agarose beads and subsequent target identification. We discuss a case study using photo-cross-linking beads later (see Sections 11.4.2–11.4.4).

11.4
Chemical Biological Research/Evaluation

In this section, we introduce case studies of target identification of six novel compounds, NPD6689/NPD8617/NPD8969, BNS-22, M-GFN, and xanthohumol (XN) (Figure 11.4) using our methodologies described earlier.

11.4.1
NPD6689/NPD8617/NPD8969

In the course of our phenotypic screening from a chemical library in NPDepo and microbial metabolites, we identified several hundreds of compounds that induced unique cell-shape changes in *src*^{ts}-NRK and HeLa cells. Among them, NPD6689, NPD8617, and NPD8969 were very potent, as evidenced by the nanomolar ranges of their effective doses.

To elucidate the mechanism of action of these compounds, we performed MorphoBase profiling [13]. Following treatment with the compounds, the resulting morphological changes were quantified by an imaging cytometer. The obtained phenotypic multiparameters were compared with the reference dataset, and the similarities in morphological changes between test samples and reference compounds were defined by "similarity ranking" and "probability scores." As a result, these compounds were predicted to perturb the microtubule system. In

Figure 11.3 *Identification of protein targets using photo-cross-linking beads.* (a) Upon UV irradiation, aryl-diazirine groups covalently introduced onto a solid matrix (e.g., agarose beads) are transformed into highly reactive carbenes, which in turn bind to or insert irreversibly into a proximal small molecule in a functionality independent manner. (b) Pull-down assays are performed by mixing compound-conjugated beads and cell lysates for target identification. Then, nonspecifically bound proteins are eliminated by extensive washing of beads. Tightly bound proteins are eluted under highly denaturing conditions and analyzed by SDS-PAGE. The protein bands, which are detected specifically on compound-conjugated beads, are identified by mass spectrometry.

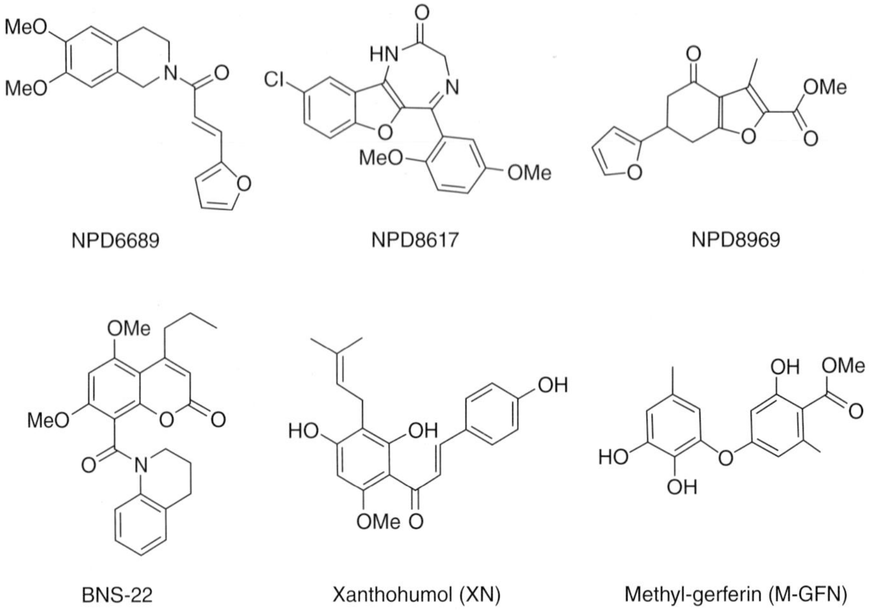

Figure 11.4 Chemical structures of unique bioactive compounds mentioned in this chapter.

addition, ChemProteoBase profiling supplementarily supported the prediction of MorphoBase profiling; NPD6689, NPD8617, and NPD8969 were clustered with typical microtubule-targeting inhibitors. Further tests using conventional techniques actually confirmed the results. The combination of these two profiling systems was able to clarify which plausible targets should be validated first, and we thus established that all three compounds inhibit microtubule integrity in both cell-free and cell-based systems immediately.

11.4.2 BNS-22

BNS-22 is a chemically synthesized derivative of the natural plant product GUT-70. GUT-70 has been reported to inhibit leukemic cell growth and induce caspase-mediated apoptosis [17]. Among more than 60 derivatives of GUT-70, BNS-22 was selected as the compound with the most robust biological activity. However, its molecular target and mechanism of action remained unknown.

Using ChemProteoBase, we analyzed HeLa cells treated with BNS-22 [18]. We compared proteomic profiles of HeLa cell treated with BNS-22 to a dataset of 42 well-characterized anticancer drugs and found that the profile of BNS-22 was similar to that of ICRF-193, a DNA topoisomerase II (TOP2) catalytic inhibitor. BNS-22 was shown to inhibit TOP2 and TOP2 poison-mediated DNA damage, consistent with other TOP2 catalytic inhibitors. By pull-down assay using photo-cross-linking beads containing BNS-22 and purified TOP2, our data suggested that BNS-22 bound to TOP2 directly, with an inhibitory mechanism different from that of ICRF-193. Thus, in this case, rapid identification of the cellular target was possible because of the optimal choice of direct and indirect methods.

11.4.3 Methyl-Gerferin

M-GFN, the methyl ester of the natural product gerfelin [19], was found to suppress osteoclastogenesis by cellular phenotype-based screening [20]. M-GFN strongly suppressed the differentiation of mouse bone-marrow-derived macrophages (BMMs) to tartrate-resistant acid phosphatase-positive (TRAP+) multinucleated osteoclasts induced by RANKL (receptor activator of NF-κB ligand) and M-CSF (macrophage colony-stimulating factor). Our proteomic analysis matched M-GFN with two V-ATPase inhibitors; however, its biological effects in a cell system were not confirmed. Using M-GFN-immobilized beads utilized in our photo-cross-linking approach described earlier, we identified glyoxalase 1 (GLO1) as a binding protein of M-GFN. GLO1 plays a critical role in the detoxification of 2-oxoaldehydes, especially the cytotoxic metabolite methylglyoxal (MG). M-GFN inhibited the enzymatic activity of GLO1 competitively, and GLO1 knockdown by siRNA suppressed RANKL-induced osteoclastogenesis, suggesting that inhibition of GLO1 results in the inhibition of osteoclastogenesis. The lack of standards for new targets in phenotype profiling analysis might lead to erroneous predictions. In such cases, it is more effective to apply a direct approach.

11.4.4 Xanthohumol

Sasazawa and coworkers [21] identified XN as an autophagy modulator by screening for a small molecule from an in-house natural product library using HeLa cells stably expressing EGFP-LC3 (enhanced green fluorescent protein-light chain 3). XN is the principal prenylated chalcone of the female inflorescences of hops, an ingredient of beer, and inhibits autophagosome maturation. Using XN-immobilized beads, they further identified valosin-containing protein (VCP) as an XN-binding protein from human epidermoid carcinoma A431 cell lysates. VCP was reported to be essential for autophagosome–lysosome fusion and formation of autolysosomes in human cell lines. XN bound directly to the N-domain of VCP, which is known to be the binding domain for the substrate and cofactor. Knockdown of VCP or treatment with XN impaired the maturation of autophagosomes. Taken together, these data suggest that XN inhibits VCP function, resulting in the inhibition of autophagosome maturation. Thus, XN was the first inhibitor that was identified to bind to the N-domain of VCP and inactivate VCP, and it may be used as a powerful tool for identifying the cofactor or substrate protein of VCP.

11.5 Conclusion

Natural products are the important source of bioprobes and pharmaceutical agents because of their significant advantages in both chemical and biological diversity. However, target identification and validation of natural bioactive compounds is often difficult, partly due to their low yield and the difficulties in total synthesis and chemical modification. So far, we have developed both direct and indirect approaches, two phenotypic profiling systems (MorphoBase and ChemProteoBase) and photo-cross-linking beads, for target identification of bioactive small molecules, specifically natural products. Each method has its own strengths and weaknesses; our direct approach offers the best method to detect the interaction between compound and target proteins without any derivatization of small molecules. However, it is often difficult to exclude nonspecific binding. Moreover, this method cannot be applied to UV-labile compounds. Indirect approaches are based on phenotypic changes in response to a compound. For these, it is difficult to identify the molecular target of the compound if the compound of interest does not induce any phenotype, if corresponding target information is not in the reference datasets, and if the compound interacts with multiple targets. We should thus make an appropriate choice in response to the chemistry and cellular phenotypes associated with small molecules, and, in most cases, supplemental combination of these methods would be helpful for "finding any tiny needle in the haystack."

References

1. Harding, M.W., Galat, A., Uehling, D.E., and Schreiber, S.L. (1989) A receptor for the immunosuppressant FK506 is a *cis-trans* peptidyl-prolyl isomerase. *Nature*, **341**, 758–760.
2. Schreiber, S.L. (1991) Chemistry and biology of the immunophilins and their immunosuppressive ligands. *Science*, **251**, 283–287.
3. Yoshida, M., Kijima, M., Akita, M., and Beppu, T. (1990) Potent and specific inhibition of mammalian histone deacetylase both in vivo and in vitro by trichostatin A. *J. Biol. Chem.*, **265**, 17174–17179.
4. Yoshida, M., Furumai, R., Nishiyama, M., Komatsu, Y., Nishino, N., and Horinouchi, S. (2001) Histone deacetylase as a new target for cancer chemotherapy. *Cancer Chemother. Pharmacol.*, **48** (Suppl 1), S20–S26.
5. Osada, H. (2010) Introduction of new tools for chemical biology research on microbial metabolites. *Biosci. Biotechnol. Biochem.*, **74**, 1135–1140.
6. Kato, N., Takahashi, S., Nogawa, T., Saito, T., and Osada, H. (2012) Construction of a microbial natural product library for chemical biology studies. *Curr. Opin. Chem. Biol.*, **16**, 101–108.
7. Osada, H. and Nogawa, T. (2012) Systematic isolation of microbial

metabolites for natural products depository (NPDepo). *Pure Appl. Chem.*, **84** (6), 1407–1420.

8. Kino, T., Hatanaka, H., Hashimoto, M., Nishiyama, M., Goto, T., Okuhara, M., Kohsaka, M., Aoki, H., and Imanaka, H. (1987) FK-506, a novel immunosuppressant isolated from a Streptomyces. I. Fermentation, isolation, and physicochemical and biological characteristics. *J. Antibiot.*, **40**, 1249–1255.
9. Yoshida, M., Nomura, S., and Beppu, T. (1987) Effects of trichostatins on differentiation of murine erythroleukemia cells. *Cancer Res.*, **47**, 3688–3691.
10. Ziegler, S., Pries, V., Hedberg, C., and Waldmann, H. (2013) Target identification for small bioactive molecules: finding the needle in the haystack. *Angew. Chem. Int. Ed.*, **52**, 2744–2792.
11. Futamura, Y., Muroi, M., and Osada, H. (2013) Target identification of small molecules based on chemical biology approaches. *Mol. BioSyst.*, **9**, 897–914.
12. Lomenick, B., Jung, G., Wohlschlegel, J.A., and Huang, J. (2011) Target identification using drug affinity responsive target stability (DARTS). *Curr. Protoc. Chem. Biol.*, **3**, 163–180.
13. Futamura, Y., Kawatani, M., Kazami, S., Tanaka, K., Muroi, M., Shimizu, T., Tomita, K., Watanabe, N., and Osada, H. (2012) Morphobase, an encyclopedic cell morphology database, and its use for drug target identification. *Chem. Biol.*, **19** (12), 1620–1630.
14. Muroi, M., Kazami, S., Noda, K., Kondo, H., Takayama, H., Kawatani, M., Usui, T., and Osada, H. (2010) Application of proteomic profiling based on 2D-DIGE for classification of compounds according to the mechanism of action. *Chem. Biol.*, **17** (5), 460–470.
15. Kanoh, N., Honda, K., Simizu, S., Muroi, M., and Osada, H. (2005) Photo-cross-linked small-molecule affinity matrix for facilitating forward and reverse chemical genetics. *Angew. Chem. Int. Ed.*, **44**, 3559–3562.
16. Kanoh, N., Kumashiro, S., Simizu, S., Kondoh, Y., Hatakeyama, S., Tashiro, H., and Osada, H. (2003) Immobilization of natural products on glass slides by using a photoaffinity reaction and the detection of protein-small-molecule interactions. *Angew. Chem. Int. Ed.*, **42**, 5584–5587.
17. Kimura, S., Ito, C., Jyoko, N., Segawa, H., Kuroda, J., Okada, M., Adachi, S., Nakahata, T., Yuasa, T., Filho, V.C., Furukawa, H., and Maekawa, T. (2005) Inhibition of leukemic cell growth by a novel anti-cancer drug (GUT-70) from calophyllum brasiliense that acts by induction of apoptosis. *Int. J. Cancer*, **113**, 158–165.
18. Kawatani, M., Takayama, H., Muroi, M., Kimura, S., Maekawa, T., and Osada, H. (2011) Identification of a small-molecule inhibitor of DNA topoisomerase II by proteomic profiling. *Chem. Biol.*, **18**, 743–751.
19. Zenitani, S., Tashiro, S., Shindo, K., Nagai, K., Suzuki, K., and Imoto, M. (2003) Gerfelin, a novel inhibitor of geranylgeranyl diphosphate synthase from Beauveria felina QN22047. I. Taxonomy, fermentation, isolation, and biological activities. *J. Antibiot.*, **56**, 617–621.
20. Kawatani, M., Okumura, H., Honda, K., Kanoh, N., Muroi, M., Dohmae, N., Takami, M., Kitagawa, M., Futamura, Y., Imoto, M., and Osada, H. (2008) The identification of an osteoclastogenesis inhibitor through the inhibition of glyoxalase I. *Proc. Natl. Acad. Sci. U.S.A.*, **105**, 11691–11696.
21. Sasazawa, Y., Kanagaki, S., Tashiro, E., Nogawa, T., Muroi, M., Kondoh, Y., Osada, H., and Imoto, M. (2012) Xanthohumol impairs autophagosome maturation through direct inhibition of valosin-containing protein. *ACS Chem. Biol.*, **7**, 892–900.

12
Activity-Based Proteasome Profiling in Medicinal Chemistry and Chemical Biology

Gerjan de Bruin, Nan Li, Guillem Paniagua, Lianne Willems, Bo-Tao Xin, Martijn Verdoes, Paul Geurink, Wouter van der Linden, Mario van der Stelt, Gijs van der Marel, Herman Overkleeft, and Bogdan Florea

12.1
Introduction

Activity-based protein profiling (ABPP) is a powerful technique to identify enzymatic activities and to study their functioning – and the effect of inhibitors on this – *in vitro*, *in situ*, and *in vivo*. In this chapter some case studies on activity-based profiling of mammalian proteasomes are discussed. Both direct and two-step bioorthogonal ABPP strategies and their merits are discussed, and the value of ABPP in the establishment of previously uncharted enzymatic activities and the direct visualization of inhibitor specificity are presented.

12.2
The Biological Problem

Proteasomes are the major cytosolic and nuclear protein degradation machineries and they are also responsible for the proteolysis of misfolded, ER-dislocated (endoplasmic reticulum) proteins [1–3]. Proteasomal protein turnover takes place in an ubiquitin-dependent manner. The proteasome-generated products – oligopeptides varying in length from 3 to up to 30 amino acid residues – are further processed by aminopeptidases. In higher vertebrates, antigenic peptides are selected from the peptide pool produced by proteasomes and downstream aminopeptidases for presentation on the outer cell surface by major histocompatibility class I (MHCI) protein complexes. In this way, proteasomes are essential factors in the detection and eradication of virally infected cells.

Proteasomes are expressed almost ubiquitously throughout the kingdoms of life, and the overall shape of the 20S core particles in which the proteolytic proteasome activities reside is highly conserved. Prokaryotic 20S proteasomes are C2-symmetrical barrel-shaped particles assembled in four stacked rings of seven proteins each (Figure 12.1). The two outer rings are composed of seven

Concepts and Case Studies in Chemical Biology, First Edition. Edited by Herbert Waldmann and Petra Janning.

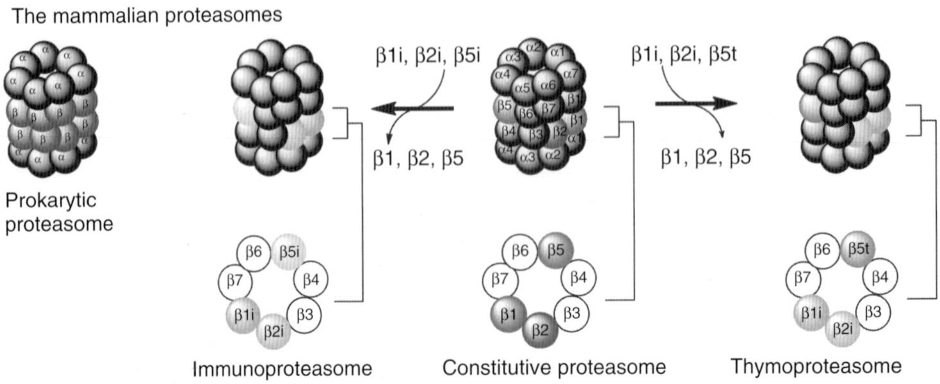

Figure 12.1 Evolution of proteasome 20S core particles.

identical α-subunits and the two inner rings contain seven identical β-subunits. The catalytic activity resides within the β-subunits, and each of these subunits contains an N-terminal threonine residue within its active site. In eukaryotic proteasomes, C2-symmetry is maintained, but each α-subunit within an outer ring has a unique sequence. The same holds true for the β-subunits; and three β-subunits out of the seven possess catalytic activity, each with a different substrate preference. Of these, β1 has a preference for processing peptide bonds containing an acidic amino acid residue at position 1 (P1 – the amino acid residue at the C-terminus of the oligopeptide product), β2 for a basic residue and β5 for a hydrophobic residue. Mammals express at least three distinct 20S core particles. Next to the ubiquitously expressed constitutive proteasome (having β1, β2, and β5 as catalytic residues), immune-competent cells can express immunoproteasomes. In the immunoproteasome 20S particles, β1, β2, and β5 are replaced by β1i, β2i, and β5i, respectively. The immunoproteasome catalytic activities display substrate preferences resembling that of their constitutive proteasome counterparts, yet subtle differences may be behind the specific contribution of immunoproteasomes to the generation of MHCI antigenic peptides. Recently, murine thymoid cortical epithelial cells were found to express thymoproteasomes, featuring next to β1i and β2i the unique β-particle, β5t. Hybrid 20S core particles containing both constitutive and immunoproteasome catalytic activities have been found in various tissues, adding to the complexity of proteasome biochemistry. Although harboring the catalytic activities, 20S core particles alone have limited physiological function, and cap structures, themselves composed of multiple proteins, associate with the α-rings to form fully functional proteasomes. Constitutive proteasome 20S particles associate with one or two 19S caps (to give 26S and 30S proteasomes, respectively). Substrate recognition (poly-ubiquitylated proteins), substrate unfolding, removal of the ubiquitin signaling tags, and ATP-dependent channeling of the unfolded polypeptides to the inner 20S core particles are executed by these 19S caps.

Immunoproteasomes associate with 11S caps, whereas the cap structures of thymoproteasomes have not been conclusively established yet.

Besides their fundamental interest, proteasomes are relevant therapeutic targets. The peptide boronic acid, bortezomib (Velcade, Figure 12.2) is the first proteasome inhibitor to have reached the clinic and is used to treat late-stage multiple myeloma [4]. Originally developed as a β5-specific inhibitor, it was later on found to target also β1, β5i, and β1i and, moreover, it became clear that exclusive inhibition of β5 is not sufficiently effective for tumor eradication. In recent years, a number of structurally distinct compounds targeting the proteasome have reached the clinic, amongst others the covalent and irreversible inhibitor carfilzomib [5]. The structure of carfilzomib is based on that of the natural product, epoxomicin, that also features the epoxyketone electrophilic trap. Indeed, numerous natural product proteasome inhibitors with a distinguishing electrophile grafted onto a peptidic core have been described over the years, including lactacystin, syringolin A (SylA), and fellutamide B. An important class of synthetic covalent proteasome inhibitors is represented by the peptide vinyl sulfones, whereas numerous noncovalent proteasome inhibitors have been discovered as well (e.g., TMC 95A).

Current research on proteasome inhibitors focuses on compounds that are truly selective for one of the seven catalytic activities: β1, β2, β5, β1i, β2i, β5i, or β5t. Such compounds would be highly valuable to downregulate in a chemical genetics setting one individual catalytic site without hampering the structural integrity of the 20S core particles. Assessment of proteasome activities is however hampered by the fact that the individual catalytic activities are only active in the context of the 20S core particles. In fact, the N-terminal threonine residues in the active sites are formed after assembly of the 20S particles by autocleavage of N-terminal extended peptides, and isolating a single catalytic β-subunit renders it inactive. Fluorogenic substrates that are designed to report on a specific β-subunit are often used but it can never be excluded that the production of fluorescence is caused by a number of catalytic β-subunits. It is for this reason that activity-based proteasome profiling has become a prominent technique to establish and assess proteasome activities.

12.3 The Chemical Approach

Activity-based probes (ABPs) are tagged covalent and irreversible enzyme inhibitors. Formation of a stable covalent bond ensures that the inhibitor will remain attached to the polypeptide after protein denaturation, after which the tag (radio-isotope, biotin, fluorophore) allows visualization and/or identification of the thus modified enzyme or enzyme family. The first proteasome ABP described comprised a tritium-labeled lactacystin analog [6, 7]. Proteasome bands are visualized on one-dimensional SDS-PAGE (sodium docecylsulfate-polyacrylamide gel electrophoresis) gel in an autoradiogram after treatment with a radiolabeled

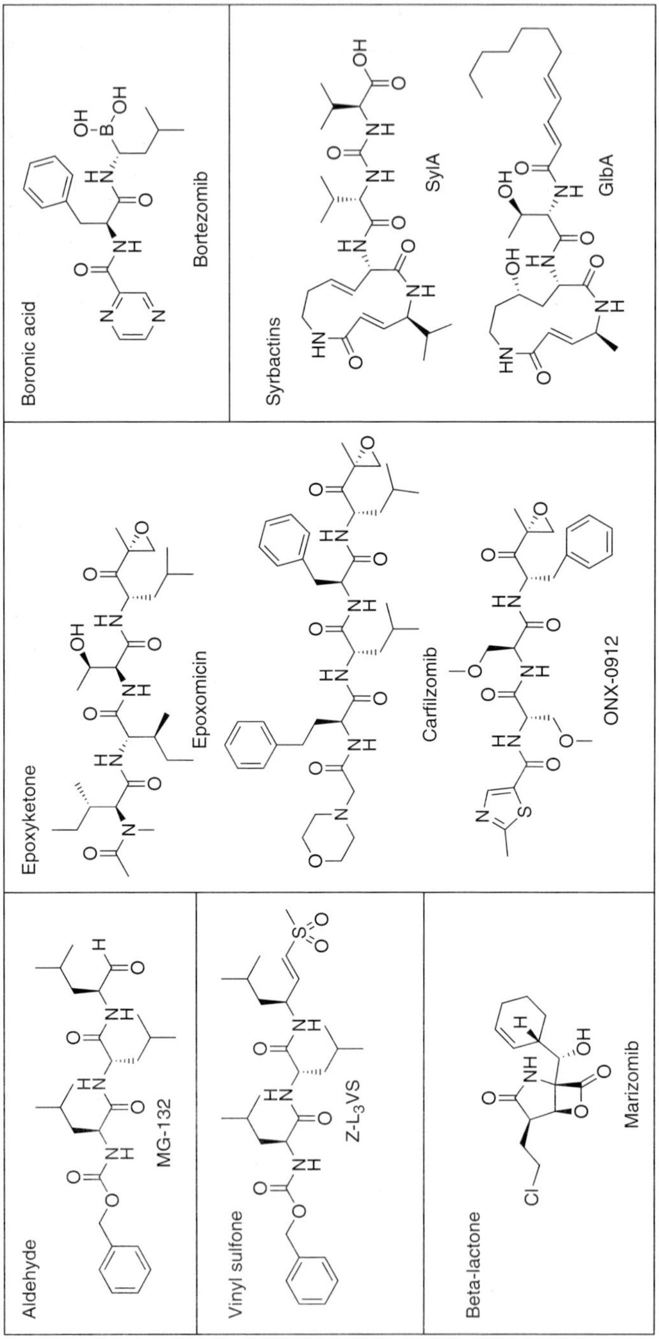

Figure 12.2 Relevant proteasome inhibitors: compound classes and clinical drugs.

covalent and irreversible inhibitor and for this purpose next to ^{3}H-lactacystin also ^{125}I-peptide vinyl sulfones have been employed in the past [8]. Conceptually related but less complicated from a practical point of view is the use of fluorescent ABPs. Biotinylated probes allow – next to in-gel detection – also for streptavidin-mediated pull down after which the modified proteins are identified by trypsinolysis and mass spectrometry analysis of the tryptic fragments. Proteasome inhibitors extended with a biotin or a fluorophore tend to have limited proteasome subunit specificity, most likely because, owing to their larger size, they resemble more closely actual proteasome substrates – polypeptides – than do small oligopeptide-based inhibitors. For the purpose of activity-based profiling of individual proteasome subunits, bioorthogonal chemistry is therefore the method of choice. In this section, the various ABPP techniques are exemplified, after which in the next section their application to some proteasome-related biological problems is discussed.

12.3.1 Comparative and Competitive Activity-Based Proteasome Profiling

Most of the proteasome ABPs reported are based on either peptide vinyl sulfones or peptide epoxyketones. In Figure 12.3a, the mechanism by which N-terminal threonines within proteasome active sites catalyze peptide bond hydrolysis is depicted. Aligning of either a vinyl sulfone or an epoxyketone at the appropriate position (i.e., the carbonyl of the scissile amide bond) allows covalent and irreversible reaction with the N-terminal threonine-OH (and $-NH_2$ in case of epoxyketones) (Figure 12.3b) and therefore employment of these electrophiles in ABP design. Of the two electrophiles, especially the epoxyketone – evolved in nature – is an intriguing electrophilic trap: it presents two electrophilic carbons to the 1,2-aminoalcohol characteristic and almost unique for proteasome active sites.

Treatment of either a tissue culture or a cell extract with Bodipy-epoxomicin MVB-003 followed by SDS-PAGE readily reveals proteasome active sites (Figure 12.3c) [9, 10].[1)] In case the constitutive proteasome is expressed exclusively, the three catalytic species β1, β2, and β5 are readily resolved. In case the treated tissue also expresses immunoproteasomes, the one-dimensional gel will resolve β1 and β2i, but only partially β1, β1i, β5, and β5i (see for a detailed experimental protocol Box 12.1 [11]). Two-dimensional gel electrophoresis, by which proteins are separated on the basis of the charge followed by mass allows for complete resolution of all six catalytic activities. Figure 12.3d represents an example of competitive ABPP. In this experiment, tissue or tissue extract is treated first with a prospective inhibitor. Ensuing incubation with

1) In this chapter, cartoons of representative SDS-PAGE gels are shown. Original data can be found in referenced articles.

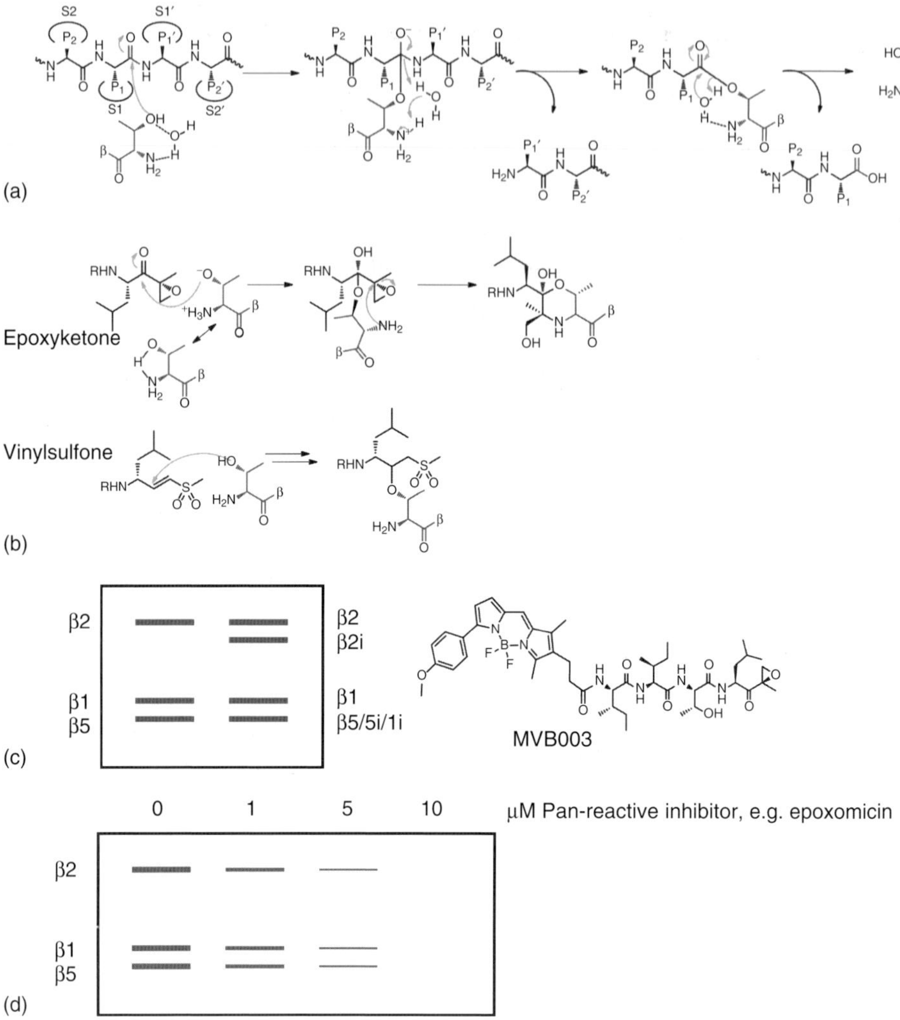

Figure 12.3 *Peptide vinyl sulfones and peptide epoxyketones in activity-based proteasome profiling.* (a) Catalytic mechanism of proteasome-mediated peptide bond cleavage. (b) Covalent adducts of proteasomes reacting with vinyl sulfones and epoxyketones. (c) One-dimensional SDS-PAGE of human constitutive proteasomes and immunoproteasomes labeled with Bodipy-epoxomicin X [9]. (d) Competitive activity-based proteasome profiling reveals a pan-reactive inhibitor.

a fluorescent ABP followed by SDS-PAGE and fluorescence scanning reveals those catalytic subunits not targeted or partially targeted by the inhibitor. In other words, the potency of a candidate-proteasome inhibitor is revealed by the amount (compared to the nontreated sample) by which ABP labeling disappears.

Box 12.1 Tagging and Resolving by One- and Two-dimensional SDS-PAGE of Proteasomes by Activity-based Protein Profiling

A protocol for proteasome labeling *in vitro* and *in situ* with broad-spectrum biotin- or Bodipy-containing peptide vinyl sulfones or peptide epoxyketones.

1) a. *In situ labeling*: Grow cells till mid log-phase. Incubate the cells with inhibitor or ABP. After incubation, lyse the cells with a mild buffer and determine the protein concentration in the sample. Continue with step 2.
   b. *In vitro labeling*: Label the proteasome in the cell extract with ABP by 1 h incubation at 37 °C. Continue with step 2.
2) *In situ* and *in vitro*:
   a. *One-dimensional SDS-PAGE*: Denature proteins and separate them on SDS-PAGE. In case of a fluorescent probe, it can be directly visualized on a fluorescence scanner. If a biotinylated probe has been used, proteins can be detected by streptavidin-HRP on Western blot.
   b. *Two-dimensional SDS-PAGE*: The denatured proteins are separated on the first-dimension isoelectric focusing (separating the proteins by pK_a), followed by second-dimension SDS-PAGE (size-based separation). The visualization procedure is the same as described earlier.

The above-mentioned examples focus on proteasomes as established targets in inhibitor studies. ABPP is however also a highly useful strategy to establish the nature of biological targets. ABPP was used in an early study to identify proteasomes as the proteins modified by the natural product, epoxomicin. By adding a biotin moiety to the peptide, epoxyketone proteasomes were readily identified after incubation of cell extracts, SDS-PAGE resolution, and Western blotting. In the same vein, in more recent years plant proteasomes were identified as the targets of SylA, a macrolactam Michael acceptor produced by plant pathogenic bacteria [12].

12.3.2 Two-Step Activity-Based Proteasome Profiling

During research on activity-based proteasome probes it became clear that proteasomes do not only allow, but in fact often favor inhibitors equipped with a biotin, fluorophore, or even both compared to their nontagged counterparts. This general finding can be rationalized by the realization that oligopeptidic proteasome inhibitors resemble in structure more closely proteasome products than substrates, and that elongation may endow some substrate-like character. However this may be, it makes direct ABPP of proteasomes sometimes complicated in establishing subunit selectivity of a prospective proteasome inhibitor. A case in point is peptide vinyl sulfone **1** (Figure 12.4a) [13]. This compound was designed on the basis of a position scanning experiment in which

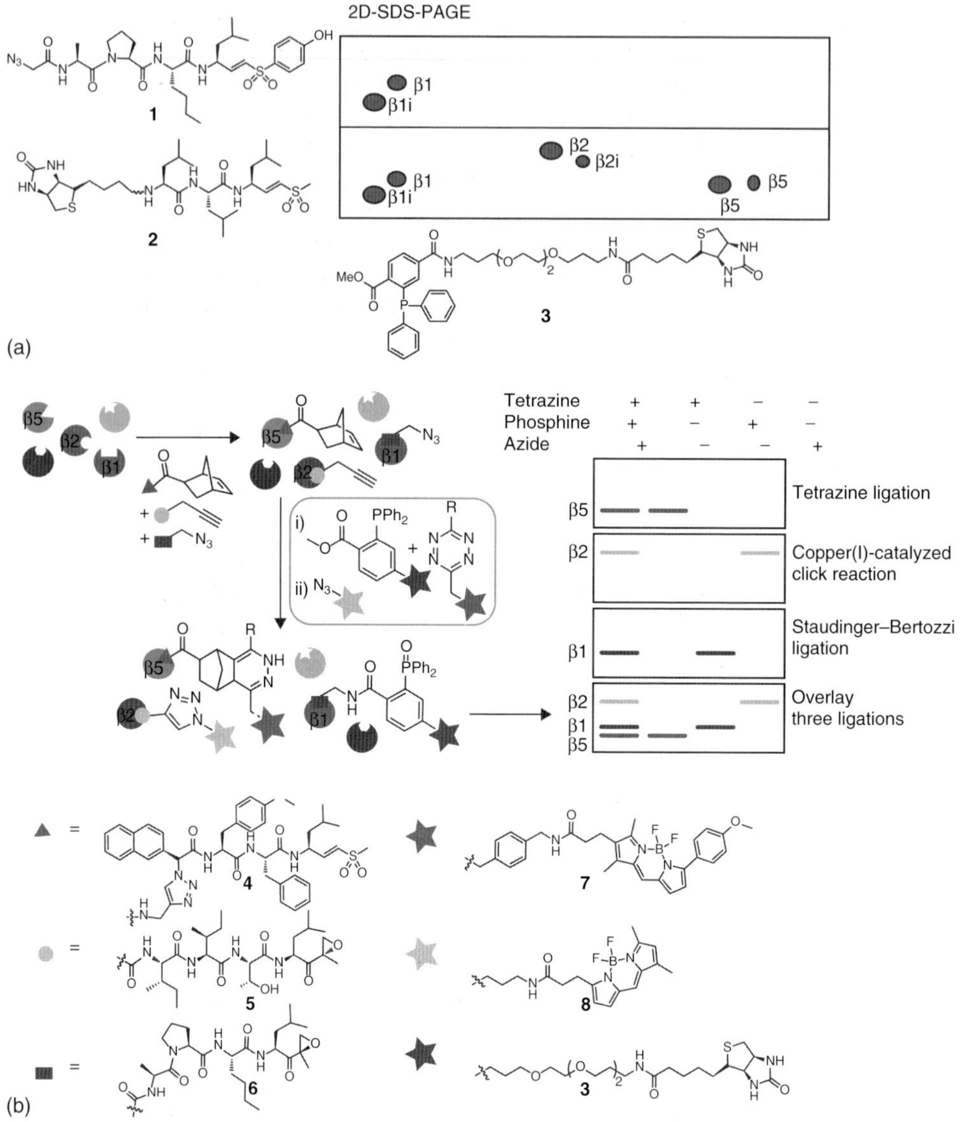

Figure 12.4 *Two-step bioorthogonal activity-based proteasome profiling.* (a) Determining the β1/β1i selectivity of peptide vinyl sulfone X featuring a proline residue at P3. (b) Independent labeling of β1, β2, and β5 in one sample using a multiplexing bioorthogonal chemistry strategy.

a library of peptide aldehydes was screened on proteasome inhibitory activity in a fluorogenic peptide substrate assay. It appeared that the combination of Leu at P1 and Pro at P3 conferred β1/β1i-specificity. However, one can never be sure which catalytic activities contribute to fluorogenic substrate hydrolysis. Direct attachment of either a biotin or a fluorophore to **1** (for instance, by bioconjugation to the N-terminal azide) delivered a proteasome probe that next to β1/β1i also targeted β5/β5i. The azide in **1** can, however, also be employed for bioorthogonal chemistry as a means to install a biotin or fluorophore after proteasome inhibition. Treatment of tissue extract expressing both constitutive proteasome and immunoproteasome with azide-modified peptide vinyl sulfone **1** was followed by incubation with biotin-phosphane **3**. Staudinger–Bertozzi ligation introduced biotin to those proteasome active sites covalently modified by ABP **1**. These were visualized by two-dimensional SDS-PAGE and the β1/β1i subunit specificity of **1** was evidenced by the appearance of only those proteins corresponding to β1/β1i, in comparison with, for instance, the six catalytic sites visualized by two-step ABPP using pan-reactive probe **2**.

Next to the Staudinger–Bertozzi ligation, a number of mechanistically distinct bioorthogonal reactions have been proposed over the past decades. These include hydrazone formation (from a ketone and a hydrazine), azide-alkyne [2+3] Huisgen cycloaddition (both in Cu(I)-catalyzed and Cu-free strain-promoted azide-alkyne cycloaddition form) and Diels–Alder ligations. Combining these bioorthogonal ligations with subunit-specific two-step proteasome ABPs allowed the demonstration that these reactions are also orthogonal with respect to each other (see Figure 12.4b and Box 12.2 for a detailed description) [14].

Box 12.2 Three Bioorthogonal Ligations in one Sample

Experimental protocol for Staudinger–Bertozzi, Cu(I)-catalyzed Huisgen alkyne-azide cycloaddition, and reverse-electron-demand Diels–Alder ligation to distinguish between β1, β2, and β5.

The simultaneous labeling of three individual enzymatic activities can be done by combining Staudinger–Bertozzi ligation, Cu(I)-catalyzed Huisgen alkyne-azide cycloaddition, and inverse-electron-demand Diels–Alder ligation in one sample, given that selective inhibitors equipped with one of three different tags are accessible. In case of the three catalytically active proteasome subunits, only selective inhibitors for β1 and β5 are available. However, after complete blocking of these subunits, the third subunit β2 can readily be targeted by a (pan-reactive) proteasome inhibitor. Hence, in a typical triple proteasome labeling experiment, cell extracts are first treated with β1-selective azide-tagged inhibitor **6** and β5-selective inhibitor **4** with a norbornene moiety using a concentration that gives complete labeling of these subunits, after which alkyne-modified inhibitor **5** is added to target β2. Having modified all three subunits, the bioorthogonal ligations can now be performed in a two-step sequence, which is necessary owing to the instability of the tetrazine reagent under click conditions and also to avoid cross-reactivity of phosphine **3** with the fluorescent azide probe **8**. Thus, first Staudinger–Bertozzi ligation and inverse-electron-demand Diels–Alder ligation are performed using biotin-phosphine **3** and red fluorescent tetrazine **7**, after which buffer exchange to a Cu(I) click buffer allows click reaction with green fluorescent azide **8**. The proteins are resolved by SDS-PAGE and visualized by fluorescence imaging followed by biotin-streptavidin Western blotting.

12.4 Biological Research/Evaluation

The ubiquitin-proteasome system (UPS), of which proteasomes form the downstream part, is highly complex and involved in numerous physiological processes in health and disease. Tools able to modulate specific events in the controlled, UPS-mediated cytosolic and nuclear protein turnover, and that at the same time identify the modulated factor, have proved to be highly useful in biochemical and cell biological studies. Proteasome ABPs are routinely used to establish proteasome activity in a given physiological setting and to determine the relative activities of the active sites present. At a first glance, one would expect a 1 : 1 : 1 ratio between β1, β2, and β5 – they are present in a 1 : 1 : 1 stoichiometric ratio in constitutive proteasome 20S core particles and inactive outside these assemblies – but this appears not the case. Moreover, relative activities appear to differ going from one tissue to the next, even if from the same organism. The situation becomes more complex in tissue expressing both constitutive proteasomes and immunoproteasomes, especially when considering that hybrid structures composed of both types of active sites likely exist. The fact that proteasome ABPs report on the number of active sites functional at a given point of time, rather than substrate turnover by the combined pool, has implications for their potential use. It becomes more and more clear that proteasome functioning is controlled by many factors. These include, next to factors involving the associating caps (19S, 11S), most likely posttranslational modifications (phosphorylation, acylation, *O*-GlcNAcylation) and perhaps events within the 20S core particles as well. Figure 12.5b, for instance, depicts a representative competitive ABPP experiment involving β1, β2, and β5 selective inhibitors. As can be seen, β2 and β5 labeling – and by this virtue activity – increases significantly (up to 50%) when partially or completely blocking β1. Similar increased labeling of nonblocked subunits is found when β2 and β5 are blocked selectively.

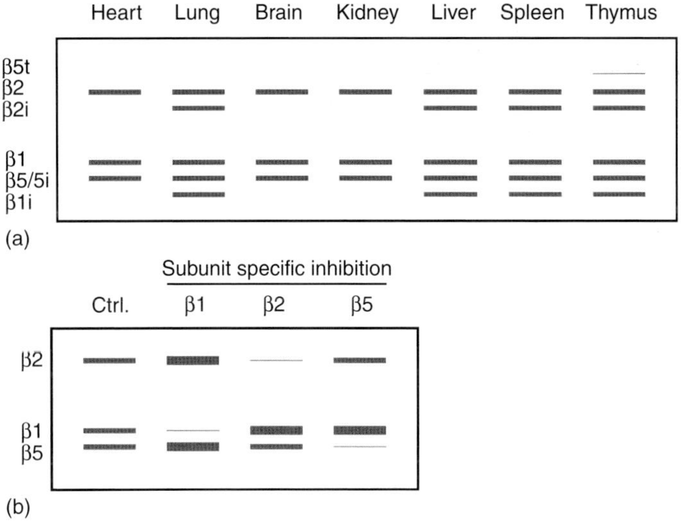

Figure 12.5 *Relative activity of the different active sites.* (a) Proteasome labeling of different mouse organs with MVB003. (b) Subunit specific inhibition result is increased activity of non-inhibited subunits.

12.4.1
Identification of Proteasome Active Sites

One of the advantages of ABPP as a means to identify or quantify proteins in a proteomics experiment is that ABPs report on active enzymes, and not on mere protein expression levels. This intrinsic property makes ABPs highly useful discovery tools both to identify previously unknown enzyme activities and to establish whether a putative enzyme is in fact reactive or not. An illustration of the latter comprises the unambiguous establishment of the reactivity of the thymoproteasome-specific β-subunit, β5t. Following the discovery of the existence of thymoproteasomes in cortical epithelial cells in the thymus, the question arose how it would contribute to T-cell selection. In the thymus, $CD4^+$ T cells that should distinguish between MHCI molecules presenting self-peptides and those presenting foreign (pathogenic) peptides are produced. During negative T-cell selection, T cells recognizing self-peptides are eradicated. In positive T-cell selection, T cells expressing T-cell receptors able to bind to MHC–peptide complexes are generated, and it is in this process that thymoproteasomes are involved. One question that arose upon discovery of the thymoproteasome-specific subunit β5t was the issue whether it possesses catalytic activity (and, if so, whether it would deviate from that of β5/β5i) or whether it would be inactive. This issue was resolved by means of ABPP, as is outlined in Figure 12.5a [15].

In the first experiment, treatment of murine thymus extracts with fluorescent peptide epoxyketone MVB-003 followed by SDS-PAGE revealed, next to the expected bands denoting the catalytic residues of constitutive proteasomes and immunoproteasomes, a seventh, relatively faint band with comparatively higher molecular weight. Literature data had already revealed that β5t is comparatively larger in size compared to β5/β5i and indeed is the largest of the murine β-subunits. Next to this, β5t is only expressed in a small subset of thymal cells, which is reflected in the relatively faint band. Although suggestive of β5t activity, these results are however not conclusive. Unambiguous proof was obtained by performing a pull-down experiment using biotin-tagged epoxomicin [16] followed by trypsinolysis and mass spectrometry analysis of the tryptic fragments. In this way (see Box 12.3 for a typical workflow), the tryptic fragment corresponding to the β5t active site fragment covalently attached to epoxomicin-based probe could be identified. Although the reactivity of β5t to ABP MVB-003 does not necessarily mean that it is capable of processing polypeptides, it is highly likely that this is in fact the case: only those β-subunits involved in protein processing are normally found to react with ABPs.

Box 12.3 Chemical Proteomics to Establish Enzymatic Activity

Workflow for the identification of the thymoproteasome β5t active site peptide.

1) Cell treatment and lysis (as described in Box 1.1)
2) ABP reaction (as described in Box 1.1)
3) C/M precipitation
   Remove the excess of probe by chloroform/methanol precipitation.
4) Reduction/alkylation.
   Denature the proteins, and open disulfide bond by DTT (dithiothreithol), and then alkylate the cysteines with iodoacetamide.
5) Pull down with paramagnetic beads
6) Wash away specific binding proteins
7) On-bead digestion by trypsin
8) Elution of the active site peptides
   After on-bead digestion, incubate beads with high-concentration biotin to release the active site peptide from the beads.
9) Desalting and LC/MS (liquid chromatography coupled to mass spectrometry) analysis
   Desalt the peptides by C18 column, and then analyze them by LC/MS. An MS^3 protocol is necessary for this work.

12.5 Conclusions

In summary, activity-based proteasome profiling has evolved from a playground for bioorganic chemistry to a mature field contributing to our understanding of proteasome biochemistry and clinical research on – and targeting of – proteasomes. A plethora of subunit-selective and pan-reactive proteasome inhibitors and ABPs are available to establish activities of individual active sites. It is now clear that proteasome subunit specificities vary throughout the kingdoms of life. The individual activities can be readily assessed by making use of the probes in a cross-species screening setting and the results may well be capitalized upon in combating, for instance, infectious diseases. Activity-based proteasome profiling has, on the other hand, also been a fruitful field in which new chemical biology tools and techniques are developed. Proteasomes were amongst the first enzymes targeted in two-step bioorthogonal ABPP experiments [17] and more recently also served as readout in the first triple bioorthogonal chemistry experiment. In this way, proteasomes are useful targets in the development of chemical biology research strategies, and in return chemical biology approaches help in establishing proteasome activities in health and disease and in determining their relevance as drug targets.

References

1. Kisselev, A.F., van der Linden, W.A., and Overkleeft, H.S. (2012) Proteasome inhibitors: an expanding army attacking a unique target. *Chem. Biol.*, **19**, 99–115.
2. Beck, P., Dubiella, C., and Groll, M. (2012) Covalent and non-covalent reversible proteasome inhibition. *Biol. Chem.*, **393** (10), 1101–1120.
3. Tanaka, K. (2009) The proteasome: overview of structure and functions. *Proc. Jpn. Acad. Ser. B Phys. Biol. Sci.*, **85**, 12–36.
4. Adams, J., Behnke, M., Chen, S., Cruickshank, A.A., Dick, L.R., Grenier, L., Klunder, J.M., Ma, Y.-T., Plamondon, L., and Stein, R.L. (1998) Potent and selective inhibitors of the proteasome: Dipeptidyl boronic acids. *Bioorg. Med. Chem. Lett.*, **8**, 333–338.
5. Demo, S.D., Kirk, C.J., Aujay, M.A., Buchholz, T.J., Dajee, M., Ho, M.N., Jiang, J., Laidig, G.J., Lewis, E.R., Parlati, F., Shenk, K.D., Smyth, M.S., Sun, C.M., Vallone, M.K., Woo, T.M., Molineaux, C.J., and Bennett, M.K. (2007) Antitumor activity of PR-171, a novel irreversible inhibitor of the proteasome. *Cancer Res.*, **67**, 6383–6391.
6. Craiu, A., Gaczynska, M., Akopian, T., Gramm, C.F., Fenteany, G., Goldberg, A.L., and Rock, K.L. (1997) Lactacystin and clasto-lactacystin β-lactone modify multiple proteasome β-subunits and inhibit intracellular protein degradation and major histocompatibility complex class I antigen presentation. *J. Biol. Chem.*, **272**, 13437–13445.
7. Fenteany, G., Standaert, R.F., Lane, W.S., Choi, S., Corey, E.J., and Schreiber, S.L. (1995) Inhibition of proteasome activities and subunit-specific amino-terminal threonine modification by lactacystin. *Science*, **268**, 726–731.
8. Bogyo, M., McMaster, J.S., Gaczynska, M., Tortorella, D., Goldberg, A.L., and Ploegh, H. (1997) Covalent modification of the active site threonine of proteasomal β subunits and the Escherichia coli homolog HslV by a new class of inhibitors. *Proc. Natl. Acad. Sci. U.S.A.*, **94**, 6629–6634.
9. Verdoes, M., Willems, L.I., van der Linden, W.A., Duivenvoorden, B.A., van der Marel, G.A., Florea, B.I., Kisselev, A.F., and Overkleeft, H.S. (2010) A panel of subunit-selective activity-based proteasome probes. *Org. Biomol. Chem.*, **8**, 2719–2727.
10. Sin, N., Kim, K.B., Elofsson, M., Meng, L., Auth, H., Kwok, B.H.B., and Crews, C.M. (1999) Total synthesis of the potent proteasome inhibitor epoxomicin: a useful tool for understanding proteasome biology. *Bioorg. Med. Chem. Lett.*, **9**, 2283–2288.
11. Li, N., Kuo, C.-L., Paniagua, G., van den Elst, H., Verdoes, M., Willems, L.I.,

van der Linden, W.A., Ruben, M., van Genderen, E., Gubbens, J., van Wezel, G.P., Overkleeft, H.S., and Florea, B.I. (2013) Relative quantification of proteasome activity by activity-based protein profiling and LC-MS/MS. *Nat. Protocols*, **8**, 1155–1168.

12. Kolodziejek, I., Misas-Villamil, J.C., Kaschani, F., Clerc, J., Gu, C., Krahn, D., Niessen, S., Verdoes, M., Willems, L.I., Overkleeft, H.S., Kaiser, M., and van der Hoorn, R.A.L. (2011) Proteasome activity imaging and profiling characterizes bacterial effector syringolin A. *Plant Physiol.*, **155**, 477–489.
13. van Swieten, P.F., Samuel, E., Hernández, R.O., van den Nieuwendijk, A.M.C.H., Leeuwenburgh, M.A., van der Marel, G.A., Kessler, B.M., Overkleeft, H.S., and Kisselev, A.F. (2007) A cell-permeable inhibitor and activity-based probe for the caspase-like activity of the proteasome. *Bioorg. Med. Chem. Lett.*, **17**, 3402–3405.
14. Willems, L.I., Li, N., Florea, B.I., Ruben, M., van der Marel, G.A., and Overkleeft, H.S. (2012) Triple bioorthogonal ligation strategy for simultaneous labeling of multiple enzymatic activities. *Angew. Chem. Int. Ed.*, **51**, 4431–4434.
15. Florea, B.I., Verdoes, M., Li, N., van der Linden, W.A., Geurink, P.P., van den Elst, H., Hofmann, T., de Ru, A., van Veelen, P.A., Tanaka, K., Sasaki, K., Murata, S., den Dulk, H., Brouwer, J., Ossendorp, F.A., Kisselev, A.F., and Overkleeft, H.S. (2010) Activity-based profiling reveals reactivity of the murine thymoproteasome-specific subunit β5t. *Chem. Biol.*, **17**, 795–801.
16. Meng, L., Mohan, R., Kwok, B.H.B., Elofsson, M., Sin, N., and Crews, C.M. (1999) Epoxomicin, a potent and selective proteasome inhibitor, exhibits in vivo antiinflammatory activity. *Proc. Natl. Acad. Sci. U.S.A.*, **96**, 10403–10408.
17. Ovaa, H., van Swieten, P.F., Kessler, B.M., Leeuwenburgh, M.A., Fiebiger, E., van den Nieuwendijk, A.M.C.H., Galardy, P.J., van der Marel, G.A., Ploegh, H.L., and Overkleeft, H.S. (2003) Chemistry in living cells: detection of active proteasomes by a two-step labeling strategy. *Angew. Chem. Int. Ed.*, **42**, 3626–3629.

13
Rational Design of Activity-Based Retaining β-Exoglucosidase Probes

Kah Yee Li, Wouter Kallemeijn, Jianbing Jiang, Marthe Walvoort, Lianne Willems, Thomas Beenakker, Hans van den Elst, Gijs van der Marel, Jeroen Codée, Hans Aerts, Bogdan Florea, Rolf Boot, Martin Witte, and Herman Overkleeft

13.1
Introduction

Activity-based protein profiling (ABPP) is one of the most visible areas of research in chemical biology where organic chemistry plays an essential role. Activity-based probes (ABPs) have been developed for numerous serine hydrolases, cysteine proteases, and threonine hydrolases (see also Chapter 12), but less frequently for other enzyme families. This chapter details the successful development and application of a number of activity-based retaining β-exoglucosidase probes. The design principles of these probes can serve as a blueprint for the development of ABPs aimed at various retaining glycosidase families, next to exoglycosidases and also endoglycosidases.

13.2
The Biological Problem

Biochemical and biological research on carbohydrates and glycoconjugates, their structure, and their function is complicated. Carbohydrates and glycoconjugates often exist only transiently, are heterogeneous in structure, and their biosynthesis is only indirectly controlled by the genetic code. The combined action of glycosyl transferases and glycosidases – enzymes that create and break glycosidic linkages, respectively – in conjunction with substrate levels determine the nature of the eventual carbohydrate structures. Therefore, the nature of the pool of carbohydrates and glycoconjugates present in a given organism (termed a *glycome*) can, in contrast to proteins and nucleic acids, not be extracted from the genetic material of this organism.

Studies toward the glycome are further complicated by its structural complexity. The chemical space covered by carbohydrate-containing compounds is vast. Compared to their biopolymer counterparts, nucleic acids and peptides/proteins,

Concepts and Case Studies in Chemical Biology, First Edition. Edited by Herbert Waldmann and Petra Janning.

which are both synthesized from a relatively small set of building blocks, carbohydrates and glycoconjugates are constructed from a large variety of monosaccharide building blocks. A limited set of monosaccharides (predominantly D-glucose, D-mannose, D-galactose, D-glucuronic acid, D-xylose, D-ribose, L-idose, D-neuraminic acid) is used to construct the glycome in humans, but, for instance, bacterial glycomes contain up to hundreds of monosaccharides differing in stereochemistry and functional group pattern. Monosaccharide building blocks can, and are, interconnected through glycosidic bonds to various positions of the core of other monosaccharides forming oligomeric structures, called *oligosaccharides* and *polysaccharides* (biomolecules composed of monosaccharide building blocks exclusively). Hybrid biomolecules composed of carbohydrates and lipids (glycolipids), carbohydrates, and peptides (glycopeptides), as well as glycoconjugates involving other biomolecules, also exist. Nucleic acids and amino acids are linked through achiral linkages (phosphodiester bonds and amide bonds, respectively), whereas the glycosidic linkages that make up oligosaccharides and glycoconjugates involve a chiral (anomeric) carbon center, increasing the structural complexity even further.

The structural complexity, together with the fact that the glycome is nontemplated encoded, limits the use of molecular biology techniques, and therefore other means of studying the glycome are often employed. One attractive and often-used strategy is to study the glycome by perturbation, which can be achieved by manipulating the corresponding glycoprocessing enzymes, the glycosyl transferases, and glycosidases. A host of natural and synthetic (fluorogenic) substrates and inhibitors that act on, predominantly, glycosidases [1] exist. With these, the activity of a given glycosidase can be monitored (fluorogenic substrate) or inhibited, yet direct insight in the presence and/or nature of a glycoprocessing enzyme in a biological sample cannot be established directly and unambiguously. ABPP does provide the means to do so but requires that suitable ABPs are available. This chapter discusses how, by rational design, suitable ABPs for retaining β-exoglucosidases can be designed.

13.3
The Chemical Approach

In designing ABPs for a specific enzyme/class of enzymes, both the nature of the substrate and the mechanism employed by the enzyme are taken into consideration. Ideally, the enzyme of interest forms a covalently bound enzyme-substrate intermediate at some point of the catalytic cycle. Analysis of such a covalent intermediate allows the design of a mechanism-based inhibitor, normally a substrate analog that undergoes part of the catalytic process as if it were a substrate, only to get stuck at the covalent intermediate stage because this covalent intermediate is (much) more stable than that of the corresponding enzyme-substrate adduct. This strategy has met with most success in the design of ABPs for hydrolytic enzymes, in particular serine hydrolases, cysteine proteases, and threonine proteases

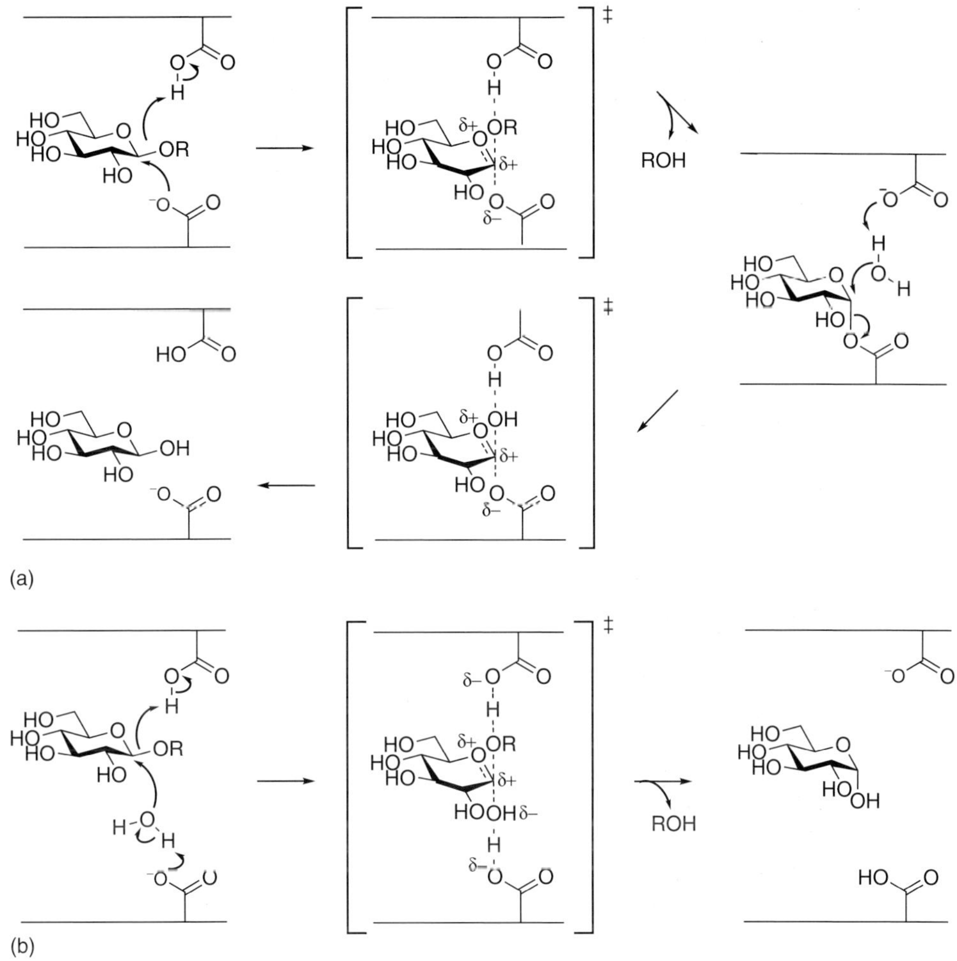

Figure 13.1 Mechanism of (a) retaining and (b) inverting β-glucosidases.

(proteasomes, see also Chapter 12). Glycosidases are a large family of hydrolytic enzymes that hydrolyze the acetal linkages that characterize oligosaccharides and glycoconjugates to form hemiacetal linkages [2]. Mechanistic studies on glycosidases revealed that many, although not all, enzymes from the glycosidase family develop covalent intermediates during glycosidase action as is exemplified for the β-exoglucosidases (Figure 13.1). In theory, such glycosidases that do form a covalent adduct are amenable to active site labeling and therefore ABPP.

Figure 13.1a depicts the classical double displacement mechanism proposed originally by Koshland as employed by *retaining* β-exoglucosidases [3]. The enzyme active site contains two carboxylic acid (aspartate or glutamate) residues: a general acid/base (carboxylic acid at the onset of enzyme catalysis) situated above the β-glucose substrate and a nucleophile (carboxylate) situated below.

Upon binding, the substrate adopts a $^{1}S_{3}$ skew-boat conformation [4]. This distorted conformation positions the aglycon in a pseudoaxial position and aligns the σ* orbital of the acetal linkage to the nucleophilic carboxylate residue, and minimizes steric hindrance by H3 and H5. After protonation of the aglycon (exocyclic oxygen of the acetal linkage) and expulsion of the leaving group, a transient oxocarbenium ion is formed, with concomitant flattening of the pyranose ring to the $^{4}H_{3}$ half-chair. This putative intermediate is trapped by the carboxylate nucleophile to form the covalently linked glucosyl-enzyme adduct with inversion of configuration at the anomeric center (alpha product). In the second step of the catalytic process, the covalent adduct is hydrolyzed via a similar oxocarbenium ion transition state to produce β-glucose with overall retention of configuration. Figure 13.1b depicts an alternative mechanism employed by *inverting* β-exoglucosidases. Although these enzymes are quite similar to *retaining* β-exoglucosidases with respect to their substrate (β-glucosides) and the composition of their enzyme active site, the overall stereochemical outcome of the hydrolysis is different: net inversion versus net retention. From a structural point of view, the two catalytic carboxylates in *inverting* β-exoglucosidases are positioned more distal (8-9 Å compared to the 4-5 Å observed in *retaining* β-exoglucosidases) and a water molecule positioned below the scissile acetal linkage can now be accommodated within the enzyme active site. The substrate binds in a distorted $^{2}S_{0}$ skew-boat and upon protonation of the aglycon, as before, the developing oxocarbenium ion can now be trapped directly by water, itself deprotonated by the alpha-carboxylate in the process, to yield alpha-glucose. From the viewpoint of ABPP, the lack of a covalent intermediate makes the development of activity-based inverting glycosidase probes rather complicated, much more so than is the case for retaining glycosidases. Although not the subject of this chapter, it should be noted that the development of probes to label enzymes that do not form a covalent enzyme–substrate intermediate mostly relies on photoreactive groups (such probes are often referred to as *photoactivatable affinity-based probes*) and a recent report describes the design of such a probe based on the competitive inhibitor, deoxynojirimycin equipped with a photoactivatable aryl azide and a bioorthogonal tag for probing inverting glycosidase activities [5].

As stated, inhibitors that proceed through the catalytic process, but form a long-lived covalent intermediate, are good leads for ABP development. With respect to retaining β-exoglucosidases, two compound classes that meet this requirement have been studied in detail in the past decades: 2-deoxy-2-fluoroglucosides (Figure 13.2a) and cyclitol epoxides (Figure 13.2b) [6, 7]. Substitution of the 2-hydroxyl by an electron-withdrawing fluorine, as in compound **1**, results in the formation of an enzyme-glucoside adduct that is comparatively more stable than that formed from the natural substrate, because the 2-deoxy-2-fluoroglucoside is comparatively less able to sustain a developing positive charge that accompanies hydrolysis of the enzyme-substrate adduct. This feature, formation of the oxocarbenium ion, is also inherent to the first step of the catalytic cycle and the

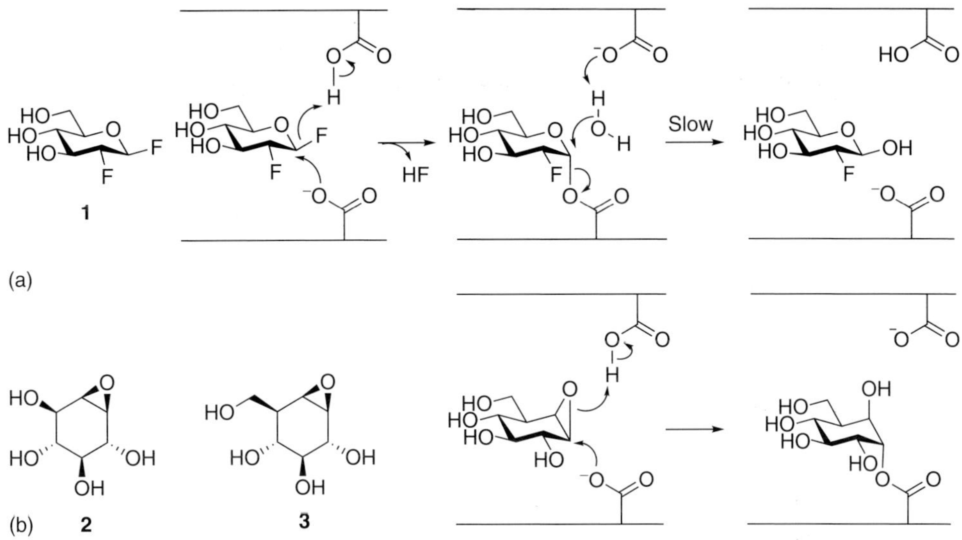

Figure 13.2 *Overview of mechanism-based retaining β-exoglucosidase inhibitors and their mode of action.* (a) 2-Deoxy-β-1,2-difluoroglucose 1 and (b) cyclophellitol 3.

fluorine residue therefore also decreases the rate of formation of the glucosyl-enzyme adduct. A good leaving group (here: fluorine) is thus a requirement to assure that the first step of the catalytic cycle proceeds uneventfully [6]. It should be noted that 2-deoxy-2-fluoroglycosides were employed by the Withers group [8] in a seminal paper demonstrating the involvement of covalent enzyme-substrate adducts in the action of retaining glycosidases and thus in proving the mechanism hypothesized by Koshland (Figure 13.1a) correct. In an alternative design, replacement of the monosaccharide core by a cyclitol analog equipped with an pseudoequatorial epoxide produces after enzyme catalysis (protonation of the epoxide followed by nucleophilic substitution) the ester adduct, comparatively more stable than the acetal formed as depicted in Figure 13.1a, thereby effectively inactivating the enzyme. As much as five decades ago, Legler and coworkers reported on the use of conduritol B epoxide (CBE) **2** (Figure 13.2b) for this purpose [9]. The most effective inhibitor of this class is the natural product, cyclophellitol **3** (Figure 13.2b), a molecule that closely resembles β-glucopyranose in configuration and substitution pattern and as such appeared a highly potent mechanism-based inhibitor of retaining β-exoglucosidases from different origins [10].

13.3.1 Development of a Human Acid Glucosylceramidase Activity-Based Probe

Human acid glucosylceramidase, or GBA (glucosidase, beta, acid), catalyzes the hydrolysis of glucosylceramide to glucose and ceramide. As such, it is responsible for the penultimate step in the turnover of glycosphingolipids, an important metabolic pathway malfunctioning of which is responsible for numerous inherited metabolic disorders. Mutations in the gene encoding GBA can lead to partial malfunctioning of the enzyme, leading to accumulation of its substrate, glucosylceramide [11, 12]. This is in a nutshell the basis of the lysosomal storage disorder, Gaucher disease. Two Gaucher therapies are practiced in the clinic. In enzyme replacement therapy, patients are treated with recombinant GBA, whereas in substrate reduction therapy glucosylceramide levels are downregulated through partial inhibition of the enzyme responsible for glucosylceramide biosynthesis: glucosylceramide synthase [13–15]. A third potential clinical strategy that received much attention in recent years is called *chemical* (or pharmacological) chaperone therapy and this strategy aims to enhance the activity of mutant GBA through stabilizing molecules [16]. Both for monitoring GBA levels in healthy and Gaucher patients and for assessment of the effect of interference in glucosylceramide metabolism, it would be advantageous to have access to potent and selective activity-based GBA probes.

With the aim of developing such tools, a comparative study was performed on the merits of the two scaffolds described earlier – 2-deoxy-2-fluoroglucosides and cyclitol epoxides – as activity-based GBA probes. Figure 13.3 depicts the four probes that were designed for this purpose: two direct probes and two probes relying on two-step bioorthogonal ligation (see for bioorthogonal chemistry in conjunction to ABPP, Chapter 12). GBA is a member of the large family of exoglycosidases, an enzyme class normally rather particular to the nature of their substrates. At the onset of the studies, it was therefore considered unlikely that attachment of a bulky group such as a fluorophore or a biotin would be accepted within the enzyme active site and thus 1,2,6-deoxy-6-azido-1,2-difluoroglucoside **4**, its click-conjugated fluorescent counterpart **5**, as well as the corresponding azido-cyclophellitol and BODIPY-cyclophellitol (boron dipyrromethene difluoride) derivatives **6** and **7**, respectively, were designed [17, 18]. Comparison of the inhibitory potency of these compounds relative to that of the known mechanism-based inhibitors, CBE (**2**) and adamantane pentyloxy deoxynojirimycin (AMP-DNM, MZ21, **8**) for both almond retaining β-exoglucosidase (ABG, the workhorse

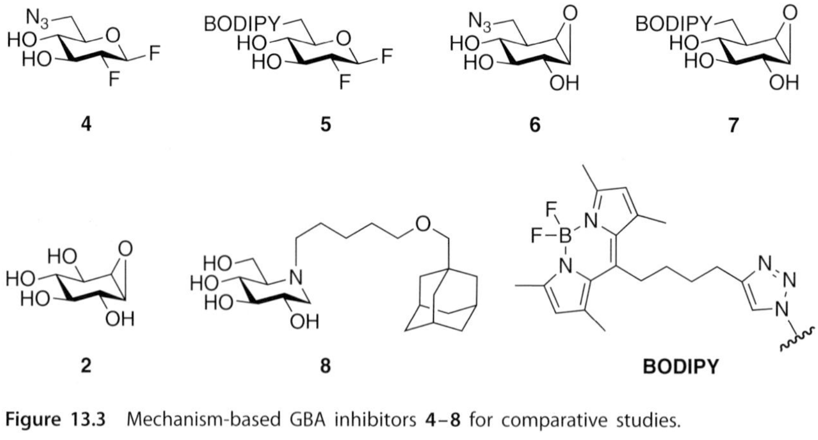

Figure 13.3 Mechanism-based GBA inhibitors **4–8** for comparative studies.

Table 13.1 Apparent IC_{50} of **2**, **4–8** for almond β-glucosidase and glucocerebrosidase.

Compound	Almond β-glucosidase IC_{50} (μM)	Glucocerebrosidase IC_{50} (μM)
2	461	9.49
4	> 10 000	1 665
5	> 1 000	785
6	27	0.120
7	56.5	0.0012
8	—	0.2

retaining β-glucosidase in the field) and GBA yielded a rather surprising result (Table 13.1). The ABG inhibitory potency of the small set of compounds was as expected. The cyclitol epoxides outperform the 2-deoxy-2-fluoroglycosides, with the close glucose mimic cyclophellitol as the most potent inhibitor, and partial to complete loss of inhibition was observed for the C6-modified compounds. In contrast, by far the most potent GBA inhibitor proved to be cyclophellitol derivative **7** equipped with a bulky fluorescent group at C6 [17, 18].

This superior inhibition becomes also evident in a comparative direct and two-step bioorthogonal ABPP experiment on the four compounds. Figure 13.4a depicts a general strategy for direct and two-step labeling on cells and cell extracts, whereas Figure 13.4b gives a representative image of the potency and specificity of the various ABPs on GBA. Labeling of GBA with BODIPY-cyclophellitol **7** is very clean both *in vitro* and *in situ*, much more so than is the case in two-step bioorthogonal labeling using copper(I)-catalyzed click reaction conditions starting with azidocyclophellitol **6**. No GBA-specific labeling was achieved with either direct or two-step probes **4** and **5** based on the 2-deoxy-2-fluoroglucoside scaffold [17]. The latter result is perhaps not so surprising as 2-deoxy-2-fluoroglucosides are rather poor glucosidase inhibitors, likely because OH-2 of the corresponding substrates is an important structural feature in binding to the enzyme active site. Another intrinsic feature of 2-deoxy-2-fluoroglucosides that sets these apart from cyclitol epoxides is their tempered reactivity as a result of the electron-withdrawing fluorine at C2 (i.e., the same effect as that in stabilizing the enzyme-glycoside adduct; see Figure 13.1). To offset this disadvantage, good anomeric leaving groups (fluoride, dinitrophenyl) are often employed. Figure 13.4c depicts a few structures that were employed to further look into the labeling activity of this class of compounds [19]. Anomeric imidate **9** proved by far the most potent of these series. Moreover, 1,2-difluoroderivative **5** labeled mutant GBA in which the acid–base residue (Glu235) is mutated for Gln equally well, whereas imidate **9** proved inactive toward this mutant. Arguably, imidate **9** is therefore a more "true" ABP that truly recruits the glucosidase active site residues. At the same time, BODIPY-cyclophellitol **7** out-competes imidate **9** by several orders of magnitude and cyclophellitol therefore appears the superior scaffold for retaining glycosidase ABP design.

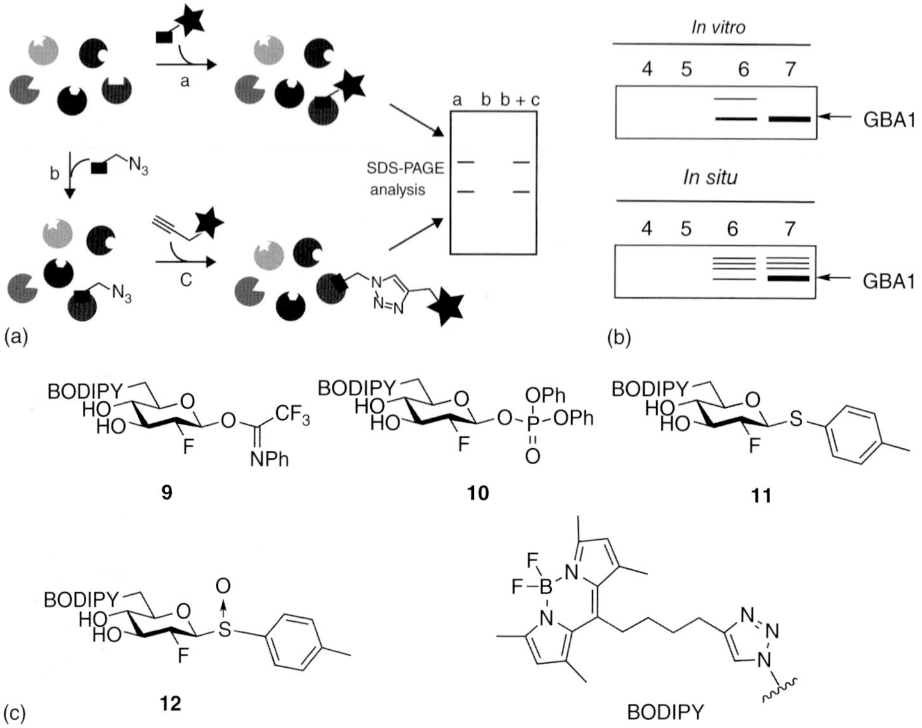

Figure 13.4 *Direct and two-step bioorthogonal labeling of GBA in cells and cell extracts.* (a) General workflow. (b) BODIPY-cyclophellitol **7** is the most effective *in vitro* and *in situ* probe. (c) Tuning of the leaving group on 2-deoxy-2-fluoroglucosides yields comparatively more potent GBA probe.

13.3.2
Cyclophellitol Aziridine Is a Broad-Spectrum Activity-Based Retaining β-Exoglucosidase Probe

The mammalian genome contains at least four retaining β-exoglucosidase genes. Next to GBA, these are the nonlysosomal retaining beta glucosidase (GBA2), another cytosolic glucosidase termed *GBA3* and *lactase-phlorizin hydrolase* – an

Figure 13.5 *Synthesis of broad-spectrum retaining β-exoglucoside probe cyclophellitol aziridine **13** and **15**.* (a) (i) CCl_3CN, DBU (1,8-diazabicyclo[5.4.0]undec-7-ene), CH_2Cl_2, 0 °C, 2 h, (ii) followed by the addition of H_2O, $NaHCO_3$, I_2, 18 h, (b) (i) 37% HCl, MeOH, 3.5 h, (ii) 37% HCl, dioxane, 60 °C, 1 h, (iii) $NaHCO_3$, MeOH, 4 days (over five steps 60%), (c) (i) Li (s), NH_3, THF (tetrahydrofurane), −60 °C, 30 min, (ii) EEDQ (ethoxycarbonyl-ethoxy-dihydroquinoline), hept-6-ynoic acid, DMF (*N,N*-dimethylformamide), 0 °C, 1 h, 20%, (d) bodipy-azide, sodium ascorbate, $CuSO_4$, DMF, 1 h, 45%, (e) EEDQ, 7-azido-octanoic acid, DMF, 0 °C, 1 h, and (f) biotin-Ahx-alkyne, sodium ascorbate, $CuSO_4$, DMF, 1 h, 17%.

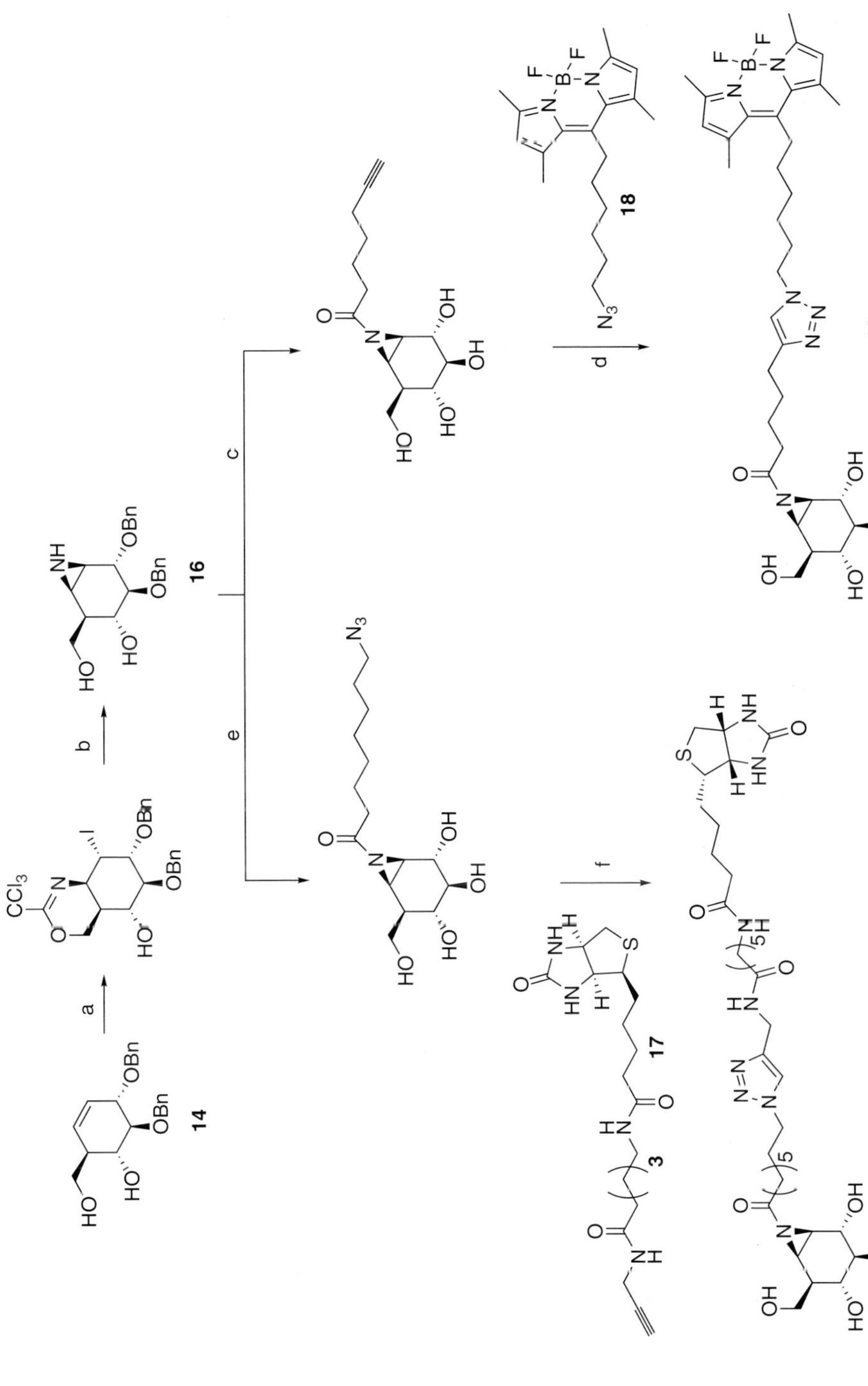

intestinal, dual-activity glycosidase containing both a β-glucosidase activity and a β-galactosidase activity [20]. Of these, only GBA and, to a lesser extent, lactase-PGH are targeted by BODIPY-cyclophellitol **7**, presumably because the bulky substituent at C6 is not accepted by the other enzymes. Indeed, and as stated before, one would expect exoglycosidases to be rather particular to the substitution pattern and configuration of the glycomimetic emulating the corresponding natural substrate. In contrast, exoglycosidases are often much less selective toward the aglycon – the anomeric leaving group – as is evident both from the range of natural substrates and artificial substrates (including fluorogenic substrates often used for glycosidase kinetics studies). Figure 13.5 depicts broad-spectrum retaining β-exoglucosidase probe **13**, the design of which is based on the thought that pointing the bulky reporter group toward the direction normally occupied by the substrate aglycon would result in a mechanism-based inhibitor accepted by all the enzymes mentioned [20].

A key aspect to ABPP studies is, next to the design of an ABP, obviously also its synthesis. This is often not an easy task and one complicating factor is that ABPs are intrinsically reactive. Their reactivity needs to be balanced such that they are – if only just – stable under physiological conditions, yet react efficiently with their target enzyme(s). The synthesis and purification of cyclophellitol aziridine **13** is depicted in Figure 13.5a. Partially protected cyclohexenol **14** is an advanced intermediate in the synthesis of the natural product, cyclophellitol, as reported by Madsen and coworkers [21]. This compound proved an ideal intermediate both in the synthesis of epoxides **6** and **7** and aziridines **13** and **15** (Figure 13.5). Installation of the acetimidate at O6 is followed by iodocyclization, acidic hydrolysis of the resulting iminal and base-induced cyclization to give aziridine **16** in a complete stereospecific manner. Global deprotection and aziridine acylation is followed by copper(I) catalyzed [2+3] azide-alkyne cycloaddition to biotin-alkyne **17** or BODIPY-azide **18**. Purification of the resulting compounds **13** and **15** has to be conducted with care, as they are both acid- and base labile. HPLC (high-performance liquid chromatography) using neutral conditions (solvent H_2O/ACN (acetonitrile)) followed by lyophilization afforded the cyclitol aziridines.

Table 13.2 presents a head-to-head comparison of GBA-specific probe **7** and aziridine **13** [20]. As expected, aziridine **13** labels all four murine retaining β-exoglucosidases depending on their expression in various tissues. As is the case with epoxide **7**, aziridine **13** is both cell permeable and tissue permeable and both probes are therefore amenable for *in vivo* labeling experiments. The question why GBA, but not the other enzymes, accepts (and, in fact, prefers) a bulky substituent at C6 remains unanswered. However, both probes react with a considerable number of bacterial glucosidases, some of which appear to have evolved from endoglucosidases and it might well be that GBA is evolutionary related to these bacterial enzymes [20].

Table 13.2 Head-to head comparison of epoxide **7** and ariridine **13**.

Glucosidases	Epoxide 7	Aziridine 13
GBA1	+	+
GBA2	−	+
GBA3	−	+
LPH	−	+
Bacterial	+	+

13.4 Biological Research/Evaluation

ABPs, in general, find various uses in biology research. They can be used to discover new enzymatic species and their active site residues (comparative ABPP) and in the evaluation of the potency and selectivity of putative inhibitors aimed at one of the enzymes targeted by the ABP (competitive ABPP). Depending on the bioavailability of the probe, these studies can be conducted in an *in vitro*, an *in situ*, or an *in vivo* research setting. Such studies are common practice with serine hydrolase, cysteine protease, and threonine protease probes (see also Chapter 12), yet only start to emerge in the field of glycobiology – this for the obvious reason that suitable glycosidase probes were until recently not available. In the following two sections, two examples of biochemical and biological studies are described to highlight the potential of activity-based glycosidase probes in chemical biology research.

13.4.1 *In situ* Monitoring of Active-Site-Directed GBA Chemical/Pharmacological Chaperones

Chemical or pharmacological chaperones form a conceptually new approach to treat inherited diseases that are characterized by point mutations in a hydrolytic enzyme that causes its partial dysfunctioning. Gaucher disease is caused by point mutations in GBA that lead to lower enzyme activity in total, and this lowered activity appears to be caused by a comparatively lower number of GBA copies that reach the lysosome, rather than a lower activity of an individual GBA protein. Indeed, probing tissues from different Gaucher type patients with epoxide 7 (Figure 13.6a) shows GBA labeling in varying intensities, corresponding to the severity of the disease (or the impact of the nature of the point mutation) [18]. This, while the intrinsic reactivity of the GBA mutants toward probe 7, is largely invariable and therefore the partial loss in lysosomal activity is thought to rely on the partial impairment of ER (endoplasmic reticulum) folding of the mutants.

Chemical/pharmacological chaperone strategies aim at correcting this impaired folding through stabilization of the enzyme in its proper fold, which can be achieved by inhibition of the enzyme active site. The caveat of this strategy

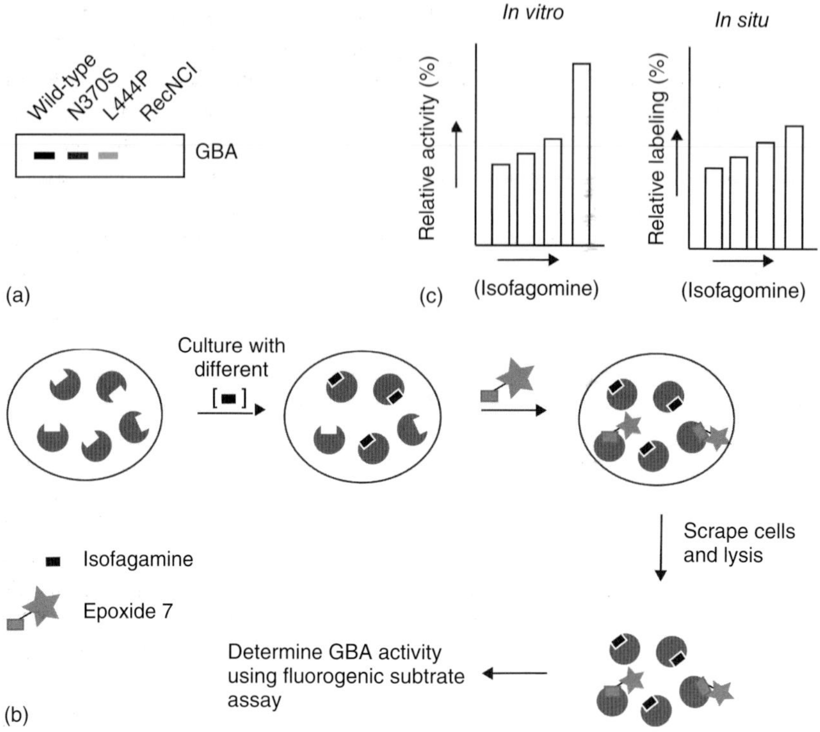

Figure 13.6 *In situ monitoring of GBA activity with epoxide 7.* (a) Wild-type versus mutant GBA from healthy and Gaucher tissue. (b) Workflow for *in vitro* and *in situ* GBA activity profiling in the presence of chemical chaperones. (c) *In vitro* and *in situ* effect of the pharmacological chaperone, isofagomine, on GBA activity.

is that in this way, a larger number of enzymes may traverse to the lysosome but these will be accompanied by their active site inhibitor. Thus, an increase in activity within lysosomes may not be the actual result. BODIPY-cyclophellitol **7** allows for the first time to probe intracellular GBA activity directly, all or not in the presence of a chemical chaperone. Isofagomine is the archetypal GBA chemical chaperone studied broadly in the field, yet almost exclusively in a setting in which, after cell culture treatment with this compound or derivatives thereof, the sample is lysed and after which GBA activity is monitored using a fluorogenic substrate assay (Figure 13.6b). For various reasons (dilution of the chemical chaperone being the most obvious one), such a research setting may not reflect the intracellular situation. As is depicted in Figure 13.6c, this appears indeed true. *In vitro* measurement reveals a marked increase in activity of mutant (N370S) GBA activity in the presence of isofagomine, whereas *in situ* measurement using BODIPY cyclophellitol **7** as the readout shows a comparatively much less pronounced activity increase. It should be noted that these assays are rather complicated and that care has to be taken in the interpretation of the results.

At the same time, this observation should serve as a warning to the field: for an enzyme active-site-directed chemical chaperone to be effective, it should bind within the ER, there stabilize the enzyme in its proper fold, and once in the endo-lysosomal compartments dissociate to become an inactive bystander. It may not be so easy to reach this result using iminosugars, intrinsically basic by nature and therefore prone to be trapped in acidic milieu. Interestingly, the Withers group recently proposed the use of 2-deoxy-2-fluoroglycosides as potentially useful alternative chemical chaperones based on their mechanism-based binding followed by slow but sure release to produce 2-deoxy-2-fluoroglucose as such an inert molecule [22].

13.4.2 Mapping of Human Retaining β-Glucosidase Active Site Residues

An intrinsic nature of ABPs is their covalent attachment to enzyme active site nucleophiles. In case the nature of these is unknown, they can, in fact, be unearthed using ABPs following the workflow as depicted in Figure 13.7a. Such studies can be executed on recombinant purified enzymes and therefore both 2-deoxy-2-fluoroglucosides and cyclitol epoxides/aziridines can be used for this purpose. Figure 13.7b–d provides a representative example. In contrast to GBA and GBA3, the active site acid base and nucleophile of GBA2 were unknown. Moreover, at least six aspartate/glutamate residues appeared suitable candidate nucleophiles. Figure 13.7e–g depicts the GBA2 active site as determined using the flow of experiments as outlined in Figure 13.7a [23].

13.5 Conclusions

In conclusion, rational design has resulted in the development of a panel of active and selective activity-based retaining β-exoglucosidase probes. Cyclophellitol is a natural product and it is therefore fair to state that, as before (see the epoxomicin-based proteasome probes in Chapter 12), nature has paved the way for these studies. The first-generation probes, represented by epoxide **7**, provided a rather surprising result: an active and highly selective probe for the Gaucher enzyme, GBA. Arguably, this design principle – the fluorophore or biotin grafted at C6 – will not meet with success when applied to other retaining glycosidases. The aziridine-based scaffold, in contrast, holds more promise, and as one can learn from Cazypedia [24], there are numerous retaining exoglycosidases that follow the general Koshland mechanism and that are, in principle, amenable to ABPP using cyclitol aziridines emulating in configuration and substitution pattern the corresponding substrate glycosides.

Another intriguing feature of the epoxide and aziridine probes is the highly potent activity they display. Next to offering a suitable electrophile (epoxides, acylaziridine) to the general acid/base, the (putative) half-chair conformation they

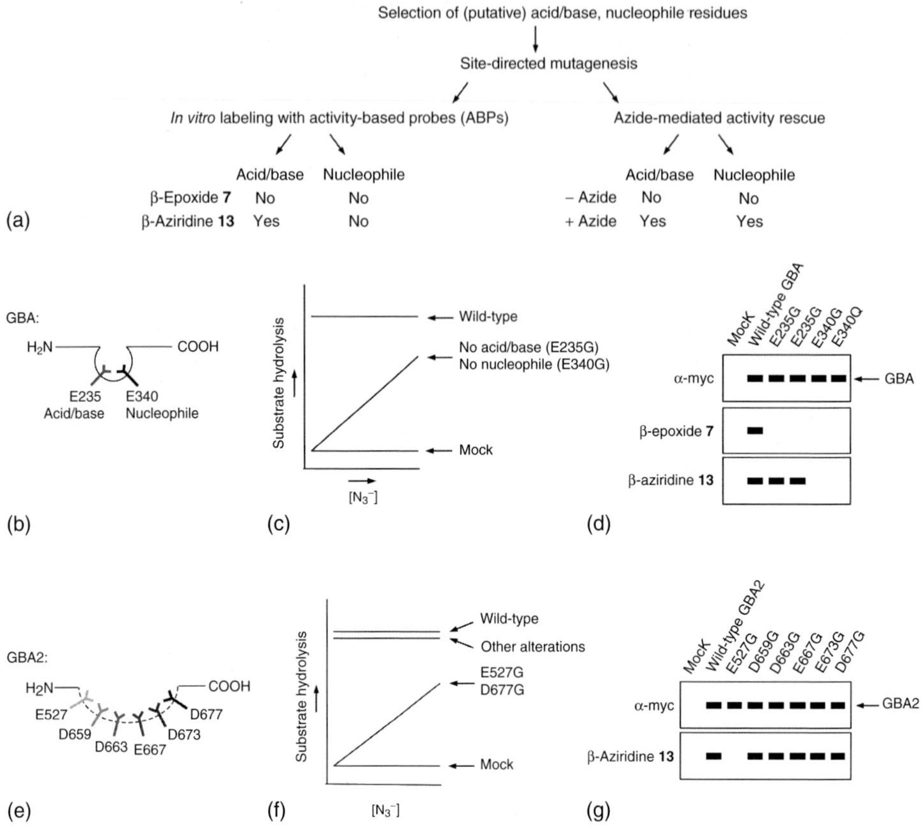

Figure 13.7 (a–g) *Retaining β-exoglucosidase active site mapping using activity-based probes.* (a) General workflow. (b) GBA2 active site as mapped by using probe **7** and **13**.

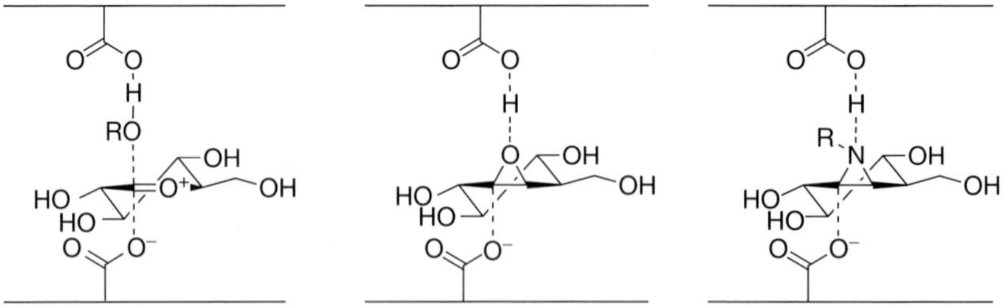

Figure 13.8 Cyclitol epoxides and aziridine may feature ideal conformational behavior for retaining β-exoglucosidases inhibition.

adopt resembles that of the developing oxocarbenium ion that is the result of aglycon protonation and may be at the basis of this activity (Figure 13.8). Not only are they highly potent, they also appear to react almost instantaneously, further suggesting that they fit exceedingly well within the active site. All this bodes well for the future development of ABPs aimed at other retaining glycosidases and perhaps – given that there are literature speculations on the covalent intermediacy of some glycosyl transferases as well – for glycosyl transferases.

References

1. Wennekes, T., van den Berg, R.J., Boot, R.G., van der Marel, G.A., Overkleeft, H.S., and Aerts, J.M.F.G. (2009) Glycosphingolipids-nature, function and pharmacological modulation. *Angew. Chem. Int. Ed.*, **48**, 8848–8869.
2. Davies, G.J. and Henrissat, B. (1995) Structures and mechanisms of glycosyl hydrolases. *Structure*, **3**, 853–859.
3. Koshland, D.E. (1953) Stereochemistry and the mechanism of enzymatic reactions. *Biol. Rev. Camb. Philos. Soc.*, **28**, 416–436.
4. Vocadlo, D.J. and Davies, G.J. (2008) Mechanistic insights into glycosidase chemistry. *Curr. Opin. Chem. Biol.*, **12**, 539–555.
5. Gandy, M.N., Debowski, A.W., and Stubbs, K.A. (2011) A general method for affinity-based proteomic profiling of exo-α-glycosidases. *Chem. Commun.*, **47**, 5037–5039.
6. Withers, S.G., Street, I., Bird, P., and Dolphin, D. (1987) 2-deoxy-2-fluoroglucosides- a novel class of mechanism-based glucosidase inhibitors. *J. Am. Chem. Soc.*, **109**, 7530–7531.
7. Atsumi, S., Umezawa, K., Iinuma, H., Naganawa, H., Nakamura, H.M., Iitaka, Y., and Takeuchi, T. (1990) Production, isolation and structure determination of a novel β-glucosidase inhibitor, cyclophellitol, from Phellinus sp. *J. Antibiot.*, **43**, 49–53.
8. Vocadlo, D.J., Davies, G.J., Laine, R., and Withers, S.G. (2001) Catalysis by hen egg-white lysozyme proceeds via a covalent intermediate. *Nature*, **412**, 835–838.
9. Legler, G. (1968) Investigations on the mechanism of action of glycoside-splitting enzymes. 3. Labelling of the active centre of a β-glucosidase from aspergillus wenti with (14C) conduritol B epoxide. *Hoppe-Seyler's Z. Physiol. Chem.*, **349**, 767–774.
10. Atsumi, S., Iinuma, H., Nosaka, C., and Umezawa, K. (1990) Biological activities of cyclophellitol. *J. Antibiot.*, **43**, 1579–1585.
11. Beutler, E. and Grabowski, G.A. (2001) in *The Metabolic and Molecular Bases of Inherited Disease*, 8th edn (eds C.R. Scriver, W.S. Sly, and D. Valle), McGraw-Hill, New York, pp. 3653–3668.
12. Aerts, J.M.F.G., Hollak, C., Boot, R., and Groener, A. (2003) Biochemistry of glycosphingolipid storage disorders: implications for therapeutic intervention. *Philos. Trans. R. Soc. Lond. B: Biol. Sci.*, **368**, 905–914.
13. Barton, N.W., Furbish, F.S., Murray, G.J., Garfield, M., and Brady, R.O. (1990) Therapeutic response to intravenous infusions of glucocerebrosidase in patient with gaucher disease. *Proc. Natl. Acad. Sci. U.S.A.*, **87**, 1913–1916.
14. Cox, T., Lachmann, R., Hollak, C., Aerts, J.M.F.G., van Weely, S., Hrebicek, M., Platt, F., Butters, T., Dwek, R., Moyses, C., Gow, I., Elstein, D., and Zimran, A. (2000) Novel oral treatment of Gaucher's disease with *N*-butyldeoxynojirimycin (OGT 918) to decrease substrate biosynthesis. *Lancet*, **355**, 1481–1485.
15. McEachern, K.A., Fung, J., Komarnitsky, S., Siegel, C.S., Chuang, W.L., Hutto, E., Shayman, J.A., Grabowski, G.A., Aerts, J.M., Cheng, S.H., Copeland, D.P., and Marshall, J. (2007) A specific and

potent inhibitor of glucosylceramide synthase for substrate inhibition therapy of Gaucher disease. *Mol. Genet. Metab.*, **91**, 259–267.

16. Benito, J.M., García Fernández, J.M., and Ortiz Mellet, C. (2011) Pharmacological chaperone therapy for Gaucher disease: a patent review. *Expert Opin. Ther. Pat.*, **21**, 885–903.
17. Witte, M.D., Walvoort, M.T.C., Li, K.-Y., Kallemeijn, W.W., Donker-Koopman, W.E., Boot, R.G., Aerts, J.M.F.G., Codee, J.D.C., van der Marel, G.A., and Overkleeft, H.S. (2011) Activity-based profiling of retaining b-glucosidases: a comparative study. *ChemBioChem*, **12**, 1263–1269.
18. Witte, M.D., Kallemeijn, W.W., Aten, J., Li, K.-Y., Strijland, A., Donker-Koopman, W.E., van den Nieuwendijk, A.M.C.H., Bleijlevens, B., Kramer, G., Florea, B.I., Hooibrink, B., Hollak, C.E.M., Ottenhoff, R., Boot, R.G., van der Marel, G.A., Overkleeft, H.S., and Aerts, J.M.F.G. (2010) Ultrasensitive in situ visualization of active glucocerebrosidase molecules. *Nat. Chem. Biol.*, **6**, 907–913.
19. Walvoort, M.T.C., Kallemeijn, W.W., Willems, L.I., Witte, M.D., Aerts, J.M.F.G., van der Marel, G.A., Codee, J.D.C., and Overkleeft, H.S. (2012) Tuning the leaving group in 2-deoxy-2-fluoroglucosides results in improved activity-based retaining b-glucosidase probes. *Chem. Commun.*, **48**, 10386–10388.
20. Kallemeijn, W.W., Li, K.-Y., Witte, M.D., Marques, A.R.A., Atemn, J., Scheij, S., Jiang, J., Willems, L.I., Voorn-Brouwer, T.M., van Roomen, C.P.A.A., Ottenhoff, R., Boot, R.G., van den Elsts, H., Walvoort, M.T.C., Florea, B.I., Codee, J.D., van der Marel, G.A., Aerts, J.M.F.G., and Overkleeft, H.S. (2012) Novel activity-based probes for broad-spectrum profiling of retaining β-exoglucosidases in situ and in vivo. *Angew. Chem. Int. Ed.*, **51**, 12529–12533.
21. Hansen, F.G., Bundgaard, E., and Madsen, R. (2005) A short synthesis of (+) cyclophellitol. *J. Org. Chem.*, **70**, 10139–10142.
22. Rempel, B.P., Tropak, M.B., Mahuran, D.J., and Withers, S.G. (2011) Tailoring the specificity and reactivity of a mechanism-based inactivator of glucocerebrosidase for potential therapeutic applications. *Angew. Chem. Int. Ed.*, **50**, 10381–10383.
23. Kallemeijn *et al.*, to be published in detail.
24. CAZypedia (2007) *http://www.cazypedia.org* (accessed 29 December 2014).

14
Modulation of ClpP Protease Activity: from Antibiotics to Antivirulence

Malte Gersch and Stephan A. Sieber

14.1
Introduction

In this chapter, we describe how modulating the activity of a single protease by two different classes of compounds gives rise to two distinct mechanisms that can be used against bacterial infections. We first look into the discovery of the acyldepsipeptides (ADEPs). These natural products activate the caseinolytic protease ClpP for uncontrolled proteolysis and thereby act as antibiotics. We then focus on synthetic β-lactones that represent ClpP-specific inhibitors showing the feasibility of an antivirulence strategy against the bacterial pathogen *Staphylococcus aureus.*

14.2
The Biological Problem

Antibacterials are chemical entities that either kill bacteria (bactericidal) or inhibit bacterial growth (bacteriostatic) and their discovery is commonly considered to be one of the greatest achievements of the twentieth century. While infectious diseases were the leading cause of death at around 1900, at present the number of people in developed countries dying from bacterial infections is significantly reduced. Although antibacterial agents span a diverse chemical space, their biological targets, however, are restricted to a comparably small subset of essential physiological processes such as nucleic acid synthesis, ribosomal function, cell-wall synthesis, plasma membrane integrity, and folate biosynthesis [1]. The fact that many compounds rely on the same molecular targets has proved to be a problem because certain structural alterations in the target can convey resistance to several antibiotics. In fact, the discovery of novel classes of antibiotics was paralleled by naturally evolving resistance mechanisms as the presence of an antibacterial agent induces a selection pressure favoring the growth of mutation-carrying clones (Figure 14.1) [2]. Currently, the emergence of multiresistant bacterial pathogens significantly challenges the treatment of bacterial infections.

Concepts and Case Studies in Chemical Biology, First Edition. Edited by Herbert Waldmann and Petra Janning.

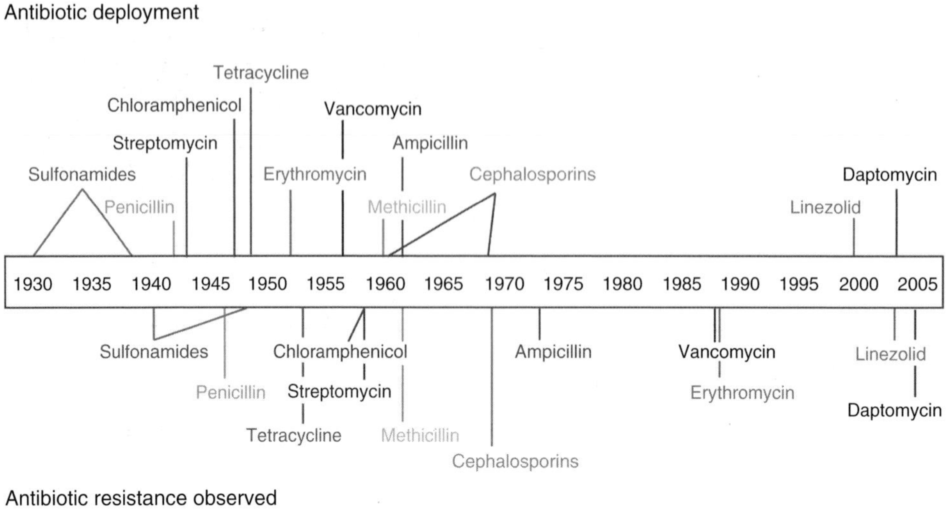

Figure 14.1 *Timeline depicting the parallel developments of novel antibiotic compound classes and antibiotic resistance.* In the upper panel, the year in which the antibiotic was first deployed is indicated. The lower panel shows the year in which the first resistant strains were observed. (Reprinted by permission from Macmillan Publisher Ltd: Nature Chemical Biology [2], copyright 2007.)

In addition, the number of hospital-acquired life-threatening infections has been steadily increasing in the past decades. Thus, great efforts are being undertaken in the search for novel chemical entities with antibacterial activity and possibly novel modes of action.

Natural products provide a wealth of bioactive and chemically diverse small molecules [3]. Although it may sound paradoxical, antibiotics of bacterial origin are commonly observed and their occurrence is assumed to regulate the composition of the microbiome in environmental niches. In 1982, a team from the pharmaceutical company Eli Lilly reported the isolation of a set of ADEPs from a fermentation broth of *Streptomyces hawaiiensis* [4]. The main constituents showed promising antibacterial activity and treatment of bacterial cells resulted in a filamentation phenotype, indicating impaired cell division (Figure 14.2). The structure of the ADEPs was then elucidated. They consist of the pentapeptide Ser–Pro–Ala–Ala–Methyl-Pro, which is cyclized through an ester linkage between the serine hydroxyl group and the C-terminal carboxylic acid (hence depsi-peptide). The amino functionality of the serine is linked to a phenylalanine to which an unsaturated acyl moiety is appended. All stereocenters are in the L-configuration, which is common for proteinogenic amino acids. One alanine is *N*-methylated, whereas the amide proton of the second alanine engages in intramolecular hydrogen bonding as revealed by a small molecule crystal structure [5]. While these ADEPs show antibacterial activity against a wide panel

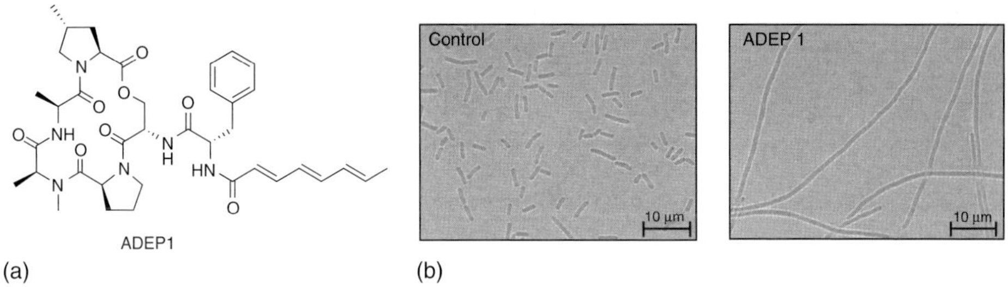

Figure 14.2 (a) Structure of the natural product ADEP1. (b) *Bacillus subtilis* develops a filamentation phenotype after incubation with ADEP1 for 5 h, as shown by phase contrast microscopy. (Reprinted by permission from Macmillan Publisher Ltd: Nature Medicine [6], copyright 2005.)

of gram-positive bacteria, they have the disadvantages of low solubility, high instability, and lack of efficacy in a mouse model of infection [6].

14.3 The Chemical Approach

To address these pharmacological issues and to provide sufficient amounts for a detailed investigation of the ADEPs' therapeutic potential, a medicinal chemistry optimization program was initiated by researchers at Bayer AG [5]. They reduced the number of double bonds in the acyl side chain, which greatly improved the stability of the compound toward light and oxygen. Moreover, the ring structure was rigidified by incorporation of pipecolic acid instead of *N*-methyl alanine. The rationale behind this change was the reduction of conformational freedom in the resulting bicyclus, and hence a lower entropy penalty (Box 14.1) upon binding to the target protein.

Box 14.1 Entropic Penalty

The binding of a ligand to a protein can be viewed in analogy to a chemical reaction whose free energy difference ΔG has to be negative in order for the reaction to proceed. ΔG can be related to the change in enthalpy (i.e., heat) during the binding event, ΔH, to the temperature, T, and to the change in entropy (i.e., disorder) during the binding event, ΔS, by the following equation:

$$\Delta G = \Delta H - T\,\Delta S$$

How well protein and ligand interact is reflected by ΔH and it becomes negative when, for instance, new hydrogen bonding contacts are made. As the absolute temperature T is always positive, reactions with a positive change in entropy (or, at most, a small negative change) are favored because the second term $-T\Delta S$ – then becomes negative (or, at most, only small and positive). Linear compounds usually have many freely rotatable bonds and can adopt many conformations. Upon protein binding, their conformational freedom is drastically reduced and they are forced into a certain conformation that fits to the binding pocket; thus, their change in entropy is negative (entropic penalty). Therefore, (poly)cyclic compounds are often desired as their degrees of freedom are already reduced in the free form and the change in entropy caused by binding is usually less pronounced. In addition, solvation effects and changes in protein conformation contribute to ΔS.

In addition, substitution of the phenyl ring by fluorine atoms was carried out. The synthesis proceeded along a previously established route, as depicted in Figure 14.3. First, a tripeptide and a dipeptide were coupled with standard peptide coupling chemistry followed by the deprotection of the serine carboxyl group. Then, this carboxyl group was activated through formation of the pentafluorophenyl ester, the boc (*N-tert*-butoxycarbonyl) group of the N-terminal amino moiety was cleaved off, and cyclization was achieved in mild basic conditions, presumably aided through intramolecular hydrogen bonds of the peptide backbone. In the next step, the orthogonally protected amino group of the serine was released, to which a phenylalanine derivative was coupled. The last steps comprised deprotection of the newly introduced amino group and its coupling to an activated, aliphatic acid [5]. This synthesis demonstrates how large natural products and improved derivatives can be built up in a stepwise manner.

14.4 The Discovery of a Novel Antibiotic Mechanism

14.4.1 Target Identification

With optimized compounds in hand, the team at Bayer set out for a full characterization of their antibiotic mechanism [6]. The new compound ADEP4 showed greatly improved solubility and stability and largely surpassed the natural product in *in vitro* potency against several pathogens with minimal inhibitory concentrations (MICs, Box 14.2) of 0.01–0.05 $mg\,l^{-1}$ (MIC of the natural product: 0.2–6.3 $mg\,l^{-1}$). ADEP4 also outperformed by far the marketed antibiotic linezolid in a mouse infection study (Figure 14.4a). Moreover, no cross-resistance of these compounds to a large panel of hospital-acquired, multiresistant pathogenic strains was observed. This is of particular importance because it strongly indicates the presence of an unprecedented molecular mode of action. Thus, several methods were applied to identify the molecular target of the ADEPs [6]. *Escherichia coli* was chosen as a model organism as

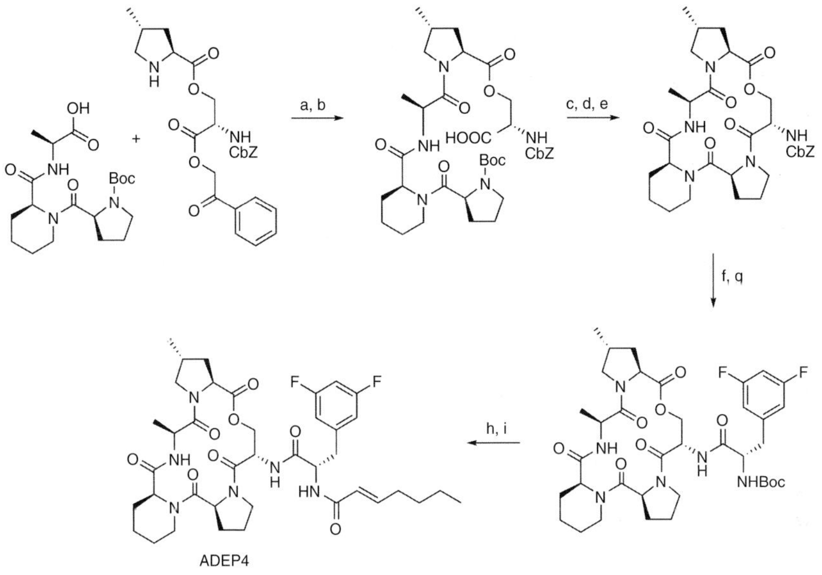

Figure 14.3 *Synthesis of the medicinal chemistry optimized ADEP4.* (a) CH_2Cl_2, HOBT (hydroxybenzotriazole), TBTU (*o*-benzotriazole-1-yl-1,1,3,3-tetramethyluronium tetrafluoroborate), *i*-Pr_2EtN, 0 °C to rt, 62%; (b) AcOH/H_2O (9 : 1), Zn, 2 h, rt, 67%; (c) CH_2Cl_2, pentafluorophenol, EDC (1-ethyl-3-(3-dimethylaminopropyl)carbodiimide), 0 °C to rt, 18 h; (d) 4N HCl in dioxane, 1 h, rt; (e) CH_2Cl_2, H_2O, $NaHCO_3$, rt, 62%; (f) MeOH, aqueous HCl, H_2 (1 bar), Pd/C, 92%; (g) 3,5-difluoro-*N*-boc-phenylalanine, DMF (*N,N*-dimethylformamide), HATU (*o*-(7-azabenzotriazole-1-yl)-1,1,3,3-tetramethyluronium hexafluorophosphate), *i*-Pr_2EtN, rt, 87%; (h) CH_2Cl_2, TFA (trifluoro acetic acid), H_2O (9 : 1), 45 min, rt, quant.; and (i) 2-hexenecarboxylic acid, DMF, HATU, *i*-Pr_2EtN, 88%; rt: room temperature; Boc: *N*-*tert*-butoxycarbonyl; CbZ: carboxybenzyl. (Figure adapted from [5] Copyright © 2006 WILEY-VCH Verlag GmbH & Co. KGaA, Weinheim.)

its genetic manipulation is easy to achieve. In a genomic approach, an ADEP-resistant *E. coli* strain was generated by cultivating bacteria on ADEP-agar plates. While under these conditions all bacteria with the canonical genome were killed, the growth of *E. coli* that comprised a spontaneous mutation conferring ADEP resistance was favored. Indeed, the isolated strain showed a much lower susceptibility toward ADEP than the parental strain (MIC > 100 vs 3 mg l^{-1}, respectively). Next, the genomic DNA of this resistant strain was isolated and partially digested. The fragments obtained were 2-4 kbp in length and were subsequently ligated into a library of vectors (Box 14.3). This library was then transformed into the parental, ADEP-susceptible strain, and those plasmids that conferred resistance were sequenced. This forward chemical genetics method (Box 14.4) allowed the identification of the ClpP protease (Box 14.5) with a single amino acid mutation (Thr182Ala) as the resistance determinant [6].

Box 14.2 Minimal inhibitory concentration (MIC)

MIC defines the smallest concentration of a chemical compound that is needed to prevent growth of bacteria in a standardized laboratory procedure.

Box 14.3 Ligation and transformation

Ligation refers to the joining of two ends of DNA fragments during molecular cloning. In order to incorporate a linear insert into a vector, the empty vector is cut linear and mixed with the fragment. With the aid of recombinant enzymes (ligases), the linear fragment is then incorporated into the vector during a ring-closing ligation. The vector backbone can thus contribute elements essential for the stable maintenance of genetic information on the fragment such as an origin of replication and an additional gene that allows for selection.

Transformation describes the uptake of DNA into bacteria, leading to their genetic alteration, that is, expansion of their genetic information. Plasmids are often transformed to become a stable, easily manipulable, and nonchromosomal (i.e., second) source of genetic information. During target identification, transformation enabled the overexpression of genes from the resistant bacterial strain. In this particular example, the overexpression of a catalytically dead mutant of the target, although still binding the compound, reduced the susceptibility of the strain.

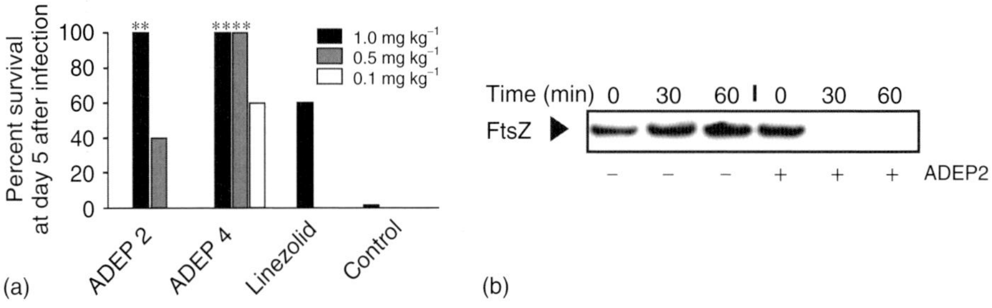

Figure 14.4 (a) *In vivo* efficacy of optimized ADEPs. Treatment of lethal systemic infections in mice caused by *Enterococcus faecalis* with a single dose of antibiotic (amounts indicated). Untreated control animals died within 24 h of infection. Survival is depicted 5 days after infection. (Reprinted by permission from Macmillan Publisher Ltd: Nature Medicine [6], copyright 2005.) (b) ADEP induces the ClpP-dependent degradation of FtsZ in bacterial cells. ADEP treatment of exponentially growing cells of *S. aureus* HG001 resulted in a decreased abundance of FtsZ over time, compared with the untreated control. Immunodetection of FtsZ was performed using a specific anti-FtsZ antibody. (Reprinted with permission from [7].)

Box 14.4 Forward and reverse chemical genetics

Forward chemical genetics approaches start with compounds that cause a particular phenotype in a biological system (e.g., ADEPs inhibit cell growth and trigger filamentation). Subsequently, the target of this small molecule and its mechanism of action are elucidated. Hence, the forward chemical genetics workflow is from compounds to gene. Usually, *in vivo* experiments are carried out ahead of *in vitro* experiments.

Reverse chemical genetics approaches start with a known gene or protein target of interest of which it is known that alteration of its function causes a certain phenotype. Subsequently, a compound collection is screened to identify small molecules that bind the target and alter its function in the desired way. Hence, the reverse chemical genetics workflow is from gene to compounds. Usually, *in vitro* experiments are carried out ahead of *in vivo* experiments.

Box 14.5 ClpP Protease

Similar to the eukaryotic proteasome, the bacterial ClpP protease degrades a wide variety of substrates. It is composed of two heptameric rings that are stacked on top of each other and form an enclosed cavity where 14 serine protease active sites reside [8]. ClpP gains proteolytic activity in complex with chaperones such as ClpX from the AAA+-class of ATPases that bind the ClpP tetradecamer (i.e., 14mer) on either or both axial sides [9]. The chaperone selects substrate proteins for degradation, unfolds them under ATP hydrolysis, and threads them into the catalytic chamber of ClpP where they are degraded to small peptides [10]. The ClpP protease contributes to protein homeostasis through the transfer-messenger ribonucleic acid (tmRNA) system. When ribosomes stall during protein synthesis, for example, owing to a defective mRNA (messenger ribonucleic acid) template, the tmRNA system effects the addition of a small, 11-amino acid tag to the C-terminus of the nascent protein chain, and its successive release from the ribosome. This SsrA tag is then recognized by a chaperone, which causes ClpP-mediated degradation of the defective protein [11, 12]. Besides its function in protein quality control, the ClpP protease controls the levels of multiple regulatory proteins. Although precise mechanisms remain to be elucidated, it is generally assumed that one of its substrates is a virulence repressor. Once ClpP activity is abrogated through inhibition, the level of this repressor increases by which the production of virulence factors would be reduced (Figure 1).

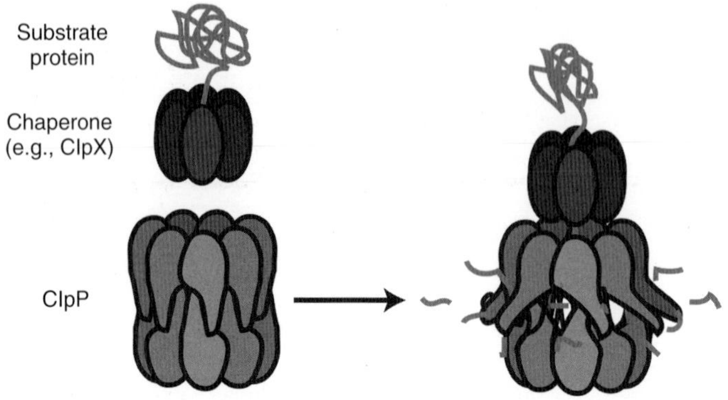

Figure 1 Degradation of a substrate protein by the ClpXP complex.

14.4.2
Target Validation

Next, ClpP was validated as target by complementary methods. As the ClpP protease is known to be nonessential for bacterial growth, ClpP-knockout strains in *Bacillus subtilis* and *S. aureus* were generated, both of which were highly ADEP resistant. This further corroborated ClpP as the target of the ADEPs, as its absence conferred resistance. Moreover, an affinity chromatography experiment was carried out to probe direct interaction between protein and compound. To this end, modified ADEP was synthesized and immobilized. When bacterial lysate was passed over this column, ClpP was found to be the only protein specifically binding to the column. Collectively, these results validated ClpP as the ADEP binding partner [6].

14.4.3
Mechanism of Action

Further experiments unraveled a unique mechanism of action. ADEPs bind to the outer side of the ClpP barrel at the interaction site of two subunits (Figure 14.5). During regular proteolysis, this binding site is usually occupied by accessory chaperones such as ClpX that select proteins to be degraded. ADEP binding to ClpP stabilizes the enzyme in its active oligomerization state, prevents ClpX-binding, and changes the conformation of the flexible N-termini that gate entry into the substrate degradation chamber. The structural changes were visualized by crystal structures showing ClpP in the ADEP-bound state that displays a marked increase in the size of the axial pore (Figure 14.5) [13]. As a consequence, not only small peptides but also larger and partially unfolded proteins can enter the ClpP interior through the axial pore and are then degraded. ADEP binding to ClpP thus abrogates the chaperone-exerted substrate control and results in an overactivated enzyme [14]. Interestingly, this complex does not degrade all cellular proteins regardless of their specific features but rather reduces the

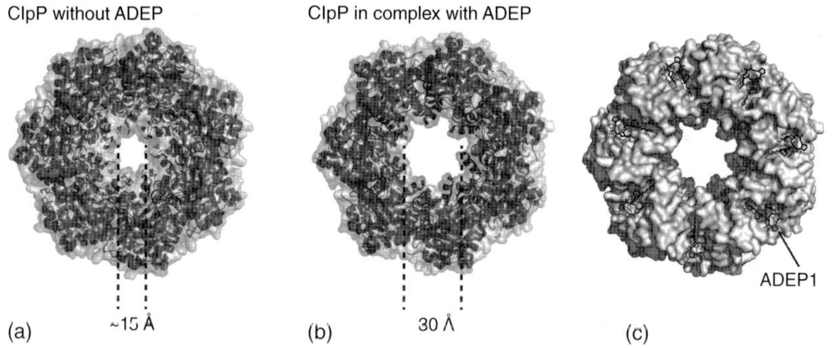

Figure 14.5 *Top views of crystal structures of ClpP from B. subtilis in ADEP-free form (a) and bound to ADEP1 (b).* ADEP binds between the ClpP subunits on the outside of the barrel in a 1 : 1 stoichiometry and induces an opening of the axial pore. (Structures published in [13] (PDB-Codes: 3KTG and 3KTI).)

amounts of selected proteins prone to degradation, one of which is FtsZ [7]. This finding was verified by Western blot analysis against FtsZ on proteome of bacteria treated with ADEP for different times (Figure 14.4b). Moreover, the ability of ClpP to degrade FtsZ upon addition of ADEP was verified by *in vitro* assays.

When bacteria multiply through cell division, FtsZ is the first protein recruited to the division site and thereby defines the geometry of binary fission. Moreover, it directs the production of a new cell wall between the separating parts of the cell and is thus absolutely essential to bacterial reproduction. A temperature-sensitive mutant strain devoid of FtsZ above a certain temperature showed a filamentation phenotype exactly as observed with ADEP-treated cells. In summary, ADEP-overactivated ClpP degrades FtsZ, which prevents cell division, causes the formation of massively elongated bacterial cells, and ultimately leads to cell death [7].

Most antibiotics act through the inhibition of essential physiological processes. The mechanism of action of ADEP-mediated cell death is special in that it relies on ClpP as a nonessential enzyme. On the one hand, this can be considered an advantage as it constitutes an unprecedented mode of action, which is responsible for the lack of cross-resistance and which opens up the possibility of developing orthogonal treatment of bacterial infections. On the other hand, ClpP, being nonessential, allows the bacterial cell to evade the ADEP-mediated selection pressure by ClpP point mutations that render the protease inactive. Consequently, resistant clones are identified with elevated frequencies in the range of 10^{-6}, which, among other reasons, led to the termination of the drug development process [6].

14.5 The Antivirulence Approach

In the second section of this chapter, we turn to a different class of compounds. While ADEPs can be comprehended as large cyclic esters, β-lactones are among

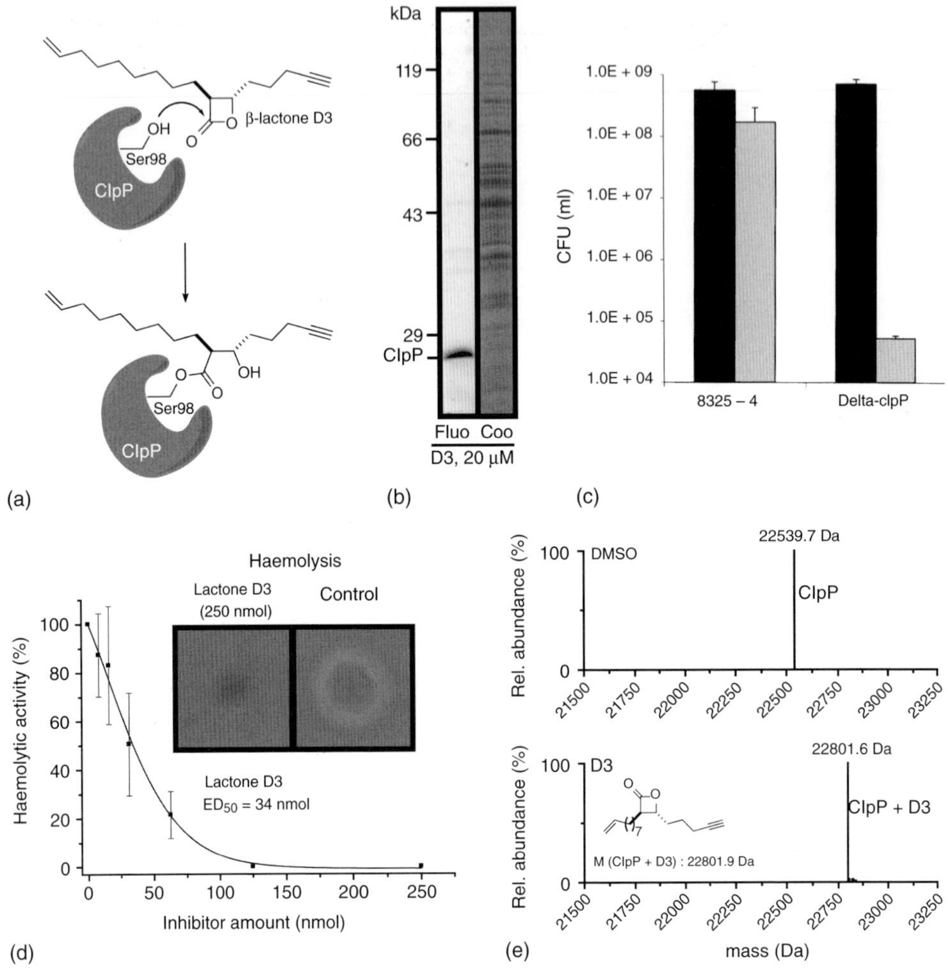

Figure 14.6 (a) Schematic of the reaction between the serine protease ClpP with its nucleophilic active site residue serine 98 and D3. (b) The fluorescence scan of an *in situ* ABPP experiment shows ClpP to be the main protein target of probe D3 in *S. aureus* cells. (Reprinted with permission from [15]. Copyright (2008) American Chemical Society.) (c) ClpX and ClpP are required for virulence in a murine skin abscess model. BALB/c mice were inoculated subcutaneously with ~10^8 cells of *S. aureus* 8325 – 4 wild type or *ClpP* mutant (solid bars) using seven mice per strain. Mice were sacrificed after seven days, and the number of bacteria recovered from skin lesions was counted (gray bars). (Reprinted with permission from [17]. Copyright (2003), John Wiley & Sons.) (d) Incubation of *S. aureus* cells with D3 leads to a reduction of hemolytic activity. (Reprinted with permission from [15]. Copyright (2008) American Chemical Society.) (e) ClpP and D3 form a covalent adduct in 1 : 1 stoichiometry as evidenced by protein mass spectrometry.

the smallest cyclic esters possible. Owing to their ring-strained nature, the electrophilicity of the carbonyl carbon atom and its reactivity toward nucleophiles is increased (Figure 14.6a). Yet, activity-based protein profiling (ABPP) (Box 14.6) of a collection of β-lactones revealed that lactone D3 almost exclusively binds the ClpP protease in living *S. aureus* cells (Figure 14.6b) [15]. Follow-up studies showed that the aliphatic substituent next to the carbonyl group of D3 occupies a deep hydrophobic pocket next to the ClpP active site, which directs the electrophilic β-lactone core in proximity to the nucleophilic active site serine [16]. Consequently, β-lactone ring opening and covalent attachment to the catalytic serine take place (Figure 14.6a). Intact protein mass spectrometry (Box 14.7) showed that all 14 ClpP active sites can be modified, which leads to inhibition of protease activity (Figure 14.6a,e).

Box 14.6 Activity-based protein profiling (ABPP)

ABPP is a forward chemical genetics technique that enables the identification of protein-binding partners of modified small molecules [18, 19]. Cells are grown and then incubated with a small molecule that contains a reactive moiety through which it covalently attaches to its binding partner. Following cell lysis, a fluorescent tag or an affinity tag such as a biotin is appended by bioorthogonal click chemistry via an alkyne handle on the small-molecule probe and an azide on the tag. Targeted proteins may then be enriched via avidin beads. After separation of the proteome via SDS-PAGE (sodium dodecylsulfate-polyacrylamide gel electrophoresis), the gel is imaged for fluorescence, indicating target proteins. These can be excised from the gel, tryptically digested and identified through peptide mass spectrometry. In the gel-free version, the enriched proteins are directly digested. The resulting mixture of peptides is then separated chromatographically and analyzed via mass spectrometry (Figure 2).

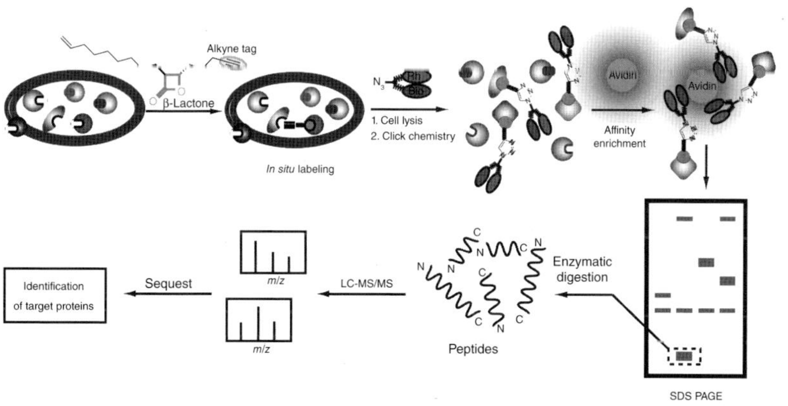

Figure 2 Schematic of a typtical ABPP workflow. Reproduced from [3] with permission from the Royal Society of Chemistry.

Box 14.7 Intact protein mass spectrometry

Intact protein mass spectrometry allows the molecular mass determination of either proteins or complexes of proteins and covalently bound ligands/other proteins. In a first step, the sample is desalted to detach from buffer components and small ions that would interfere through noncovalent complexes in the gas phase. Next, the isolated protein is ionized, for example, by electrospray ionization (ESI). The acid in the eluent causes protonation of the protein at basic sites, particularly lysine and arginine residues, so that m/z values of multiple species with different charges can be measured in a mass spectrometer. These data are then combined during the deconvolution process to yield the mass of the protein or complex.

A ClpP-knockout strain of *S. aureus* was previously assessed in a murine infection model and it was found to be severely impaired in its ability to establish an infection (Figure 14.6c) [17]. The number of ClpP-knockout bacteria isolated from an infection was more than three orders of magnitude lower than with the wild-type strain, suggesting reduced pathogenicity of the ClpP-knockout strain. This result suggests a prominent role of ClpP in virulence regulation and it prompted the researchers to test the β-lactone compounds for their antivirulence potential (Box 14.8) [2]. Indeed, treatment of living *Staphylococci* with β-lactones caused a reduction in the secretion of several virulence factors such as red-blood-cell lysing hemolysins and extracellular proteases (Figure 14.6d). These results show how a chemical knockout can be achieved with suitable compounds. Moreover, they show ClpP to have a promising role as an antivirulence target. The strength of this approach lies in its insensitivity toward antibiotic resistance mechanisms, which are frequently observed in multiresistant *Staphylococci* such as MRSA (methicillin-resistant *Staphylococcus aureus*). Future research has to focus on an optimization of the pharmacokinetic properties of the β-lactones similar to the ADEPs in order to progress in the development of a novel therapeutic option.

Box 14.8 Antivirulence Approach

The therapy of bacterial infections through antibiotics is inextricably connected with the problem of resistance. The antivirulence approach thus targets bacterial virulence rather than viability [2]. It aims at reducing the pathogenic potential of bacteria either through direct inhibition of virulence factors (such as secreted toxins, proteases, inflammation-triggering agents) or through alteration of the pathways that regulate virulence factor expression. The main advantage of this approach is that it lacks a direct feedback loop between resistance to a compound and growth advantage, because, ideally, all bacteria stay alive and can be cleared by the immune system. It would also allow for therapy of certain bacterial infections, where the exposure to antibiotics leads to a massive increase in toxin production such as EHEC (enterohemorrhagic *Escherichia coli*) and where conventional therapy is counterproductive. A common example for antivirulence therapy is the vaccination against tetanus. Here, the vaccine triggers the formation of antibodies directed against the tetanus toxin rather than against the producing strain *Clostridium tetani.*

14.6 Conclusions

Finding novel and innovative ways to target bacterial infections is a key challenge in chemical biology research. This chapter delineates two forward chemical genetic approaches that took inspiration from natural products and converged at the same target, the ClpP protease. While the ADEP class of compounds acts through ClpP activation and thereby as antibiotics, the β-lactone class of compounds inhibits ClpP and thus confers an antivirulence phenotype. The research discussed in this chapter is a prime example of how the precise modulation of protease activity leads to distinct, but complementary phenotypes. It is a remarkable illustration of the versatility of biological systems and for the contribution of chemical synthesis through suitable tool compounds that allow the dissection of complex biological processes. Finally, this chapter emphasizes the central and powerful role proteases play in physiology and shows how their activity can be exploited for the treatment of disease.

References

1. Kohanski, M.A., Dwyer, D.J., and Collins, J.J. (2010) How antibiotics kill bacteria: from targets to networks. *Nat. Rev. Microbiol.*, **8**, 423–435.
2. Clatworthy, A.E., Pierson, E., and Hung, D.T. (2007) Targeting virulence: a new paradigm for antimicrobial therapy. *Nat. Chem. Biol.*, **3**, 541–548.
3. Gersch, M., Kreuzer, J., and Sieber, S.A. (2012) Electrophilic natural products and their biological targets. *Nat. Prod. Rep.*, **29** (6), 659–682.
4. Michel, K.H., Kastner, R.E., and Eli Lilly and Company (1982) A54556 Antibiotics and process of production thereof. US patent 4492650.
5. Hinzen, B., Raddatz, S., Paulsen, H., Lampe, T., Schumacher, A., Habich, D., Hellwig, V., Benet-Buchholz, J., Endermann, R., Labischinski, H., and Brötz-Oesterhelt, H. (2006) Medicinal chemistry optimization of acyldepsipeptides of the enopeptin class antibiotics. *ChemMedChem*, **1** (7), 689–693.
6. Brötz-Oesterhelt, H., Beyer, D., Kroll, H.P., Endermann, R., Ladel, C., Schroeder, W., Hinzen, B., Raddatz, S., Paulsen, H., Henninger, K., Bandow, J.E., Sahl, H.G., and Labischinski, H. (2005) Dysregulation of bacterial proteolytic machinery by a new class of antibiotics. *Nat. Med.*, **11** (10), 1082–1087.
7. Sass, P., Josten, M., Famulla, K., Schiffer, G., Sahl, H.G., Hamoen, L., and Brötz-Oesterhelt, H. (2011) Antibiotic acyldepsipeptides activate ClpP peptidase to degrade the cell division protein FtsZ. *Proc. Natl. Acad. Sci. U.S.A.*, **108** (42), 17474–17479.
8. Wang, J., Hartling, J.A., and Flanagan, J.M. (1997) The structure of ClpP at 2.3 Å resolution suggests a model for ATP-dependent proteolysis. *Cell*, **91** (4), 447–456.

9. Sauer, R.T. and Baker, T.A. (2011) AAA + proteases: ATP-fueled machines of protein destruction. *Annu. Rev. Biochem.*, **80**, 587–612.
10. Ortega, J., Singh, S.K., Ishikawa, T., Maurizi, M.R., and Steven, A.C. (2000) Visualization of substrate binding and translocation by the ATP-dependent protease, ClpXP. *Mol. Cell*, **6** (6), 1515–1521.
11. Weber-Ban, E.U., Reid, B.G., Miranker, A.D., and Horwich, A.L. (1999) Global unfolding of a substrate protein by the Hsp100 chaperone ClpA. *Nature*, **401** (6748), 90–93.
12. Gottesman, S., Roche, E., Zhou, Y., and Sauer, R.T. (1998) The ClpXP and ClpAP proteases degrade proteins with carboxy-terminal peptide tails added by the SsrA-tagging system. *Genes Dev.*, **12** (9), 1338–1347.
13. Lee, B.-G., Park, E.Y., Lee, K.-E., Jeon, H., Sung, K.H., Paulsen, H., Rübsamen-Schaeff, H., Brötz-Oesterhelt, H., and Song, H.K. (2010) Structures of ClpP in complex with acyldepsipeptide antibiotics reveal its activation mechanism. *Nat. Struct. Mol. Biol.*, **17** (4), 471–478.
14. Kirstein, J., Schlothauer, T., Dougan, D.A., Lilie, H., Tischendorf, G., Mogk, A., Bukau, B., and Turgay, K. (2006) Adaptor protein controlled oligomerization activates the AAA + protein ClpC. *EMBO J.*, **25** (7), 1481–1491.
15. Böttcher, T. and Sieber, S.A. (2008) Beta-lactones as specific inhibitors of ClpP attenuate the production of extracellular virulence factors of Staphylococcus aureus. *J. Am. Chem. Soc.*, **130** (44), 14400–14401.
16. Gersch, M., Gut, F., Korotkov, V.S., Lehmann, J., Bottcher, T., Rusch, M., Hedberg, C., Waldmann, H., Klebe, G., and Sieber, S.A. (2013) The mechanism of caseinolytic protease (ClpP) inhibition. *Angew. Chem. Int. Ed.*, **52** (10), 3009–3014.
17. Frees, D., Qazi, S.N., Hill, P.J., and Ingmer, H. (2003) Alternative roles of ClpX and ClpP in Staphylococcus aureus stress tolerance and virulence. *Mol. Microbiol.*, **48** (6), 1565–1578.
18. Liu, Y., Patricelli, M.P., and Cravatt, B.F. (1999) Activity-based protein profiling: the serine hydrolases. *Proc. Natl. Acad. Sci. U.S.A.*, **96** (26), 14694–14699.
19. Fonovic, M. and Bogyo, M. (2008) Activity-based probes as a tool for functional proteomic analysis of proteases. *Expert Rev. Proteomics*, **5** (5), 721–730.

15
Affinity-Based Isolation of Molecular Targets of Clinically Used Drugs

Shin-ichi Sato and Motonari Uesugi

15.1
Introduction – The Biological/Medicinal Problem

Recent drug development efforts have increasingly focused on discovery of so-called molecular medicines, searching for small molecules with affinity for specific proteins that are relevant to particular human diseases or conditions. The targets of these molecular medicines are already known or preselected, so that there is no apparent need for later target identification. However, isolation of targets of drugs or drug candidates is often conducted during the drug discovery process for three reasons [1]. First, molecular targets of lead molecules discovered by phenotype-based screening are usually unknown or unclear, which poses disadvantages in the later stages of development including optimization, patent acquisition, and clinical trials. Second, isolation of "off-targets" of approved drugs or drug candidates can lead to improved drug efficacy and prediction of potential side effects. Isolation of off-targets of "dropped" or failed drugs can allow removal of side effects and rekindle interest in those drugs. Third, isolation of *bona fide* molecular targets provides a rationale for drug repositioning, that is, allows expanded application of drug candidates to other unintended diseases [2]. Such unexpected indications are sometimes perceived during clinical trials or long-term clinical uses.

Genetic, genomic, proteomic, and biochemical approaches have been developed to determine modes of action of bioactive small molecules. Of these, isolation of cellular protein targets using affinity resins is the classic and, perhaps, the most straightforward approach. This chapter focuses on affinity-based target isolation of clinically used drugs.

15.2
The Chemical Approach

Biochemical isolation of protein targets of a clinically used drug requires the chemical synthesis of its affinity resins. Preparation of the affinity resins is facilitated by information about the structure–activity relationships of the small

Concepts and Case Studies in Chemical Biology, First Edition. Edited by Herbert Waldmann and Petra Janning.

molecule of interest. Structure–activity relationships of clinically used drugs have usually been extensively studied, and often suggest sites appropriate for linker modification, which can be used to design the affinity resins. Once a modification site is identified, the drug is covalently attached to an appropriate linker and bound, covalently (Figures 15.1 and 15.2) or noncovalently (Figure 15.3), to solid supports. The proteins that bind directly to the drug are purified from cell lysates by affinity chromatography, separated by sodium docecylsulfate-polyacrylamide gel electrophoresis (SDS-PAGE), and excised from the gel. The isolated proteins are identified by partial tryptic digestion, mass sequencing of the digested peptides, and database search of the sequences. The identified proteins must then be validated as the actual molecular targets by independent molecular and cell biology experiments, to confirm that one or more of them is responsible for part or all of the effects of the small molecule. A major problem with affinity-based

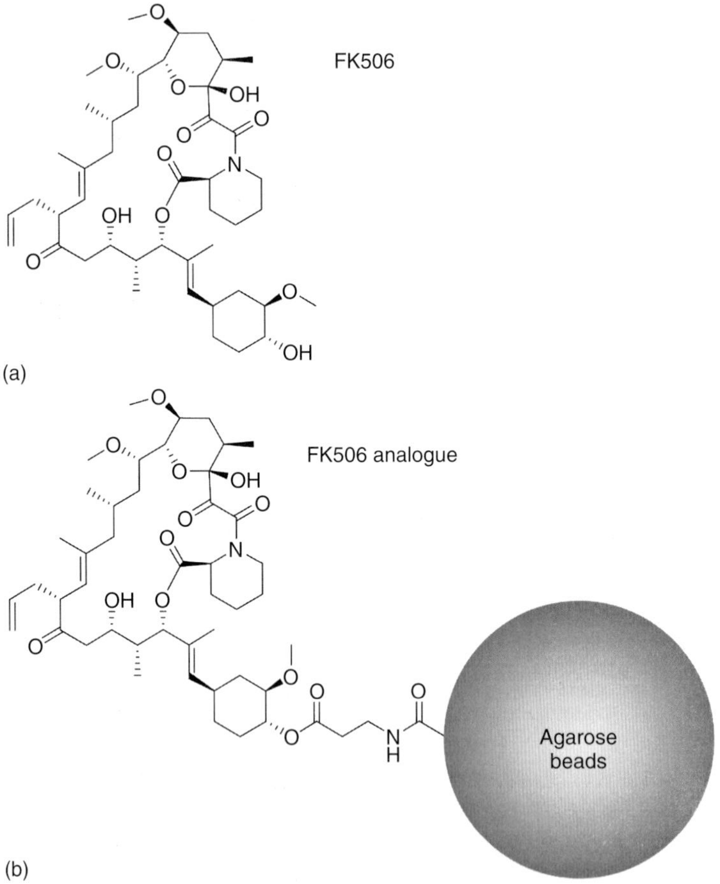

(a)

(b)

Figure 15.1 *Structure of FK506 (a) and FK506-conjugated agarose beads (b).* FK506 is directly immobilized to agarose beads through a linker.

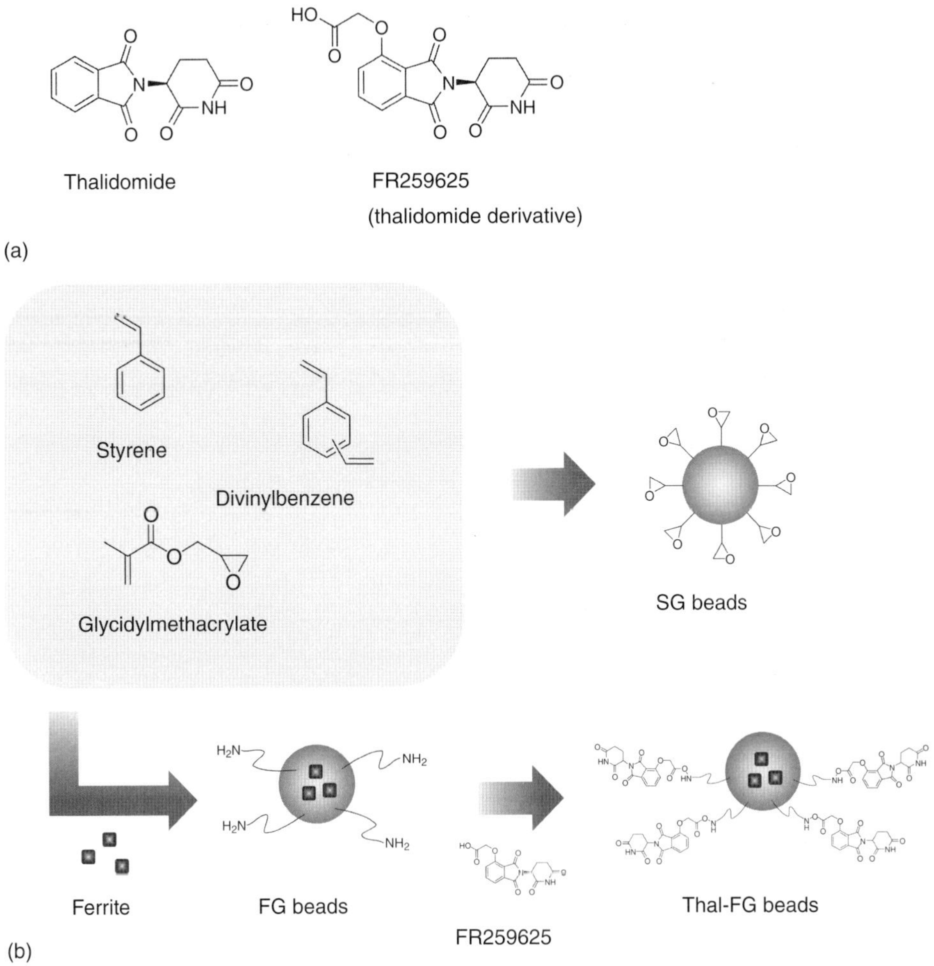

Figure 15.2 (a) Structure of thalidomide and its derivative, FR259625. (b) Scheme of thalidomide immobilization to FG beads is shown.

target identification is the existence of nonspecific binding proteins, which are hard to remove completely, even after careful washing of the affinity resin. Nonspecific or less specific proteins have often confused scientists in academia and industry, and a number of chemical approaches have been developed to exclude them.

Two main approaches are used to prepare affinity resins of bioactive molecules: (i) direct covalent conjugation of the molecule of interest to affinity resins and (ii) preparation of a biotinylated version of the molecule, followed by its noncovalent binding to avidin resins. The covalently conjugated resins exhibit the high load capacity of the molecule (a millimolar range in slurry), permitting capture of binding proteins with relatively low affinities. However, nonspecific adsorption

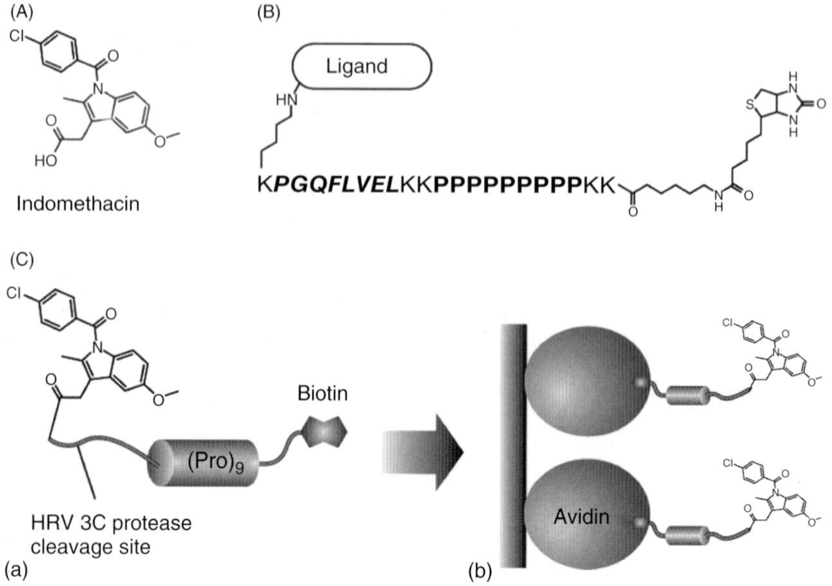

Figure 15.3 (A) Structure of indomethacin. (B) The amino acid sequence of a polyproline linker. (C) The model structure of biotinylated indomethacin with a polyproline linker (a) and its affinity resin (b).

of proteins and other biopolymers is a problem. One potential chemical solution would be development of polymer resins with reduced protein adsorption properties and their magnetic bead versions for efficient washing during protein purification.

For example, Handa and coworkers developed such a low-adsorption matrix and its magnetic bead version, starting with a glycidylmethacrylate (GMA)-covered GMA-styrene copolymer core (SG[1)] beads), originally used for the affinity purification of DNA-binding proteins (Figure 15.2) [3]. A divalent epoxide, ethyleneglycoldiglycidylether (EGDE), was introduced as a spacer, following aminolysis of epoxy groups on the surface of the beads. The lack of pores and hydrophilic surfaces (due to the GMA and the EGDE spacer arm) of these SGNEGDE[2)] beads provided efficient removal of residual proteins during washing, and reduced nonspecific protein interactions in comparison with the commonly used matrix. Handa and coworkers further developed a magnetic version of SGNEGDE beads, constructed by admicellar polymerization, with a uniform core/shell/shell nanostructure carrying 40 nm magnetite particles (Figure 15.2).

The biotinylation approach suffers less than the covalent conjugation method from nonspecific protein adsorption. However, the biotinylation method offers limited load capacity of the molecule, which can result in lower recovery of the

1) Glycidylmethacrylate (GMA)-covered GMA-styrene copolymer core
2) Ethyleneglycoldiglycidylether-coupled epoxide on SG

binding proteins. One chemical solution for this limitation is optimization of the linker moiety of the biotinylated molecules. A variety of polymethylene linkers and polyethylene glycol (PEG) linkers with different lengths is commercially available and many have been used for this purpose. Hydrophilic PEG linkers are generally preferable to polymethylene linkers, because the PEG conjugates exhibit more desirable physical properties and usually reduce the binding of nonspecific proteins.

Length is also an important factor in determining the usefulness of a specific linker. Our laboratory examined the effects of linker length on recovery rates of target proteins from an affinity resin. PEG-based linkers with different lengths (11–32 Å) were inserted between a small-molecule bait and biotin, and target recovery was compared for the resulting conjugates. Longer PEG linkers tend to exhibit higher recovery of the molecular target from cell lysates. Furthermore, elongation of the linker, by insertion of a long, rigid polyproline helix between the small-molecule bait and the biotin tag, boosted the capacity of affinity purification (Figure 15.3) [4, 5]. The rigid polyproline helix might project the small-molecule bait away from the biotin–avidin complex, enhancing its interaction with protein targets. Another improvement that we have tested is insertion of a cleavage site of HRV3C protease, a highly specific protease that digests the peptide at 4 °C, between the polyproline linker and the small-molecule bait. The mild cleavage condition of HRV3C allows selective elution of the binding proteins by proteolytic cleavage. The polyproline-rod approach has now been used in target identification programs by pharmaceutical companies and academic researchers.

15.3 Chemical Biological Research

As mentioned earlier, there are three major motivations for target protein isolation of clinically used drugs: identification of unknown molecular targets, identification of off-targets for side effects, and drug repositioning. In this section, we provide one successful example of each category with a distinct chemical approach, with the attempt to extract lessons from the representative examples.

15.3.1 Lessons from Isolation of FK506-Binding Protein (FKBP) Using FK506

The most prominent example of successful target isolation of a clinically used drug might be the isolation of FKBP with immunosuppressant FK506 (tacrolimus) by Schreiber and coworkers. FK506, a highly potent immunosuppressive drug, was discovered by Fujisawa Pharmaceutical (now Astellas Pharma) in 1984 from the fermentation broth of a Japanese soil sample that contained the bacterium, *Streptomyces tsukubaensis*. FK506 had been used mainly for organ transplantation and atopic dermatitis, but its molecular target was unknown.

To isolate the target of FK506, Schreiber and coworkers [6] covalently linked FK506 directly to agarose beads, through a linker attached to a biologically inert hydroxyl group of FK506 (Figure 15.1). The resulting affinity beads were treated with human cell lysates. SDS-PAGE analysis of the proteins that bound to the affinity beads showed a single 14 kDa band, which turned out be a band of the protein, FKBP. Although the isolation of FKBP had profound impacts on both biology and medicine, from a technical point of view, this successful target isolation seemed to rely heavily on the fact that the target protein was particularly well suited for affinity-based isolation: (i) FK506 exerts its biological activity in the picomolar range, and its interaction with FKBP is very tight; (ii) FKBP is highly abundant in cells; and (iii) FKBP is highly soluble in cell lysates.

Tight interactions and abundant targets are the most favorable conditions for target isolation. In general, the higher the affinity of a small molecule for the target protein, the more successful the target isolation will be. High-affinity complexes tend to be maintained after extensive washing, which reduces the amount of nonspecific binding proteins in the sample. Assuming that more potent molecules have higher affinities for their targets, the molecule with the lowest EC_{50} (effective concentration 50) or IC_{50} (inhibitor concentration 50) value is usually considered best suited for target identification. Although a high-affinity ligand is advantageous for isolating its complexes, its target protein is not necessarily easy to identify. A very low effective concentration, for example, in the picomolar range, might reflect a low abundance of the target protein, making target identification difficult.

Another point to note in the case of FKBP isolation is that later studies showed that the FK506/FKBP complex binds to and inhibits a less abundant protein, calcineurin [7]. In fact, calcineurin is the *bona fide* target of FK506 and responsible for its immunosuppressive activity. Thus, in some instances, the complex of a drug with an abundant target interacts with another protein target that is actually responsible for the biological activity of the drug. Another example of this is chromeceptin, an inhibitor of IGF signaling, which binds to the abundant protein, MFP-2 (membrane fusion protein), to interact with its *bona fide* target, ACC1 [8]. Similarly, fusicoccin, a fungal phytotoxin, binds to the abundant protein, 14-3-3, to recruit other proteins to exert its biological activity [9].

15.3.2
Lessons from Isolation of Cereblon (CRBN) Using Thalidomide

Thalidomide is a sedative or hypnotic drug that was released in the market in West Germany in 1957 by Grünenthal GmbH. Shortly after its launch, thalidomide was used to treat nausea and alleviate morning sickness in pregnant women, resulting in severe birth defects when used during the first trimester of pregnancy [10–12]. The drug was withdrawn from use in 1961, and the molecular mechanism of the teratogenetic effects was a mystery for 50 years, until it was resolved by Handa and coworkers.

To purify thalidomide-binding proteins, Handa and coworkers used covalent conjugation of a thalidomide derivative to a magnetic version of low-adsorption

SGNEGDE beads, called *FG*[3] *beads* (Figure 15.2) [3]. The carboxylic thalidomide derivative, FR259625, was covalently conjugated to the beads and incubated with human HeLa cell extracts. Two polypeptides were specifically eluted by adding free thalidomide, and were subsequently identified as CRBN (cereblon) and damaged DNA-binding protein 1 (DDB1). In this case, successful target isolation appears to have depended on the suppressed recovery of nonspecific proteins by the employment of the low-adsorption beads, and by the elution of the specific targets with the free ligand.

Human CRBN encodes a 442-amino acid protein that had been reported to interact with DDB1 in a proteomic analysis [13]. However, the functional relevance of this interaction was unclear at that time. DDB1 is a component of E3 ubiquitin ligase complexes [14]. Molecular biological and biochemical studies ultimately showed that thalidomide binds to CRBN and inhibits the associated ubiquitin ligase activity. From extensive experimentation with zebrafish and chicks, Handa and coworkers demonstrated that these interactions are responsible for the teratogenic effects of thalidomide.

In 1996, thalidomide was reapproved for treatment of leprosy, and in 2006, the US Food and Drug Administration granted accelerated approval for use of thalidomide in the treatment of newly diagnosed multiple myeloma patients. Identification of the off-targets of thalidomide is expected to contribute greatly to the development of new thalidomide derivatives without teratogenic activity [11, 12].

15.3.3
Lessons from Isolation of Glyoxalase 1 (GLO1) Using Indomethacin

Off-targets of drugs occasionally provide surprising benefits, the so-called beneficial side effects. For example, the anti-inflammatory drug, indomethacin, has a beneficial side effect found in oncology [15, 16]. This highly popular drug is known to exert its anti-inflammatory activity by inhibiting cyclooxygenase (COX). However, epidemiological studies have demonstrated that indomethacin also enhances the anticancer effects of anticancer drugs. We have succeeded in isolating a secondary target protein of indomethacin that is probably responsible for this beneficial side effect.

We used a polyproline-rod approach with biotinylated indomethacin to isolate the secondary target. A biotin molecule was conjugated with a 9-mer of prolines, an HRV-C3 protease site, and indomethacin (Figure 15.3). To improve water-solubility, two lysine residues were introduced between a biotin molecule and a polyproline linker. As expected, affinity purification from mammalian cell lysates isolated COX-1, a known target of indomethacin. In addition to COX-1, we isolated a second target of indomethacin, human GLO1 (glyoxalase 1), an abundant metabolic enzyme that catalyzes the conversion of methylglyoxal to D-lactate. Results of further biochemical and cell-based experiments suggested

3) Ferrite SG

that inhibition of GLO1's enzymatic activity is responsible for the clinically observed synergy between indomethacin and anticancer drugs [4].

The case of indomethacin and GLO1 demonstrates the importance of the balance between protein abundance and effective concentration in the success of affinity-based target isolation. The higher the affinity and the more abundant the target, the more likely is successful isolation. Although the K_D value of its interaction with indomethacin was in the low micromolar range, the high abundance of GLO1 made purification and identification of the target possible. Isolation of a low-abundance target with a low-affinity molecule would be extremely challenging [4, 17].

15.4
Conclusion

Identifying molecular targets remains a major technical challenge in phenotype-based drug discovery, overcoming side effects, and drug repositioning. In this chapter, we briefly summarized chemical approaches to affinity-based target identification and factors influencing their success. Although the affinity-based biochemical approach needs to be combined with nonbiochemical approaches to reduce the risk of incorrect target identification, the classic affinity-based approach is powerful. Recent advances in mass spectrometry, genomics, and analytical techniques have greatly facilitated biochemical isolation and validation of molecular targets. What we lack are methods for isolating target proteins that are inactive, insoluble, or in low abundance in cell lysates, and methods for isolating, identifying, and validating nonprotein molecular targets. Continued method development will lead to successful identification of additional molecular targets.

References

1. Stockwell, B.R. (2000) Chemical genetics: ligand-based discovery of gene function. *Nat. Rev. Genet.*, **1**, 116–125.
2. Ashburn, T.T. and Thor, K.B. (2004) Drug repositioning: identifying and developing new uses for existing drugs. *Nat. Rev. Drug Discovery*, **3**, 673–683.
3. Sakamoto, S., Kabe, Y., Hatakeyama, M., Yamaguchi, Y., and Handa, H. (2009) Development and application of high-performance affinity beads: toward chemical biology and drug discovery. *Chem. Rec.*, **9**, 66–85.
4. Sato, S., Kwon, Y., Kamisuki, S., Srivastava, N., Mao, Q., Kawazoe, Y., and Uesugi, M. (2007) Polyproline-rod approach to isolating protein targets of bioactive small molecules: isolation of a new target of indomethacin. *J. Am. Chem. Soc.*, **129**, 873–880.
5. Sato, S., Murata, A., Orihara, T., Shirakawa, T., Suenaga, K., Kigoshi, H., and Uesugi, M. (2011) Marine natural product aurilide activates the OPA1-mediated apoptosis by binding to prohibitin. *Chem. Biol.*, **18**, 131–139.
6. Harding, M.W., Galat, A., Uehling, D.E., and Schreiber, S.L. (1989) A receptor for the immunosuppressant FK506 is a *cis-trans* peptidyl-prolyl isomerase. *Nature*, **341**, 758–760.

7. Liu, J., Farmer, J.D. Jr., Lane, W.S., Friedman, J., Weissman, I., and Schreiber, S.L. (1991) Calcineurin is a common target of cyclophilin-cyclosporin A and FKBP-FK506 complexes. *Cell*, **66**, 807–815.
8. Jung, D., Abu-Elheiga, L., Ayuzawa, R., Gu, Z., Shirakawa, T., Fujiki, Y., Nakatsuji, N., Wakil, S.J., and Uesugi, M. (2012) Mislocalization and inhibition of acetyl-CoA carboxylase 1 by a synthetic small molecule. *Biochem. J.*, **448**, 409–416.
9. Takahashi, M., Kawamura, A., Kato, N., Nishi, T., Hamachi, I., and Ohkanda, J. (2012) Phosphopeptide-dependent labeling of 14–3–3 ζ proteins by fusicoccin-based fluorescent probes. *Angew. Chem. Int. Ed.*, **51**, 509–512.
10. Miller, M.T. and Strömland, K. (1999) Teratogen update: thalidomide: a review, with a focus on ocular findings and new potential uses. *Teratology*, **60**, 306–321.
11. Melchert, M. and List, A. (2007) The thalidomide saga. *Int. J. Biochem. Cell Biol.*, **39**, 1489–1499.
12. Knobloch, J. and Rüther, U. (2008) Shedding light on an old mystery: thalidomide suppresses survival pathways to induce limb defects. *Cell Cycle*, **7**, 1121–1127.
13. Higgins, J.J., Pucilowska, J., Lombardi, R.Q., and Rooney, J.P. (2004) A mutation in a novel ATP-dependent Lon protease gene in a kindred with mild mental retardation. *Neurology*, **63**, 1927–1931.
14. Lee, J. and Zhou, P. (2007) DCAFs, the missing link of the CUL4-DDB1 ubiquitin ligase. *Mol. Cell*, **26**, 775–780.
15. Duffy, C.P., Elliott, C.J., O'Connor, R.A., Heenan, M.M., Coyle, S., Cleary, I.M., Kavanagh, K., Verhaegen, S., O'Loughlin, C.M., NicAmhlaoibh, R., and Clynes, M. (1998) Enhancement of chemotherapeutic drug toxicity to human tumour cells *in vitro* by a subset of non-steroidal anti-inflammatory drugs (NSAIDs). *Eur. J. Cancer*, **34**, 1250–1259.
16. Hull, M.A., Gardner, S.H., and Hawcroft, G. (2003) Activity of the non-steroidal anti-inflammatory drug indomethacin against colorectal cancer. *Cancer Treat. Rev.*, **29**, 309–320.
17. Sato, S., Murata, A., Shirakawa, T., and Uesugi, M. (2010) Biochemical target isolation for novices: affinity-based strategies. *Chem. Biol.*, **17**, 616–623.

16
Identification of the Targets of Natural-Product-Inspired Mitotic Inhibitors

Kamal Kumar and Slava Ziegler

16.1
Introduction

Proteins are major drug targets, and small molecules can bind to proteins and modulate their functions. In particular, naturally occurring compounds represent very potent protein modulators. A biologically active compound does not necessarily bind to only one protein and can interact with many proteins. To understand the mechanism of action of a given substance in modulating a biological function in living organisms, it is essential to have detailed information on its molecular targets. This knowledge will also guide the synthesis of more potent and more specific derivatives. Here, we describe the design and synthesis of a natural-product-inspired compound collection that was analyzed in a phenotypic screen to identify inhibitors of mitosis and the biological characterization of the hit compounds as well as the identification of their molecular targets.

16.2
The Biological Problem

16.2.1
Mitosis and Modulation of Mitosis by Small Molecules

Cells need to divide in order to ensure growth and propagation of organisms. Cell division in eukaryotes is regulated by the cell division cycle, which consists of interphase and M phase (Box 16.1). During interphase, cells prepare for division. The M phase is divided in mitosis (i.e., the process of equal separation of genetic material of a parental cell into two daughter cells) and cytokinesis (i.e., the separation of the cytoplasm, resulting in two daughter cells). The progression through the cell cycle is unidirectional and irreversible and must be tightly regulated in time and space to ensure the error free segregation of chromosomes. This is achieved by the temporal activation and inactivation of cyclin-dependent kinases (CDKs) in complex with defined cyclins, whereby the presence of cyclins

Concepts and Case Studies in Chemical Biology, First Edition. Edited by Herbert Waldmann and Petra Janning.

oscillates during the cell cycle. Control mechanisms called *checkpoints* allow for the entry in the next cell cycle phase only after proper completion of the previous one (e.g., G1, G2, and spindle-assembly checkpoint, Box 16.1).

Box 16.1 Cell (Division) Cycle

Cell cycle is the repeated series of events in cells that lead to division of a parental cell into two daughter cells. The cell cycle is divided into interphase and M phase. The interphase encompasses the G1 phase (gap 1, cell growth), S phase (synthesis, DNA replication), G2 phase (gap 2, cell growth and preparation for mitosis). During the M phase, mitosis and cytokinesis take place. After completion of cell division, cells may enter the G1 phase for a next round of cell division. Alternatively, cells may exit the cell cycle and remain in a resting state (also called G0 phase). In this quiescent state, cells can remain for a long period of time or even indefinitely (Figure 1).

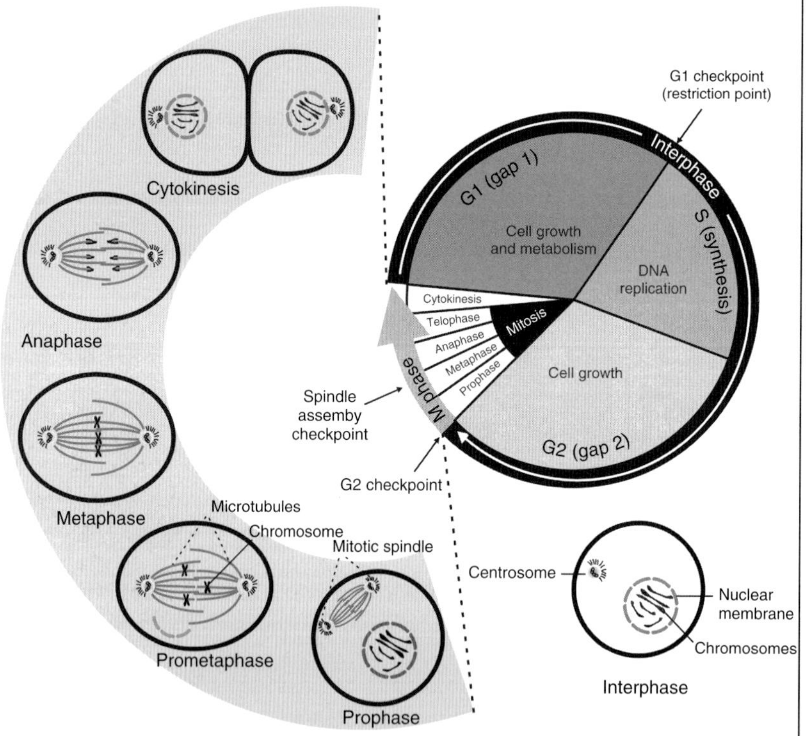

Figure 1 Cell division cycle.

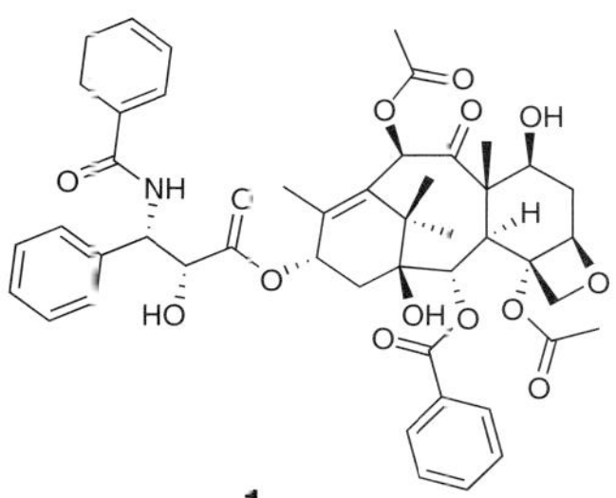

1

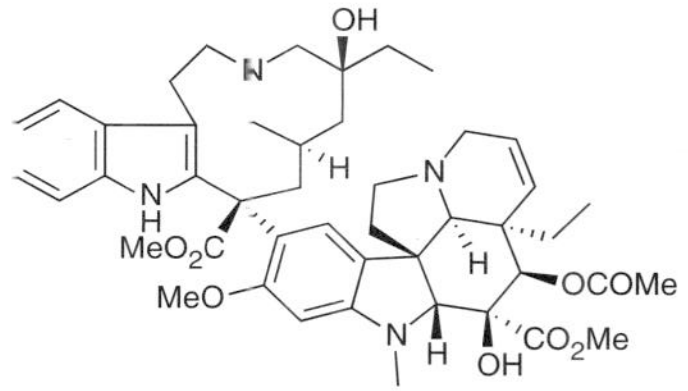

2

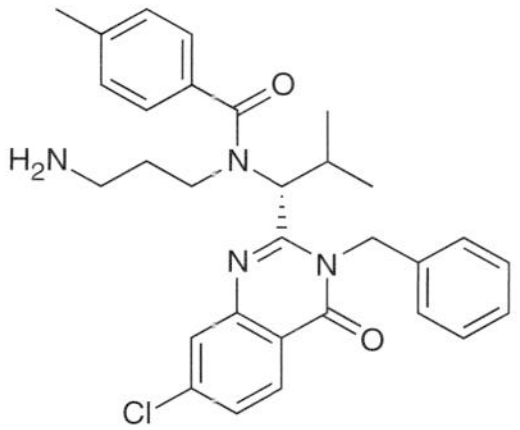

3

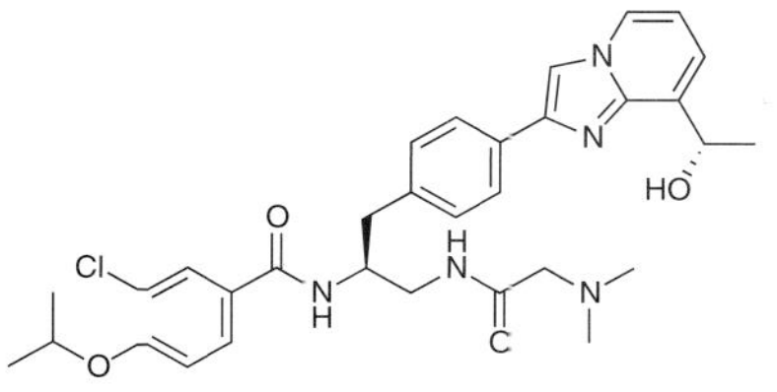

4

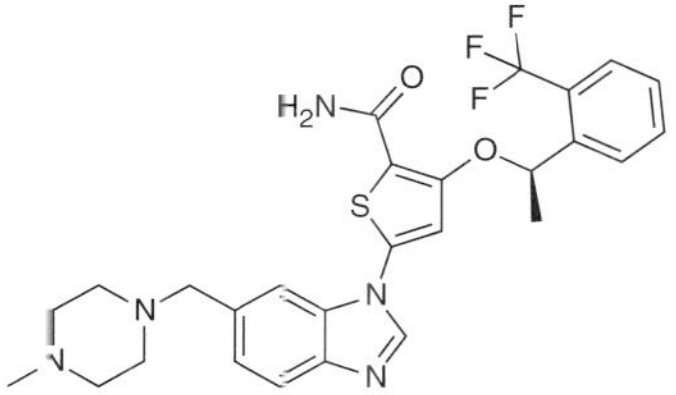

5

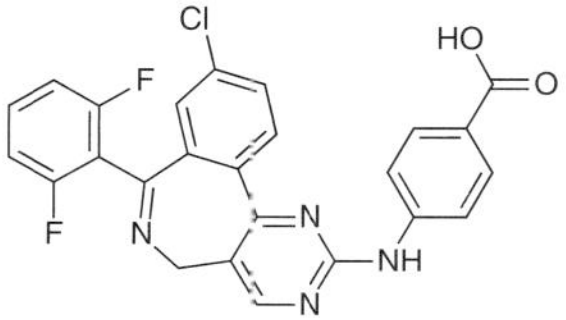

6

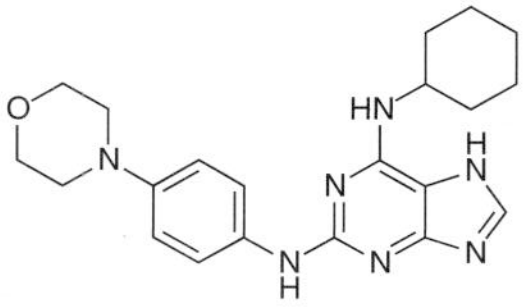

7

Figure 16.1 *Chemical structures of some mitotic inhibitors.* Microtubule modulators taxol (**1**) and vinblastine (**2**), Eg5 inhibitor ispinesib (**3**), CENP-E inhibitor GSK923295 (**4**), Plk1 inhibitor GSK 461364 (**5**), Aurora A inhibitor MLN8054 (**6**), and Mps1 inhibitor reversine (**7**).

Cancer is one of the leading causes of death worldwide. There are several hallmarks of cancer, for instance, sustained proliferative signaling that deregulates cell division [1]. Mitosis is amenable to modulation by small molecules and many of the current anticancer drugs target mitosis (e.g., taxol (**1**) and vinblastine (**2**) (Figure 16.1), which disturb microtubule dynamics [2]). However, these agents often suffer from severe adverse effects that call for the identification of novel antimitotic compounds. Although numerous proteins play important roles in the process of mitosis, only a few of them (tubulin, centromeric protein (CENP-E), Eg5, Plk, Aurora A and B, and Mps1 kinases) have been modulated by small molecules (Figure 16.1) [3].

Owing to redundant functions, it is difficult to predict which protein(s) involved in mitosis would be good drug target candidates. Therefore, cell-based, rather than target-based, screening [4] may uncover novel druggable protein targets for modulation of mitosis (Box 16.2).

Box 16.2 Chemical Genetics Approaches

Target-based screening, also called the reversed chemical genetics approach, aims to identify small-molecule modulators for a known protein of interest. The identified compounds are then employed to study protein function *in cellulo* or *in vivo*.

Cell-based screening, also called the forward chemical genetics approach, first analyzes the modulation of living systems (cells or organisms) by compounds, for example, by means of a reporter-gene activity, a fluorescence signal, or phenotypic changes detected by means of imaging methods, and only subsequently are the biological targets identified (Figure 2).

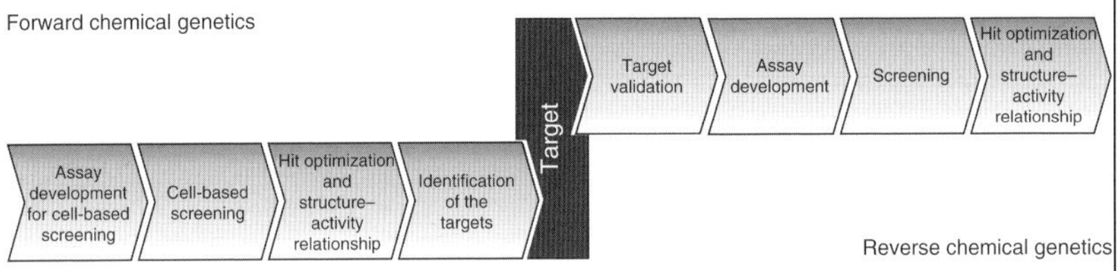

Figure 2 Chemical genetics approaches.

Modified according to Terstappen *et al.* [5].

16.2.2 Phenotypic Screening

Phenotypic (or cell-based) screens monitor the influence of a compound on a complex living system in its entirety [6]. Compounds identified in such assays have the proven ability to modulate such complex systems in the desired manner, a property not necessarily shared by hits obtained from target-based screens that need to be validated in cellular assays first (Box 16.2, Figure 2) [7].

Usually, a compound collection is subjected to a cellular assay of interest to find substances that interfere with a certain readout. This initial screening will define hit compounds and may provide a first structure–activity relationship (SAR, Box 16.3) given that structurally related compounds were included in the compound collection. Often, it requires a few rounds of organic synthesis and screening to define the sites in the hit compound that are required for activity as well as the sites that can be modified without any loss of activity. This information is essential for the attachment of a linker (spacer) with a functionality that will enable the isolation of potential target proteins by means of affinity-based proteomics (Box 16.3). Depending on the employed phenotypic assay, additional biological characterization of the hit compounds may help narrow down or exclude potential target proteins.

Box 16.3 SAR and Affinity-based Proteomics

SAR is the correlation of the chemical structure of a compound to its biological activity. SAR allows defining the chemical groups that are required for the activity of the substance and guides the design of derivatives with the objective of increasing the potency of the compound or introducing modifications without impairment of its activity.

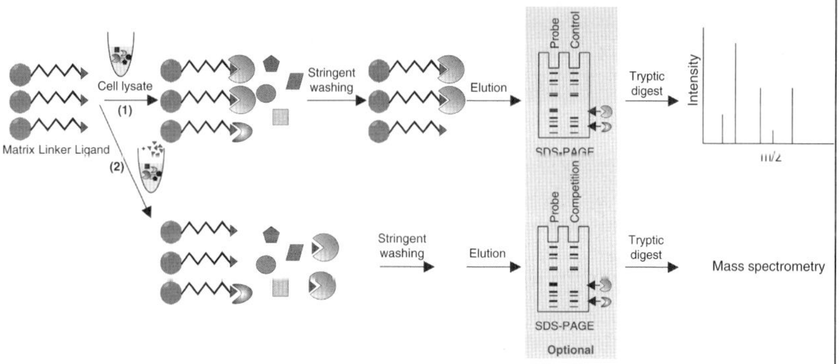

Figure 3 Identification of targets for biologically active small molecules by means of affinity-based proteomics.

Affinity-based proteomics. In affinity-based proteomics, a compound of interest is immobilized to the solid surface by means of functional groups (e.g., NH_2) or using the biotin–streptavidin interaction. Cell lysate that contains the target proteins is then incubated with the immobilized compound and this leads to enrichment of proteins on the matrix (**1**). Stringent washing removes proteins that unspecifically bind to the solid surface. Bound proteins are then released from the matrix by elution (e.g., with an excess of unmodified compound) or by heating in denaturing buffer. Proteins are identified by means of sodium dodecyl sulfate (SDS) polyacrylamide gel electrophoresis (SDS-PAGE, sodium dodecyl sulfate polyacrylamide gel electrophoresis), tryptic in-gel digest (or, alternatively, on matrix tryptic digest) followed by mass spectrometry. In a competition approach, an excess of unmodified compound is added to the cell lysate and it is expected to reduce the amount of target proteins that are enriched on the matrix (**2**) (Figure 3).

16.2.3
Target Identification and Confirmation

The most demanding step in forward-chemical genetics is the identification of the molecular targets of a compound of interest. Several methodologies have been established and so far there is no generic approach available [7]. One of the most widely applied methodologies is the affinity isolation of proteins that bind to immobilized compounds of interest. Proteins are identified after analysis using mass spectrometry and need to be further confirmed as the molecular target(s). Typically, the target confirmation includes detection of the direct binding of the small molecule to the protein(s) by means of biophysical methods (e.g., fluorescence polarization (FP), isothermal calorimetry, surface plasmon resonance, fluorescence life-time imaging microscopy), overexpression or knockdown of the protein (which can sensitize or desensitize cells for the compound or may phenocopy the compound's activity), localization studies using a derivative that is labeled with a fluorophore and target-specific assays (e.g., enzymatic activity).

16.3
The Chemical Approach

16.3.1
Design and Synthesis of Natural-Product-Inspired Compound Collections

Natural products represent a rich source of biologically active substances and have been used for thousands of years to treat diseases without the detailed knowledge of the active components. Advances in separation and analytical techniques and organic synthesis methods allowed for the isolation and characterization of biologically active natural products. Numerous natural products are successfully applied as anticancer agents (e.g., taxol (**1**), vinblastine (**2**), (Figure 16.1)) [8]. However, they usually suffer from lack of selectivity toward cancer over normal cells and the development of targeted therapies (i.e., antibodies or substances that block the growth and spreading of tumor cells by specifically interfering with molecules involved in tumor growth and progression) is on the rise. This has made these mostly cytotoxic agents less attractive [9]. Pharmaceutical companies have put a lot of effort exploring combinatorial libraries. However, combinatorial compound collections failed to meet the expectations and could not enrich the pipelines with hit and lead molecules. Drug approvals by the US Food and Drug Administration (FDA) have continued to fall from the levels of the 1990s and therefore a revival in the use of natural products for drug discovery has already begun [10]. Natural products and natural-product-derived compounds represent a major part of current anticancer small molecules. The scaffolds of natural products are regarded as prevalidated frameworks because they are created by nature and during synthesis they have experienced binding to proteins of the biosynthetic machineries [11]. Therefore, compound collections based on natural product scaffolds continue to be of particular interest for finding novel biologically active agents. However, generation of structural complexity of the level of natural products in a compound collection calls for development of efficient synthesis methods. In particular, cascade reactions wherein many chemical reactions happen consecutively in a sequence and molecular complexity is rapidly built up are gaining attention (see Chapter 27 for details) [12]. A concise and efficient 12-step cascade synthesis of tetracyclic tetrahydroindolo[2,3-*a*]quinolizines, which embody the core scaffold of numerous polycyclic indole alkaloids, was developed (Scheme 16.1) [13]. In this one-pot cascade synthesis of the centrocountins, a toluene solution of 3-formyl-chromones (**8**), acetylenedicarboxylates (**9**), and triphenylphosphine at 80 °C was slowly treated with tryptamine derivatives (**13**) followed by addition of camphorsulfonic acid (CSA). After 5–30 min, the centrocountins (**24**) were purified by either flash column chromatography or by precipitation and crystallization. A very simple and practical method to generate natural product-like complex molecules, however it is the longest known cascade reaction and involves many individual steps and chemical reactions happening one after another before centrocountins are generated. The mechanism of the reaction sequence is illustrated in Scheme 16.1 and was supported by isolating and characterizing the key intermediates (in dotted boxes) appearing in the cascade reaction sequence. Thus, the cascade sequence commenced with the [4+2] annulation of 3-formylchromones **8** with alkynes **9** catalyzed by triphenylphosphine to yield tricyclic benzopyrones **12** (Scheme 16.1) [14].

Tryptamine derivatives **13** undergo conjugate addition to ring C of the tricyclic benzopyrones **12** accompanied by ring opening in which a phenol moiety serves as a leaving group and thus forming intermediate **14**. The phenol **14** could add again to the newly generated α,β-unsaturated carbonyl moiety, and in turn pushes another pyran ring opening to generate intermediates **15**, supporting an enamine and an α-keto ester in close proximity. A 6π electrocyclization of triene **16**, the isomeric form of the **15** leads to α-hydroxy-dihydropyridine **17**. Avoiding acidic conditions, this intermediate could be isolated and characterized. Under very mild acidic conditions, **17** quickly dehydrate to form the tricyclic dihydropyridines **18**. Acid-promoted opening of the chromone ring generates the pyridiinium salt **19** to which phenolate adds to yield the tricyclic dienes **20** and thereby set the stage for a sigmatropic aza-Claisen rearrangement providing iminoesters **21**. A Pictet–Spengler cyclization of indole ring with the activated imine generates

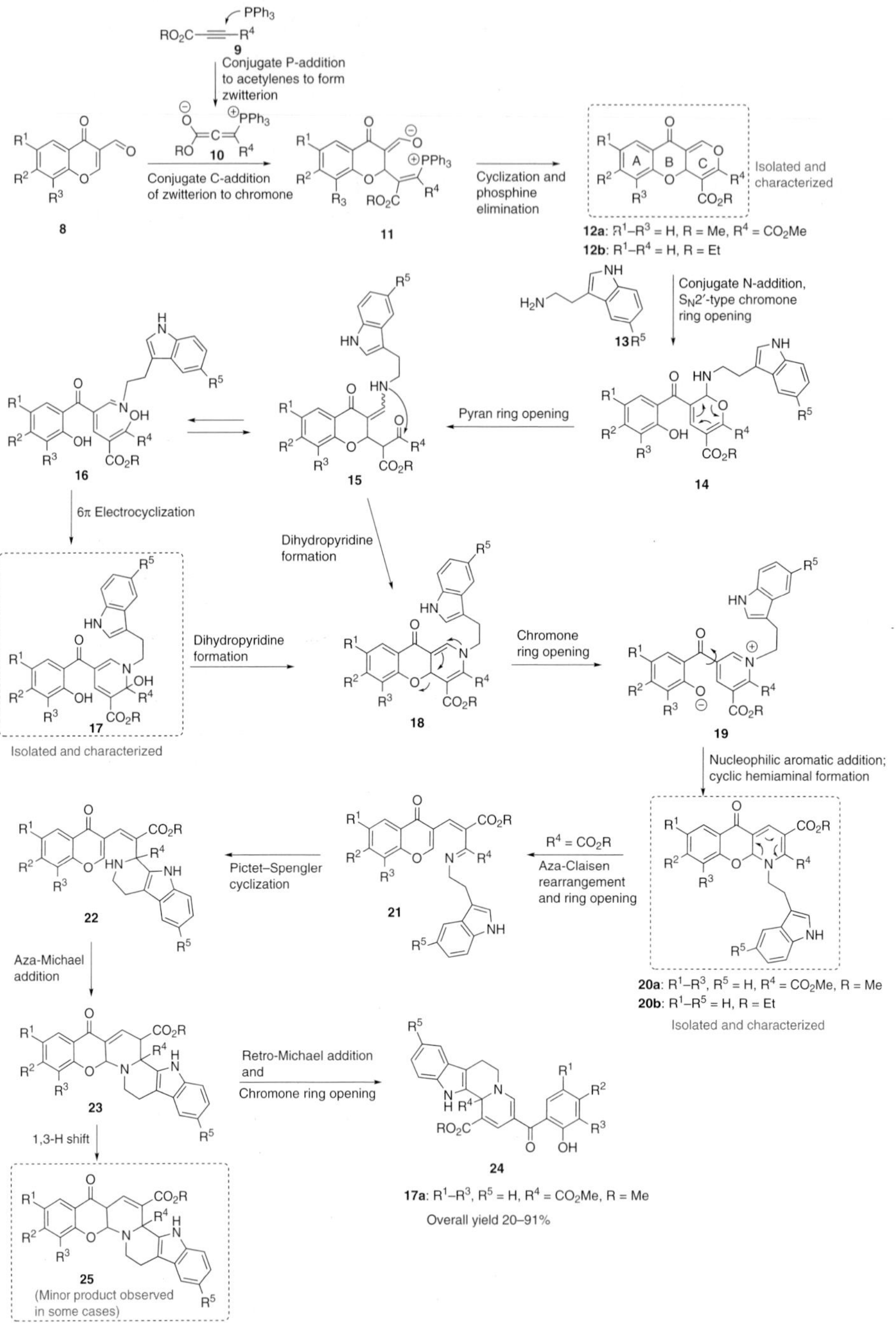

Scheme 16.1 Cascade synthesis of centrocountins.

secondary amines **22**. The final steps of the sequence consist of conjugate aza-Michael addition of the secondary amines to the doubly vinylogous esters to yield addition products **23**. A final acid-mediated pyran ring opening with phenol, serving again as a leaving group, culminates in the formation of indoloquinolizines **24**. Isolation and characterization of hexacyclic molecules **25** that should originate after a 1,3-*H* shift happens in **23** further corroborated the proposed final steps of this long cascade sequence. The synthesis was surprisingly very high yielding and led to a focused library of more than 60 indoloquinolizines [15].

16.4 Chemical Biological Evaluation

16.4.1 Phenotypic Screen for Mitotic Inhibitors

To identify novel small molecules that inhibit mitosis, a high-content cell-based assay was established to monitor changes in the cytoskeleton (Box 16.4). For this purpose, the African green monkey BSC-1 cells were treated for 24 h with compounds before immunostaining of the cells for tubulin, actin, and DNA (Figure 16.2 and Box 16.4). Compounds that cause changes in cell morphology related to mitosis are considered as hits. Mitotic inhibitors typically disturb the proper alignment of chromosomes in the equatorial plane during metaphase (also called metaphase plate) and activate the spindle-assembly checkpoint. Cells become arrested in mitosis and can remain in this phase for a long period of time. This causes the accumulation of round-shaped cells with intense staining of the DNA and microtubules, which can be detected by fluorescence staining using antibodies or dyes. The spindle-assembly checkpoint in cells that are arrested in mitosis is inactivated only after proper error correction. Alternatively, cells commit suicide by apoptosis.

Box 16.4 High-content Screen and Immunostaining

High-content screen (HCS) is a phenotypic screening that monitors multiple cellular parameters simultaneously. HCS employs fluorescence-based reagents (antibodies, dyes that bind or localize to a given cell component, sensors) to generate a multicolor fluorescence readout that is usually recorded using automated optical image acquisition devices.

Immunostaining refers to the visualization of cellular components in cells (immunocytochemistry) or tissues (immunohistochemistry) by employing specific antibodies or dyes. Cells are usually seeded on glass bottom surfaces (e.g., glass coverslips). To preserve cellular structures and localization of cellular components, proteins are cross-linked using formaldehyde. To allow detection of intracellular targets using antibodies, cells need to be permeabilized with a detergent (e.g.,

Triton X-100). Alternatively, cells can be fixed and permeablized with organic solvents such as methanol and acetone. The type of fixative depends on the cellular components that need to be detected and the employed antibodies. Cellular components are then detected either directly by means of dyes (e.g., fluorescent dyes that bind to and stain DNA) or antibodies that are labeled with fluorophores. Alternatively, an indirect detection can be performed by means of primary antibody that specifically binds to a protein of interest and a secondary antibody that recognizes the primary antibody and is labeled for detection (usually with a fluorophore). Finally, samples are analyzed by means of fluorescence microscopy.

The focused library of around 60 indoloquinolizines was subjected to the phenotypic assay for mitosis modulators at a concentration of 30 μM and this led to the identification of **24a** (centrocountin 1, Scheme 16.2a) as a mitotic inhibitor.

After treatment with centrocountin 1, a dose-dependent increase (from 1 to 25 μM, Box 16.5) in the number of mitotic cells with misaligned chromosomes was detected (Figure 16.3b). At higher concentrations of 25–50 μM, the number of mitotic cells with multipolar mitotic spindles also increased with concentration (Figure 16.3c). Treatment with centrocountin 1 caused fivefold prolonged mitosis compared to the control, which is indicative of an activated spindle-assembly checkpoint and mitotic arrest. An increase in apoptosis was observed as a consequence of this arrest.

This screening could also establish a moderate SAR. While the substitutions on the indole ring were well tolerated without significant loss of biological activity, surprisingly, modifications on any other part rendered the molecules inactive (Scheme 16.2b). Moreover, the (*R*)-enantiomer of **24a** and other active centrocountins were found to exhibit significantly more activity than (*S*)-enantiomers. On the basis of this SAR, pulldown probes **26–28** were prepared wherein **26** was

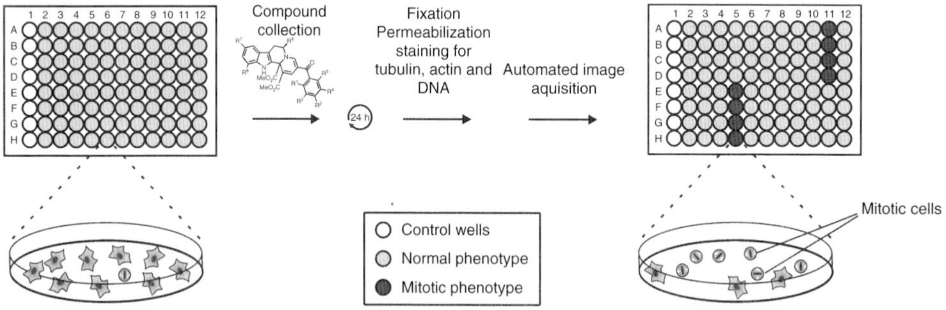

Figure 16.2 *Phenotypic assay for the identification of mitosis modulators.* Cells are seeded in a 96-multiwell plate before treatment with compounds for 24 h. Cells are then fixed and stained for tubulin, actin, and DNA. Mitotic inhibitors are identified as compounds that cause the accumulation of round-shaped cells using automated high-content imaging.

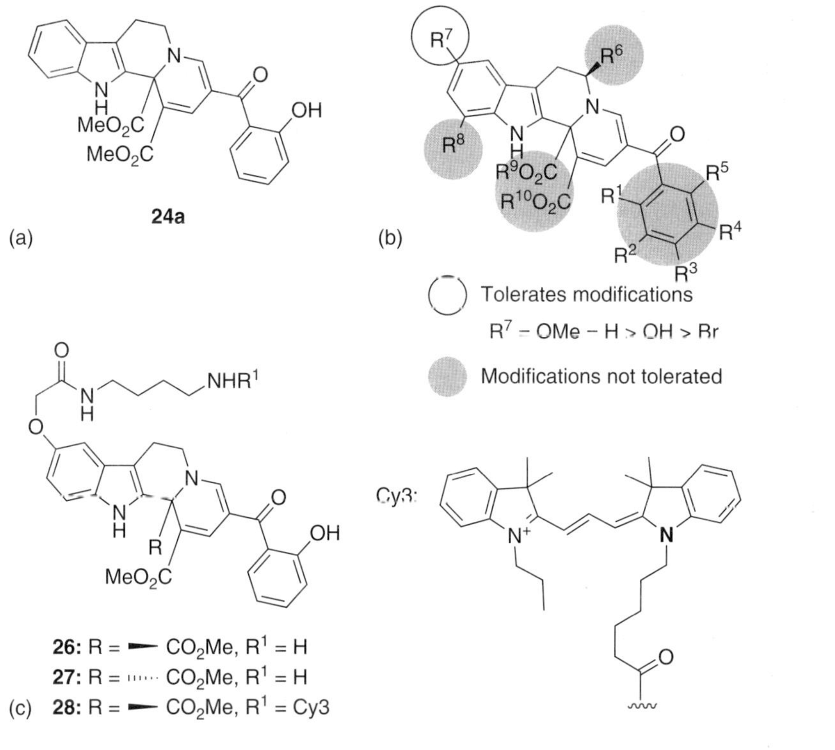

Scheme 16.2 (a) Chemical structure of centrocountin 1 (**24a**). (b) Structure–activity relationship for centrocountin derivatives. (c) Chemical structures of the employed probes (**26–28**).

the active probe and **27** was its negative control (Scheme 16.2c). Probe **28** embodying fluorescent Cy3 dye was synthesized for FP and fluorescence lifetime imaging microscopy (FLIM) experiments.

Box 16.5 Dose Dependent
The action of a given compound is dose dependent if the influence of the compound in the studied system (e.g., on cells) changes when the concentration of the compound is changed.

16.4.2
Identification of the Target Protein(s) of Centrocountin 1

Several potential proteins could be excluded as targets of centrocountin 1. Characterization of the compound revealed no inhibition of various kinases involved in mitosis regulation. As most of the mitotic inhibitors target tubulin/

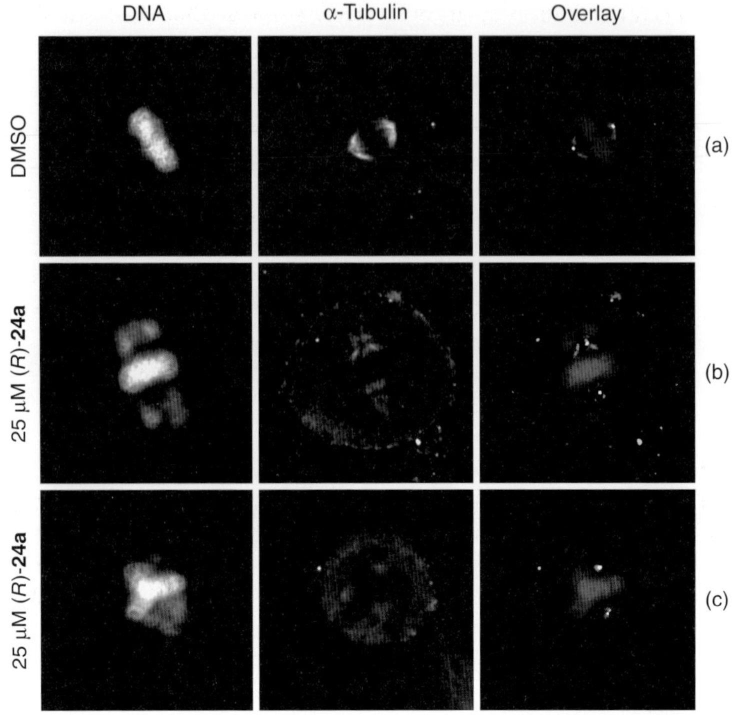

Figure 16.3 (a–c) Influence of centrocountin 1 ((*R*)-**24a**) on chromosome alignment and mitotic spindle assembly in HeLa cells. HeLa cells were treated with centrocountin 1 or dimethyl sulfoxide (DMSO) as control for 18 h. DNA and α-tubulin were visualized my means of immunostaining.

microtubules, centrocountin 1 was subjected to tubulin polymerization studies *in vitro* and in cells. However, no inhibition of tubulin dynamics was observed. To elucidate the mechanism of action, we aimed to isolate the proteins that bind to centrocountin 1 using an affinity-based approach (also called pulldown, Box 16.3). The results from the phenotypic screening of the indoloquinolizine library allowed correlating structure to activity (Scheme 16.2b). According to the SAR, the pulldown probe **26** and the corresponding inactive enantiomer **27** were synthesized and employed in a pulldown using HeLa cell lysate. Probes **26** and **27** were immobilized on NHS-activated sepharose. After incubation with the cell lysate and removal of unspecific binding using stringent washing, proteins that remained bound to the matrix were eluted with 10-fold excess of centrocountin 1 (*R*)-**24a**. Elution samples were subjected to SDS-PAGE followed by tryptic digest and identification of the enriched proteins using MS/MS analysis (Box 16.6). Comparison of proteins isolated with **26** and the inactive enantiomer **27** identified the nucleolar and centrosome-associated protein nucleophosmin (NPM) as a promising target candidate. Interestingly, knockdown of NPM by siRNA phenocopies the influence of centrocountin 1 in HeLa cells [16]. A similar phenotype was also observed for the NPM-binding partner Crm1 (exportin 1) [17, 18].

Box 16.6 Principle of Tandem (MS/MS) Mass Spectrometry to Identify Proteins After Tryptic Digest

To analyze complex protein mixtures (e.g., proteins that were enriched on solid phase after binding to a small molecule), proteins are first digested with a protease to obtain peptides. Usually, trypsin is employed for enzymatic digest. Trypsin is a protease that cleaves peptide bonds C-terminally of lysine and arginine with the exception of a following proline. The resulting peptides are partially separated using reversed-phase nanoHPLC. The HPLC is coupled online with a mass spectrometer via a nano-electrospray ion source. In the mass spectrometer, the mass-to-charge ratio and the charge state of the peptides are determined. In a further step, these peptides are fragmented in the mass spectrometer to get partial sequence information of the peptides, the so-called peptide sequence tags. This combined information, the so-called peptide fragmentation fingerprint, is used for a database search. In a protein database, the protein digest and the fragmentation of the resulting peptides are done *in silico* and the experimental results are compared to these theoretical data (Figure 4).

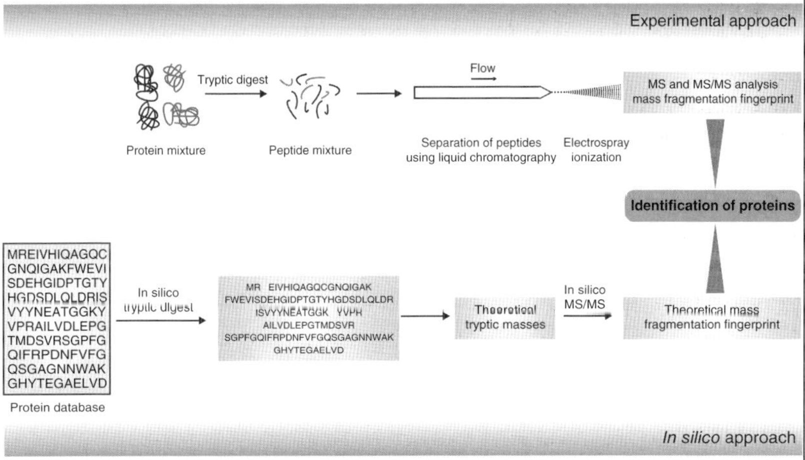

Figure 4 Identification of proteins using tandem (MS/MS) mass spectrometry.

16.4.3
Confirmation of the Target Candidates

To confirm NPM and/or Crm1 as targets of centrocountin 1, proteins enriched during the pulldown were subjected to immunoblotting to detect NPM and Crm1

(Box 16.7). Pulldown probe **26** enriched both proteins, whereas the presence of increasing concentrations of centrocountin 1 ((*R*)-**24a**) during the pulldown (competition approach) abolished the binding of NPM and Crm1 to the probe (Figure 16.4a). This competition analysis provides strong evidence for NPM and Crm1 as target proteins of centrocountin 1.

Box 16.7 Immunoblotting, Fluorescence Polarization, FLIM and Fluorescent Proteins

Immunoblotting or Western blotting refers to the transfer of proteins from a gel to a solid support matrix (also called membrane) and the subsequent detection of proteins of interest by means of specific antibodies. Protein samples are first separated according to their molecular weight by means of SDS-PAGE. Proteins are then transferred from the gel to a polyvinylidene difluoride (PVDF) or nitrocellulose membrane using electroblotting. The employed membranes have high affinity for proteins and blocking of the membrane is required to prevent unspecific binding of antibodies to the free sites on the membrane. Usually, the blocking solution contains skimmed milk or bovine serum albumin (BSA). In the next step, the membrane is incubated with an antibody that specifically binds to the protein of interest (also called primary antibody) and is typically not detectable (i.e., it lacks an appropriate label). Rinsing the membrane several times removes unbound primary antibody before incubation with a secondary antibody that specifically binds to the primary antibody and enables the indirect detection of the protein of interest. The secondary antibody can be linked to an enzyme to enable protein detection by means of an enzymatic reaction (e.g., horseradish peroxidase (HRP) and a chemiluminescence detection using X-ray films or charge-coupled device (CCD) camera imaging devices). Alternatively, the secondary antibody can be labeled with a fluorophore, which allows the detection of proteins using fluorescence imager.

Fluorescence polarization or fluorescence anisotropy is a property of fluorescent molecules and provides information on the orientation and mobility of a fluorophore using polarized light. When a sample that contains fluorescent molecules is excited using linearly polarized light, the degree of polarization of the emitted light depends on the motility and thus the size of the fluorophore. A small fluorescent molecule usually has a high degree of rotation and upon excitation using polarized light will reduce the degree of polarization of the emitted light. An increase in the molecular size of the fluorescent molecule, for example, when a protein binds to the fluorophore, will reduce its rotation and increase the degree of polarization of the emitted light. This approach is independent of the concentration of the fluorophore. FP is employed in the determination of binding constants of reactions that cause a change in motility of fluorescent molecules, for example, the binding of a compound that is labeled with a fluorophore to a protein.

FLIM employs the reduction in lifetime of a given donor fluorophore when located in close proximity to an acceptor fluorophore to allow a fluorescence resonance energy transfer (FRET). In FRET, upon excitation, a donor fluorophore can transfer its energy to a suitable acceptor fluorophore without emission of radiation. The efficiency of the energy transfer is inversely dependent on the distance between the two fluorophores. The fluorescence lifetime (τ) determines how long a fluorophore remains in excited state upon excitation and is dependent on microenvironment and is concentration independent. For example, FLIM can be employed to detect the interaction of proteins or compound and protein given that they are coupled to fluorophores that enable FRET. The binding of the two proteins (or of the compound to the protein) brings the fluorophores in close proximity. Excitation of the donor fluorphore will then promote energy transfer to the acceptor fluorophore, thus causing reduction in the fluorescence lifetime of the donor.

Citrine. It is a member of the green fluorescent protein (GFP) family. Mutations in GFP result in new fluorescent probes. Citrine is a type of yellow fluorescent protein (YFP) with increased acidic stability and photostability.

Enhanced yellow fluorescent protein (EYFP). One of the brightest fluorescent proteins.

In an FP experiment (Box 16.7) employing a centrocountin 1-derivative **28** that was labeled with the fluorophore Cy3, the dissociation constants for binding to NPM and Crm1 were determined (Figure 16.4b). Furthermore, the binding of **28** to citrine-NPM or EYFP-Crm1 was also analyzed using FLIM (Box 16.7) and reduced lifetime of the donor fluorophores citrine and EYFP was detected upon addition of **28** (Figure 16.4c,d). This finding is indicative of the close proximity of the donor and acceptor fluorophore and thus of direct interaction of NPM and Crm1 with **28**.

Besides the role of NPM in the nucleolus and of Crm1 in the nuclear transport, both proteins have been reported to regulate the duplication of centrosomes. According to the proposed mechanism, NPM associates with the centrosome through binding to Crm1 and thus prevents the duplication of the centrosome. Phosphorylation of NPM by the CDK2 displaces NPM from the centrosome, which allows for the duplication of the centrosomes during S-G2 phase. Before entry into mitosis NPM reassociates with the two centrosomes to prevent their further duplication and ensures the proper formation of a bipolar mitotic spindle [19]. Treatment of cells with centrocountin 1 results in the accumulation of mitotic cells with defective mitotic spindles. By means of specific markers for centrosomes (γ-tubulin) and centrioles (cep135), the composition of the spindle poles was examined. Mitotic cells with properly assembled spindle poles were detected. However, cells in mitosis that contain more than two centrosomes and thus containing mostly more than two spindle poles were observed. In these cells, the centrosomes were properly assembled (i.e., containing γ-tubulin and cep135) or were lacking one or both centrosomal markers. This finding demonstrates

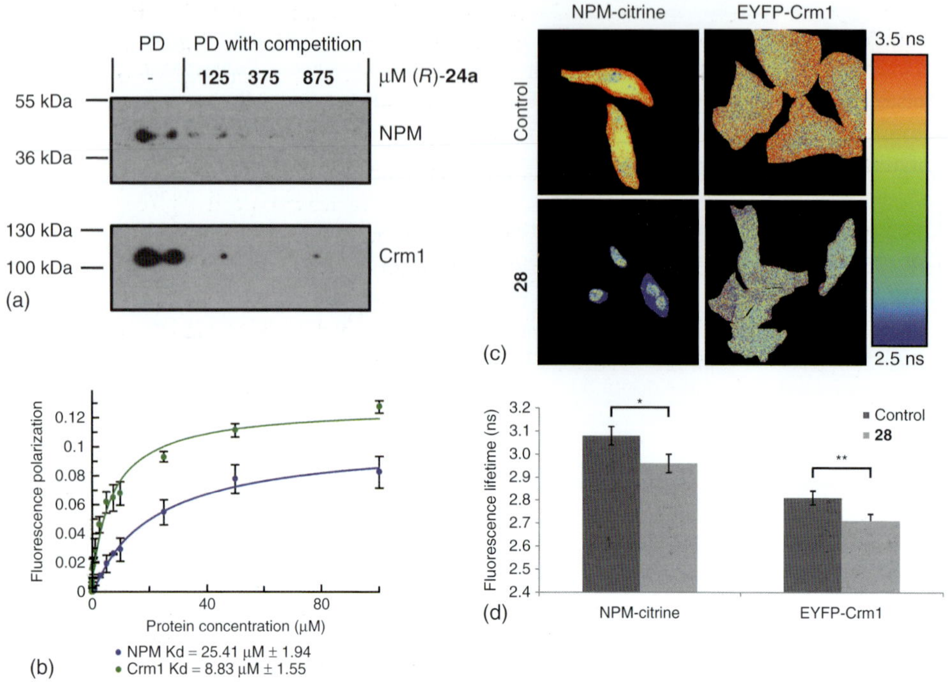

Figure 16.4 Confirmation of NPM and Crm1 as targets of centrocountins. (a) Detection of NPM and Crm1 by means of immunoblotting after enrichment by affinity chromatography using **26** in the absence or presence of different concentrations of centrocountin 1 ((*R*)-**24a**) (competition approach). (b) Binding of **28** to NPM or Crm1 as determined by means of fluorescence polarization. (c,d) Binding of **28** to NPM-citrine or EYFP-Crm1 as determined by means of fluorescence lifetime imaging microscopy. Images represent lifetime maps of cells (c). The graph (d) shows the decrease in the donor lifetime in the presence of **28** as compared to the control.

that centrocountin 1 impairs the centrosome duplication cycle, which ultimately results in the overduplication and fragmentation of centrosomes and formation of acentrosomal spindle poles.

16.5 Conclusion

The synthesis of the natural-product-inspired indoloquinolizine compound collection, the HCS for mitotic inhibitors and the elucidation of the mode of action of centrocountin 1 is a demonstrative example of the forward-chemical genetics approach. It illustrates the workflow for the identification of biologically active small molecules in cells and target deconvolution. Centrocountin 1 was the most potent hit compound in a screen for mitotic inhibition and impairs the proper chromosome congression in cells. This results in chromosomal misalignment and mitotic spindle defects. Correlation of structure to biological activity of the indoloquinolizines led to the design and synthesis of probes for target identification and target confirmation. By means of affinity-based proteomics approach coupled to LC-MS/MS analysis NPM was identified as a target protein of centrocountin 1. NPM and Crm1, which binds to NPM, were confirmed as direct targets by means of immunodetection of the proteins after target enrichment, FP, and FLIM. NPM and Crm1, besides their functions in the nucleolus and nucleus, respectively, regulate the centrosome duplication cycle. Binding of centrocountin 1 to NPM and Crm1 most likely disturbs this cycle and causes mitotic spindle defects, mitotic arrest, and finally cell death. The above-outlined strategy to unravel the molecular targets of biologically active small molecules thus provides great insights for advancing the basic biology and drug discovery research.

References

1. Hanahan, D. and Weinberg, R.A. (2011) Hallmarks of cancer: the next generation. *Cell*, **144**, 646–674.
2. Dumontet, C. and Jordan, M.A. (2010) Microtubule-binding agents: a dynamic field of cancer therapeutics. *Nat. Rev. Drug Discovery*, **9**, 790–803.
3. Manchado, E., Guillamot, M., and Malumbres, M. (2012) Killing cells by targeting mitosis. *Cell Death Differerntiation*, **19**, 369–377.
4. Connor, C.J.O., Laraia, L., and Spring, D.R. (2011) Chemical genetics. *Chem. Soc. Rev.*, **40**, 4332–4345.
5. Terstappen, G.C., Schlupen, C., Raggiaschi, R., and Gaviraghi, G. (2007) Target deconvolution strategies in drug discovery. *Nat. Rev. Drug Discovery*, **6**, 891–903.
6. Yarrow, J.C., Feng, Y., Perlman, Z.E., Kirchhausen, T., and Mitchison, T.J. (2003) Phenotypic screening of small molecule libraries by high throughput cell imaging. *Comb. Chem. High Throughput Screen.*, **6**, 279–286.
7. Ziegler, S., Pries, V., Hedberg, C., and Waldmann, H. (2013) Target identification for small bioactive molecules: finding the needle in the haystack. *Angew. Chem. Int. Ed.*, **52**, 2744–2792.
8. Newman, D.J. and Cragg, G.M. (2007) Natural products as sources of new drugs over the last 25 years. *J. Nat. Prod.*, **70**, 461–477.
9. Yue, Q.X., Liu, X.A., and Guo, D.A. (2010) Microtubule-binding natural products for cancer therapy. *Planta Med.*, **76**, 1037–1043.
10. Carter, G.T. (2011) Natural products and pharma 2011: strategic changes spur new opportunities. *Nat. Prod. Rep.*, **28**, 1783–1789.
11. Shelat, A.A. and Guy, R.K. (2007) Scaffold composition and biological relevance of screening libraries. *Nat. Chem. Biol.*, **3**, 442–446.
12. Tietze, L.F. (1996) Domino reactions in organic synthesis. *Chem. Rev.*, **96**, 115–136.
13. Dückert, H., Pries, V., Khedkar, V., Menninger, S., Bruss, H., Bird, A.W., Maliga, Z., Brockmeyer, A., Janning, P., Hyman, A., Grimme, S., Schürmann, M., Preut, H., Hübel, K., Ziegler, S., Kumar, K., and Waldmann, H. (2012) Natural product-inspired cascade synthesis yields modulators of centrosome integrity. *Nat. Chem. Biol.*, **8**, 179–184.
14. Waldmann, H., Khedkar, V., Dückert, H., Schürmann, M., Oppel, I.M., and Kumar, K. (2008) Asymmetric synthesis of natural product inspired tricyclic benzopyrones by an organocatalyzed annulation reaction. *Angew. Chem. Int. Ed.*, **47**, 6869–6872.
15. Eschenbrenner-Lux, V., Dückert, H., Khedkar, V., Bruss, H., Waldmann, H., and Kumar, K. (2013) Cascade syntheses

routes to the centrocountins. *Chem. Eur. J.*, **19**, 2294–2304.

16. Amin, M.A., Matsunaga, S., Uchiyama, S., and Fukui, K. (2008) Nucleophosmin is required for chromosome congression, proper mitotic spindle formation, and kinetochore-microtubule attachment in HeLa cells. *FEBS Lett.*, **582**, 3839–3844.
17. Liu, Q., Jiang, Q., and Zhang, C. (2009) A fraction of Crm1 locates at centrosomes by its CRIME domain and regulates the centrosomal localization of pericentrin. *Biochem. Biophys. Res. Commun.*, **384**, 383–388.
18. Wang, W., Budhu, A., Forgues, M., and Wang, X.W. (2005) Temporal and spatial control of nucleophosmin by the Ran-Crm1 complex in centrosome duplication. *Nat. Cell Biol.*, **7**, 823–830.
19. Okuda, M., Horn, H.F., Tarapore, P., Tokuyama, Y., Smulian, A.G., Chan, P.K., Knudsen, E.S., Hofmann, I.A., Snyder, J.D., Bove, K.E., and Fukasawa, K. (2000) Nucleophosmin/B23 is a target of CDK2/Cyclin E in centrosome duplication. *Cell*, **103**, 127–140.

17
Finding a Needle in a Haystack. Identification of Tankyrase, a Novel Therapeutic Target of the Wnt Pathway Using Chemical Genetics

Atwood K. Cheung and Feng Cong

17.1
Introduction

To stem the tide of declining therapeutic breakthroughs of novel mechanism a fundamental new approach is required, one that takes into account the complexity of human biology and disease, and capitalizes on the technological advances of the past few decades, such as the sequencing of the human genome, and improvements in mass spectrometry (MS).

One approach is the exploitation of molecular signaling pathways in cells to model human biology and disease. Using small molecules, the functions of these complex collections of genes and proteins can be probed in an efficient and disease-relevant manner. This approach, termed *chemical genetics*, in practice, utilizes high-throughput screens to identify small-molecule probes that can modulate disease-relevant signaling pathways [1, 2]. However, as an unbiased approach to drug and target discovery, chemical genetics has the disadvantage of the challenging deconvolution of the mechanism of action (MOA) and identification of the efficacy targets of the drugs. Chemical proteomics (small-molecule affinity chromatography followed by mass spectrometric protein identification) has emerged as one solution for resolving this target identification problem, as it is an unbiased, large-scale method enabling target discovery from a complex protein mixture. Recent advances in MS technology have made this a viable approach [3].

We chose to apply chemical genetic and chemical proteomic approaches to new drug and target discovery to the Wnt signaling pathway where dysregulation of this pathway is linked to many human diseases [4].

Concepts and Case Studies in Chemical Biology, First Edition. Edited by Herbert Waldmann and Petra Janning.

17.2 The Biological Problem

17.2.1 Modulating the Wnt Signaling Pathway for Cancer Therapeutics

The evolutionarily conserved Wnt/β-catenin signal transduction cascade controls many biological processes [5]. A key feature of the Wnt/β-catenin pathway is the regulated proteolysis of the downstream effector β-catenin by the β-catenin destruction complex. The principal components of the β-catenin destruction complex are adenomatous polyposis coli (APC), Axin, and glycogen synthase kinase-3 beta (GSK3β). In the absence of Wnt pathway activation, cytosolic β-catenin is constitutively phosphorylated by GSK3β and targeted

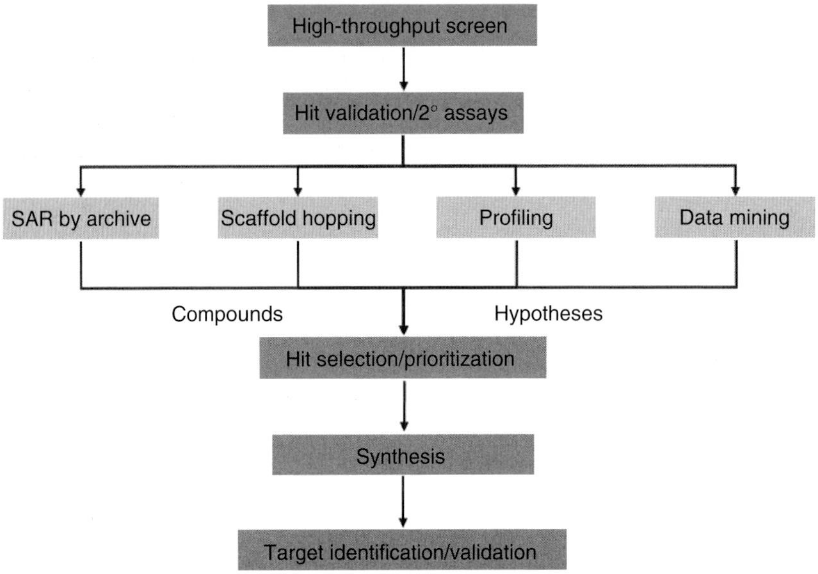

Figure 17.1 *A schematic depicts the flowchart of chemical genetics screening approach.* A primary cell-based assay that captures pathways or phenotypic readouts is established and validated to screen a compound library. Owing to the frequent off-target effects of primary screen compounds, it is essential to implement counter screens and secondary screens to filter nonspecific hits to arrive at a group of high-confidence hit compounds. *In silico* methods for scaffold hopping and compound similarity searching can be utilized to select groups of similar molecules to generate SAR data to better understand the relevant "war head." In parallel, profiling and data mining can also arrive at hypotheses and facilitate hit selection and prioritization. Next, chemistry is initiated to expand SAR for the hit, and identify sites for linker modification or prepare chemical probes. Target identification is conducted with the compound-linked beads by affinity purification of interacting proteins followed by protein identification and quantification by LC-MS/MS (liquid chromatography-tandem mass spectrometry) or utilizing other chemical probes. The final step is target validation through genetic, biochemical, or biophysical means.

for degradation. Upon Wnt stimulation, the β-catenin destruction complex dissociates, leading to the accumulation of nuclear β-catenin and transcription of Wnt-pathway-responsive genes.

Inappropriate activation of the pathway has been observed in many cancers [6]. Notably, truncating mutations of the tumor suppressor APC are the most prevalent genetic alterations in colorectal carcinomas [7]. The efficient assembly of the multiprotein destruction complex is dependent on the steady-state levels of its main components. Axin has been reported to be the concentration-limiting factor in the formation of the β-catenin destruction complex [8] and overexpression of Axin induces β-catenin degradation in cell lines expressing truncated APC [9]. Thus, Axin protein levels are tightly regulated to ensure proper Wnt pathway function. However, the molecular mechanisms that regulate protein homeostasis of destruction complex components and complex assembly remain elusive. The lack of tractable drug targets to antagonize the Wnt pathway makes this pathway especially difficult to drug utilizing a conventional small-molecule drug discovery approach. Thus, targeting this pathway requires a nontraditional approach such as chemical genetics. To identify novel targets and small-molecule modulators of the Wnt pathway, we envisaged the following steps: cell-based high-throughput screening, counter-screening, hit analysis and selection, exploration of structure–activity relationship (SAR) and generation of the affinity chromatography reagent, affinity enrichment, protein identification, and finally target validation (Figure 17.1).

17.3 The Chemical Approach

17.3.1 Screening Approach

We opted to find small-molecule inhibitors of the Wnt/β-catenin pathway from a cellular high-throughput screen using a Wnt-responsive Super-Topflash (STF) luciferase reporter assay in HEK293 cells (Figure 17.2). The reporter assay reads out the binding of β-catenin to its target transcription site. In principle, small-molecule inhibitors of any of the known, as well as, unknown steps in the signaling cascade could be identified in this unbiased or catch-all cellular assay (Box 17.1).

17.3.2 Chemical Proteomics Target Identification

Chemical proteomics consists of the classical drug-affinity chromatography and modern high-resolution MS analysis for protein identification [3, 11]. The procedure typically involves immobilization of the compound of interest to a solid support through a spacer arm, and the affinity matrix is then used to purify specific interacting proteins from cellular lysate. The complex proteomic mixture is then proteolytically digested, and the resulting peptides are sequenced

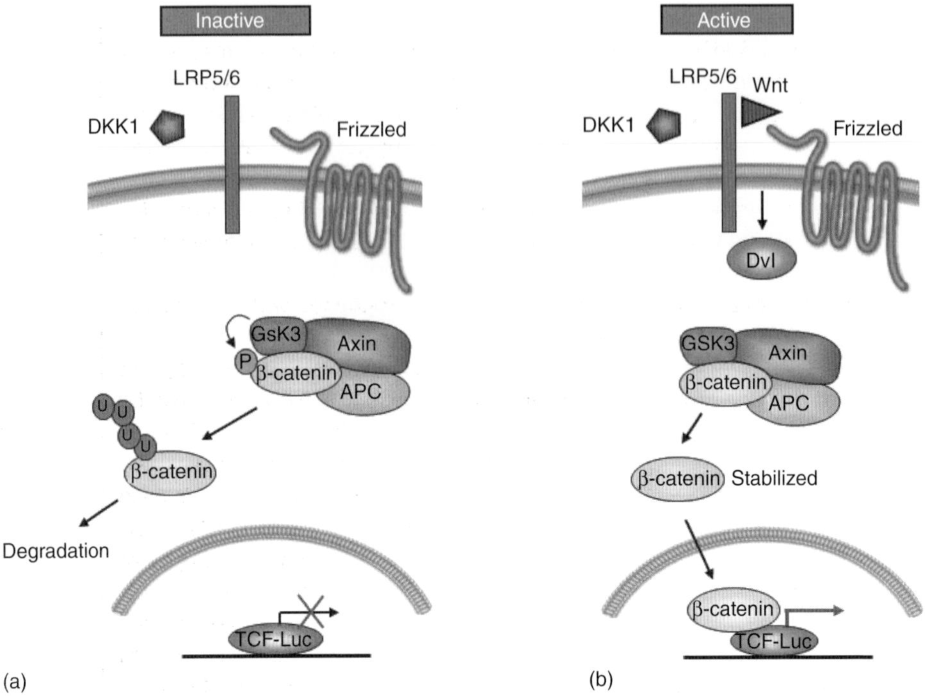

Figure 17.2 (a,b) *β-catenin is regulated by proteolysis mediated by a regulatory complex termed the "β-catenin destruction complex" that includes APC, Axin, and GSK3β*. In the absence of Wnt ligand, β-catenin, a transcriptional coactivator, associates with the destruction complex, and is phosphorylated by GSK3β, which leads to its degradation through ubiquitin-mediated proteolysis. When Wnt ligand binds to its extracellular receptor, a signal that inhibits β-catenin phosphorylation by GSK3β is transduced. β-catenin is stabilized and accumulates in the cell, then it translocates to the nucleus where it activates Wnt pathway genes. The engineered TCF-luc reporter signals binding of β-catenin to its target transcription site through the generation of luciferase.

through MS. Recent development of quantitative MS methods is critical for distinguishing specific and nonspecific binders such as stable isotope labeling by amino acids in cell culture (SILAC) [12] and chemical labeling of tryptic peptides with isobaric tags for relative and absolute quantitation (iTRAQ) [13]. iTRAQ labeling allows for multiplexing of parallel experiments to compare effects of different compound treatments such as dose response and active versus inactive compound comparisons.

Box 17.1 Considerations for Compound Screening and Hit Selection

The first step in chemical-genetics-based target finding is to perform cellular phenotypic screening to identify compounds that elicit a desired phenotype. An array of orthogonal secondary assays is then performed to remove nonspecific compounds and identify compounds with desired properties for mechanistic studies. As target deconvolution is a challenging and time-consuming process, each hit compound should be carefully evaluated. For example, common frequent hitters in cellular screens include compounds affecting general cellular processes, including tubulin destabilizers which could suppress reporter expression, or histone deacetylase inhibitors that relieve general transcriptional repression to activate pathways. Carefully designed secondary assays can remove these nonspecific compounds.

After nonselective compounds are removed, compounds with the desired activity profile need to be prioritized for target deconvolution. The performance of the compounds in past assays, *in silico* target prediction tools [10], and biochemical assay profiling can provide clues as to the target class of the hit (e.g., kinase inhibitor) or MOA (e.g., microtubule stabilizer). Physicochemical properties of the hits are important factors for hit prioritization. Solubility and permeability will effect downstream activities including generation of SAR, and hinder future MOA studies. For example, a compound with low solubility or low permeability can provide false-negative data and confound the development of SAR. Thus, prioritization of hits with good physiochemical properties will focus chemistry on establishing SAR, improving potency, and identifying the best target-identification tool compound, and not preparing compounds with inconsistent or uninterpretable data due to lack of solubility and permeability.

During target deconvolution, known targets in the pathway or targets from predictive tools should be first considered. Once these target hypotheses are eliminated, target deconvolution, the most challenging step of chemical-genetics-based target finding, can begin.

The most significant limitation for chemical proteomics is specificity. The cellular mixture is extremely complex with 10^6 dynamic range of protein number [14]. Small molecules interact with an overabundance of proteins in the cell. The high-affinity interaction between a compound and its efficacy target, often a low-abundance protein, is the most relevant interaction. However, low-affinity interactions with highly abundant proteins often create significant noises during purification. Multiple strategies are used to reduce background binding. "Sticky" proteins tend to have low affinity for the hydrophobic surface of the linker–drug complex, and modifications of the linker [15, 16], and novel matrices have been developed [17, 18] for lower background binding. Even when background binding is minimized, methods to distinguish specific and nonspecific binders are still needed. In-solution competition is one effective strategy to differentiate specific and nonspecific binders. In this approach, cell lysates are pretreated with active molecules or vehicle before the affinity matrix is added. Capture of specific binders by the affinity matrix is effectively blocked by active molecules in solution, so comparative analysis of parallel purifications using MS should reveal specific binders [19]. We describe this approach later to find novel regulatory proteins of Wnt pathway signaling.

17.3.3
Target Validation

The goal of target validation is to determine whether or not the drug target mediates the biological activity of the compound in cellular assays. Key methods include RNAi (ribonucleic acid interference) depletion of the target to phenocopy compound effects or sensitize cells to compound treatment. It should be noted that many proteins have functional homologs, which may be targeted by the probe compound, and should be considered in validation efforts even if it was not identified from drug-affinity experiments. Also, the compound may have "gain-of-function" activity, and elimination of its target would suppress, instead of phenocopy, the compound's activity. In addition, cDNA overexpression can help establish compound–target relationships by overexpression of the drug target to suppress activity of the compound.

Once the putative target is validated in a functional assay, a quantitative measurement of the binding affinity between the small molecule and the target can be done using techniques such as surface plasmon resonance or isothermal calorimetry using recombinant or purified proteins [20, 21]. If the putative target is an enzyme, a biochemical assay can be set up to measure the compound's effect on enzyme activity. Structurally related compounds with various degrees of cellular activities should be tested in the binding assay or enzymatic assay to establish the SAR. Ultimately, rigorous validation by NMR (nuclear magnetic resonance) or cocrystallization experiments should be performed to determine the three-dimensional structure of the compound–target complex to allow for future chemical optimization of the potential drug candidate.

17.4
Chemical Biological Research/Evaluation

17.4.1
Identification of XAV939 as a Wnt Pathway Inhibitor

The screening of over 1 000 000 compounds ultimately led to the identification of XAV939, which strongly inhibited Wnt3a-stimulated STF activity in HEK293 cells but did not affect CRE (cyclic AMP response element), NF-κB (nuclear factor kappa B), or TGFβ (transforming growth factor beta) luciferase reporters satisfying our selectivity requirement [22] (Scheme 17.1, Figure 17.3). LDW643, a close structural analog to XAV939, had no activity on the Wnt3a-induced STF reporter, providing an early SAR clue. XAV939 treatment blocked Wnt3a-induced accumulation of β-catenin in HEK293 cells, indicating that the compound modulates Wnt signaling upstream of β-catenin. Interestingly, XAV939 also inhibited STF activity

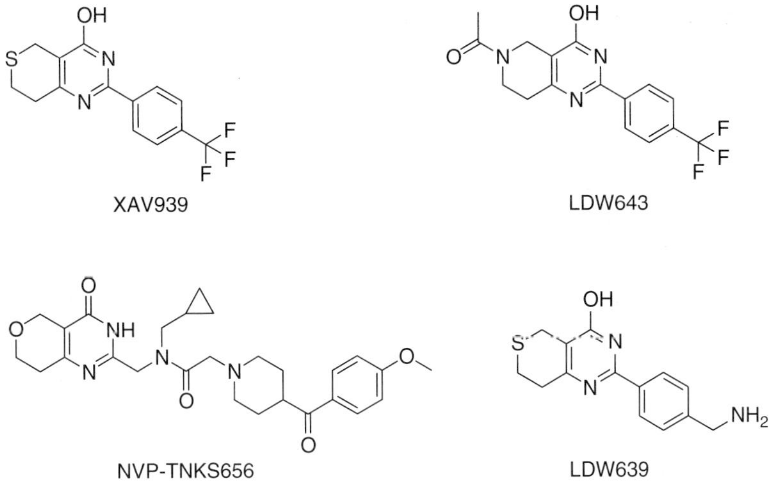

Scheme 17.1 Structures of XAV939, LDW643, LDW639, and NVP-TNKS656, respectively.

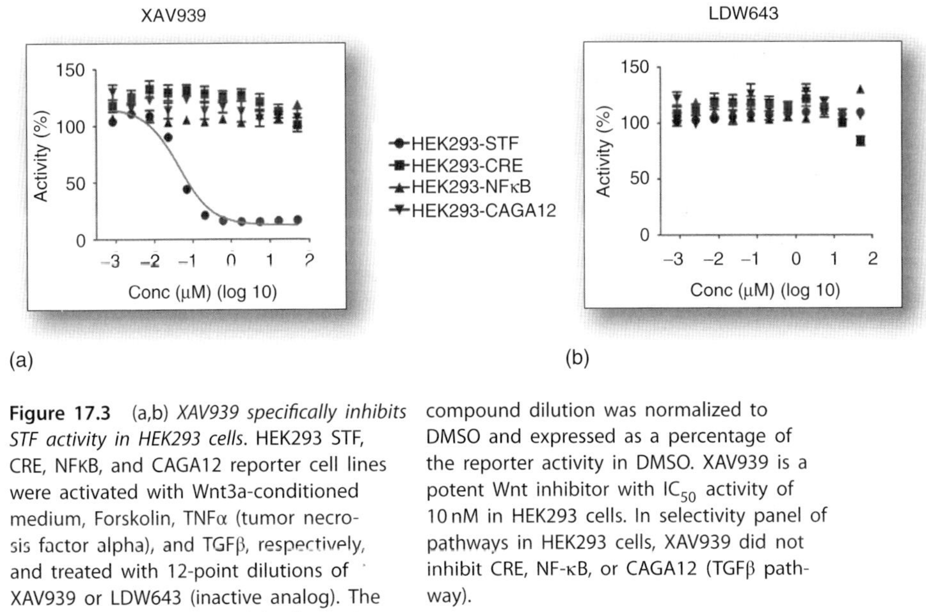

Figure 17.3 (a,b) *XAV939 specifically inhibits STF activity in HEK293 cells.* HEK293 STF, CRE, NFκB, and CAGA12 reporter cell lines were activated with Wnt3a-conditioned medium, Forskolin, TNFα (tumor necrosis factor alpha), and TGFβ, respectively, and treated with 12-point dilutions of XAV939 or LDW643 (inactive analog). The corresponding reporter activity for each compound dilution was normalized to DMSO and expressed as a percentage of the reporter activity in DMSO. XAV939 is a potent Wnt inhibitor with IC_{50} activity of 10 nM in HEK293 cells. In selectivity panel of pathways in HEK293 cells, XAV939 did not inhibit CRE, NF-κB, or CAGA12 (TGFβ pathway).

in SW480 cells, a colorectal cancer cell line harboring truncated APC. XAV939 decreased β-catenin abundance but significantly increased β-catenin phosphorylation (S33/S37/T41) in SW480 cells, suggesting that XAV939 promotes the phosphorylation-dependent degradation of β-catenin by increasing the activity of the destruction complex.

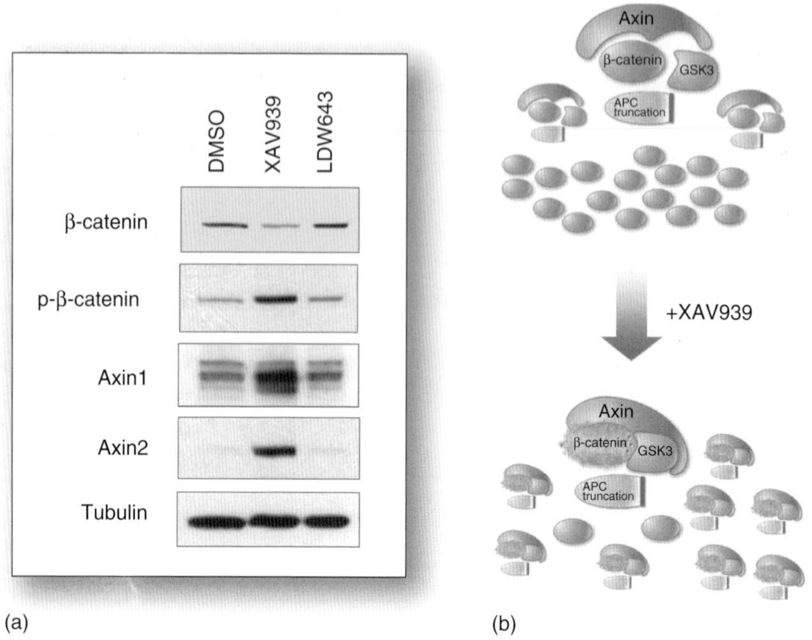

Figure 17.4 (a) SW480 cells were treated overnight with 1 μM XAV939 or LDW643, fractionated for cytosolic proteins, and immunoblotted with the indicated antibodies. XAV939 decreases the abundance of β-catenin and increases the abundance of Axin and phospho-β-catenin. Phosphorylation by GSK3β leads to E3 ligase binding and degradation of β-catenin through proteasome pathway. (b) These data suggest that the destruction complex with the truncated form of APC is weakly active. Therefore, by increasing or stabilizing Axin (thought to be the concentration-limiting component of the destruction complex) XAV939 acts to increase the number of destruction complexes, and thus increase the turnover of β-catenin.

To explore how XAV939 increased the activity of the destruction complex, we examined the levels of known Wnt pathway components. Strikingly, the protein, but not mRNA (messenger ribonucleic acid) levels of Axin1 and Axin2 were strongly increased after XAV939 treatment, as was the Axin-GSK3β complex (Figure 17.4). Similar effects of XAV939 were also observed in DLD-1 cells, another colorectal cancer cell line with truncated APC. Together, these findings support the hypothesis that XAV939 increases the concentration of Axin–GSK3β complex, and thereby promotes phosphorylation and degradation of β-catenin. Next, we turned to chemical proteomics to elucidate the efficacy target(s) of XAV939.

17.4.2
XAV939 Regulates Axin Protein Levels by Inhibiting Tankyrases

Affinity chemical proteomics relies on the affinity capture of the cellular efficacy targets onto solid support. Therefore, it is critical to introduce a linker onto the small molecule in such a way that target binding is not disrupted. To prepare the optimal small-molecule affinity reagent, more than 50 analogs of XAV939 were synthesized to develop SAR that led to linker analog LDW639 which had comparable activity to the lead molecule. To identify the cellular efficacy target(s) through which XAV939 upregulates Axin protein levels, we performed a 3-channel iTRAQ quantitative chemical proteomics experiment with in-solution competition. The immobilized, bioactive analog of XAV939 was used to affinity capture cellular proteins from HEK293 cell lysates spiked with an excess amount (20 μM) of XAV939, the inactive analog LDW643, or DMSO (dimethyl sulfoxide). Specific binding to the immobilized compound should be competed with XAV939 but not with LDW643. Of 699 proteins quantified, 18 proteins were significantly and specifically competed-off (>65%, $>2\sigma$ of the mean) with soluble XAV939, including the poly(ADP-ribose) polymerases PARP1, PARP2, tankyrase 1 (TNKS1), tankyrase 2 (TNKS2), and several known PARP1 substrates, presumably co-purified with PARP1 [22].

17.4.3 Validation of Tankyrase as the Target for XAV939

To determine which PARP family member(s) are the actual efficacy targets of XAV939, we assessed their siRNA-mediated loss-of-function phenotypes [22]. Co-depletion of TNKS1 and TNKS2 phenocopied the effect of XAV939 by increasing the protein levels of Axin1 and Axin2, whereas combinatorial PARP1/2 knockdown did not. In addition, ABT-888, a potent PARP1 and PARP2 inhibitor [23] that has minimal activity on TNKS1 and TNKS2, did not affect the protein levels of Axin and TNKS. Collectively, these results suggest that TNKS1 and TNKS2 are the cellular efficacy targets of XAV939.

Using additional siRNAs, we further demonstrated that co-depletion of TNKS1 and TNKS2 increases β-catenin phosphorylation, decreases β-catenin abundance, and inhibits the transcription of β-catenin target genes in SW480 cells [22]. Notably, depletion of TNKS1 or TNKS2 alone did not increase Axin1/2 protein levels, indicating that TNKS1 and TNKS2 function redundantly in regulating Axin protein levels. Co-depletion of TNKS1 and TNKS2 also phenocopied the pharmacological effect of XAV939 in HEK293 and DLD-1 cells.

Using Cy5-labeled XAV939 and recombinant PARP proteins, we found that XAV939 binds tightly to the catalytic (PARP) domains of TNKS1 and TNKS2 (K_d of 0.099 and 0.093 μM, respectively) using fluorescence polarization [22]. XAV939 also binds to recombinant PARP1, although with a significantly lower binding affinity (K_d of 1.2 μM). TNKS1 and TNKS2 modify their substrates through the addition of multiple ADP-ribose units, referred to as *poly-ADP ribosylation (PARsylation)* [24]. In biochemical activity assays measuring depletion of nicotinamide adenine dinucleotide (NAD^+), XAV939 strongly inhibited TNKS1 and TNKS2, with respective IC_{50} (inhibitor concentration 50) values of 0.011 and 0.004 μM. Auto-PARsylation of TNKS has been reported to promote its own degradation through the ubiquitin–proteasome pathway. We found that XAV939 treatment of SW480 cells led to a significant increase in TNKS protein levels, suggesting that XAV939 also inhibits TNKS auto-PARsylation *in vivo*.

A study by Chen and coworkers described compounds that increase Axin protein levels and inhibit Wnt signaling [25]. The authors show that IWR-1-endo increased Axin protein levels, while its close analog IWR-1-exo had much weaker activity. We showed that IWR-1-endo strongly inhibited TNKS1 and TNKS2 in biochemical assays, with IWR-1-exo being approximately 10-fold less active, consistent with their potency in Axin stabilization assays. Consistent with their activity in TNKS1/2 biochemical assays, IWR-1-endo, but not IWR-1-exo, significantly stabilized endogenous TNKS1, TNKS2, and Axin2, suggesting that IWR-1-endo inhibited auto-PARsylation of TNKS *in vivo*. These results suggest that IWR-1-endo stabilizes Axin through TNKS inhibition.

17.4.4
XAV939 Inhibits TNKS-Mediated Ubiquitination and PARsylation of Axin

Further experiments were performed to understand how TNKS and XAV939 regulate Axin protein levels. TNKS associates with a small *N*-terminal region of Axin1 (amino acid 19–30) through its ankyrin repeat domain, and this interaction is critical for TNKS-mediated degradation of Axin [22, 26]. Supporting the significance of this interaction, single amino acid mutation in the TNKS binding motif of Axin2 leads to decreased Wnt signaling and embryonic lethality in mice [27]. Auto-PARsylation of TNKS leads to its own ubiquitination and degradation [28]. Thus, Axin degradation may be facilitated through direct PARsylation by TNKS. *In vitro*, TNKS was able to PARsylate Axin, which was completely inhibited by XAV939 treatment. In cells, TNKS promoted PARsylation and ubiquitination of Axin, and this was completely blocked by XAV939. The E3 ubiquitin ligase that mediates TNKS-dependent degradation of Axin has recently been identified. It has been shown that E3 ubiquitin ligase RNF146 interacts with PARsylated Axin through its WWE domain, and mediates ubiquitination and degradation of Axin [29, 30]. Together, these findings suggest that TNKS promotes the ubiquitination and degradation of Axin through the direct PARsylation of Axin, and XAV939 treatment modulates the protein stability of Axin by preventing its poly-ubiquitination (Figure 17.5).

17.4.5
TNKS Inhibitor Blocks the Growth of Colon Cancer Cells

Because XAV939 inhibited β-catenin signaling even in APC-deficient cells, we examined whether this compound could inhibit the proliferation of APC-deficient colorectal cancer cells. Under low serum growth conditions, XAV939, but not inactive analog LDW643, significantly inhibited colony formation of DLD-1 cells, and the growth inhibitory activity was rescued by Axin1/2 siRNA [22]. Follow-up studies from other laboratories demonstrate that TNKS inhibitors can inhibit the

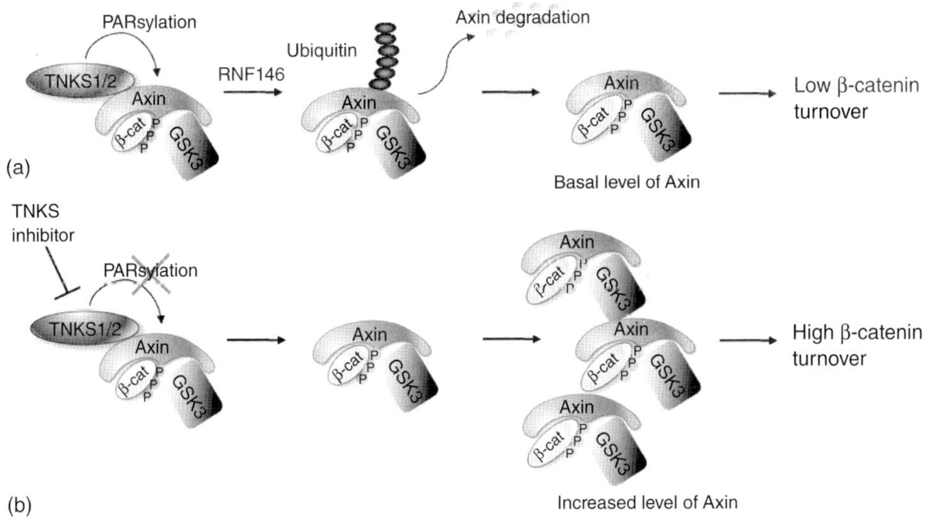

Figure 17.5 *The mechanism of regulating Axin stability by TNKS.* (a) TNKS promotes the ubiquitination and degradation of Axin through the direct PARsylation of Axin, and PARsylated Axin is ubiquitinated by RNF146. (b) The addition of TNKS inhibitor prevents PARsylation and ubiquitination of Axin, and thereby increases Axin levels and promotes β-catenin turnover.

growth of colorectal cancer in xenograft models and tumor growth in conditional APC mutant mice [31, 32].

17.4.6
Crystal Structure of XAV939 and TNKS1

As additional validation that XAV939 binds to TNKS and inhibits its enzymatic activity, crystal structure of the catalytic PARP domain of TNKS1 in complex with XAV939 was resolved [33]. The structure revealed that XAV939 binds to the nicotinamide crevice in the NAD^+-binding site. The pyrimidine ring of XAV939 stacks with Tyr1224, and hydrogen bonds between hydroxyl of Ser1221 and Gly1185 with the pyrimidine hydroxyl and nitrogen, respectively, are evident (Figure 17.6). The D-loop of TNKS1, which binds in the NAD^+-binding pocket in the unliganded crystal structure, is displaced by the molecule, and trifluoromethyl moiety of XAV939 makes nonpolar interactions, allowing for replacement with a methyl amine linker for use in affinity purification. In a comparison of the TNKS2 and PARP1 crystal structures, Karlberg and coworkers observed that aside from the hydrogen-bonding interactions detailed earlier, XAV939 makes mostly nonpolar interactions with the TNKS NAD^+ pocket, whereas in PARP1 the NAD^+ pocket is lined by more polar side chains [34]. Moreover, the trifluoromethyl group of XAV939 makes nonpolar interactions with TNKS2 side chains near the pocket opening; in contrast, the trifluoromethyl group would sterically clash

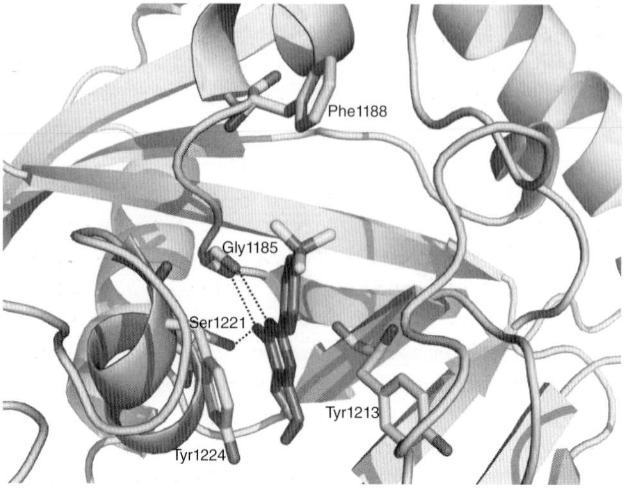

Figure 17.6 *Structure of TNKS1 in complex with XAV939.* The inhibitor molecule occupies the nicotinamide-binding site of the NAD$^+$-binding groove. Reprinted with permission from IUCr.

with the regulatory domain of PARP1. These differences in the crystal structures suggest an explanation of the selectivity of XAV939 for the tankyrases over other PARPs.

With the aid of the co-crystal structure of TNKS1 and XAV939, the highly potent and selective, orally bioavailable TNKS inhibitor, NVP-TNKS656, has been reported [35]. NVP-TNKS656 is the first reported compound to interact with three pockets in the NAD$^+$-binding site, and displays potent antagonism of Wnt pathway activity in the MMTV-Wnt1 (MMTV, mouse mammary tumor virus) mouse xenograft model.

17.5 Conclusion

Chemical genetics provides a solution to one of the pharmaceutical industry's major challenges, that of identifying novel disease-relevant targets for drug discovery. The growth of cell-based high-throughput screening has been well complemented with chemical proteomics to provide a robust and powerful platform for the identification of targets of bioactive small molecules. We have applied this approach to the Wnt/β-catenin signaling pathway, where there are few druggable targets for small-molecule inhibitors. Using a chemical genetics approach, we discovered tankyrases as novel targets for Wnt pathway inhibition and identified a novel mechanism to promote β-catenin degradation through inhibition of tankyrases and stabilization of Axin. Underappreciated steps in the chemical genetic approach to drug and target discovery are the selection of a potent and *selective* hit compound, and robust synthetic chemistry efforts needed to define the SAR of the hit and prepare a valid linker-incorporated derivative. Both steps were instrumental in the discovery of tankyrase's involvement in Wnt signaling that may pave the way for mechanism-based treatments of Wnt-dependent cancers.

TNKS inhibitors serve as versatile tools to probe the function of Wnt/β-catenin signaling in various biological processes and diseases, and are likely to find their utility in diseases other than cancer. For example, increased Wnt/β-catenin signaling is responsible for the inability to form new myelin after neonatal hypoxia, and TNKS inhibitors reverse the "differentiation block" of oligodendrocyte linage cells and can potentially be used to treat demyelination diseases [36]. Moreover, TNKS inhibition blocks aberrant Wnt signaling in pulmonary fibrosis, and thus may serve as a novel therapeutic approach for fibrotic disorders [37].

Although a small-molecule screen in theory should identify all tractable nodes in a signaling pathway, the tractable node with the most robust effect on signaling output tends to be identified more frequently. This point is illustrated by the identification of TNKS inhibitors as inhibitors of the Wnt/β-catenin signaling pathway. Axin is a concentration-limiting factor for the formation of the β-catenin destruction complex [8]. Mathematical modeling suggests that the most efficient way to inhibit Wnt signaling is to increase the concentration of Axin [8]. Consistent with this prediction, the most potent and specific inhibitors from Wnt reporter screens performed in different laboratories are all Axin stabilizers and TNKS inhibitors [22, 25, 32]. Therefore, TNKS inhibition represents the most robust and tractable mechanism to inhibit Wnt signaling.

References

1. Cong, F., Cheung, A.K., and Huang, S.M. (2012) Chemical genetics-based target identification in drug discovery. *Annu. Rev. Pharmacol. Toxicol.*, **52**, 57–78.
2. Cheung, A.K. and Jain, R.K. (2010) Accelerating the discovery of new drug targets with chemical proteomics. *IDrugs: Invest. Drugs J.*, **13** (12), 862–868.
3. Han, X., Aslanian, A., and Yates, J.R. III, (2008) Mass spectrometry for proteomics. *Curr. Opin. Chem. Biol.*, **12** (5), 483–490.
4. Zimmerman, Z.F., Moon, R.T., and Chien, A.J. (2012) Targeting Wnt pathways in disease. *Cold Spring Harbor Perspect. Biol.*, **4** (11).
5. Clevers, H. (2006) Wnt/beta-catenin signaling in development and disease. *Cell*, **127** (3), 469–480.
6. Polakis, P. (2007) The many ways of Wnt in cancer. *Curr. Opin. Genet. Dev.*, **17** (1), 45–51.
7. Miyaki, M., Konishi, M., Kikuchi-Yanoshita, R., Enomoto, M., Igari, T., Tanaka, K., Muraoka, M., Takahashi, H., Amada, Y., Fukayama, M., Maeda, Y., Iwama, T., Mishima, Y., Mori, T., and Koike, M. (1994) Characteristics of somatic mutation of the adenomatous polyposis coli gene in colorectal tumors. *Cancer Res.*, **54** (11), 3011–3020.
8. Lee, E., Salic, A., Kruger, R., Heinrich, R., and Kirschner, M.W. (2003) The roles of APC and Axin derived from experimental and theoretical analysis of the Wnt pathway. *PLoS Biol.*, **1** (1), E10.
9. Kishida, M., Koyama, S., Kishida, S., Matsubara, K., Nakashima, S., Higano, K., Takada, R., Takada, S., and Kikuchi, A. (1999) Axin prevents Wnt-3a-induced

accumulation of beta-catenin. *Oncogene*, **18** (4), 979–985.

10. Bender, A., Young, D.W., Jenkins, J.L., Serrano, M., Mikhailov, D., Clemons, P.A., and Davies, J.W. (2007) Chemogenomic data analysis: prediction of small-molecule targets and the advent of biological fingerprint. *Comb. Chem. High Throughput Screening*, **10** (8), 719–731.
11. Domon, B. and Aebersold, R. (2006) Mass spectrometry and protein analysis. *Science*, **312** (5771), 212–217.
12. Ong, S.E., Blagoev, B., Kratchmarova, I., Kristensen, D.B., Steen, H., Pandey, A., and Mann, M. (2002) Stable isotope labeling by amino acids in cell culture, SILAC, as a simple and accurate approach to expression proteomics. *Mol. Cell. Proteomics*, **1** (5), 376–386.
13. Ross, P.L., Huang, Y.N., Marchese, J.N., Williamson, B., Parker, K., Hattan, S., Khainovski, N., Pillai, S., Dey, S., Daniels, S., Purkayastha, S., Juhasz, P., Martin, S., Bartlet-Jones, M., He, F., Jacobson, A., and Pappin, D.J. (2004) Multiplexed protein quantitation in Saccharomyces cerevisiae using amine-reactive isobaric tagging reagents. *Mol. Cell. Proteomics*, **3** (12), 1154–1169.
14. Rix, U. and Superti-Furga, G. (2009) Target profiling of small molecules by chemical proteomics. *Nat. Chem. Biol.*, **5** (9), 616–624.
15. Fonovic, M., Verhelst, S.H., Sorum, M.T., and Bogyo, M. (2007) Proteomics evaluation of chemically cleavable activity-based probes. *Mol. Cell. Proteomics*, **6** (10), 1761–1770.
16. Speers, A.E. and Cravatt, B.F. (2005) A tandem orthogonal proteolysis strategy for high-content chemical proteomics. *J. Am. Chem. Soc.*, **127** (28), 10018–10019.
17. Shimizu, N., Sugimoto, K., Tang, J., Nishi, T., Sato, I., Hiramoto, M., Aizawa, S., Hatakeyama, M., Ohba, R., Hatori, H., Yoshikawa, T., Suzuki, F., Oomori, A., Tanaka, H., Kawaguchi, H., Watanabe, H., and Handa, H. (2000) High-performance affinity beads for identifying drug receptors. *Nat. Biotechnol.*, **18** (8), 877–881.
18. Shiyama, T., Furuya, M., Yamazaki, A., Terada, T., and Tanaka, A. (2004) Design and synthesis of novel hydrophilic spacers for the reduction of nonspecific binding proteins on affinity resins. *Bioorg. Med. Chem.*, **12** (11), 2831–2841.
19. Bantscheff, M., Eberhard, D., Abraham, Y., Bastuck, S., Boesche, M., Hobson, S., Mathieson, T., Perrin, J., Raida, M., Rau, C., Reader, V., Sweetman, G., Bauer, A., Bouwmeester, T., Hopf, C., Kruse, U., Neubauer, G., Ramsden, N., Rick, J., Küster, B., and Drewes, G. (2007) Quantitative chemical proteomics reveals mechanisms of action of clinical ABL kinase inhibitors. *Nat. Biotechnol.*, **25** (9), 1035–1044.
20. Kuettel, S., Mosimann, M., Maser, P., Kaiser, M., Brun, R., Scapozza, L., and Perozzo, R. (2009) Adenosine Kinase of T. b. Rhodesiense identified as the putative target of 4-[5-(4-phenoxyphenyl)-2*H*-pyrazol-3-yl]morpholine using chemical proteomics. *PLoS Negl. Trop. Dis.*, **3** (8), E506.
21. Morohashi, K., Yoshino, A., Yoshimori, A., Saito, S., Tanuma, S., Sakaguchi, K., and Sugawara, F. (2005) Identification of a drug target motif: an anti-tumor drug NK109 interacts with a PNxxxxP. *Biochem. Pharmacol.*, **70** (1), 37–46.
22. Huang, S.M., Mishina, Y.M., Liu, S., Cheung, A., Stegmeier, F., Michaud, G.A., Charlat, O., Wiellette, E., Zhang, Y., Wiessner, S., Hild, M., Shi, X., Wilson, C.J., Mickanin, C., Myer, V., Fazal, A., Tomlinson, R., Serluca, F., Shao, W., Cheng, H., Shultz, M., Rau, C., Schirle, M., Schlegl, J., Ghidelli, S., Fawell, S., Lu, C., Curtis, D., Kirschner, M.W., Lengauer, C., Finan, P.M., Tallarico, J.A., Bouwmeester, T., Porter, J.A., Bauer, A., and Cong, F. (2009) Tankyrase inhibition stabilizes axin and antagonizes Wnt signalling. *Nature*, **461** (7264), 614–620.
23. Donawho, C.K., Luo, Y., Luo, Y., Penning, T.D., Bauch, J.L., Bouska, J.J., Bontcheva-Diaz, V.D., Cox, B.F., DeWeese, T.L., Dillehay, L.E., Ferguson, D.C., Ghoreishi-Haack, N.S., Grimm, D.R., Guan, R., Han, E.K., Holley-Shanks, R.R., Hristov, B., Idler, K.B., Jarvis, K., Johnson, E.F., Kleinberg, L.R., Klinghofer, V., Lasko, L.M., Liu, X., Marsh, K.C., McGonigal,

T.P., Meulbroek, J.A., Olson, A.M., Palma, J.P., Rodriguez, L.E., Shi, Y., Stavropoulos, J.A., Tsurutani, A.C., Zhu, G.D., Rosenberg, S.H., Giranda, V.L., and Frost, D.J. (2007) ABT-888, an orally active poly(ADP-ribose) polymerase inhibitor that potentiates DNA-damaging agents in preclinical tumor models. *Clin. Cancer Res.*, **13** (9), 2728–2737.

24. Hsiao, S.J. and Smith, S. (2008) Tankyrase function at telomeres, spindle poles, and beyond. *Biochimie*, **90** (1), 83–92.
25. Chen, B., Dodge, M.E., Tang, W., Lu, J., Ma, Z., Fan, C.W., Wei, S., Hao, W., Kilgore, J., Williams, N.S., Roth, M.G., Amatruda, J.F., Chen, C., and Lum, L. (2009) Small molecule-mediated disruption of Wnt-dependent signaling in tissue regeneration and cancer. *Nat. Chem. Biol.*, **5** (2), 100–107.
26. Morrone, S., Cheng, Z., Moon, R.T., Cong, F., and Xu, W. (2012) Crystal structure of a Tankyrase-Axin complex and its implications for Axin turnover and Tankyrase substrate recruitment. *Proc. Natl. Acad. Sci. U.S.A.*, **109** (5), 1500–1505.
27. Qian, L., Mahaffey, J.P., Alcorn, H.L., and Anderson, K.V. (2011) Tissue-specific roles of Axin2 in the inhibition and activation of Wnt signaling in the mouse embryo. *Proc. Natl. Acad. Sci. U.S.A.*, **108** (21), 8692–8697.
28. Yeh, T.Y., Meyer, T.N., Schwesinger, C., Tsun, Z.Y., Lee, R.M., and Chi, N.W. (2006) Tankyrase recruitment to the lateral membrane in polarized epithelial cells: regulation by cell-cell contact and protein poly(ADP-ribosyl)ation. *Biochem. J.*, **399** (3), 415–425.
29. Zhang, Y., Liu, S., Mickanin, C., Feng, Y., Charlat, O., Michaud, G.A., Schirle, M., Shi, X., Hild, M., Bauer, A., Myer, V.E., Finan, P.M., Porter, J.A., Huang, S.M., and Cong, F. (2011) RNF146 is a poly(ADP-ribose)-directed E3 ligase that regulates axin degradation and Wnt signalling. *Nat. Cell Biol.*, **13** (5), 623–629.
30. Callow, M.G., Tran, H., Phu, L., Lau, T., Lee, J., Sandoval, W.N., Liu, P.S., Bheddah, S., Tao, J., Lill, J.R., Hongo, J.A., Davis, D., Kirkpatrick, D.S., Polakis, P., and Costa, M. (2011) Ubiquitin ligase RNF146 regulates tankyrase and Axin to promote Wnt signaling. *PLoS One*, **6** (7), e22595.
31. Lau, T., Chan, E., Callow, M., Waaler, J., Boggs, J., Sandoval, W.N., Liu, P.S., Bheddah, S., Tao, J., Lill, J.R., Hongo, J.A., Davis, D., Kirkpatrick, D.S., Polakis, P., and Costa, M. (2013) A novel tankyrase small-molecule inhibitor suppresses APC mutation-driven colorectal tumor growth. *Cancer Res.*, **73** (10), 3132–3144.
32. Waaler, J., Machon, O., Tumova, L., Dinh, H., Korinek, V., Wilson, S.R., Paulsen, J.E., Pedersen, N.M., Eide, T.J., Machonova, O., Gradl, D., Voronkov, A., von Kries, J.P., and Krauss, S. (2012) A novel tankyrase inhibitor decreases canonical Wnt signaling in colon carcinoma cells and reduces tumor growth in conditional APC mutant mice. *Cancer Res.*, **72** (11), 2822–2832.
33. Kirby, C.A., Cheung, A., Fazal, A., Shultz, M.D., and Stams, T. (2012) Structure of human tankyrase 1 in complex with small-molecule inhibitors PJ34 and XAV939. *Acta Crystallogr. Sect. F Struct. Biol. Cryst. Commun.*, **68** (Pt 2), 115–118.
34. Karlberg, T., Markova, N., Johansson, I., Hammarstrom, M., Schütz, P., Weigelt, J., and Schüler, H. (2010) Structural basis for the interaction between tankyrase-2 and a potent Wnt-signaling inhibitor. *J. Med. Chem.*, **53** (14), 5352–5355.
35. Shultz, M.D., Cheung, A.K., Kirby, C.A., Firestone, B., Fan, J., Chen, C.H., Chen, Z., Chin, D.N., Dipietro, L., Fazal, A., Feng, Y., Fortin, P.D., Gould, T., Lagu, B., Lei, H., Lenoir, F., Majumdar, D., Ochala, E., Palermo, M.G., Pham, L., Pu, M., Smith, T., Stams, T., Tomlinson, R.C., Touré, B.B., Visser, M., Wang, R.M., Waters, N.J., and Shao, W. (2013) Identification of NVP-TNKS656: the use of structure-efficiency relationships to generate a highly potent, selective, and orally active tankyrase inhibitor. *J. Med. Chem.*, **56** (16), 6495–6511.

36. Fancy, S.P., Harrington, E.P., Yuen, T.J., Silbereis, J.C., Zhao, C., Baranzini, S.E., Bruce, C.C., Otero, J.J., Huang, E.J., Nusse, R., Franklin, R.J., and Rowitch, D.H. (2011) Axin2 as regulatory and therapeutic target in newborn brain injury and remyelination. *Nat. Neurosci.*, **14** (8), 1009–1016.

37. Ulsamer, A., Wei, Y., Kim, K.K., Tan, K., Wheeler, S., Xi, Y., Thies, R.S., and Chapman, H.A. (2012) Axin pathway activity regulates in vivo pY654-beta-catenin accumulation and pulmonary fibrosis. *J. Biol. Chem.*, **287** (7), 5164–5172.

18
The Identification of the Molecular Receptor of the Plant Hormone Abscisic Acid

Julian Oeljeklaus and Markus Kaiser

18.1
Introduction

Plants constitute one of the five kingdoms of life and their roughly 500 000 different species are found on land, in oceans, and in freshwater. They are multicellular, eukaryotic organisms endowed with the capacity to perform photosynthesis, which allows them to generate food supplies such as carbohydrates from light energy. Owing to this unique trait, plants represent the basis of the food chain for animals and humans, and crops have been since centuries the major energy source of human nutrition.

In contrast to organisms such as animals or bacteria, crops and all other higher plants are sessile. Consequently, they cannot escape if challenged from adverse environmental influences. During their lifetime, they are therefore steadily exposed to many different unfavorable conditions, and plants have consequently evolved many different and highly complex regulatory systems that allow them to survive and even prosper in swiftly changing and often detrimental environments. A better understanding of these regulatory processes is therefore not only of scientific interest but owing to their role in basic nutrition also of socioeconomic importance. In fact, molecular studies on model plant organisms such as *Arabidopsis thaliana* have in the past years helped to massively enlarge our knowledge of the unique biology of plants (Box 18.1).

These studies have shown that plant hormones (also known as *phytohormones*) play an important role in most plant biological processes and influence almost all facets of plant life (Box 18.2). One of these plant hormones is (+)-abscisic acid (ABA or (+)-ABA, **1**, Figure 18.1), a sesquiterpenoid natural product first discovered in the 1960s [1, 2]. The name *abscisic acid* derives from its originally deduced bioactivity, which is the induction of leaf abscission. Nowadays, it is known that this process is induced in only some plant species and ABA is more recognized for its capability to regulate many other vital plant physiological processes, ranging from nonstress responses such as seed maturation and bud dormancy to adaptive

Concepts and Case Studies in Chemical Biology, First Edition. Edited by Herbert Waldmann and Petra Janning.

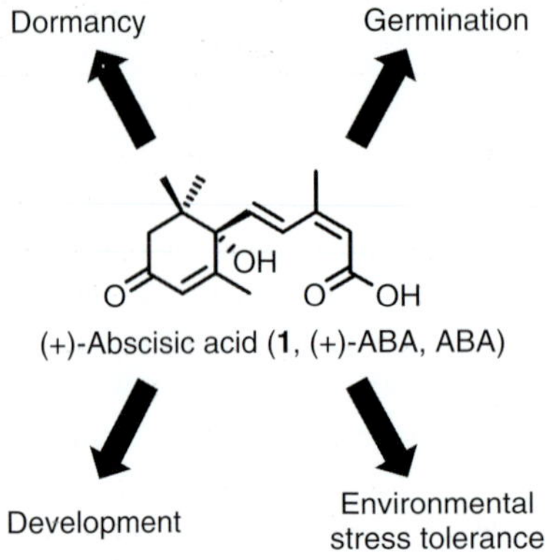

Figure 18.1 ABA controls many different physiological processes in plants.

stress responses (e.g., toward heat, drought, and salinity). For example, ABA leaf concentrations significantly increase when plants dry out or get too cold, thereby activating gene translation of a variety of genes involved in stress responses. The hormone thereby curtails water loss, induces seed dormancy until suitable conditions for germination are reached, and inhibits root and other vegetative growth. Owing to its essential function in plant physiology (and particular stress resistance), a proper (e.g., agrochemical) modulation of the ABA signaling pathway therefore holds enormous promise for agricultural applications. To this end, a detailed and mechanistic understanding of the ABA signaling network, including the molecular mechanism of ABA sensing and signal transduction is however required. In contrast to other plant hormones, this, however, turned out to be a major challenge that was only recently solved via the use of a chemical biology approach.

Box 18.1 *Arabidopsis thaliana*

A. thaliana (or thale cress) is a commonly occurring plant species from the family of *Brassicaceae.* It has no importance in agriculture but has been used since the 1940s as the major model organism in plant science. Consequently, the *Arabidopsis* genome was the first fully sequenced plant genome and comprises around 157 mega base pairs coding for 27 000 genes. These traits turn *Arabidopsis* into a plant with a rather small genome, thereby facilitating genetic mapping. Further reasons for the use of *Arabidopsis* as a model organism are a relatively short generation cycle of only 6 weeks, the relative small size of this plant which is beneficial for cultivation in laboratories, and abundant seed production. Moreover, the long-term storage of seeds is quite simple and well established. Genetic manipulation of *Arabidopsis* can be easily achieved *via Agrobacterium tumefaciens* transfection and is nowadays a routine technique in plant research. Finally, the physiology of *A. thaliana* matches quite well with most other higher plant species.

Box 18.2 Plant Hormones

Plant hormones (also referred to as *phytohormones*) are endogenous signal molecules produced by plants. Although peptidic or proteinaceous plant hormones are also known, plants mainly utilize small molecule hormones. These compounds move easily through the plant's tissues, for example, by diffusion, although they are produced and often exert their bioactivity locally within the plant. Plant hormones regulate a vast number of physiological processes, for instance, the time point of flowering, breeding of stems, leaves, and the development and ripening of fruits. Plants, contrary to animals, lack corresponding glands that produce and secrete hormones. Instead, each plant cell is theoretically capable of producing hormones. Plant hormones are vital for cell division and differentiation and thus critically regulate all facets of plant growth and morphology. Classically, five major classes of plant hormones are differentiated, of which most consist of more than one chemical. The five classes are the auxins, the cytokinins, ethylene, gibberellins, and the abscisic acids. The plant hormones are each gathered together into one of these classes based on their structural similarities and on their effects on plant physiology. Each hormone class has positive as well as inhibitory influences, and they mostly work in a well-orchestrated team play with each other to affect plant growth regulation.

18.2 The Biological Problem

For many plant hormones, the use of forward genetic approaches has enabled the identification of the direct target proteins (also known as *receptors*) and other components of their signaling pathway. Along these lines, researchers screened for plant mutants that were insensitive to the corresponding phytohormone and – after analysis of these mutants – were able to identify key components of the corresponding signaling pathway [2].

Naturally, such an approach was also applied in ABA research. In fact, a phenotypic screen for mutants that germinate despite the presence of the "germination inhibitor" ABA resulted in the identification of a set of mutants commonly denoted as *abi* (for ABA insensitive). Among them, two mutants, that are *abi1* and *abi2*, turned out to be particularly interesting because they featured mutations in two related type 2C phosphatases genes, thereby suggesting a role of these enzymes in ABA signal transduction [3]. Despite intense research in the following years, their exact mechanistic involvement in ABA signaling however remained elusive. Although additional *abi* mutants were also isolated afterwards, the "standard" genetic approach continued to fail to deliver the desperately sought direct ABA-binding protein capable of "sensing" cellular ABA levels. In fact, later studies showed that there are many reasons for the identification of such a mutant being hampered [1, 2]. For example, the multitude and diversity of ABA gene responses found in plants strongly impedes the efficiency of a forward genetics screen. In *Arabidopsis* seedlings, roughly 10% of the whole genome is regulated by ABA (with a nearly one-to-one ratio between induced and repressed genes). At that, a thorough analysis furthermore revealed that the observed changes in gene transcription heavily depend on cell type and developmental stage, resulting in the finding that there is no "universal" set of ABA-regulated genes (although the expected preferences become visible, that is, ABA-induced genes are generally enriched for those encoding proteins involved in stress tolerance, while ABA-repressed genes are enriched for proteins associated with growth). In some plant tissues, ABA is even able to neutralize the effect of growth-promoting phytohormones such as auxin or the gibberellins. Thus, ABA is a rather pervasive hormone that simultaneously affects many genes and pathways that sometimes even overlap, thereby making it difficult to screen for and characterize mutant plants with clear-cut defects in ABA responses. Moreover, the ABA receptor(s) most probably turned out to be functionally redundant, a trait often observed in plants because extensive genome duplications have occurred multiple times in their evolutionary history. Therefore, most plant genes are part of whole gene families and single knockout mutant plants often lack a clear phenotypic response because the missing knocked-out enzyme activity is compensated for by other members of the gene family.

Consequently, researchers subsequently turned their attention to biochemical instead of genetic approaches for discovering the molecular ABA receptor(s). These studies culminated in the identification of several proteins that bound ABA *in vitro* with nanomolar affinities [2]. These proteins however turned out to be rather "unusual" ABA-binding proteins and this finding immediately led to the controversial question if they represent the physiologically relevant ABA receptor proteins. In fact, cellular ABA levels in ABA-sensitive cells are often in the low-micromolar range that may even increase further under stress conditions, thereby questioning if proteins that bind ABA in the nanomolar range may serve as biologically relevant "sensors" of ABA levels; moreover, the identified proteins failed to explain the many different physiological effects that occur upon ABA application [4]. In fact, these discrepancies persuaded most researchers that the biochemically identified proteins also did not represent the biologically relevant ABA receptor and further efforts to elucidate the "real" ABA receptor were still required.

This challenge was therefore taken up by the Cutler group, which started an alternative, chemical genetics approach and finally succeeded in the discovery of the ABA receptor [5].

18.3 The Chemical Genetics Approach

18.3.1 Identification of a Synthetic ABA-Agonist Using a Chemical Genetics Screen

In contrast to the "classical" genetic approach, a chemical genetic approach uses small molecules to directly modulate protein functions, thereby causing

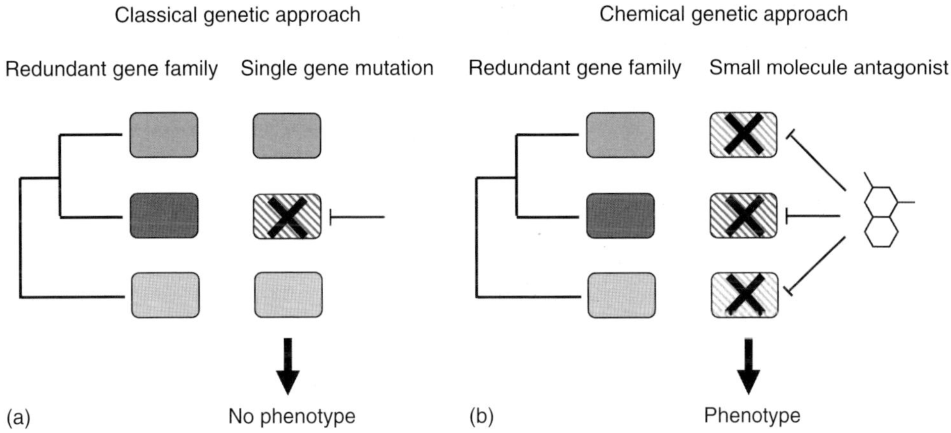

Figure 18.2 *"Classical" genetics (a) versus chemical genetics (b) in the presence of gene redundancy*. Single genetic knock-outs of one gene from a redundant gene family might lack a phenotypic response because other family members will functionally substitute its role. In contrast, a small molecule antagonist might be able address all members of a gene family due to conserved structural elements (e.g., conserved binding pockets), thereby chemically knocking out all members of the redundant gene family and causing a phenotypic response.

an observable phenotypic effect that often resembles a mutant (e.g., knockout) phenotype, thereby justifying the term *chemical genetics* (Figure 18.2). Such an approach holds many advantages, for example, chemical interference can be dosed (via concentration series) or introduced at any time point and at any developmental stage of an organism. Of particular interest for ABA receptor discovery, chemical genetics is furthermore often capable of overcoming problems associated with gene redundancy as proteins encoded by one gene family often share conserved binding sites susceptible to the same small molecule inhibitor (and thus cause a gene family-wide chemical knockout). Small molecule chemical genetics screens may therefore reveal phenotypes in cases where simple gene mutations fail.

With these advantages in mind, the Cutler group therefore devised a chemical genetics approach, aiming to identify chemical modulators of the ABA signaling pathway (Figure 18.3). To this end, a chemical library of 3600 small molecules were screened for compounds that perturbed seed maturation of the model plant *A. thaliana* (subtype Col-0), thereby using a phenotypic readout similar to the bioactivity of ABA. These efforts resulted in the elucidation of a synthetic naphthalene sulfonamide germination inhibitor which they called *pyrabactin* (compound **2**, Figure 18.3).

A subsequent structure–activity analysis of structurally closely related derivatives revealed that a substitution of the pyridine moiety in pyrabactin (compound **2**, Figure 18.4) by a phenyl residue resulted in an inactive derivative, named apyrabactin (compound **4**, Figure 18.4).

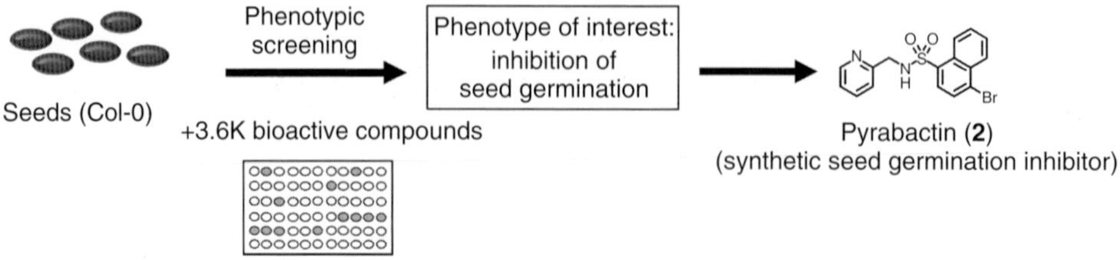

Figure 18.3 *Workflow of the employed phenotypic screen to identify potential ABA-agonists.* Wildtype (WT) seeds of *Arabidopsis* were incubated with a library of 3600 different small molecules. Compounds that prevented seed-germination were considered as hits. Validation then confirmed the compound pyrabactin (**2**) as a synthetic seed germination inhibitor.

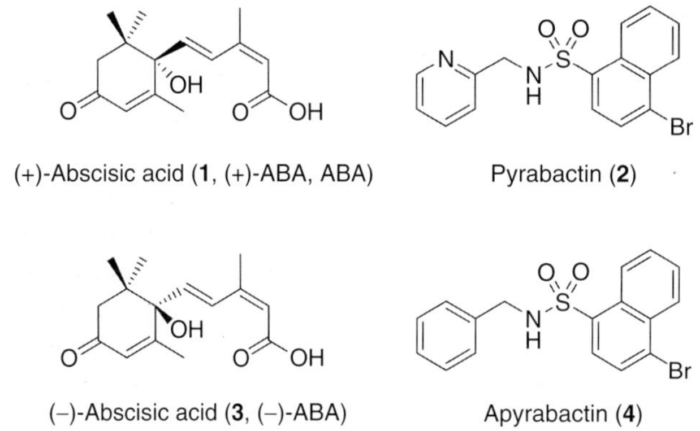

Figure 18.4 Chemical structures of (+)-abscisic acid (**1**), (−)-abscisic acid (**3**), the ABA agonist pyrabactin (**2**), and the inactive control compound apyrabactin (**4**).

Microarray-based analyses then demonstrated that ABA (**1**) and pyrabactin (**2**) induced highly correlated transcriptomal responses in seeds. In a control experiment, three previously known, but structurally unrelated, germination inhibitors were also analyzed. These compounds however failed to induce similar ABA responses, thereby illustrating that the ABA agonistic activity of pyrabactin (**2**) in seeds was not the result of a simple, "indirect" inhibitory germination effect. In an additional experiment, the Cutler group then investigated the transcriptional responses of ABA and pyrabactin in seedlings and found – in contrast to the seed assay – a lower correlation. These findings therefore pinpointed that pyrabactin was affecting some but not all of the pathways regulated by ABA.

18.3.2
Target Gene Identification of Pyrabactin

In order to identify the molecular target of the ABA agonist pyrabactin, a forward genetic screen was next employed aiming at the isolation of pyrabactin-

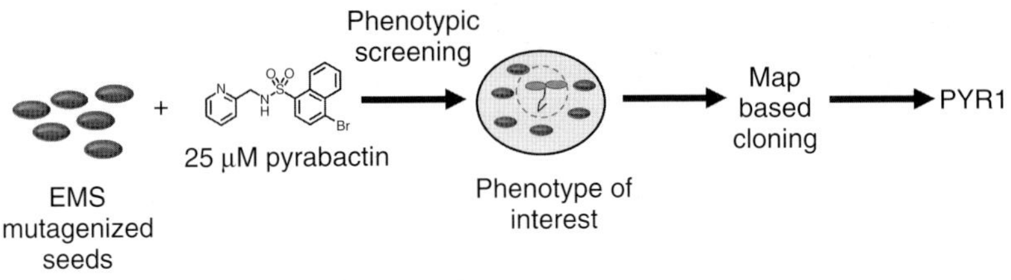

Figure 18.5 *A second phenotypic screen was used to identify pyrabactin-resistant mutant lines.* EMS-mutagenized seeds were incubated for 4 days with 25 μM of pyrabactin. Seedlings that germinated within this timeframe into plants with fully developed cotyledons were considered as resistant mutants. Subsequent map based cloning then identified that most of these mutants had defects in the PYR1 coding gene.

resistant mutants. To this end, ethylmethane sulfonate (EMS)-mutagenized seeds were generated and subsequently screened for germination in presence of 25 μM pyrabactin (Figure 18.5). This screen revealed 16 seedlings with fully developed cotyledons, which were therefore considered as pyrabactin resistant.

By using map-based cloning, 12 of these pyrabactin resistance 1 (PYR1) mutants were subsequently determined to contain mutations in the gene At4g17870, a gene encoding PYR1, a member of the PYR/PYL/RCAR (PYL, pyrabactin resistance like; RCAR, regulatory component of ABA receptor) proteins family, a subfamily of polyketide cyclase-like proteins. Sequence alignments revealed that 13 genes in the *Arabidopsis* genome shared distinct similarities with PYR1 and were therefore designated as Pyl1 to Pyl13 (for PYR1-like). Noteworthy, this 14-membered gene family has also almost simultaneously been independently identified as ABI1

Table 18.1 Nomenclature and locus of the 14 members of the PYR/PYL/RCAR family of ABA receptors.

Nomenclature		Locus
PYR/PYL	RCAR	
PYR1	RCAR11	At4g17870
PYL1	RCAR12	At5g46790
PYL2	RCAR14	At2g26040
PYL3	RCAR13	At1g73000
PYL4	RCAR10	At2g38310
PYL5	RCAR8	At5g05440
PYL6	RCAR9	At2g40330
PYL7	RCAR2	At4g01026
PYL8	RCAR3	At5g53160
PYL9	RCAR1	At1g01360
PYL10	RCAR4	At4g27920
PYL11	RCAR5	At5g45860
PYL12	RCAR6	At5g45870
PYL13	RCAR7	At4g18620

(Abelson interactor 1)-interacting proteins (one of the previously via standard-genetics-identified ABA-resistance mutants) and likewise designated as RCAR1 to RCAR14 [6]. The reports of the elucidation of the PYR/PYL and RCAR proteins at the same time therefore led to two sets of nomenclature for this family (Table 18.1).

Because the *pyr1* mutant responded normally to ABA (i.e., displayed impaired seed maturation), functional redundancy from other members of this gene family might mask PYR1's role in ABA signaling. To investigate whether PYR/PYLs display functional redundancy in ABA signaling, *Pyr1*, *Pyl1*, *Pyl2*, and *Pyl4* insertion alleles were isolated and multilocus mutants were constructed. These four alleles were chosen because they showed the highest expression levels in seeds (Figure 18.6a). While the triple (*pyr1*; *pyl1*; *pyl4*) and quadruple (*pyr1*; *pyl2*; *pyl1*; *pyl4*) mutant lines showed reduced sensitivity to ABA and displayed reduced ABA-mediated transcriptional responses, the double mutant (*pyr1*; *pyl4*) as before the *pyr1*-mutant did not, suggesting that a triple homozygous mutant is minimally required for ABA insensitivity (Figure 18.6b). Interestingly, the triple and quadruple mutants could be rescued by introducing PYR1 or PYL4-expressing transgenes, thereby proving that the PYR/PYL genes act redundantly in ABA signaling. Monitoring of the kinase activity of SnRK2, a downstream

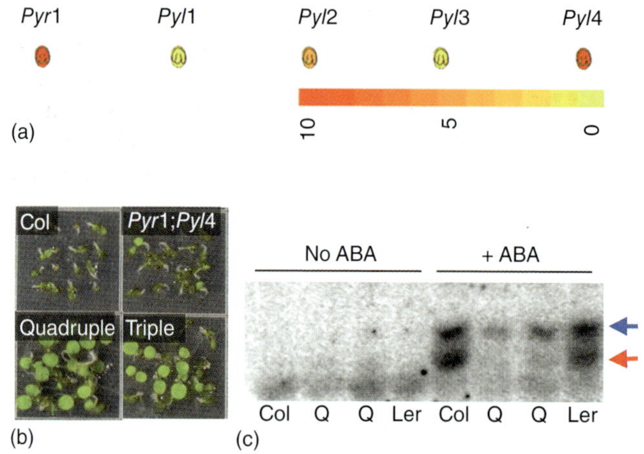

Figure 18.6 *Relevance of PYR/PYL proteins for ABA signaling.* (a) PYR1 and PYL1 to PYL4 expression levels in seeds; the heatmap shows normalized microarray expression values according to the color scale. (b) The PYR/PYL genes act redundantly. Deletion mutants were germinated on medium containing 0.9 μM ABA for 7 days. Col = wt *Arabidopsis* (Col-0), *pyr1*; *pyl4* = double mutant, triple = *pyr1*; *pyl1*; *pyl4* triple mutant, quadruple = *pyr1*; *pyl1*; *pyl2*; *pyl4* quadruple mutant. (c) The PYR/PYLs are necessary for SnRK2 kinase activity. The in-gel kinase assays were conducted on extracts made from ABA-treated plants. Active, phosphorylated SnRK2-kinase was visualized by ^{32}P-radioactivity detection (arrows indicate different types of SnRK2 kinases, namely SnRK2.2 and SnRK2.3 (red arrow) and SnRK2.6 (blue arrow)). Col = wt *Arabidopsis* (Col-0); Q = quadruple mutant; Ler = wt *Arabidopsis* Landsberg erecta ecotype. (From [5]. Reprinted with permission from AAAS.)

effector known to become autophosphorylated and activated upon ABA signaling [7], furthermore demonstrated that PYR1 and PYLs are also required for normal ABA-induced perception (Figure 18.6c).

Altogether, these findings demonstrated that PYR1 and PYLs are required for mediating multiple ABA responses *in vivo* and display differential binding affinities for natural and synthetic agonists.

18.4 The Chemical Biology Approach

18.4.1 Elucidation of the Functional ABA-Receptor Complex

Although the previous findings indicated that PYR1 and PYLs play critical roles in mediating ABA responses, no studies on a protein level or mechanistic studies that demonstrate how ligand binding is transmitted into ABA signaling had until now been performed. To overcome this and as a working hypothesis, the Cutler group therefore hypothesized that PYR1 might act as a binding protein for ABA or pyrabactin, thereby promoting the formation of a protein–protein complex between PYR1 and a so far unidentified downstream effector. To test and potentially validate this hypothesis on a protein level, they therefore employed a yeast-2-hybrid (Y2H) assay in the presence of either 10 μM pyrabactin or ABA, aiming to identify potential PYR1 interactors (of note, the authors called this assay a *Y2H approach* in their original article; in literature, such assays are however often denoted as yeast-3-hybrid assays because it is built up from three different components, that is, PYR1, the interacting protein and the small molecule ligand. For consistency with the original literature, we will however maintain the original nomenclature) (Box 18.3). For this purpose, 2 million prey cDNA clones were screened against the PYR1 Y2H bait. This experiment revealed an enzyme called *HAB1* that was interacting with PYR1 in the presence of pyrabactin and ABA, while no interaction was observed if apyrabactin was used. HAB1 belongs to a group of type 2C protein phosphatases (PP2Cs), which consists of nine partially redundant members. It is noteworthy that these findings also correlate with the results from the traditional genetic screens performed in 1994, which found two PP2C mutants called *ABI1* and *ABI2* as regulators of ABA signaling [3].

Box 18.3 Yeast-2-hybrid Screening

Y2H screening (also known as a *yeast-two-hybrid system*) is a chemical biology technique widely used to discover protein–protein interactions of a protein of interest.

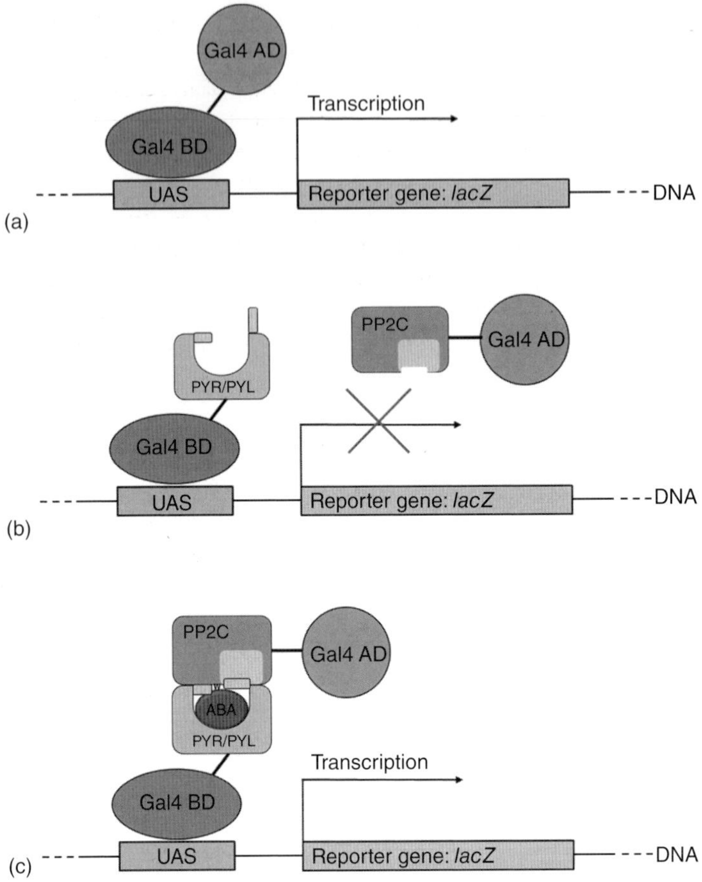

Figure 1 *The principle of yeast-2-hybrid screening using the example of PYR/PYL interacting with PP2C in the presence of ABA.* (a) The yeast Gal4 transcription factor consists of two domains (BD and AD), which are both required for a functional transcription of the reporter gene (*lacZ*), giving a blue stain after substrate treatment. (b) Two fusion proteins are generated, Gal4/BD-PYR/PYL and Gal4/AD-PP2C. None of them alone is sufficient to initiate the transcription and the absence of ABA hampers the physical interaction between PYR/PYL and PP2C. (c) When both fusion proteins interact with each other, the Gal4-AD domain is recruited to the Gal4-BD domain, resulting in reconstitution of the transcription factor, and transcription is initiated as exemplified here by ABA binding to PYR/PYL, which enables interaction with PP2C resulting in transcription of the reporter.

To this end, a genetically modified yeast strain is used in which the activation of a downstream reporter gene, for example, the *lacZ* gene, is followed. Initiation of transcription relies on a binding of a transcription factor (e.g., Gal4 that is frequently used) onto an upstream activating sequence (UAS). In Y2H screens, this transcription factor is split into two separate fragments, called the *binding domain* (*BD*) and *activating domain* (*AD*). The BD binds to the DNA, while the AD controls the activation of transcription, turning the Y2H screen into a protein-fragment complementation assay. For identifying protein interactors, the protein of interest, also known as the *bait* (here: PYR/PYL) is fused to a BD of transcription factor such as Gal4, resulting in the construct Gal4-BD. Simultaneously, each yeast cell expresses a distinct Gal4-AD-fused "prey" that originates from a cDNA library which was produced through reverse transcription of fully transcribed mRNA (messenger ribonucleic acid) and therefore contains all expressed genes of an organism (here: *A. thaliana*). If a yeast strain now contains the "matching partners," this is a protein that forms a protein–protein interaction with the protein of interest fused to the Gal4-BD; a functional transcription factor is reconstituted, resulting in activation of the downstream reporter gene (Figure 1).

Next, the defective *pyr1* mutants from the previous PYR screen were tested for an ABA-responsive interaction with HAB1. $PYR1^{P88S}$ and the $PYR1^{S152L}$ mutants disrupted the ABA-induced PYR1–HAB1 interaction, whereas the $PYR1^{R157H}$ did not (Figure 18.7). Subsequent Co-IP experiments using either mock or ABA-treated living plants confirmed that the (+)-ABA-mediated interaction between PYR1 and HAB1 is not only occurring in the yeast test system (Figure 18.8a and Box 18.4). Finally, the corresponding interaction could also be affirmed in biochemical experiments, which demonstrated that recombinantly produced PYR1 and HAB1 proteins only interacted in presence of ABA (Figure 18.8b).

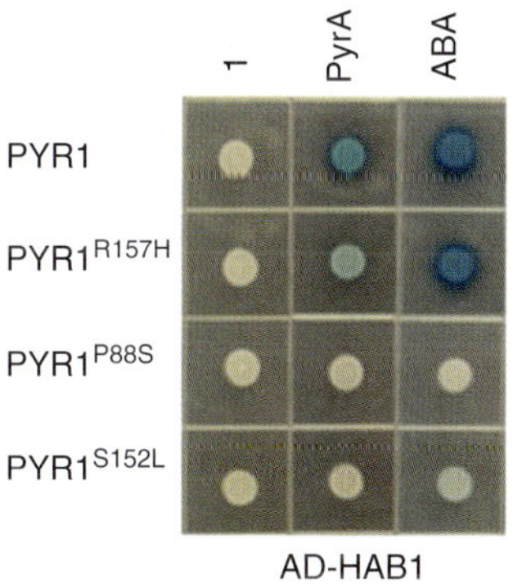

Figure 18.7 *Results of the yeast-2-hybrid assay.* PYR1 and three different pyrabactin-insensitive mutants were constructed as binding domain (BD) fusion proteins and tested for their interaction with activation domain (AD)-fused HAB1 in the presence of pyrabactin or ABA. Blue staining then indicates a protein–protein interaction between HAB1 and the different wild type or mutant PYR1s. Either 10 mM pyrabactin (PyrA) or 10 mM ABA were used. (From [5]. Reprinted with permission from AAAS.)

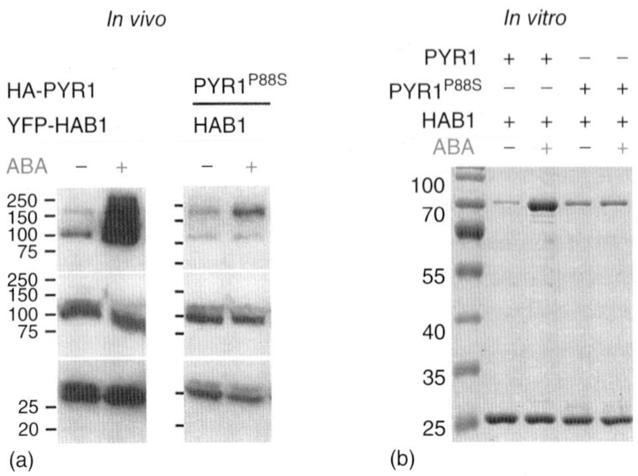

Figure 18.8 *ABA-dependent formation of PYR1–PP2C protein complexes.* (a) ABA promotes PYR1 binding to the PP2C phosphatase HAB1 *in vivo*. Protein extracts were made from plant leaves, transformed with the indicated constructs, immunoprecipitated with an antibody against HA-agarose and immunodetected with an antibody against yellow fluorescent protein or HA. The YFP-HAB1 fusion protein migrates at 100 kDa and HA-PYR1 fusion protein at 25 kDa. (b) Reconstitution of ABA responses *in vitro*. Pull-down assays with recombinant, purified glutathione S-transferase (GST)-tagged HAB1 (~80 kDa) and His6-tagged PYR1 (~25 kDa) were conducted in presence or absence of 10 mM ABA. (From [5]. Reprinted with permission from AAAS.)

Box 18.4 Co-immunoprecipitation

Co-immunoprecipitation (Co-IP) is a widely used technique to elucidate physiologically important protein–protein interactions of a target protein by using a target protein-specific antibody to "indirectly" trap proteins binding to the target protein. These protein complexes can be subsequently analyzed to elucidate new protein-binding partners, binding affinities, the kinetics of binding, and the function of the target protein. The concept of immunoprecipitation is very simple but ingenious. First, an antibody (monoclonal or polyclonal) against a specific target protein is used to form a selective immune complex with that target protein in a complex protein mixture such as, for example, a cell lysate. The immune complex is then captured (or precipitated) on a solid support, such as a bead, to which an antibody-binding protein is immobilized (such as protein A or G). Any proteins that are not bound to the beads via the antibody are then washed away. Finally, the antigen (target protein) is eluted from the solid support and analyzed by, for example, gel electrophoresis (SDS-PAGE, sodium dodecyl sulfate polyacrylamide gel electrophoresis), often followed by Western blot detection to confirm the identity of the target protein (Figure 2).

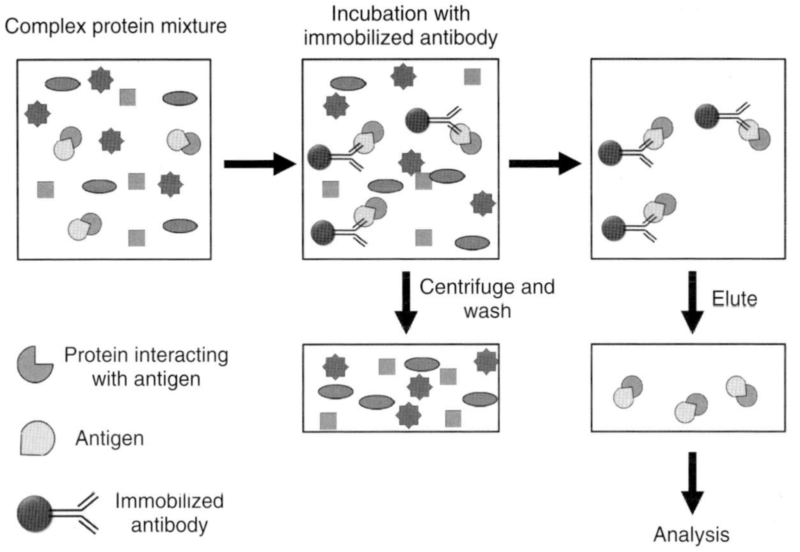

Figure 2 Principle of co-immunoprecipitation experiments.

The observed functional redundancy on a genetic level for the double, triple, and quadruple mutants was then further examined using, again, a Y2H assay. Therefore, a series of 12 PYR/PYLs was screened using different ligands. These experiments showed that ABA mediated an HAB1-interaction for the proteins PYR1 to PYL4, whereas pyrabactin promoted an HAB1 interaction between PYR1, PYL1, and PYL3. Of these three proteins, only PYR1 is highly transcribed in seeds, thereby explaining why *pyr1* mutants showed insensitivity toward pyrabactin. These results furthermore indicated that the five PYR/PYLs do have different ligand selectivity and do not bind HAB1 in response to nonnatural ligands (Figure 18.9). Interestingly, additional experiments demonstrated that PYL2, PYL3, and PYL4 also respond to the nonnatural (−)-ABA stereoisomer, turning these proteins into candidates for dual-stereoisomer receptors and that PYL12 also interacted in an ABA-dependent manner with PP2CA/AHG3.

In summary, at least six proteins from the 14-membered PYR/PYL gene family showed ABA responsiveness in yeast. The observed selectivity within the *pyr/pyl* genes toward ABA-triggered formation of protein–protein complexes with HAB1 indicates that the ABA pathway may be dissected using selective PYR/PYL small-molecule agonists.

Next, the binding of ABA to ^{15}N-labeled PYR1 and PYR1^{P88S} was examined using heteronuclear single-quantum coherence (HSQC) nuclear magnetic resonance (NMR) spectroscopy. This technique allows the detection of ligand binding to proteins by monitoring the chemical shifts of protein amide-NH bonds that shift on ligand binding because of a changing chemical environment. As PYR1

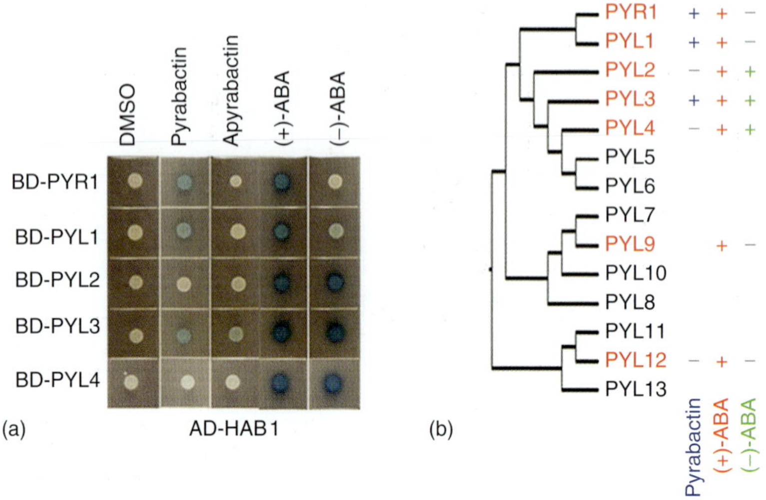

Figure 18.9 *PYR/PYL proteins have different selectivity for responses to different small-molecule ligands.* (a) A panel of PYR/PYL genes were generated as BD fusion proteins and tested in yeast-2-hybrid assays for interactions with HAB1 in the presence of (+)-ABA, (−)-ABA, pyrabactin, apyrabactin (all at a concentration of 10 mM) or dimethyl sulfoxide (DMSO) (carrier solvent, 1%). Blue staining indicates formation of a protein–protein complex. (b) ABA response activity is distributed throughout the PYR/PYL family. Shown is a phylogenetic tree of the PYR/PYL family, which is correlated with ligand selectivity data derived from the yeast-2-hybrid assays. (+)-ABA-responsive PYR/PYLs are colored red (PYL9 red coloring is performed on the basis of data from Ma *et al.* [6]). (From [5]. Reprinted with permission from AAAS.)

belongs to the protein family of START-like proteins (StAR-related lipid transfer proteins) that contain a conserved hydrophobic ligand-binding cavity, it was rationalized that ABA binding should selectively disturb the NMR signals from this binding site on the protein. The NMR experiments confirmed this hypothesis as they revealed that addition of ABA to PYR1 caused an alteration of the HSQC signals in the expected protein region. A complete saturation of the binding cavity was achieved using a 2.5-fold excess of ABA. Even more surprising was the finding that the PYR1^{P88S} mutant was not defective in ABA binding, whereas it was malfunctioning in ABA signaling. Instead, this mutant failed to bind HAB1 after ABA binding, thereby again confirming the developed ABA receptor model (Figure 18.10).

PP2Cs such as HAB1 have previously been recognized as negative regulators of ABA perception. Consequently, the Cutler group rationalized that (+)-ABA-mediated formation of the PYR/PYL-PP2C-protein interaction might result in inhibition of HAB1's phosphatase activity. To evaluate this hypothesis, they subsequently investigated the effects of (+)-ABA on HAB1's PP2C enzyme activity using a biochemical assay consisting of purified PYR1 or PYR^{P88S} and HAB1 together with *para*-nitrophenylphosphate, a synthetic phosphatase substrate. These assays showed that ABA application resulted in potent phosphatase

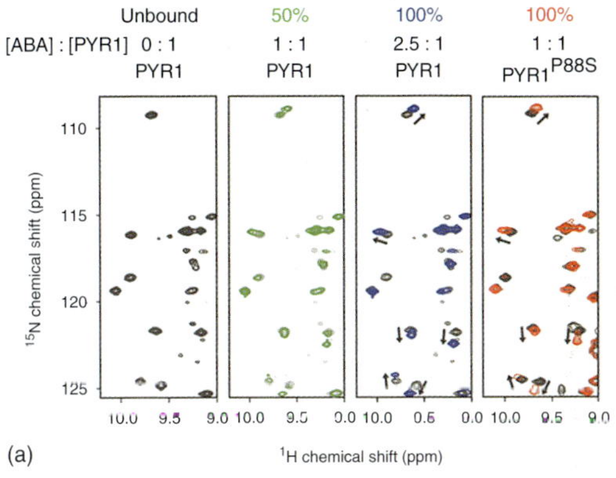

Figure 18.10 *ABA binds to PYR1 and inhibits PP2C phosphatase activity.* (a) Depicted are selected subregions of HSQC spectra for ^{15}N-labeled PYR1 and PYR1^{P88S} in the presence of increasing concentrations of ABA. Arrows indicate amide protons that shift upon ABA binding. (b) PYR1 inhibits PP2C phosphatase activity in the presence of ABA. Initial reaction velocities of recombinant GST-HAB1 were determined in the presence of PYR1 or the PYR1^{P88S} mutant and differing ABA concentrations using the colorimetric substrate *para*-nitrophenyl phosphate (pNPP) and used to calculate the IC_{50} values of 125 nM for PYR1 and 50 mM for PYR1^{P88S}, respectively. (From [5]. Reprinted with permission from AAAS.)

inhibition with an IC_{50} (inhibitor concentration 50) value of 125 nM if PYR1 was present. As expected, in case the PYR^{P88S} mutant was used, no inhibition was found as the mutation prevented HAB1 binding.

Taken together, the experimental data set revealed the following mechanistic model of ABA sensing in living plants: PYR1 first acts as a (+)-ABA receptor. This binding then induces the formation of a protein–protein interaction with PP2Cs such as HAB1, thereby inhibiting their phosphatase activity. Moreover, the different ligand-binding specificity of the diverse PYR/PYLs control which ligands trigger PP2C interactions.

18.4.2
Validation and Further Structural Studies on the ABA–Receptor Mechanism

After the identification of the essential elements of the ABA–receptor complex, several additional research groups undertook efforts to validate the proposed ABA perception model. Among the many studies, of particular relevance is a report that demonstrated that the ABA pathway could be reconstituted *in vitro* by combining the ABA receptor PYR1 with the PP2C phosphatase ABI1 and the downstream effectors SnRK2.6/OST1, a serine/threonine protein kinase, and the transcription factor ABF2/AREB1 [8]. Their experiments furthermore elucidated that the PP2Cs usually interact with and dephosphorylate the SnRK2 kinases, thereby keeping them in an inactive state. This interaction is interrupted by addition of ABA, which instead results in the formation of PYR/PYL-PP2C complexes, thereby activating the SnRK2 kinases that finally phosphorylate and thus activate the transcription factor ABF2/AREB1. Altogether, this study therefore not only established the minimal core set of protein components required for ABA signaling but also confirmed the previously established ABA receptor model (Figure 18.11).

Subsequently performed crystallographic studies from various groups have furthermore helped to complete the picture on the mechanism of ABA recognition and perception [9–14]. Structures of several PYR/PYL proteins in different functional states have been obtained so far and crystallographic structures of PYL2 in a ligand-free, ligand-bound, and ligand/phosphatase-bound state enabled a detailed view of conformational changes that PYL2 undergoes upon ABA and phosphatase binding. These studies revealed that ABA binding to PYR/PYLs is associated with "closing" of a gating loop over the ABA-binding cavity [9]. Thus, in the absence of ABA, the PYR/PYL proteins exist in an open form with an accessible, water-filled binding pocket. Two flexible surface loops, referred to as *gate* and *latch* flank the ligand entry site of the hydrophobic binding pocket (Figure 18.12). When this binding pocket is occupied by ABA, the gating loop undergoes a conformational

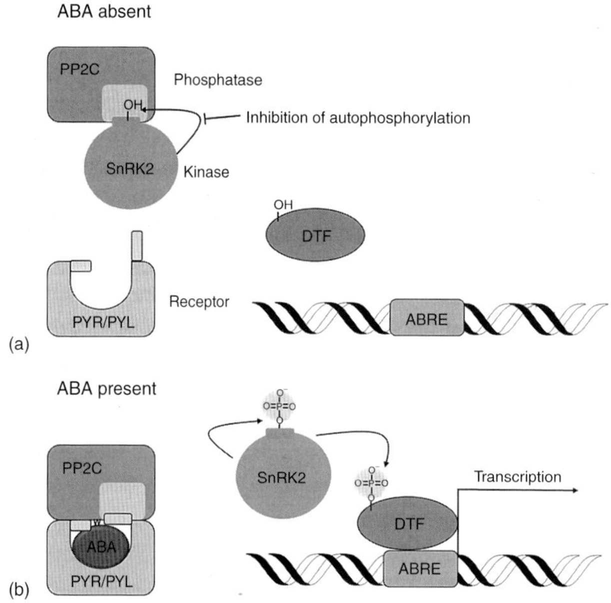

Figure 18.11 *Schematic presentation of the ABA signaling pathway.* (a) In the absence of ABA, the phosphatase PP2C binds to and thus inhibits autophosphorylation of SnRK2 kinases, a family of downstream ABA effectors. Consequently, signaling and thus gene transcription is suppressed. (b) The presence of ABA enables the PYR/PYL proteins to bind to and inhibit PP2C phosphatase. This relieves phosphatase inhibition of the kinase, thereby enabling autophosphorylation and thus activation of the kinase domain. Active SnRK2 kinases subsequently activate further downstream transcription factors (DTFs), thereby starting transcription at ABA-responsive promoter elements (ABREs).

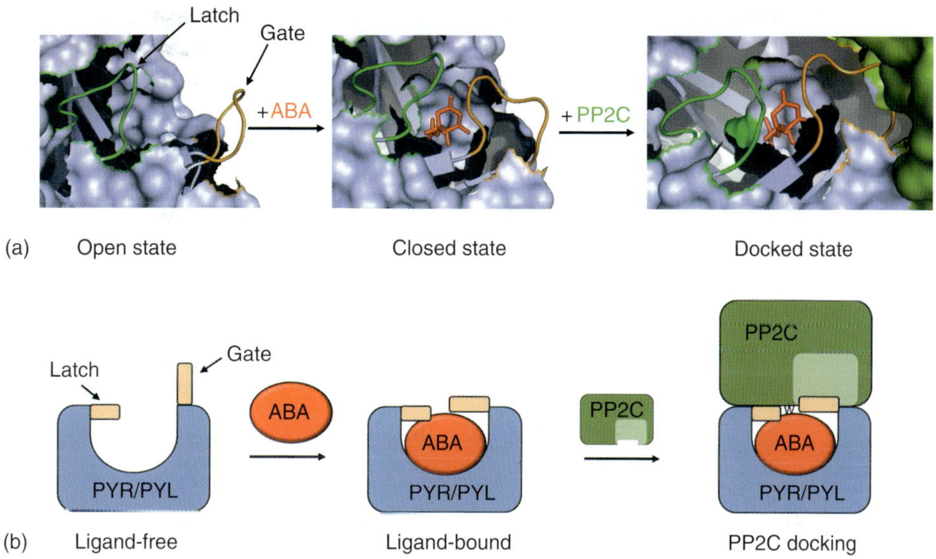

Figure 18.12 *Mechanism of ligand binding.* (a) Crystal structures reveal ABA-induced conformational changes of the PYR1 receptor. In the absence of ABA, the PYR1 protein adopts an open conformation at the "gate" loop. Binding of ABA induces a closure of the "gate," thereby sealing (+)-ABA within the cavity. Simultaneously, a functional interface that enables binding of the PP2C phosphatase is created. A conserved tryptophan residue of PP2C thereby inserts next to the gating loop, further locking the closed conformation. The interaction of PYR1 and PP2C thereby blocks access to the phosphatase-substrate-binding pocket, resulting in its inhibition. (This figure was made using Protein Data Bank coordinates for PYL2 (3KAZ), ABA-bound PYL2 (3KB0), and ABA/PP2C-bound PYL2 (3KB3) from Melcher *et al.* [9].) (b) A schematic depiction of the above-described binding events.

change and seals ABA in the binding pocket. This "closed" conformation then acts as a complementary interface that allows docking into the PP2C active site, thus inhibiting PP2C activity by blocking substrate access. A conserved tryptophan residue of PP2Cs was thereby found to play a particular role in this mechanism: in the closed state, it inserts its side chain between the gate and the latch, thereby functioning as a molecular seal and allowing the gating loop to closely interact with the substrate-binding site of the phosphatase, thereby blocking its ability to bind and thus inhibit the autophosphorylation of its substrate, the downstream effector kinase SnRK2. Of note, so far most structures of PYR/PYL proteins indicate that these proteins in the absence of ABA form dimers. It is however not clear if this dimerization also occurs *in vivo* and is biologically relevant.

x Recently, the structural basis of the interaction of SnRK2 kinases and PP2C phosphatases has also been elucidated [14]. These studies confirmed the previously established ABA signal transduction models and revealed an astonishing similarity in PP2C recognition by SnRK2 kinases and the PYR/PYL receptors in which a locking of the conserved tryptophan residue of PP2C is involved in either an interaction with the SnRK2 kinases or the ABA receptors of the PYR/PYL family. The autoactivation of the downstream effector SnRK2 kinases is thereby inhibited by two mechanisms: on the one hand, the interaction of the SNRK2 kinase with the PP2C phosphatases leads to a dephosphorylation of the activation loop of SnRK2. On the other hand, the physical interaction furthermore blocks the access of potential substrates to the kinase active site. Recruitment of PP2C by the ABA-receptor proteins of the PYR/PYL family then occurs at the same binding site. Thus, the ABA-bound PYR/PYL proteins as well as the SnRK2 kinases use a type of molecular mimicry to "compete" for binding to PP2Cs [14].

18.5
Conclusion

The present case study demonstrates the power of chemical genetics and other chemical biology techniques to uncover the molecular targets of natural products such as ABA. This combination approach has allowed elucidation of the family of PYR/PYLs proteins as the direct protein targets of ABA and has validated them as the long-sought protein family of ABA receptors. Their previous discovery has been hampered by the redundancy in the *pyr/pyl* genes. The chemical genetic approach using pyrabactin, which targets only a subset of proteins from the PYR/PYL family, however circumvented this genetic redundancy, which masked the ABA phenotype in single knockout mutants. The identification of the synthetic ABA agonist pyrabactin was thereby found to be crucial as this small molecule has the capacity to bind selectivity only a subset of the different members of the PYR/PYL family (Figure 18.13).

The target identification of pyrabactin and the subsequent validation experiments then allowed not only elucidating the direct ABA target proteins but also

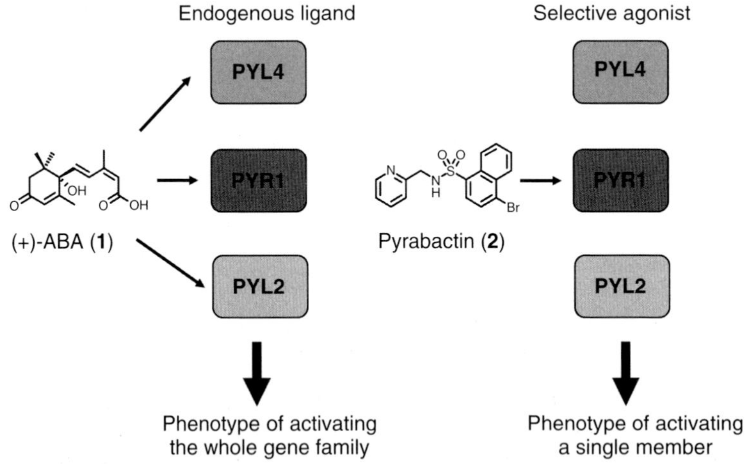

Figure 18.13 *Different target selectivities of ABA and pyrabactin.* The identification of selective agonists allows dissection of the different functions of PYR/PYL proteins.

identifying and establishing the core ABA signaling system. These studies have paved the way for further crystallographic studies that finally revealed atomic insights into the molecular mechanism by which ABA is sensed *via* the PYR/PYL proteins and how binding is transduced into a signal *via* the PP2C phosphatases and SnRK2 kinases.

References

1. Cutler, S.R., Rodriguez, P.L., Finkelstein, R.R., and Abrams, S.R. (2010) Abscisic acid: emergence of a core signaling network. *Annu. Rev. Plant Biol.*, **61**, 651–679.
2. McCourt, P. and Creelman, R. (2008) The ABA receptors – we report you decide. *Curr. Opin. Plant Biol.*, **11**, 474–478.
3. Meyer, K., Leube, M.P., and Grill, E. (1994) A protein phosphatase 2c involved in ABA signal-transduction in *Arabidopsis thaliana*. *Science*, **264**, 1452–1455.
4. Pennisi, E. (2009) Plant biology. Stressed out over a stress hormone. *Science*, **324**, 1012–1013.
5. Park, S.Y., Fung, P., Nishimura, N., Jensen, D.R., Fujii, H., Zhao, Y., Lumba, S., Santiago, J., Rodrigues, A., Chow, T.F., Alfred, S.E., Bonetta, D., Finkelstein, R., Provart, N.J., Desveaux, D., Rodriguez, P.L., McCourt, P., Zhu, J.K., Schroeder, J.I., Volkman, B.F., and Cutler, S.R. (2009) Abscisic acid inhibits type 2C protein phosphatases via the PYR/PYL family of START proteins. *Science*, **324**, 1068–1071.
6. Ma, Y., Szostkiewicz, I., Korte, A., Moes, D., Yang, Y., Christmann, A., and Grill, E. (2009) Regulators of PP2C phosphatase activity function as abscisic acid sensors. *Science*, **324**, 1064–1068.
7. Fujii, H., Verslues, P.E., and Zhu, J.K. (2007) Identification of two protein kinases required for abscisic acid regulation of seed germination, root growth, and gene expression in Arabidopsis. *Plant Cell*, **19**, 485–494.
8. Fujii, H., Chinnusamy, V., Rodrigues, A., Rubio, S., Antoni, R., Park, S.Y., Cutler, S.R., Sheen, J., Rodriguez, P.L., and Zhu, J.K. (2009) *In vitro* reconstitution of an abscisic acid signalling pathway. *Nature*, **462**, 660–664.
9. Melcher, K., Ng, L.M., Zhou, X.E., Soon, F.F., Xu, Y., Suino-Powell, K.M., Park, S.Y., Weiner, J.J., Fujii, H., Chinnusamy, V., Kovach, A., Li, J., Wang, Y., Peterson, F.C., Jensen, D.R., Yong, E.L., Volkman, B.F., Cutler, S.R., Zhu, J.K., and Xu, H.E. (2009) A gate-latch-lock mechanism for hormone signalling by abscisic acid receptors. *Nature*, **462**, 602–608.
10. Miyazono, K., Miyakawa, T., Sawano, Y., Kubota, K., Kang, H.J., Asano, A., Miyauchi, Y., Takahashi, M., Zhi, Y., Fujita, Y., Yoshida, T., Kodaira, K.S., Yamaguchi-Shinozaki, K., and Tanokura, M. (2009) Structural basis of abscisic acid signalling. *Nature*, **462**, 609–614.
11. Santiago, J., Dupeux, F., Round, A., Antoni, R., Park, S.Y., Jamin, M., Cutler, S.R., Rodriguez, P.L., and Marquez, J.A. (2009) The abscisic acid receptor PYR1 in complex with abscisic acid. *Nature*, **462**, 665–668.
12. Nishimura, N., Hitomi, K., Arvai, A.S., Rambo, R.P., Hitomi, C., Cutler, S.R., Schroeder, J.I., and Getzoff, E.D. (2009) Structural mechanism of abscisic acid binding and signaling by dimeric PYR1. *Science*, **326**, 1373–1379.
13. Yin, P., Fan, H., Hao, Q., Yuan, X.Q., Wu, D., Pang, Y.X., Yan, C.Y., Li, W.Q., Wang, J.W., and Yan, N. (2009) Structural insights into the mechanism of abscisic acid signaling by PYL proteins. *Nat. Struct. Mol. Biol.*, **16**, 1230–1236.
14. Soon, F.F., Ng, L.M., Zhou, X.E., West, G.M., Kovach, A., Tan, M.H., Suino-Powell, K.M., He, Y., Xu, Y., Chalmers, M.J., Brunzelle, J.S., Zhang, H., Yang, H., Jiang, H., Li, J., Yong, E.L., Cutler, S., Zhu, J.K., Griffin, P.R., Melcher, K., and Xu, H.E. (2012) Molecular mimicry regulates ABA signaling by SnRK2 kinases and PP2C phosphatases. *Science*, **335**, 85–88.

19
Chemical Biology in Plants: Finding New Connections between Pathways Using the Small Molecule Sortin1

Chunhua Zhang, Glenn R. Hicks, and Natasha V. Raikhel

19.1
Introduction

Chemical genomics combines chemical library screening and genomics approaches to dissect biological processes. It is especially useful in studying essential cellular processes such as endomembrane trafficking pathways. This chapter describes a small molecule – Sortin1 – that connects plant vacuolar trafficking defects and flavonoid metabolism, both of which play important roles in plant growth and development. Sortin1 was obtained by a chemical library screen aimed at finding small molecules that disrupt plant vacuolar trafficking based on the evolutionary conservation between plants and *Saccharomyces cerevisiae* in vacuolar trafficking pathways. The subsequent mutant screening in *Arabidopsis* identified flavonoid metabolism mutants that link flavonoids to vacuole biogenesis and vacuolar trafficking.

19.2
The Biological Problem

In plant cells, as in other eukaryotic cells, the endomembrane system is composed of multiple organelles with distinct morphology and functions. The compartmentalized endomembrane system ensures proper processing and trafficking of macromolecules to the sites of function. A normally functioning endomembrane trafficking machinery is essential for plant growth and development [1].

Vacuoles are organelles that are present in all plant and fungal cells. Plant vacuoles are multifunctional organelles that can take up to 90% of the total cell volume. Vacuoles are storage organelles for ions, sugars, polysaccharides, pigments, proteins, and flavor compounds of fruits and vegetables (reviewed in [1, 2]). The high solute concentration in the vacuole creates the turgor pressure, which serves as the driving force for cell expansion during plant growth. Vacuoles are highly dynamic organelles and there is a large diversity in the forms, sizes, and contents of vacuoles in different plant tissues and organs, and even the same cell

Concepts and Case Studies in Chemical Biology, First Edition. Edited by Herbert Waldmann and Petra Janning.

under different biotic/abiotic conditions/stimulations. Severe vacuole biogenesis mutants, such as *vcl1* and *amsh3*, cause either embryonic or seedling lethality [3, 4].

Vacuolar proteins are synthesized at the endoplasmic reticulum (ER) and then delivered to the vacuole through the vesicle trafficking pathway, which includes the Golgi apparatus, trans-Golgi network (TGN)/early endosomes, and late endosomes/prevacuolar compartments (reviewed in [5]). The interaction between the proteins' vacuolar sorting sequences, such as the amino-terminal propeptides (NTPPs) and carboxyl-terminal propeptides (CTPPs), and the vacuolar sorting receptors (VSRs) directs proteins to clathrin-coated vesicles. These vesicles deliver cargos to the prevacuolar compartments, which then fuse with the vacuole (reviewed in [1, 5]). In plants, there is also a Golgi-independent pathway for trafficking of protein storage vacuoles, where proteins are delivered from ER to the vacuole *via* large vesicles rather than Golgi (reviewed in [2]). Transport of other contents such as ions and pigmented anthocyanins to the vacuole involve ion channels, tonoplast-bound ABC (ATP-binding cassette) transporters and solute/H^+ antiporters.

We have been interested in finding other regulators of protein vacuolar trafficking pathways. However, it has been challenging to study vacuolar trafficking using conventional genetic knockout mutants because of lethality or genetic redundancy, as we mentioned earlier. That is why it has been interesting to the cell biologists to develop chemical genomics tools to study essential and dynamic cellular processes. Chemical genomics uses small molecules to disrupt cellular pathways and provide tunable and reversible tools in addressing critical cellular process such as membrane trafficking.

19.3
The Chemical Approach

The strategy allowed the identification of chemicals that disrupt the trafficking of carboxypeptidase Y (CPY), a vacuolar protein that contains an NTPP vacuolar targeting sequence, to the vacuole through the endomembrane system in yeast. The selected compounds were then tested in *Arabidopsis* for vacuolar trafficking and vacuole morphology effects. Although there are differences in the vesicle trafficking pathways between plants and yeast (reviewed in [6]), many protein complexes that mediate these pathways are homologous to one another. Thirteen compounds were effective in yeast and three of them also disrupt vacuolar trafficking in *Arabidopsis*.

19.3.1
Chemical Library Screening

A wild-type yeast strain *S. cerevisiae* INVSc1 (*his3-Δ1, leu2, trip1-289, ura3-52,* Invitrogen) that does not display secretion of CPY under normal growth condition was used for chemical library screening. A chemical library containing 4800 compounds (DIVERSetE, Chembridge) was screened. CPY secretion was monitored by the accumulation of CPY protein in the growth medium with different compounds using a dot-blot approach. The compounds that cause the secretion of CPY to the growth medium after treatment were selected to further test their effects on yeast vacuolar morphology. The positive compounds were then tested in plant cells for vacuolar morphology and vacuolar trafficking. These further confirmed compounds that were positive in the plant system were named Sortins for protein SORTing Inhibitors [7]. Sortin1 is one of the compounds found to affect vacuolar trafficking and vacuole morphology in both yeast and *Arabidopsis*. The structure of Sortin1 is shown in Figure 19.1a. Sortin1 induces CPY secretion in a dosage-dependent manner, with the maximum amount of secreted CPY protein detected in the yeast growth medium in the presence of 25 mg l^{-1} Sortin1 (Figure 19.1b). The protein secretion was dramatically reduced with increased Sortin1, which indicates that it might cause lethality at higher dosage.

Sortin1 also causes secretion of *Arabidopsis* carboxypeptidase Y (AtCPY) protein in cultured cells (Figure 19.1c). To test the effect of Sortin1 in *Arabidopsis* vacuole morphology, a transgenic line carrying enhanced green fluorescent protein (EGFP)-tagged δTIP (delta-tonoplast intrinsic protein), a plant tonoplast intrinsic protein expressed abundantly in vacuolated cells in vegetative organs [8], was used as a vacuole marker. When grown in the presence of Sortin1, the vacuoles of EGFP-δTIP seedlings are highly fragmented compared with untreated control (Figure 19.1d,e). The specificity of Sortin1 to the vacuole was confirmed by examining cellular localization of different organelle markers after Sortin1 treatment [9]. Sortin1 also stimulates AtCPY secretion in whole seedlings, as revealed by increased AtCPY localization in apoplast after Sortin1 treatment by immunoelectron microscopy analysis [7]. In order to test whether Sortin1 only affects proteins with NTPP sorting signal, Rosado and coworkers generated a CLV3-secreted peptide fused with the CTPP vacuole sorting signal from barley lectin (CLV3 : T7 : $CTPP_{BL}$) [9]. Immunoelectron microscopy shows that CLV3 : T7 : $CTPP_{BL}$ was secreted to the apoplast upon Sortin1 treatment as well [9]. This shows that in plants, Sortin1 affects vacuolar trafficking of proteins containing both NTPP and CTPP sorting signals. Different phenotypes indicate that Sortin1 affects vacuolar trafficking and vacuolar morphology in both yeast and plant.

19.3.2
Identification of Pathway(s) that are Targeted by Sortin1

In order to find the related genetic pathways that were affected by Sortin1, chemical genomics screening was carried out to identify *Arabidopsis* mutants that were hypersensitive to Sortin1 treatment. An M2 segregating mutant population was created by ethyl methanesulfonate (EMS) treatment of seeds from an EGFP-δTIP-expressing *Arabidopsis* line. The EMS-treated seeds were then

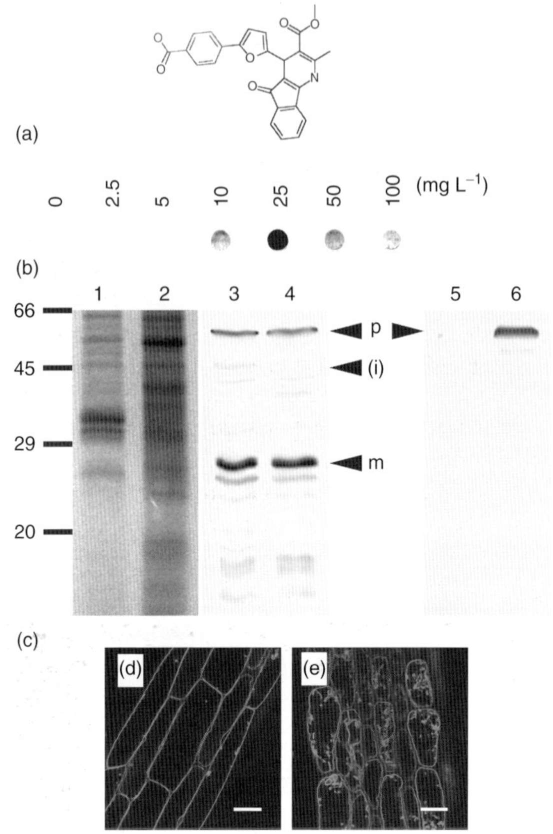

Figure 19.1 *Sortin1 affects vacuolar trafficking of CPY in both yeast and Arabidopsis and affects Arabidopsis vacuole morphology.* (a) Sortin1 chemical structure. (b) A dot-blot detecting CPY in yeast growth media after treatment with different concentrations of Sortin1 showing that Sortin1 stimulates the secretion of CPY in yeast. (c) Proteins that are secreted into the growth media of *Arabidopsis* cultured cells that were untreated (lane 1) or treated with Sortin1 (lane 2) were concentrated and analyzed by SDS-PAGE (sodium docecylsulfate-polyacrylamide gel electrophoresis). The AtCPY processing pattern was examined in cell pellets in untreated (lane 3) or treated with Sortin1 (lane 4). The arrowheads point to the positions of the 60-kDa precursor (p), 48-kDa intermediate (i), and 24-kDa mature (m) forms of AtCPY in Western blots. Western blot detection of AtCPY in the growth media of *Arabidopsis* cultured cells under untreated condition or treated with Sortin1. (d,e) Vacuolar morphology in hypocotyls epidermal cells of 1-week-old seedlings expressing EGFP-δTIP under untreated (d) or Sortin1-treated (e) conditions. Reprinted with permissions from [7], Copyright (2004) National Academy of Sciences, U.S.A.

grown to produce the next generation. A sublethal dosage of Sortin1 was used to select for the mutants that showed a strong hypersensitive response to Sortin1 (11 μM) compared to the wild-type seedlings that were not affected significantly. Six strong Sortin1 hypersensitive (*s1h*) *Arabidopsis* mutants were identified after three rounds of screens [9].

19.3.3
Sortin1-Hypersensitive Mutants Link Vacuolar Trafficking to Flavonoids Metabolism

The seed coat color of some *s1h* mutants vary from the wild-type, indicating that the mutant genes might be involved in flavonoid biosynthesis and transport (Figure 19.2a). In order to examine the flavonoid accumulation phenotype in *s1h* mutants, chemical staining in seed coats using aromatic aldehyde *p*-dimethylaminocinnamaldehyde (DMACA), which specifically reacts with proanthocyanidins, flavan-3,4-diols, and flavan-3-ols [10], was carried out. As shown in Figure 19.2b, flavonoid vacuolar accumulation was significantly reduced in the seed coats of three *s1h* mutants (*s1h2-50, s1h2-21, s1h1-27*) compared to the wild-type and three other mutants. One of the *s1h* mutants, *s1h2-50*, is allelic to the flavonoid biosynthesis mutant *tt4* and has vesiculated

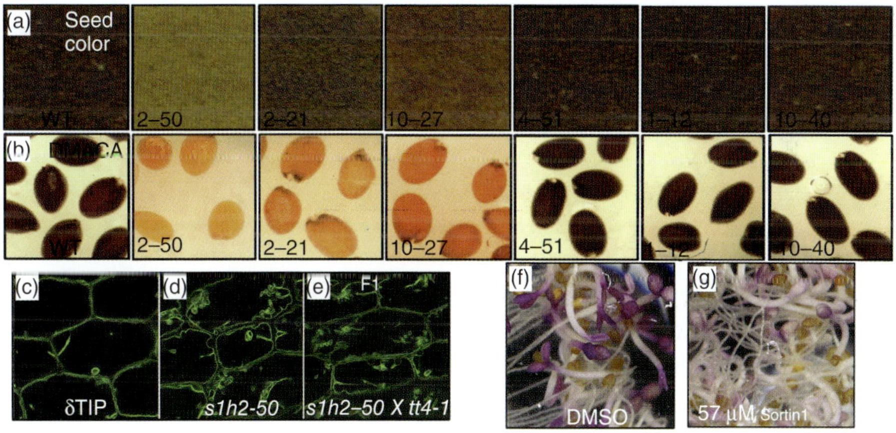

Figure 19.2 *Sortin1 affects flavonoid transport to the vacuole.* (a) Seed coat colors of different *s1h* mutants. (b) DMACA staining shows that some *s1h* mutants (2–50, 2–21, and 10–27) have defects in vacuolar accumulation of proanthocyanidins in the seed coats. (c–e) The recessive mutant *s1h2-50* is allelic to *tt4*, a flavonoid biosynthesis mutant that also affects vacuole morphology. The localization of EGFP-δTIP in wild-type (c), *s1h2-50* homozygous mutant (d), or the F1 seedling (e) from the cross between *s1h2-50* and *tt4*. (f,g) Vacuolar accumulation of pigmented anthocyanins in seedlings untreated (f) or treated with Sortin1 (g) under anthocyanin-inductive conditions. Reprinted with permissions from [9]. Copyright (2011), with permission from Elsevier.

vacuoles even without Sortin1 treatment (Figure 19.2c–e) [9]. Reduced vacuolar flavonoid accumulation was also observed in seedlings treated with Sortin1 under anthocyanin-inductive conditions (AICs) [11], when the accumulation of pigmented anthocyanins in the vacuole was detected (Figure 19.2f,g). Thorough quantification of pigmented anthocyanin accumulation after different dosages of Sortin1 treatment further confirmed the effect of Sortin1 in flavonoid vacuole accumulation [9]. Reduced flavonoid accumulation in the vacuole upon Sortin1 treatment is consistent with the fact that flavonoid biosynthesis mutants are hypersensitive to Sortin1 treatment.

19.3.4
Sortin1 Resembles the Effects of Buthionine Sulfoximine (BSO)

Sortin1 inhibition of flavonoid accumulation in vacuoles provoked Rosado and coworkers [9] to investigate whether it acts through inhibition of ABC-type tonoplast transporters, which function in secondary metabolite transport [12]. The authors compared the effects of Sortin1 and the ABC-type transporter inhibitor BSO on root growth, anthocyanin accumulation, and vacuole vesiculation (Figure 19.3a–i) [9]. Sortin1 and BSO induced similar phenotypes in all the assays tested, which included reduced root hair elongation, heterogeneous accumulation of pigmented anthocyanins, and tonoplast vesiculation. Furthermore, neutral red staining bodies (NRSBs), which have been described as an indicator of dynamic subvacuolar structures that accumulate under light and oxidative stress and in the mutants with enhanced autophagy [11], were formed in cotyledons after treatment with both Sortin1 and BSO.

The notion of Sortin1 resembling BSO was further supported by the transcription profiles induced by the two chemicals. Sortin1 and BSO induce transcription of similar groups of genes such as P450-dependent monooxygenases, glutathione (GSH) S-transferases, and glucosyltransferases and others [9]. Similar classes of genes were also induced by another plant oxidative stress generator xenobiotic trinitrotoluene (TNT) [9].

19.3.5
Substructures Required for Sortin1 Bioactivity

It was also noticed that Sortin1 is not stable in *Arabidopsis* growth media and probably decomposes to produce a compound that is active [9]. The authors analyzed different Sortin1 substructures for their induction of vacuolar morphology and anthocyanin vacuolar accumulation defects [9]. It was found that the substructure in Figure 19.3j is required for induction of vacuolar morphology defects and the substructure in Figure 19.3k is required for the induction of the anthocyanin vacuolar accumulation defect.

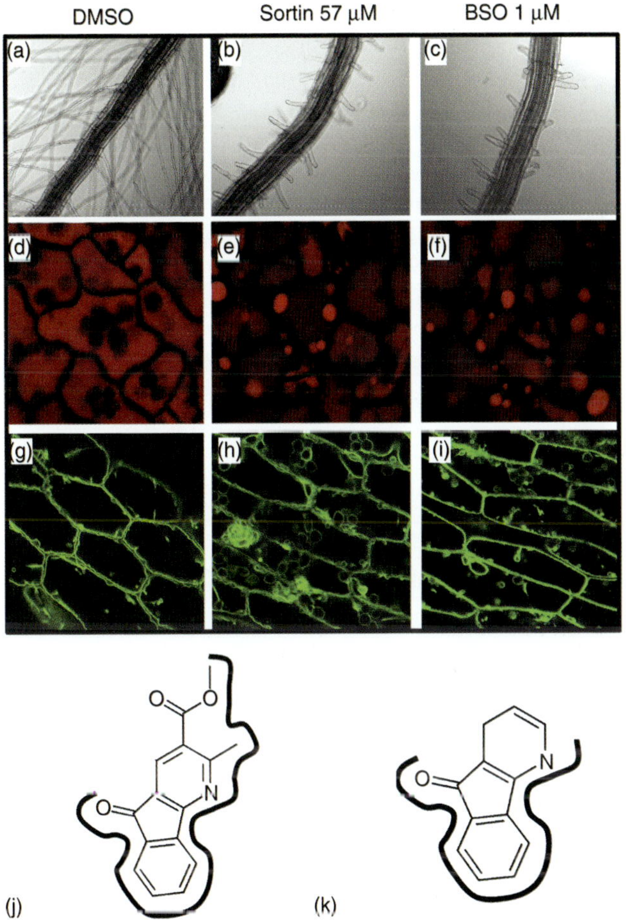

Figure 19.3 *Sortin1-induced phenotypes resemble those of the glutathione biosynthesis inhibitor BSO.* (a–c) Both Sortin1 (b) and BSO (c) inhibit root hair elongation compared with the control (a). (d–f) Fluorescence accumulation in neutral red staining bodies in untreated (d), Sortin1-treated (e), or BSO-treated (f) seedlings. The seedlings were grown under AIC conditions and stained with Naringenin. The samples were excited with a 543 nm laser line and the fluorescence was detected in the 565–620 nm range. (g–i) Both Sortin1 and BSO induce tonoplast vesiculation. Six-day-old seedlings expressing EGFP-δTIP were incubated for 24 h with DMSO (dimethyl sulfoxide) (g), Sortin1 (h), or BSO (i) and the fluorescence was analyzed using confocal microscopy. (j) Minimal bioactive Sortin1 substructure that is required for vacuolar biogenesis disruption. (k) Minimal bioactive Sortin1 substructure that is required for vacuolar-anthocyanin accumulation disruption. Reprinted with permissions from [9]. Copyright (2011), with permission from Elsevier.

19.4 Biological Research/Evaluation

19.4.1 Chemicals That Disrupt Yeast Vacuolar Trafficking also Target Plant Vacuolar Trafficking Pathway

We started chemical screening using yeast as a system, because it is easy to grow and suitable for large-scale and automated screening. Although vacuoles are dispensable for yeast survival compared to plant cells, there are some homologous vesicle trafficking genes between the two systems. It turned out that Sortin1, one of the chemicals that disrupts yeast vacuolar CPY trafficking, also affects plant vacuolar trafficking and vacuole morphology. This provides clear evidence that chemicals derived from phenotypic screens are translatable between different systems.

19.4.2 Sortin1 Disrupts Vacuolar Trafficking of both Proteins and Flavonoids

As mentioned earlier, vacuoles are storage sites for different cellular materials, such as proteins and secondary metabolites. We showed that Sortin1 affects not only CPY protein but also flavonoid transport. Flavonoids are secondary metabolites that are not essential for plant survival but are involved in multiple cellular processes such as auxin transport, defense, modulating levels of reactive oxygen species, and providing different flower colors for attracting pollinators (reviewed in [13]). The enzymes for flavonoid biosynthesis are localized in the cytoplasm, ER, and nucleus, the proposed sites of flavonoid biosynthesis [14]. Flavonoids are located in different subcellular locations such as the vacuole, the cytosol, the ER, and the apoplast (reviewed in [15]). Flavonoids are transported from sites of synthesis to storage *via* two proposed models (reviewed in [15]). One model is called *vesicle-trafficking-mediated flavonoid transport*, wherein flavonoids are transported through the same pathway as protein trafficking and does not need specialized flavonoid carriers or tonoplast transporters. The other model suggests that flavonoid transport occurs through ABC-type transporters or the multidrug and toxic compound extrusion (MATE) antiporters, both of which need energy (reviewed in [15]).

Sortin1 affects the vacuolar trafficking of both proteins and flavonoids, indicating that flavonoid transport to the vacuole could share the same membrane-mediated pathway as vacuolar proteins. But this does not rule out the possibility that one of the trafficking defects is direct, whereas the other one is indirect. The fact that Sortin1 mimics BSO to produce an oxidized cell environment to affect ABC transporter activity supports a model that a transporter-mediated flavonoid transport defect induces miss-trafficking of other vacuolar contents such as the proteins. On the other hand, mutants defective in flavonoid metabolism, such as *tt4*, show hypersensitivity to Sortin1 and have altered vacuolar morphology without treatment. This indicates that a properly regulated flavonoid pathway participates in the maintenance of plant vacuole integrity.

19.4.3 Mechanism of Sortin1 Action

Sortin1 decomposes in plant growth media where one of the products resembles the effects of BSO, a GSH biosynthesis inhibitor. This indicates that a series of observed Sortin1 effects might be due to an oxidized cellular environment that it creates. It could be that Sortin1 or an active decomposition product sequesters GSH or inhibits GSH production. This may induce the transcription of cytochrome P450-dependent monooxygenases and mixed function oxidases, as shown in the transcription profile after Sortin1 treatment. Reduced GSH level then affects flavonoid transport.

The mechanism of flavonoid regulation of vacuolar biogenesis is not understood. However, the existence of vacuole biogenesis and flavonoid accumulation phenotypes in different mutants, such as *mtv6*, *aha10*, *tt12*, and *tds4* [9, 10, 16, 17] indicates that there is a connection between the two pathways. Sortin1 might serve as a tool in solving this question.

As reported by Rosado and coworkers [9], not all the *s1h* mutants have flavonoid accumulation/transport defects; this indicates that there are flavonoid-independent pathways that involve the regulation Sortin1-targeted vacuolar transport and vacuole biogenesis.

19.5 Conclusion

Chemical genomics is a valuable tool for cell biologists to study essential cellular processes such as endomembrane trafficking. Chemicals that are identified in simpler organisms can be translated to more complex systems. Chemical genomics in combination with reverse genetics can identify and integrate different cellular pathways. In the case of Sortin1, it connects three cellular processes: protein vacuolar trafficking, flavonoid vacuolar trafficking and metabolism, and vacuole biogenesis. Further identification of target proteins that directly interact with Sortin1 or its active decomposition product will reveal proteins that serve as a linker or hub of three different cellular processes. This will significantly expand our knowledge about the mechanism of pathway interaction during plant growth and development.

Acknowledgment

The research is supported by the Division of Chemical Sciences, Geosciences, and Biosciences, Office of Basic Energy Sciences of the US Department of Energy through Grant DE-FG02-02ER15295 to N.V.R.

References

1. Surpin, M. and Raikhel, N. (2004) Traffic jams affect plant development and signal transduction. *Nat. Rev. Mol. Cell Biol.*, **5**, 100–109.
2. Marty, F. (1999) Plant vacuoles. *Plant Cell*, **11**, 587–600.
3. Isono, E., Katsiarimpa, A., Müller, I.K., Anzenberger, F., Stierhof, Y.D., Geldner, N., Chory, J., and Schwechheimer, C. (2010) The deubiquitinating enzyme AMSH3 is required for intracellular trafficking and vacuole biogenesis in *Arabidopsis thaliana. Plant Cell*, **22**, 1826–1837.
4. Rojo, E., Gillmor, C.S., Kovaleva, V., Somerville, C.R., and Raikhel, N.V. (2001) VACUOLELESS1 is an essential gene required for vacuole formation and morphogenesis in *Arabidopsis. Dev. Cell*, **1**, 303–310.
5. Xiang, L., Etxeberria, E., and Van den Ende, W. (2012) Vacuolar protein sorting mechanisms in plants. *FEBS J.*, **280**, 979–993.
6. Bassham, D.C. and Raikhel, N.V. (2000) Plant cells are not just green yeast. *Plant Physiol.*, **122**, 999–1001.
7. Zouhar, J., Hicks, G.R., and Raikhel, N.V. (2004) Sorting inhibitors (Sortins): chemical compounds to study vacuolar sorting in *Arabidopsis. Proc. Natl. Acad. Sci. U.S.A.*, **101** (7), 9497–9501.
8. Karlsson, M., Johansson, I., Bush, M., McCann, M.C., Maurel, C., Larsson, C., and Kjellbom, P. (2000) An abundant TIP expressed in mature highly vacuolated cells. *Plant J.*, **21**, 83–90.
9. Rosado, A., Hicks, G.R., Norambuena, L., Rogachev, I., Meir, S., Pourcel, L., Zouhar, J., Brown, M.Q., Boirsdore, M.P., Puckrin, R.S., Cutler, S.R., Rojo, E., Aharoni, A., and Raikhel, N.V. (2010) Sortin1-hypersensitive mutants link vacuolar-trafficking defects and flavonoid metabolism in *Arabidopsis* vegetative tissues. *Chem. Biol.*, **18**, 187–197.
10. Abrahams, S., Lee, E., Walker, A.R., Tanner, G.J., Larkin, P.J., and Ashton, A.R. (2003) The *Arabidopsis* TDS4 gene encodes leucoanthocyanidin dioxygenase (LDOX) and is essential for proanthocyanidin synthesis and vacuole development. *Plant J.*, **35**, 624–636.
11. Poustka, F., Irani, N.G., Feller, A., Lu, Y., Pourcel, L., Frame, K., and Grotewold, E. (2007) A trafficking pathway for anthocyanins overlaps with the endoplasmic reticulum-to-vacuole protein-sorting route in *Arabidopsis* and contributes to the formation of vacuolar inclusions. *Plant Physiol.*, **145**, 1323–1335.
12. Yazaki, K. (2006) ABC transporters involved in the transport of plant secondary metabolites. *FEBS Lett.*, **580**, 1183–1191.
13. Buer, C.S., Imin, N., and Djordjevic, M.A. (2010) Flavonoids: new roles for old molecules. *J. Integr. Plant Biol.*, **52**, 98–111.
14. Saslowsky, D.E., Warek, U., and Winkel, B.S. (2005) Nuclear localization of flavonoid enzymes in *Arabidopsis. J. Biol. Chem.*, **280**, 23735–23740.
15. Zhao, J. and Dixon, R.A. (2010) The 'ins' and 'outs' of flavonoid transport. *Trends Plant Sci.*, **15**, 72–80.
16. Baxter, I.R., Young, J.C., Armstrong, G., Foster, N., Bogenschutz, N., Cordova, T., Peer, W.A., Hazen, S.P., Murphy, A.S., and Harper, J.F. (2005) A plasma membrane H+-ATPase is required for the formation of proanthocyanidins in the seed coat endothelium of *Arabidopsis thaliana. Proc. Natl. Acad. Sci. U.S.A.*, **102**, 2649–2654.
17. Marinova, K., Pourcel, L., Weder, B., Schwarz, M., Barron, D., Routaboul, J.M., Debeaujon, I., and Klein, M. (2007) The *Arabidopsis* MATE transporter TT12 acts as a vacuolar flavonoid/H+-antiporter active in proanthocyanidin-accumulating cells of the seed coat. *Plant Cell*, **19**, 2023–2038.

20
Selective Targeting of Protein Interactions Mediated by BET Bromodomains

Susanne Müller, Hannah Lingard, and Stefan Knapp

20.1
Introduction

Transcription of genes is regulated by a complex network of transcription factors and transcriptional regulators that are recruited to acetylated euchromatin by protein interaction modules of the bromodomain family, a conserved helical domain that specifically recognizes ε-*N*-acetylated lysine residues. Here we describe the discovery and characterization of inhibitors of the bromo and extra terminal (BET) family, which consists of BRD2, BRD3, BRD4, and BRDT in humans. Evaluation of specific inhibitors targeting the acetyl-lysine site validated the targeting of BET bromodomains in cancer and we discuss the major biological findings that have been elaborated by BET specific chemical probes.

20.2
The Biological Problem

Transcriptional regulators and chromatin modifiers are deregulated in a large variety of diseases and have emerged as attractive therapeutic targets. For instance, inhibitors that target transcriptional regulators of the nuclear hormone receptor family have been developed into highly efficient drugs, and chromatin-modifying enzymes of the histone deacetylase family (HDACs) have been extensively targeted in oncology. However, protein interaction modules that mediate transcriptional processes are considered challenging sites to target and few inhibitors have been developed so far.

Histone acetylation is a hallmark of open and accessible chromatin structure, which results in activation of gene transcription. Histone acetylation leads to the recruitment of a number of bromodomain-containing proteins (BRDs) that exclusively recognize ε-*N*-acetylated lysine residues. These specific protein interaction modules, also called *epigenetic reader domains*, are evolutionarily highly conserved. The human proteome encodes 61 bromodomain interaction domains present in 42 proteins; comprising chromatin modifying enzymes (histone

Concepts and Case Studies in Chemical Biology, First Edition. Edited by Herbert Waldmann and Petra Janning.

acetyl transferases, methyl lysine transferases), ATP (adenosine triphosphate)-dependent chromatin remodeling complexes, transcriptional coactivators, and adapter proteins [1]. Dysfunction of BRDs has been linked to the development of many diseases [2], suggesting that selective inhibitors targeting the interaction of bromodomains with their specific cellular targeting sites might be disease modifying.

The BET family of bromodomain proteins comprises four members in mammals (BRD2, BRD3, BRD4, and BRDT), each containing two conserved *N*-terminal bromodomains and an extra terminal (ET) protein interaction motif. BET proteins specifically recognize acetylated lysine residues in histones H3 and H4, play critical roles in cellular proliferation and cell cycle progression, and are particularly required for efficient expression of several growth-promoting and anti-apoptotic genes. The BET family member BRD4 specifically interacts with the positive transcription elongation factor b (P-TEFb), a complex of the kinase Cdk9 and its activator cyclin T. When recruited to mitotic chromatin by BRD4, P-TEFb phosphorylates the C-terminal heptad repeat region of RNA polymerase II, stimulating transcriptional elongation by releasing the promoter-proximal paused polymerase. Transcription of highly expressed genes is driven by enhancers that interact with the transcription apparatus at the core promoter. BRD4 is known to interact with the mediator complex present at gene enhancers controlling transcriptional elongation by RNA polymerase II [3, 4]. Importantly, specific regulation of certain target genes is highly dependent on stimulating interactions with enhancer elements. A recent study revealed that phosphorylation of serine 10 at histone H3 (S10H3) at the FOSL1 enhancer by the serine/threonine kinase PIM1 results in the recruitment of the histone acetyltransferase MOF *via* a 14-3-3-mediated interaction. This complex subsequently leads to acetylation of histone H4 at lysine 16 (H4K16ac) at promoter regions and the recruitment of BRD4/P-TEFb and hence amplification of FOSL1 transcription [5] (Figure 20.1).

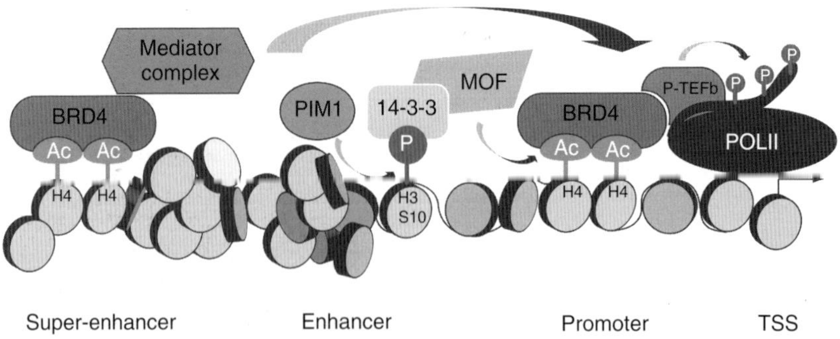

Figure 20.1 *BRD4-mediated transcriptional regulation.* BRD4 is recruited to enhancer/super-enhancer regions as well as to the promoter of highly transcribed genes. Recruitment of transcriptional regulators to the promoter is regulated by a sequence of signaling events involving the kinase PIM1 and the acetyltransferase MOF. Polyacetylated histone H4 in the promoter region recruits the pTEFb complex promoting transcriptional elongation at the transcriptional start site (TSS).

Intriguingly, recent publications suggest that a small number of lineage-specific survival genes are regulated by the so-called super-enhancer regions that differ from enhancers in size, transcription factor density, and their ability to activate transcription. Super-enhancers are present in loci of key oncogenic drivers and BRD4 is particularly enriched in these critical control regions [6]. Chemical inhibition or genetic knockdown of BRD4 leads to transcriptional downregulation of a number of growth-promoting (c-Myc, Aurora B, FOSL1, CDK6) and antiapoptotic (BCL2) genes, providing a compelling case for selective targeting of BRD4 in cancer [6, 7]. Indeed, a recent study showed that BRD4 is critically required for disease maintenance in acute myeloid leukemia (AML) [8]. Interestingly, expression of proinflammatory cytokines is also regulated by BET proteins. Inhibition of BET proteins results in suppression of inflammation and protection against lipopolysaccharide-induced endotoxic shock and bacteria-induced sepsis [9].

In addition, translocations of the BET locus have been linked to development of aggressive cancers: genetic rearrangement of the BRD4 and BRD3 loci gives rise to an aggressive form of squamous carcinoma in which in-frame chimeric proteins of the tandem *N*-terminal bromodomains of BRD4 or BRD3 fused with the protein NUT (nuclear protein in testis) lead to the development of nuclear protein in testis midline carcinoma (NMC), an incurable, uniformly fatal subtype of squamous carcinoma [10]. Thus, the key role of BET proteins in regulating transcription of key drivers of cancer growth and inflammation makes these BRDs interesting targets for the development of protein–protein interaction inhibitors with anti-inflammatory and anticancer activity.

20.2.1
Druggability of the BET Acetyl-Lysine-Binding Pocket

Bromodomains share a conserved fold that comprises a left-handed bundle of four alpha helices (αZ, αA, αB, αC). The four canonical helices form a large central cavity flanked by highly diverse loop regions (ZA and BC loops) that determine specificity of the interaction of bromodomains with their recognition sequences. The acetyl-lysine of bromodomain interaction sites is typically anchored by a hydrogen bond to a conserved asparagine residue and water-mediated interactions with a conserved tyrosine residue. Interestingly, for some bromodomains of the BET family, high affinity substrate recognition requires poly-acetylated sequences. Crystal structures of BET complexes with di-acetylated histone H4 tail peptides show that the first bromodomain of BET bromodomains requires two appropriately spaced acetyl-lysines for high-affinity interaction [1].

Acetylation of the lysine side chains neutralizes the charge of this residue. As a consequence, the acetyl-lysine binding pocket in bromodomains is populated by mainly hydrophobic and aromatic residues, suggesting that inhibitors with suitable pharmacokinetic properties can be developed. The acetyl-lysine binding site of BETs is sufficiently large to accommodate an inhibitor of around 400 Da. The pocket is deep with a good level of enclosure, and analysis of BET bromodomains revealed very favorable druggability scores [11]. Several conserved and tightly

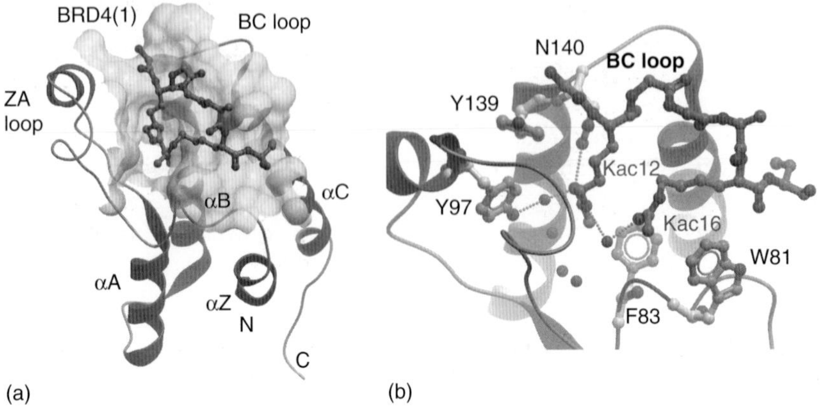

Figure 20.2 *Architecture of the acetyl-lysine binding pocket.* (a) Ribbon diagram of the first bromodomain of BRD4 [BRD4(1), pdb-code: 3uvx] in complex with the diacetylated histone tail of histone H4 (H4Kac12Kac16). Main structural elements are labeled. (b) Detailed view of the BRD4(1) acetyl-lysine binding site and interactions formed by the histone peptide. Water molecules are shown as spheres and main binding site residues and the interacting peptide are shown in ball-and-stick representation.

bound water molecules are present at the bottom of the acetyl-lysine pocket, suggesting that these structural waters should be treated as integral parts of the binding site (Figure 20.2).

20.3 The Chemical Approach

20.3.1 Development of High-Throughput Assays

Interaction of bromodomains with their target sequences is generally weak ($K_D \sim 1-50\,\mu M$), making the development of *in vitro* screening assays challenging. In order to identify the most suitable peptide sequence for assay development, we systematically screened libraries of acetylated histone peptides against members of the human bromodomain family [1]. This effort identified a large number of sequences that interact with human bromodomains. Highest affinities of BET family members were measured with poly-acetylated sequences in histone H4. Crystal structures revealed the binding mode of these interactions and isothermal titration calorimetry (ITC) determined dissociation constants (K_ds) in the low micromolar region [1].

We developed an *a*mplified *l*uminescence *p*roximity *h*omogeneous *a*ssay (AlphaScreen) for *in vitro* screening. This assay detects the proximity of donor

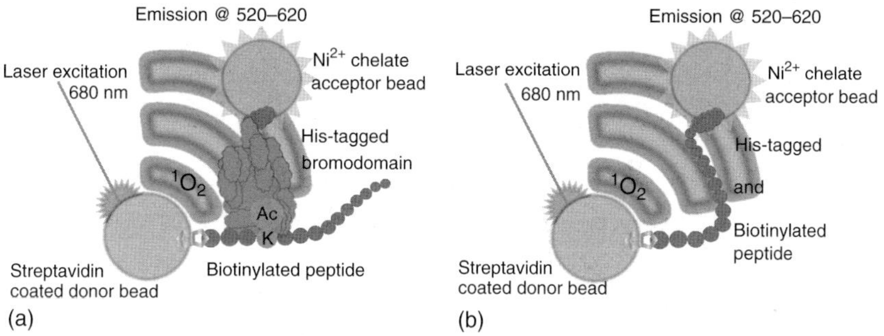

Figure 20.3 Schematic representation of (a) an AlphaScreen assay and (b) the counter screen for the identification of false positive hits. See text for details.

and acceptor beads conjugated to biomolecular binding partners by a chemiluminescent reaction triggered by singlet oxygen (Figure 20.3) [12]. AlphaScreen has been widely used for the development of protein interaction assays using a number of different formats. For the development of BET-specific assays, we used donor beads coated with streptavidin, which allow tight conjugation to terminally biotinylated tetra-acetylated H4 peptide, together with nickel-chelator-coated acceptor beads for immobilization of His-tagged bromodomain proteins [13]. In AlphaScreen assays, donor beads are excited with a laser (680 nm) resulting in the conversion of ambient triplet oxygen to singlet oxygen. Owing to the reactivity of this oxygen species, the acceptor bead needs to be in close proximity (<200 nm) to the donor beads. If this is the case, a chemiluminescent reaction is initiated by the singlet oxygen, which emits light in the 520–620 nm range from the acceptor bead. If an acetyl-lysine competitive inhibitor interacts with the bromodomain, the inhibitor will release the peptide from the binding site, resulting in dissociation of the donor– acceptor bead complex. The singlet oxygen thus returns to its ground state and no chemiluminescent signal is detected. Because of the multiplicity of binding sites on the AlphaScreen beads, the AlphaScreen assay format is particularly useful for screening weak interactions because the signal is considerably amplified. AlphaScreen assays have been shown to generate reproducible IC_{50} (inhibitor concentration 50) values for tightly binding inhibitors (as low as 50 nM) as well as very weakly binding fragments and solvent molecules [12]. Owing to the reactive oxygen chemistry and potential interference of metal chelating agents with the His-tag-mediated bromodomain-acceptor bead interaction, AlphaScreen assays tend to generate a large number of false-positive hits. In order to identify compounds that produce false-positive results in the AlphaScreen assay, we screen all compounds in an assay not containing a bromodomain but a biotinylated His-tagged peptide. In this case, the peptide directly links the donor and acceptor beads, resulting in a chemiluminescent signal. Inhibitors that still show activity in this assay are therefore likely to be false-positive hits.

20.3.2 Secondary Screening Assays

Independent confirmation of primary hits is performed by a variety of reliable secondary assays. We use a number of direct binding assays such as biolayer interferometry (BLI) or (for strongly binding inhibitors) ITC as well as indirect binding assays such as thermal shift assays for hit validation.

20.3.3 Cellular Testing

One of the most direct ways of measuring an "on-target" effect of an inhibitor is by means of the fluorescence recovery after photobleaching (FRAP) assay. The assay has been used to describe dynamics of nuclear proteins including transcription factors and high mobility group box proteins and histones in living cells [14]. Chromatin-associated proteins typically show low mobility owing to their tight immobilization on chromosomes. If the interaction is principally mediated by bromodomain proteins, acetyl-lysine competitive inhibitors are expected to release the bromodomain from chromatin, resulting in increased mobility of the target protein. In FRAP, this increase in mobility of the target protein is measured by comparing the speed of diffusion of a fluorescent-labeled target protein [e.g., GFP-BRD4 (GFP, green fluorescent protein)] into a photobleached area on chromatin in the presence and absence of an inhibitor. This assay works very well for BET bromodomains [13], but owing to the presence of other chromatin and/or DNA-binding domains in other bromodomain targets the development of cellular FRAP assays may require deletion of additional interaction domains from the target protein. Typically, the half-recovery time is plotted, which is inversely proportional to bromodomain inhibition (Figure 20.4). While FRAP assays are suitable assays for measuring the displacement of bromodomain proteins from chromatin, they provide no insight into the biological consequence of the removal of the target protein from its nuclear interaction sites. FRAP assays are therefore usually complemented by well-established assays measuring genome-wide changes in gene transcription (e.g., microarray studies or RNA-seq) or by quantitative reverse transcriptase polymerase chain reaction assays (qPCR) that monitor expression changes of specific genes.

20.3.4 Discovery of Acetyl-Lysine Competitive Inhibitors

20.3.4.1 Acetyl-Lysine Mimetic Fragments Crystallized with Bromodomains

The compounds mimic the acetyl-lysine head group and share similar interactions with BET bromodomains (Figure 20.5). Acetyl-lysine (**1**) has an IC_{50} value of about 7 mM with BRD4(1). A number of solvent molecules showed similar binding potency. These molecules include dimethylsulfoxide (**2**) (280 mM) and *N*-methylpyrrolidin-2-one (**3**) (6 mM) [12]. Among the identified fragments are

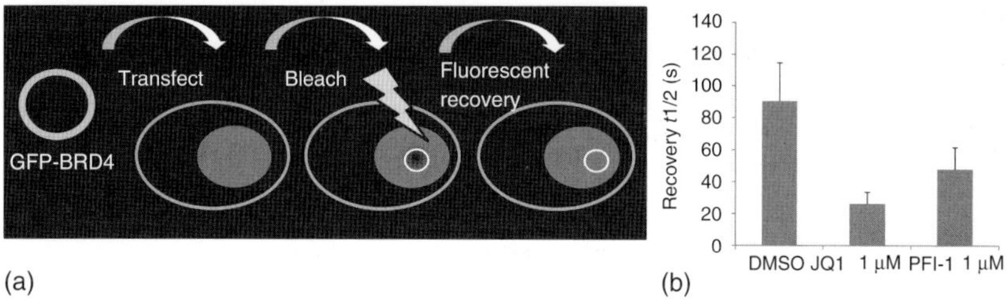

Figure 20.4 *Principle of FRAP assay.* (a) A GFP (green fluorescent protein)-tagged protein is transfected into a mammalian cell. The ectopically expressed protein is targeted to the nucleus and binds to acetyl-lysine sites on chromatin. A small area of the nucleus is bleached using a laser and the recovery of the photobleached region (circle) is measured as a function of time. (b) Data evaluation. Half recovery time of GFP-tagged BRD4 after photobleaching is shown without inhibitor treatment (DMSO) and for two different BRD4 inhibitors (JQ1 and PFI-1) at 1 μM compound concentration. Both inhibitors significantly decrease half recovery time, with JQ1 having a stronger effect compared to PFI-1. The graph shows average values of 10 bleaching experiments and the calculated error.

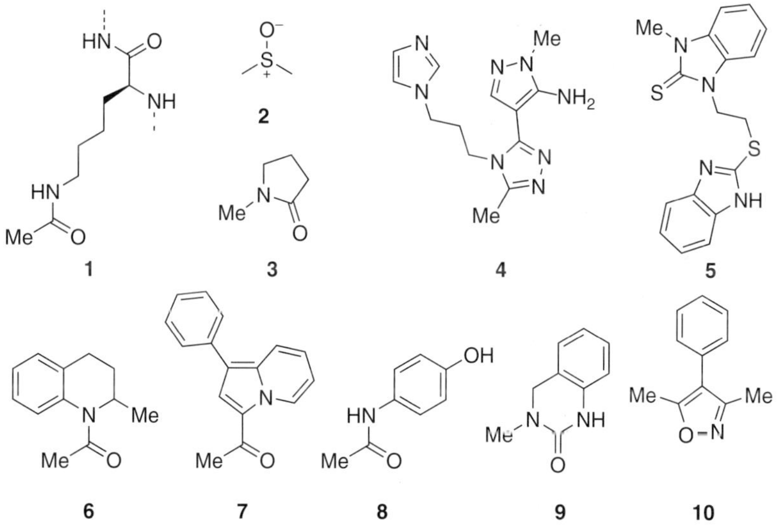

Figure 20.5 *Cocrystallized acetyl-lysine mimetic fragments.* Shown are acetyl-lysine (**1**), dimethylsulfoxide (DMSO) (**2**), *N*-methylpyrrolidin-2-one (NMP) (**3**), methyl-triazolo (**4**), benzimidazole (**5**), *N*-acetyl-2-methyltetrahydroquinoline (**6**), 1-(1-(pyridin-2-yl)indolizin-3-yl)ethanone (**7**), acetaminophen (paracetamol) (**8**), 3-methyl-3,4-dihydroquinazolin-2(1*H*)-one (**9**), and 4-phenyl 3,5-dimethyl isoxazole (**10**).

methyl-triazolo 4-benzimidazole (**5**), *N*-acetyl-2-methyltetrahydroquinoline (**6**), 1-(1-(Pyridin-2-yl)indolizin-3-yl)ethanone (7), acetaminophen (paracetamol) (**8**), 3-methyl-3,4-dihydroquinazolin-2(1*H*)-one (**9**), and 4-phenyl 3,5-dimethyl isoxazole (**10**). This set of fragments represents a rich source of chemical starting points for the development of acetyl-lysine mimetic inhibitors [15].

20.3.4.2 Discovery of Benzo- and Thienodiazepines

Highly potent and selective BET inhibitors of the benzodiazepine class (BZDs) have been discovered by GlaxoSmithKline using cellular phenotypic assays together with a chemoproteomic approach that identified the cellular targets as BET bromodomain proteins [9]. During the same time, Mitsubishi Pharmaceuticals disclosed a series of structurally related thienodiazepine inhibitors with strong growth inhibitory activity on an array of cancer cell lines (WO/2009/084693). This patent gave rise to the development of a novel thieno-triazolo-1,4-diazepine, JQ1, with the (*S*) enantiomer, (+)-JQ1, binding to BRD4(1) with an IC_{50} of 77 nM according to AlphaScreen assay. ITC confirmed these data, revealing low nanomolar dissociation constants for all BET bromodomains [13]. The (*R*) enantiomer, (−)-JQ1, is sterically excluded from the acetyl-lysine binding site ($IC_{50} > 10\,000$ nM) and functions therefore as a negative control molecule. Robust synthetic routes to these two molecules were developed (Figure 20.6). (−)-JQ1 was accessed in a nearly identical manner, the only difference being the use of Fmoc-D-Asp(O*t*Bu)-OH instead of the naturally derived protected amino acid in the second step.

(+)-JQ1 shows high selectivity for BET bromodomains and excellent shape complementarity with the BET acetyl-lysine binding site. Differential scanning fluorimetry (DSF) identified no significant interaction of (+)-JQ1 outside the BET family. To confirm the binding mode, cocrystal structures using racemic JQ1 and purified BRD4(1) and BRD2(2) were obtained. Only the (+) isomer of JQ1 was observed to bind directly in the acetyl-lysine binding site. A hydrogen bond formed in each case between the conserved asparagine residue [Asn140 in BRD4(1), Asn429 in BRD2(2)], and the triazole ring in (+)-JQ1, mimicking the interaction with acetyl-lysine. The binding pocket was fully occupied by the ligand and hydrophobic interactions with conserved BET residues in the ZA- and BC loops further stabilized the binding. Examination of (+)-JQ1 in U2OS cells using FRAP with GFP-BRD4 showed that 500 nM (+)-JQ1 was able to fully displace BRD4 from chromatin, while (−)-JQ1 showed no effect in this assay.

The successful targeting of BET bromodomains by the GSK (glucogen synthase kinase) inhibitor iBET and (+)-JQ1 gave rise to the development of a number of related BZD and thienodiazepine molecules (Figure 20.6). These molecules include the benzotriazepines (BzTs) [16], the JQ1 methylester MS417 [17] as well as a number of patent applications (Figure 20.7).

20.3.4.3 Other BET Inhibitors

On the basis of the methyltriazoloacetyl-lysine mimetic moiety, other isosteric ring systems have been explored as BET inhibitors. One of the most explored heterocycles is the 3,5-dimethylisoxazole moiety, which has been identified as a versatile scaffold for the development of BET inhibitors [18]. Decoration of the isoxazole ring at the four position with aromatic ring systems led to selective BET inhibitors with good ligand efficiency. Using the dimethylisoxazole template led to the development of GSK1210151A (I-BET151), a potent and highly selective BET inhibitor with optimized pharmacokinetic properties (Figure 20.8) [7].

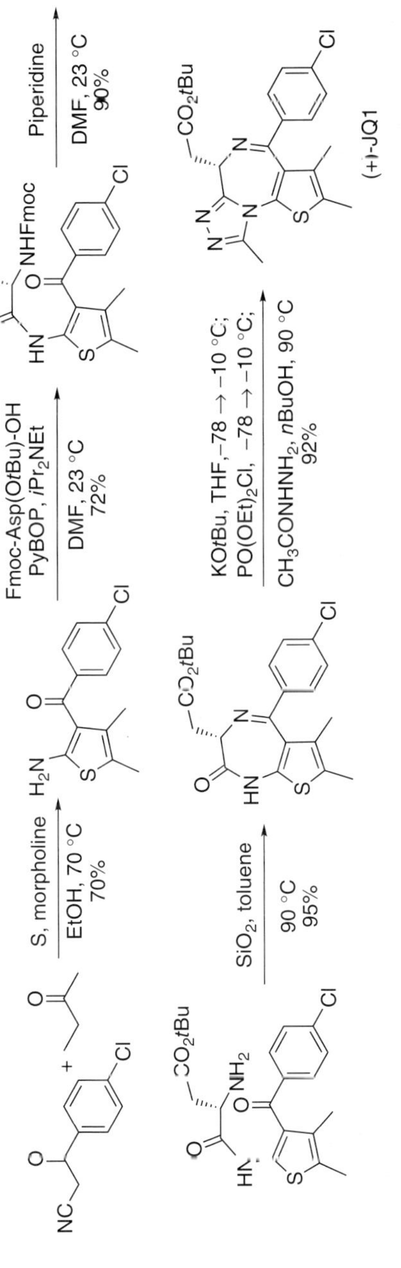

Figure 20.6 Synthetic route for (+)-JQ1.

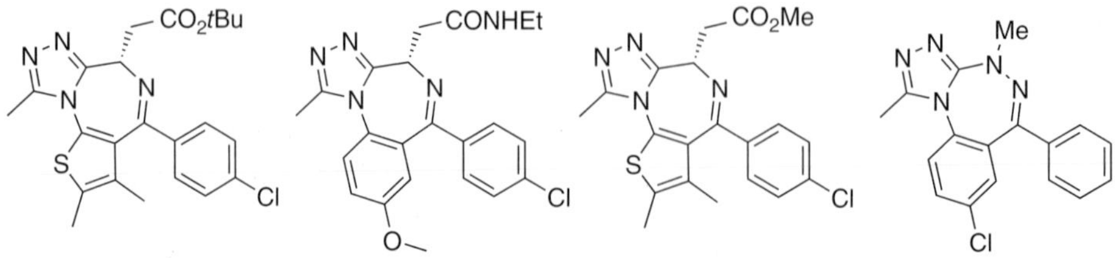

(+)-JQ1 I-BET MS417 BzT-7

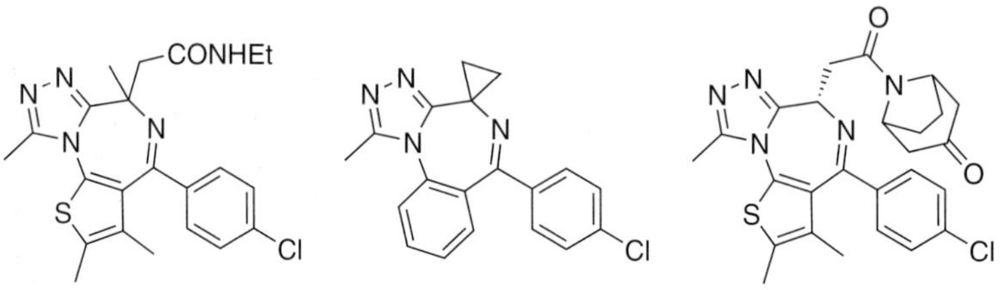

WO 2012075456 A1
Constellation

WO 2013030150 A1
Bayer

Figure 20.7 *Inhibitors based on the benzodiazepine (iBET) and thienodiazepine (JQ1).* The compounds include the benzotriazepine BzT-7 [16], the JQ1 methylester MS417 [17], and several examples disclosed by Constellation (WO 2012075456 A1) and Bayer (WO 2013030150 A1).

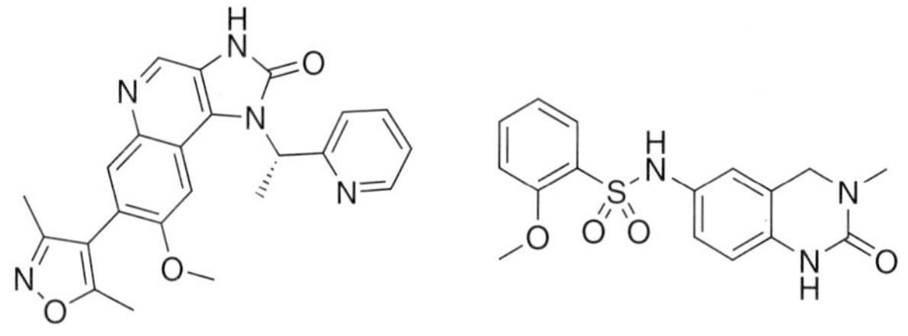

(a) **I-BET151** (b) **PFI-1**

Figure 20.8 Structures of the BET-specific inhibitors I-BET151 (a) and PFI-1 (b), respectively.

The quinazolinone fragment hit (Figure 20.5) led to the structurally orthogonal BET inhibitor PFI-1 (Figure 20.8) [19]. The inhibitor was developed on the basis of optimizing a series of sulfonamides and finally reverse sulfonamides to yield a selective BRD4 inhibitor with a dissociation constant of 136 nM for BRD4(1) and 303 nM for BRD4(2), respectively. Cocrystal structures with BRD4(1) confirmed the acetyl-lysine mimetic binding mode of PFI-1 which forms two hydrogen bonds with the conserved Asn140 and a water-mediated H-bond to the conserved tyrosine Tyr97.

20.4
Chemical/Biological Investigations

The development of BET-specific inhibitors provided a unique tool for our understanding of gene transcription and potentially new treatment options for aggressive cancers and inflammation. FRAP assays demonstrated that BET inhibitors efficiently displace BET transcriptional regulators from chromatin, resulting in tissue-specific changes in gene expression. The first tumor type studied was NMC, an aggressive tumor with no current treatment options, which directly involves BRD3 and BRD4 bromodomains in the principle driver oncogene BRD-NUT. Because these tumors highly depend on the BRD-NUT oncogene for survival, exceptional response rates were achieved in animal models of the disease. On the basis of these data, an optimized BET inhibitor entered clinical testing. Surprisingly, a large diversity of tumor types also seems to be highly dependent on BET family members, and in particular BRD4, for survival. Bromodomain inhibitors have been found to be highly efficacious in diverse subtypes of leukemia, including AML [8], mixed linage leukemia (MLL) [7], and acute lymphoid leukemia (ALL) [19]. Profound efficacy was also observed in myeloma, lung cancer, neuroblastoma, and glioblastoma; and it is likely that in the future, other tumor types that are sensitive to BET inhibition will be identified. However, to date, the exact mechanisms that determine the sensitivity of tumors to BET inhibitors are still not completely understood. In inhibitor-sensitive tumors, key oncogenes and survival genes have been found to be strongly downregulated but the molecular reasons for why this only happens in a subset of tumors remain to be elucidated. A possible explanation is that key oncogenes are exclusively driven by BET dependent (super)-enhancers, as suggested by Loven and coworkers [6].

Outside the oncology area, BET inhibitors showed promising results for the treatment of inflammatory disease as well as viral infection. It is therefore likely that the availability of selective BET inhibitors will initiate many more research activities in these disease areas and others in the future.

20.5
Conclusion

The discovery of potent and selective BET inhibitors validated this binding pocket for the development of protein interaction inhibitors that target epigenetic reader domains and thereby directly modulate gene transcription. This mode of action also allows indirect targeting of key oncogenic drivers, such as c-MYC, that are difficult to target directly. The BET acetyl-lysine binding site represents an attractive pocket for the development of inhibitors with favorable pharmacological properties. Fragment and inhibitor screening identified a diversity of acetyl-lysine isosteres that can be developed further into diverse inhibitors. In addition, the sequence diversity of bromodomains facilitates development of selective compounds that can be utilized as valuable tools for target validation.

To date, most efforts have been dedicated to the development of BET inhibitors. However, the human bromodomain family contains 42 diverse proteins and 61 bromodomains. This offers a large space for potential targets. The phenotypic consequences of bromodomain inhibition in BRDs that comprise multiple domains are however difficult to predict, which represents a major challenge in this field in the future. This issue can only be addressed by comprehensive evaluation of highly specific inhibitors in an array of cellular-disease-relevant assays. However, given the strong disease association of many bromodomain proteins that have been identified by genome-wide association studies and deep sequencing of cancers, epigenetic reader domains offer attractive targets for future evaluation of these potential drug targets.

References

1. Filippakopoulos, P., Picaud, S., Mangos, M., Keates, T., Lambert, J.P., Barsyte-Lovejoy, D., Felletar, I., Volkmer, R., Muller, S., Pawson, T., Gingras, A.C., Arrowsmith, C.H., and Knapp, S. (2012) Histone recognition and large-scale structural analysis of the human bromodomain family. *Cell*, **149** (1), 214–231.
2. Muller, S., Filippakopoulos, P., and Knapp, S. (2011) Bromodomains as therapeutic targets. *Expert Rev. Mol. Med.*, **13**, e29.
3. Jang, M.K., Mochizuki, K., Zhou, M., Jeong, H.S., Brady, J.N., and Ozato, K. (2005) The bromodomain protein Brd4 is a positive regulatory component of P-TEFb and stimulates RNA polymerase II-dependent transcription. *Mol. Cell*, **19** (4), 523–534.
4. Jiang, Y.W., Veschambre, P., Erdjument-Bromage, H., Tempst, P., Conaway, J.W., Conaway, R.C., and Kornberg, R.D. (1998) Mammalian mediator of transcriptional regulation and its possible role as an end-point of signal transduction pathways. *Proc. Natl. Acad. Sci. U.S.A.*, **95** (15), 8538–8543.
5. Zippo, A., Serafini, R., Rocchigiani, M., Pennacchini, S., Krepelova, A., and Oliviero, S. (2009) Histone crosstalk between H3S10ph and H4K16ac generates a histone code that mediates transcription elongation. *Cell*, **138** (6), 1122–1136.
6. Loven, J., Hoke, H.A., Lin, C.Y., Lau, A., Orlando, D.A., Vakoc, C.R., Bradner, J.E., Lee, T.I., and Young, R.A. (2013) Selective inhibition of tumor oncogenes by disruption of super-enhancers. *Cell*, **153** (2), 320–334.
7. Dawson, M.A., Prinjha, R.K., Dittmann, A., Giotopoulos, G., Bantscheff, M., Chan, W.I., Robson, S.C., Chung, C.W., Hopf, C., Savitski, M.M., Huthmacher, C., Gudgin, E., Lugo, D., Beinke, S., Chapman, T.D., Roberts, E.J., Soden, P.E., Auger, K.R., Mirguet, O., Doehner, K., Delwel, R., Burnett, A.K., Jeffrey, P., Drewes, G., Lee, K., Huntly, B.J., and Kouzarides, T. (2011) Inhibition of BET recruitment to chromatin as an effective treatment for MLL-fusion leukaemia. *Nature*, **478** (7370), 529–533.
8. Zuber, J., Shi, J., Wang, E., Rappaport, A.R., Herrmann, H., Sison, E.A., Magoon, D., Qi, J., Blatt, K., Wunderlich, M., Taylor, M.J., Johns, C., Chicas, A., Mulloy, J.C., Kogan, S.C., Brown, P., Valent, P., Bradner, J.E., Lowe, S.W., and Vakoc, C.R. (2011) RNAi screen identifies Brd4 as a therapeutic target in acute myeloid leukaemia. *Nature*, **478** (7370), 524–528.
9. Nicodeme, E., Jeffrey, K.L., Schaefer, U., Beinke, S., Dewell, S., Chung, C.W., Chandwani, R., Marazzi, I., Wilson, P.,

Coste, H., White, J., Kirilovsky, J., Rice, C.M., Lora, J.M., Prinjha, R.K., Lee, K., and Tarakhovsky, A. (2010) Suppression of inflammation by a synthetic histone mimic. *Nature*, **468** (7327), 1119–1123.

10. French, C.A. (2012) Pathogenesis of NUT midline carcinoma. *Annu. Rev. Pathol.*, 7, 247–265.
11. Vidler, L.R., Brown, N., Knapp, S., and Hoelder, S. (2012) Druggability analysis and structural classification of bromodomain acetyl-lysine binding sites. *J. Med. Chem.*, **55** (17), 7346–7359.
12. Philpott, M., Yang, J., Tumber, T., Fedorov, O., Uttarkar, S., Filippakopoulos, P., Picaud, S., Keates, T., Felletar, I., Ciulli, A., Knapp, S., and Heightman, T.D. (2011) Bromodomain-peptide displacement assays for interactome mapping and inhibitor discovery. *Mol. Biosyst.*, 7 (10), 2899–2908.
13. Filippakopoulos, P., Qi, J., Picaud, S., Shen, Y., Smith, W.B., Fedorov, O., Morse, E.M., Keates, T., Hickman, T.T., Felletar, I., Philpott, M., Munro, S., McKeown, M.R., Wang, Y., Christie, A.L., West, N., Cameron, M.J., Schwartz, B., Heightman, T.D., La Thangue, N., French, C.A., Wiest, O., Kung, A.L., Knapp, S., and Bradner, J.E. (2010) Selective inhibition of BET bromodomains. *Nature*, **468** (7327), 1067–1073.
14. Mueller, F., Mazza, D., Stasevich, T.J., and McNally, J.G. (2010) FRAP and kinetic modeling in the analysis of nuclear protein dynamics: what do we really know? *Curr. Opin. Cell Biol.*, **22** (3), 403–411.
15. Chung, C.W., Dean, A.W., Woolven, J.M., and Bamborough, P. (2012) Fragment-based discovery of bromodomain inhibitors part 1: inhibitor binding modes and implications for lead discovery. *J. Med. Chem.*, **55** (2), 576–586.
16. Filippakopoulos, P., Picaud, S., Fedorov, O., Keller, M., Wrobel, M., Morgenstern, O., Bracher, F., and Knapp, S. (2012) Benzodiazepines and benzotriazepines as protein interaction inhibitors targeting bromodomains of the BET family. *Bioorg. Med. Chem.*, **20** (6), 1878–1886.
17. Zhang, G., Liu, R., Zhong, Y., Plotnikov, A.N., Zhang, W., Zeng, L., Rusinova, E., Gerona-Nevarro, G., Moshkina, N., Joshua, J., Chuang, P.Y., Ohlmeyer, M., He, J.C., and Zhou, M.M. (2012) Down-regulation of NF-kappaB transcriptional activity in HIV-associated kidney disease by BRD4 inhibition. *J. Biol. Chem.*, **287** (34), 28840–28851.
18. Hewings, D.S., Wang, M., Philpott, M., Fedorov, O., Uttarkar, S., Filippakopoulos, P., Picaud, S., Vuppusetty, C., Marsden, B., Knapp, S., Conway, S.J., and Heightman, T.D. (2011) 3,5-dimethylisoxazoles act as acetyl-lysine-mimetic bromodomain ligands. *J. Med. Chem.*, **54** (19), 6761–6770.
19. Picaud, S., Da Costa, D., Thanasopoulou, A., Filippakopoulos, P., Fish, P.V., Philpott, M., Fedorov, O., Brennan, P., Bunnage, M.E., Owen, D.R., Bradner, J.E., Taniere, P., O'Sullivan, B., Muller, S., Schwaller, J., Stankovic, T., and Knapp, S. (2013) PFI-1, a highly selective protein interaction inhibitor, targeting BET bromodomains. *Cancer Res.*, **73** (11), 3336–3346.

21
The Impact of Distant Polypharmacology in the Chemical Biology of PARPs

Albert A. Antolín and Jordi Mestres

21.1
Introduction

The ability of small molecules to activate, inhibit, or modulate the function of macromolecules has long been used to probe the biological role of those proteins. However, in recent years, it has become increasingly evident that small molecules are seldom selective but tend to bind to multiple proteins, a property usually referred to as *polypharmacology* (Box 21.1). In some cases, the use of chemical probes (Box 21.1) having polypharmacology across distantly related proteins involved in the same biological processes can result in unsuspected confounding effects that could ultimately lead to wrong conclusions. In this chapter, the impact of the distant polypharmacology recently uncovered in a small molecule widely used in chemical biology to probe the biological role of poly(ADP-ribose)polymerases (PARPs) is presented. We then continue on a discussion of how these findings may affect the development of some of the PARP inhibitors currently in clinical trials. Finally, we learn how molecular informatics can be applied to identifying novel targets of chemical probes, as a knowledge-based strategy to de-risk chemical biology.

21.2
The Biological Problem

21.2.1
Studying the Function of Proteins Using Chemical Probes with Unknown Polypharmacology

One of the main goals in chemical biology is to develop bioactive small molecules (also referred to as *chemical probes, chemical tool compounds,* or *standard inhibitors*) to interrogate and study the effects of biomolecules (generally proteins) in biological processes or disease models [1–3]. Chemical probes are thus essential to complement more invasive techniques that eliminate the target of the

Concepts and Case Studies in Chemical Biology, First Edition. Edited by Herbert Waldmann and Petra Janning.

Box 21.1 Glossary of Key Concepts

Chemical Probe. Small molecule used to study the function of a certain protein. Synonyms: standard inhibitor, chemical tool compound.
Protein Target. The protein at which the small molecule is mainly directed and to which it was often designed for high affinity.
Polypharmacology. The binding of a small molecule to multiple proteins.
Synthetic Lethality. Two proteins are synthetic lethal pairs when they are nonessential for the survival of a given cell alone, but they are lethal if they are both suppressed at the same time.
IC_{50}/pIC_{50}. The half-maximal inhibitor concentration is a measure of the capacity of a compound to inhibit a biological function by measuring the concentration at which the compound inhibits half of the biological effect. It is an indirect measure of affinity. The higher the IC_{50}, the lower the effectiveness of the compound. It is usually reported in logarithmic scale as pIC_{50}. In this logarithmic scale, the higher the pIC_{50}, the higher the effectiveness of the compound.
Ki/pKi. The inhibition constant is a direct measure of the affinity of a reversible inhibitor for its target by measuring the equilibrium constant of the dissociation process.
Chemical similarity. A quantitative measure of how closely related two chemical structures are.
Molecular Descriptor. A mathematical representation of a chemical structure.
In silico Target Profiling. Computational prediction of the affinity of a given small molecule across a panel of proteins.

system to study its function, such as RNA interference (RNAi) or gene knockouts. They are also the first step in the development of new small-molecule drugs [2]. However, it is often assumed in chemical biology that chemical probes interact selectively with the protein under study (the target) and little attention is usually paid to the fact that this selectivity depends critically on the concentration of chemical probe used and that it may be dangerously lost at higher concentrations.

In this respect, many chemical probes have been made available in recent years and many research studies have used these chemical probes to drive new discoveries without accurate control of the concentration used. Currently, the development of novel chemical probes continues to be essential to unravel the unknown biological function of many proteins [4]. However, the assumption that chemical probes interact specifically with one single target at any concentration is starting to be challenged [5]. As highlighted earlier, an increasing number of publications suggest that drugs generally bind to multiple targets, a promiscuous behavior referred to as *polypharmacology* [6–8]. This behavior is likely to affect chemical probes also [5].

Accordingly, the current situation is that the unknown polypharmacology of chemical probes could compromise many of the conclusions achieved under the assumption that chemical probes interact selectively with one single target. Therefore, it is essential to clarify the exact targets of chemical probes to de-risk their utilization in chemical biology. Generally, to address their selectivity, small molecules are screened *in vitro* across a diverse panel of targets selected as representatives [2]. However, this approach can cover a rather small proportion of the entire proteome. Here, we illustrate how computational tools can help address this issue by screening *in silico* a chemical probe across thousands of targets at a reduced cost. Predicting the interaction of chemical probes with unsuspected, potentially confounding, off-targets should be regarded as a de-risking strategy in chemical biology.

21.2.2
Development of Poly(ADP-Ribose)Polymerase-1 (PARP-1) Chemical Probes and Follow-on Drugs

Poly(ADP-ribose) (PAR) is a negatively charged branched polymer that serves as a posttranslational modification of proteins [9]. The majority of the PAR in cells is produced by the catalytic activity of PARP-1, the founder of the 17-membered PARPs family, also known as *ADP-ribosyltransferases* (*ARTDs*). PARP-1, the most studied member of the family, has key functions in DNA repair, transcription, and cellular signaling, among others [9]. Its role in DNA repair makes PARP-1 a key target in oncology, both as chemopotentiator of radiation or chemotherapeutics and as stand-alone therapy in patients carrying defects in DNA repair genes [10]. This last application was a breakthrough in cancer therapy as it was the first demonstration that specific mutations in cancer cells created specific dependencies on those cells that could be exploited to selectively kill them, a concept referred to as *synthetic lethality* (Box 21.1). The first small molecule identified to

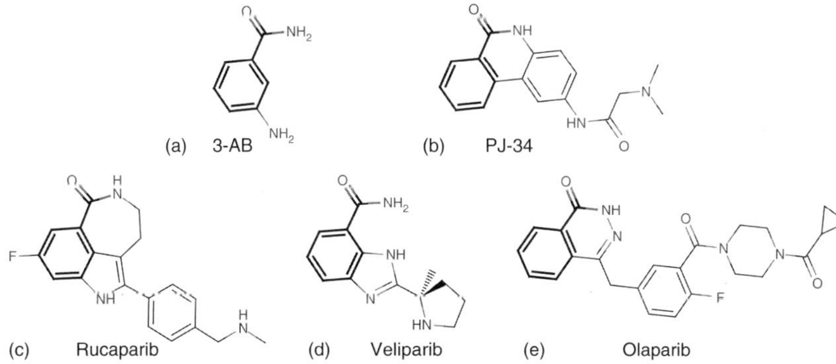

Figure 21.1 *The evolution of PARP inhibitors.* (a) 3-Aminobenzamide (3-AB), (b) PJ34, (c) rucaparib, (d) veliparib, and (e) olaparib. The benzamide core structure of all PARP inhibitors is highlighted in bold.

study the functions of PARP-1 was 3-aminobenzamide (3-AB), a close analog of the natural inhibitor nicotinamide (Figure 21.1) [10]. Despite the low micromolar potency of 3-AB (PARP-1 $IC_{50} = 30\,\mu M$, Box 21.1), its use was key for the first proof of concept studies of PARP-1 inhibition in the 1980s [10]. In the 1990s, there was an explosion in campaigns, both in industry and academia, aimed at improving the potency of a first generation of PARP inhibitors that yield the current chemical probes and drug candidates. One of those inhibitors, which is now a reference compound to probe the biological role of PARP-1, is PJ34 (PARP-1 $IC_{50} = 0.02\,\mu M$) [10], used in more than 150 publications since its discovery (Figure 21.1). More recently, the structures of the first drug candidates entering clinical trials were disclosed (Figure 21.1). Currently, there are great expectations about the results of PARP drug candidates in late-stage clinical trials, the more advanced ones being olaparib, veliparib, and rucaparib [10].

21.2.3 Unexpected Differential Effects between PARP Inhibitors

PJ34 was used for over 10 years at quite high concentrations to drive conclusions on the effects of PARP-1 under the belief that it was a specific PARP-1 inhibitor and despite the fact that its selectivity over other members of the PARP family had never been evaluated. In 2011, two publications reported that PJ34 produced different cellular effects when compared to other PARP-1 inhibitors and PARP-1 siRNA (small interfering ribonucleic acid), suggesting that PJ34 might be inhibiting more targets than just PARP-1 [11]. When the selectivity of PJ34 over 13 members of the PARP family was finally investigated, it was discovered that PJ34 was also inhibiting PARP-2 ($IC_{50} = 0.03\,\mu M$), PARP-5A/TNKS1 ($IC_{50} = 0.6\,\mu M$), PARP-3 ($IC_{50} = 0.8\,\mu M$), PARP-4 ($IC_{50} = 2\,\mu M$), and PARP-5B/TNKS2 ($IC_{50} = 20\,\mu M$), in addition to residual affinities to PARP-14, PARP-15, and PARP-16 [12]. Therefore, some of the effects ascribed to PARP-1 using PJ34 could be indeed mediated through other PARPs being inhibited by PJ34. However, these novel PARP targets could not explain many of the differential effects produced by PJ34 at the cellular level [11]. Therefore, it was essential to gain a wider understanding of the target profile of PJ34 beyond the members of the PARP protein family.

21.3 The Chemical Approach

21.3.1 Molecular Informatics

The recent development of high-throughput "omics" technologies is transforming the type and quantity of information available, profoundly affecting many disciplines and creating the necessity to develop novel tools to handle big data. To this aim, molecular informatics has gradually developed into a field that uses computers to facilitate the collection, storage, manipulation, and analysis of large quantities of data at the interface between chemistry and biology [13]. An essential aspect of this discipline resides in having access to electronic sources that store all information on small molecules, proteins, and their interaction (pharmacological data) and make all of it accessible to the broad scientific community (Box 21.2) [13, 14]. Overall, these databases host information on hundreds of thousands of small molecules interacting with thousands of targets, representing nowadays an invaluable source to understand and predict the complex polypharmacology of small molecules [14]. All these resources offer the possibility to be mined directly from their respective websites, although recently developed external applications allow for performing searches in a more integrated manner [15].

Box 21.2 Representative Public Sources of Ligand–target Interaction Data

Database	Web address	Data description
PubChem	*http://pubchem.ncbi.nlm.nih.gov/*	250 000 compounds, 2500 bioassays
ChEMBL	*http://www.ebi.ac.uk/chembl*	>700 000 small molecules with >2.7 million bioactivity data points
DrugBank	*http://www.drugbank.ca/*	4800 drug entries including >1350 FDA-approved drugs
IUPHAR-DB	*http://www.iuphar-db.org*	2000 compounds 616 protein targets
Binding DB	*http://www.bindingdb.org*	>271 000 compounds, >620 000 binding affinities against 5526 protein targets

21.3.2 *In silico* Target Profiling

Target profiling can be defined as the evaluation of the affinity of a given small molecule across a panel of targets. Ideally, this affinity should be evaluated using *in vitro* assays, but both the number of molecules and proteins for which the interaction should be evaluated make it an unmanageable task experimentally. Recent developments in molecular informatics are making it possible to predict *in silico* the affinity of a small molecule across an increasing number of targets [16]. There are many methods to predict the target of a small molecule and some of them are openly available through web services [14]. In general, these methods can be divided in structure-based methods and ligand-based methods. Structure-based methods use information on the target three-dimensional (3D) structure to calculate the possibility of a given molecule to bind to that target. However, not all target families have a representative set of protein members with known 3D structure deposited in publicly available sources, limiting these kinds of approaches [14]. Ligand-based methods, in contrast, use information on the chemical structure and its affinity to a given target. The increasing number of molecules for which pharmacological data is becoming available in open resources (Box 21.2) [14] offers a wealth of possibilities for developing ligand-based methods to target profile prediction. In this respect, the simplest approaches exploit the basic principle that similar molecules should exhibit similar biological activities and thus use chemical similarity (Box 21.1) to infer potential affinities to other targets [17].

In our research, we use the target-profiling approach implemented in the PredictFX software [18]. Given the two-dimensional structure of a molecule (smiles or sd/mol file), PredictFX returns the predicted affinities for those targets for which ligand information is available in public sources of pharmacological data [14]. Three ligand-based methods are implemented in the applied version of PredictFX that rely on descriptor-based similarities, fuzzy fragment-based mapping, and target cross-pharmacology.

Descriptor-based similarities are calculated using three types of two-dimensional descriptors, namely, PHRAG (pharmacophoric fragment), FPD (feature-pair distribution), and SHED (Shannon entropy descriptor) [19, 20], each one of them characterizing chemical structures with a different degree of fuzziness and thus complementing each other in terms of structural similarity and hopping abilities. PHRAGs are all possible fixed-length segments of five-atom features that can be extracted from the topology of a molecule. In contrast, FPDs capture the overall spreading of pairs of atom-centered features at different predefined bond lengths. Finally, SHEDs are derived from simplified FPD, in which, instead of using the actual feature-pair counts at each path length, the variability within all possible FPDs is quantified using the concept of Shannon entropy [19]. When using PHRAG and FPD, the similarity between two molecules corresponds to the overlapping fraction of their respective profiles [20], whereas with SHED, Euclidean distances are calculated instead [19]. All three descriptors were assessed on their ability to discriminate active from random compounds for all targets chemically represented in publicly available sources. As a result of this validation analysis, compounds below similarity values of 0.76 and 0.87 for PHRAG and FPD, respectively, and above a distance value of 0.52 for SHED were considered to be outside the applicability domain of these descriptors. For each of the 4681 targets for which a PredictFX model was available, the ensemble of PHRAG, FPD, and SHED molecular descriptors (Box 21.1) of all known ligands represents a mathematical description of the target from a chemical perspective. On this basis, the affinity of a compound for a given target can be estimated by inverse distance weighting interpolation from the affinity landscape defined by all neighboring molecules according to the descriptors and similarity/distance metrics used [19, 20].

Fuzzy fragment-based mapping exploits the fact that when a substantial chemical coverage is available for a given target, key interaction points can be revealed from the presence of specific chemical series with analogous scaffolds and multiple functionalities. Common trends within the same chemical series can be considered "primary" features, while the variable functionalities can be considered as "secondary" features. In this context, given a biological target, a simplest active subgraph (SAS) can be defined, which contains the minimum set of primary features required to achieve activity within a congeneric set of compounds. In order to generate a SAS model for a given biological target, all molecules with affinities below 1 μM are sorted according to their chemical complexity. Then, the simplest active molecule (SAM) is selected and molecules containing it to a certain degree of similarity are assigned to it. When all molecules have been processed, the next available SAM is selected and the process is iterated until all molecules are related to a SAM. The SAS identification protocol is not restricted to identical subgraphs. Instead, similar topologies can be identified, allowing a reasonable degree of scaffold hopping. Once the SAS model for a given biological target has been generated, it represents an alternative mathematical description of this target from a fuzzier ligand perspective and can be used for virtual screening purposes.

Finally, the target cross-pharmacology index (XPI) between two targets A and B ($XPI_{A,B}$) is defined as the fraction of compounds experimentally known to be active (pACT $\geq$ 5.5) on target A and target B at the same time relative to all known ligands active on target A. If, for a given compound, an affinity to target A is predicted on the basis of a SAS model, all cross-pharmacologically related targets B are identified for A. If no similarity-based or SAS-based affinity can be predicted for B, interaction affinities can be inferred for target B by using the corresponding cross-pharmacology index $XPI_{A,B}$ as a weighting factor on the predicted affinity for target A. If several targets A* are related to B, then the inferred affinity for target B is the weighted average of all $XPI_{A^*,B}$ derived affinity values.

The method has been successfully validated retrospectively, on its ability to predict the entire experimental interaction matrix between 13 antipsychotic drugs and 34 protein targets (Box 21.1) [20] and also prospectively on its capacity to correctly anticipate the affinity profile of the drug cyclobenzaprine [21].

21.4 Chemical Biological Research/Evaluation

21.4.1 *In silico* Identification and *In Vitro* Confirmation of Novel Targets for PJ34

We used PredictFX to predict *in silico* the target profile of PJ34. Apart from recovering many of the already known interactions on PARPs, the results anticipated novel affinities for two serine/threonine kinases, namely, Pim1 and Pim2, based on the similarity of PJ34 to a high-affinity Pim1 and Pim2 inhibitor (CHEMBL572783: ki-Pim1 = 0.008 μM, ki-Pim2 = 0.003 μM, Box 21.1) (Figure 21.2) [11]. Subsequent *in vitro* testing confirmed that PJ34 was indeed a competitive Pim1 and Pim2 inhibitor with IC50s of 3.7 and 16 μM for Pim1

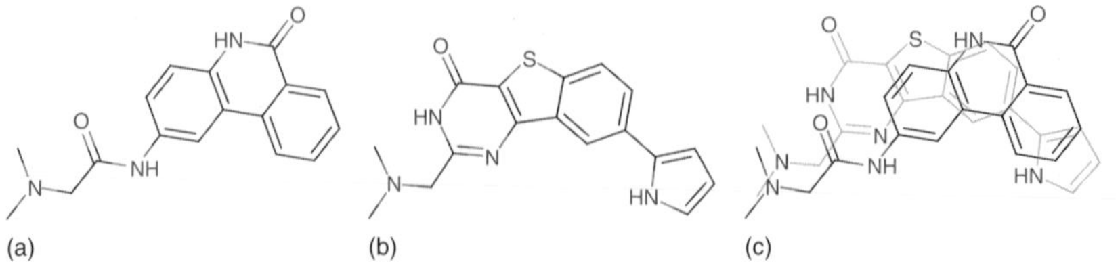

Figure 21.2 (a) PJ34, (b) CHEBL572783, and (c) model of the superposition of PJ34 (in black) and CHEMBL572783 (in gray).

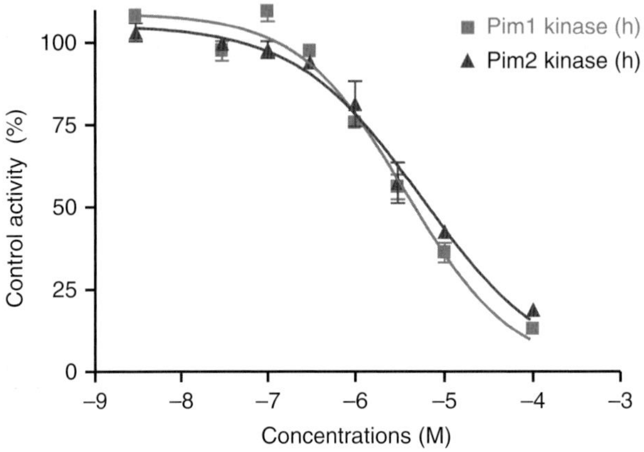

Figure 21.3 Dose–response curves of PJ34 against Pim1 and Pim2 kinases.

and Pim2, respectively (Figure 21.3) [11]. This is a good example of a prospective validation of *in silico* target predictions that enabled to expand the panel of known targets for PJ34 beyond the PARP protein family. These results also show that profiling a chemical probe only across members of the protein family of their known target (PARPs) is not sufficient guarantee for its safe use in chemical biology, as proteins from distantly related families can also bind to the same small molecule. For the particular case of PJ34, Figure 21.4 illustrates the evolution of the knowledge of its target profile over the years.

21.4.2
Implications for the Use of PJ34 and Follow-on Drugs

Pim kinases are a protein family composed of three members (Pim1, Pim2, and Pim3) that were originally discovered because of their role as oncogenes and their overexpression in a wide range of cancer types [11]. Specifically, their biological roles include protein transcription and translation, regulation of cell cycle progression, and the regulation of survival signaling, all of them being overlapping functions with PARPs. Therefore, the risks of confounding effects when using PJ34 to study the role of PARPs need to be considered, as some of the functions

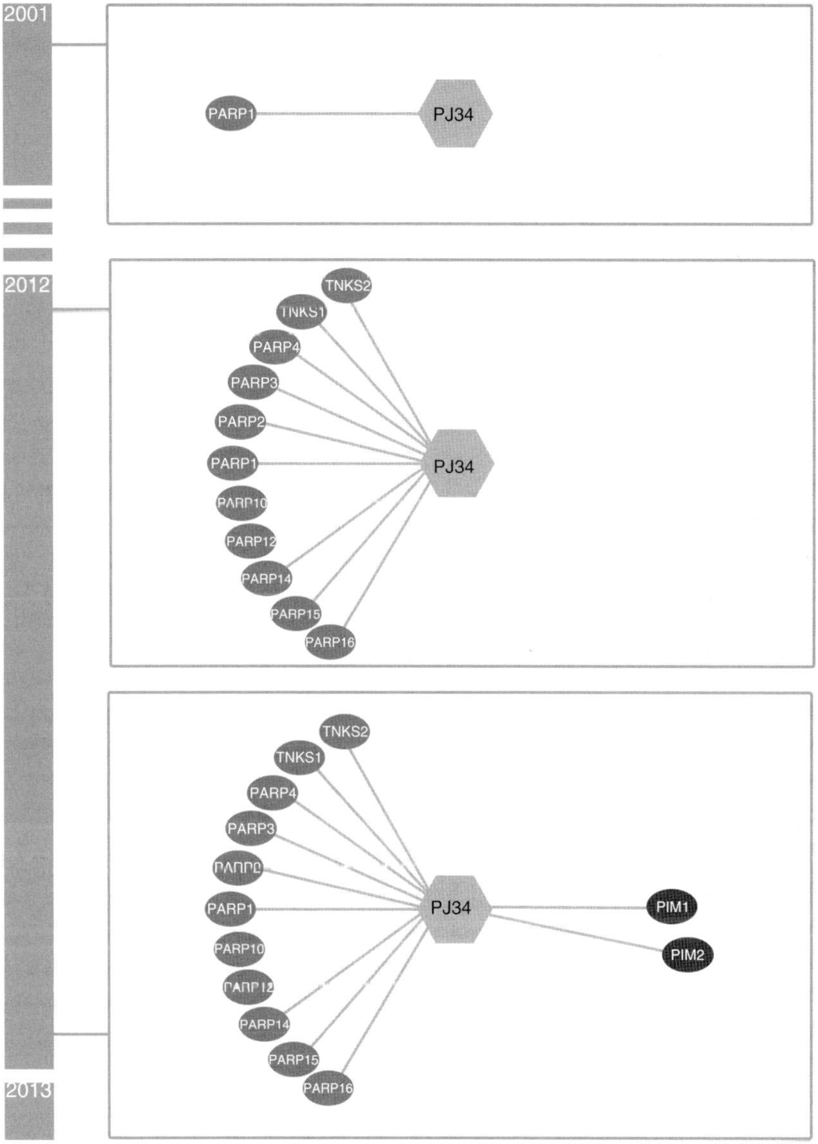

Figure 21.4 Evolution of the knowledge on the target profile of PJ34 over time.

attributed to PARP-1 using PJ34 could be due to Pim kinase inhibition. However, there is enough difference in affinity between Pims and PARPs to use PJ34 safely to probe specifically the role of PARPs. The PJ34 affinity for PARP-1 is 0.02 μM, whereas for Pim1 it is 3.7 μM, so there is an affinity gap of almost three orders of magnitude to avoid any potential confounding effects coming from Pim1. As the

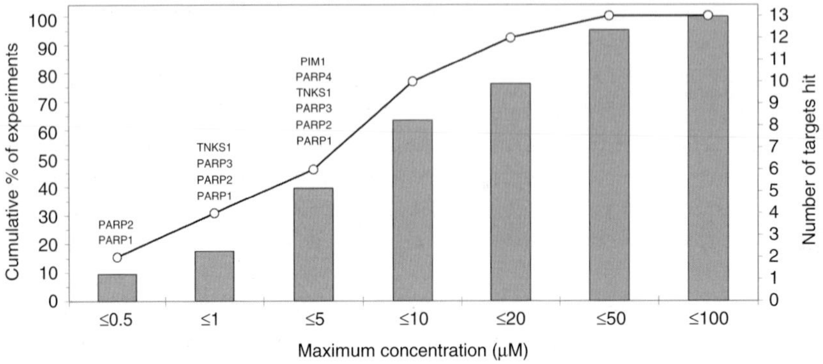

Figure 21.5 *Distribution of the percentage of cellular experiments found in the past 2 years using PJ34 concentrations below certain ranges.* Also plotted is the number of targets potentially hit by PJ34 at increasing concentration levels. Over 60% of all experiments used PJ34 concentrations above 5 μM and at this concentration PJ34 is likely to interact with at least six targets, Pim1 among them. Reprinted with permission from [11]. Copyright (2012) American Chemical Society.

IC_{50} is the concentration of the chemical probe at which half of the catalytic activity of the protein is being inhibited, at 1 μM very little Pim1 catalytic activity is being inhibited, whereas the catalytic activity of PARP-1 is being totally inhibited. Although this is the situation *in vitro*, as *in vivo* there are issues such as membrane transport and subcellular localization that need to be considered, we could assume that the use of PJ34 at concentrations lower than 1 μM would specifically inhibit members of the PARP family. However, on inspection of the publications using PJ34 in the years 2010 and 2011 to investigate the biological role of PARPs, it was surprisingly observed that PJ34 was being used at rather high concentrations. In fact, in 60% of cases PJ34 was used at concentrations higher than 5 μM (Figure 21.5) [11]. At these high concentrations, PJ34 can potentially interact with six targets, Pim1 kinase among them, and thus the risk of confounding effects should be considered (Figure 21.5). In general, scientists use siRNA to control that the effect of the small molecule is due to the inhibition of that specific target. However, if both targets are involved in the same pathways, as with Pim kinases and PARPs, the synergistic or antagonistic effects that could be occurring could pass undetected even in the siRNA experiment. Therefore, there is a need to revise the use of PJ34 as a chemical probe for PARP-1 and PARP-2, avoiding concentrations higher than 1 μM [11].

Along the same lines, a recent report highlighted a misunderstanding originated by the use of a chemical tool at high concentrations without a complete understanding of its target profile, demonstrating the utility of *in silico* target profiling (Box 21.1) [22]. Since 2002, the PARP pathway was believed to be a crucial element of tumor necrosis factor (TNF)-mediated necroptosis. However, it was recently discovered that PARP-1 and TNF represent two distinct and independent pathways and that only PJ34 (and not the rest of PARP inhibitors) was able to reduce TNF-induced necroptosis [22]. This way, the warning that PJ34 had other targets outside the PARP protein family [11] facilitated the revision and clarification of this assumption achieved using a promiscuous tool compound.

Finally, chemical probes are also important because they are the first step in further campaigns to discover new drugs. If the polypharmacology of a chemical probe is not fully understood, it can affect the development of follow-on drugs inspired by an early chemical probe. In the case of PARPs, some publications are starting to point out that different PARP drug candidates have different cellular effects despite the fact that they were supposed to be acting through the same mechanism of action [23]. Accordingly, it should be stressed that the conclusions of clinical trials involving one PARP inhibitor drug candidate might not be directly transferable to other PARP drug candidates, questioning some of the currently ongoing clinical trials. On the more positive side, additional targets discovered for PARP drug candidates could represent novel opportunities for the specific clinical development of PARP inhibitors.

21.5 Conclusions

In this chapter, we have shown how polypharmacology can impact the chemical biology of PARPs. In this case, the use of *in silico* target profiling was key to unraveling the polypharmacology of PJ34 beyond members of the PARP family. We have also discussed how these new affinities of PJ34 for Pim kinases could have confounding effects in PARP biology, promoting the attribution of functions to PARP-1 while they could actually be due to Pim kinase inhibition, as in TNF-mediated necroptosis. Finally, we have seen how the use of PJ34 at high concentrations may lead to confounding effects because of polypharmacology.

From the example provided, there are some general lessons that can be extracted for the practice of chemical biology. First of all, when a chemical tool is used, it has to be kept in mind that it might be inhibiting other yet unknown targets. Therefore, the use of small molecules to probe the biological role of proteins should ideally be done after gathering the widest possible knowledge on the affinity profile of those chemical probes across a large panel of proteins. If limited information on the target profile of the chemical probe has been obtained, then strict control on the concentration used in experiments should be imposed to limit the possibility of confounding effects masking the results and thus the conclusions drawn. Accordingly, as a general guideline, the use of the lower concentration possible can minimize the impact of unknown targets, as the number of targets being modulated by small molecules increases with the concentration used (Figure 21.5). Control experiments with siRNA are highly desirable but they cannot distinguish synergistic from antagonistic effects because of polypharmacology. Therefore, they should be complemented with the use of other chemical probes whenever possible. Finally, the use of *in silico* methods to identify novel targets that could produce confounding effects when chemical probes are used emerges as an efficient de-risking strategy in chemical biology that should be added to the toolbox of chemical biologists.

References

1. Frye, S.V. (2010) The art of the chemical probe. *Nat. Chem. Biol.*, **6**, 159–161.
2. Workman, P. and Collins, I. (2010) Probing the probes: fitness factors for small molecule tools. *Chem. Biol.*, **17**, 561–577.
3. Garcia-Serna, R. and Mestres, J. (2011) Chemical probes for biological systems. *Drug Discovery Today*, **16**, 99–106.
4. Knapp, S., Arruda, P., Blagg, J., Burley, S., Drewry, D.H., Edwards, A., Fabbro, D., Gillespie, P., Gray, N.S., Kuster, B., Lackey, K.E., Mazzafera, P., Tomkinson, N.C., Willson, T.M., Workman, P., and Zuercher, W.J. (2013) A public-private partnership to unlock the untargeted kinome. *Nat. Chem. Biol.*, **9**, 3–6.
5. Peterson, R.T. (2008) Chemical biology and the limits of reductionism. *Nat. Chem. Biol.*, **4**, 635–638.
6. Hopkins, A.L. (2008) Network pharmacology: the next paradigm in drug discovery. *Nat. Chem. Biol.*, **4**, 682–690.
7. Mestres, J., Gregori-Puigjané, E., Valverde, S., and Solé, R.V. (2008) Data completeness – The Achilles heel of drug-target networks. *Nat. Biotechnol.*, **26**, 983–984.
8. Jalencas, X. and Mestres, J. (2013) On the origins of drug polypharmacology. *Med. Chem. Commun.*, **4**, 80–87.
9. Gibson, B.A. and Kraus, W.L. (2012) New insights into the molecular and cellular functions of poly(ADP-ribose) and PARPs. *Nat. Rev. Mol. Cell Biol.*, **13**, 411–424.
10. Ferraris, D.V. (2010) Evolution of poly(ADP-ribose) polymerase-1 (PARP-1) inhibitors. From concept to clinic. *J. Med. Chem.*, **53**, 4561–4584.
11. Antolín, A.A., Jalencas, X., Yélamos, J., and Mestres, J. (2012) Identification of pim kinases as novel targets for PJ34 with confounding effects in PARP biology. *ACS Chem. Biol.*, 7, 1962–1967.
12. Wahlberg, E., Karlberg, T., Kouznetsova, E., Markova, N., Macchiarulo, A., Thorsell, A.-G., Pol, E., Frostell, Å., Ekblad, T., Öncü, D., Kull, B., Robertson, G.M., Pellicciari, R., Schüler, H., and Weigelt, J. (2012) Family-wide chemical profiling and structural analysis of PARP and tankyrase inhibitors. *Nat. Biotechnol.*, **30**, 283–288.
13. Wishart, D.S. (2012) Small molecules and diseases. *PLoS Comput. Biol.*, **8**, e1002805.
14. Koutsoukas, A., Simms, B., Kirchmair, J., Bond, P.J., Whitmore, A.V., Zimmer, S., Young, M.P., Jenkins, J.L., Glick, M., Glen, R.C., and Bender, A. (2011) From in silico target prediction to multi-target drug design: current databases, methods and applications. *J. Proteomics*, **74**, 2554–2574.
15. Carrascosa, M.C., Massaguer, O.L., and Mestres, J. (2012) PharmaTrek: a semantic web explorer for open innovation in multitarget drug discovery. *Mol. Inform.*, **31**, 537–541.
16. Ekins, S., Mestres, J., and Testa, B. (2007) In silico pharmacology for drug discovery: methods for virtual ligand screening and profiling. *Br. J. Pharmacol.*, **152**, 9–20.
17. Keiser, M.J., Setola, V., Irwin, J.J., Laggner, C., Abbas, A.I., Hufeisen, S.J., Jensen, N.H., Kuijer, M.B., Matos, R.C., Tran, T.B. *et al.* (2009) Predicting new molecular targets for known drugs. *Nature*, **462**, 175–181.
18. PredictFX v12.3 (2012) Chemotargets SL, Barcelona (Catalonia).
19. Gregori-Puigjané, E. and Mestres, J. (2006) SHED: Shannon entropy descriptors from topological feature distributions. *J. Chem. Inf. Model.*, **46**, 1615–1622.
20. Vidal, D., Garcia-Serna, R., and Mestres, J. (2011) Ligand-based approaches to in silico pharmacology. *Methods Mol. Biol. (Clifton)*, **672**, 489–502.
21. Mestres, J., Seifert, S.A., and Oprea, T.I. (2011) Linking pharmacology to clinical reports: cyclobenzaprine and its possible

association with serotonin syndrome. *Clin. Pharmacol. Ther.*, **90**, 662–665.

22. Sosna, J., Voigt, S., Mathieu, S., Lange, A., Thon, L., Davarnia, P., Herdegen, T., Linkermann, A., Rittger, A., Chan, F.K.-M., Kabelitz, D., Schütze, S., and Adam, D. (2013) TNF-induced necroptosis and PARP-1 mediated necrosis represent distinct routes to programmed necrotic cell death. *Cell. Mol. Life Sci.* doi: 10.1007/s00018-013-1381-6
23. Chuang, H.-C., Kapuriya, N., Kulp, S.K., Chen, C.-S., and Shapiro, C.L. (2012) Differential anti-proliferative activities of poly(ADP-ribose) polymerase (PARP) inhibitors in triple-negative breast cancer cells. *Breast Cancer Res. Treat.*, **134**, 649–659.

22
Splicing Inhibitors: From Small Molecule to RNA Metabolism

Tilman Schneider-Poetsch and Minoru Yoshida

22.1
Introduction

The removal of intronic sequences from primary transcripts constitutes a hallmark of eukaryotic gene expression. The ability to include and exclude different exons from one gene into the final transcript allows for increased genetic complexity without actually increasing the number of genes involved. In fact, about 90% of human genes are believed to undergo alternative splicing [1, 2].

Naturally, the splicing process and especially alternative splicing require tight regulation and control to maintain high-fidelity gene expression. Missing the correct splice site (ss) by even one nucleotide would introduce a frame shift mutation into the final transcript, resulting in a nonfunctional, if not deleterious, gene product.

While a lot of detail has been worked out about the splicing process by means of biochemistry and molecular biology, it had not received much attention from the field of chemical biology.

Before the discovery of spliceostatin A (SSA) and pladienolide B, the spliceosome had not been seriously considered as a potential drug target [3, 4]. Naturally, identification of the first inhibitors allowed further dissection of the splicing apparatus and investigation of the cellular consequences of splicing dysfunction.

22.2
The Biological Problem

22.2.1
Splicing

Without inhibition, the removal of intronic sequences proceeds in a well-ordered, highly regulated manner. The spliceosome itself is composed of five distinct small ribonuclear particles (snRNPs), named U1, U2, U4, U5, and U6, as well as many associated proteins, which are not part of the snRNPs themselves [5–7].

Concepts and Case Studies in Chemical Biology, First Edition. Edited by Herbert Waldmann and Petra Janning.

The coordinated binding of the snRNPs to their target sites on the pre-mRNA (messenger ribonucleic acid) allows the precise excision of intronic sequences from the transcript.

The pre-mRNA contains several defined regions to which the spliceosome will bind. Besides the 5′ and 3′ splice sites (5′ss and 3′ss), introns contain a branch point sequence (BPS) some 18–40 nucleotides downstream of the 3′ss, as well as a polypyrimidine tract (PPT) between BPS and 3′ss.

The splicing process (Figure 22.1) begins with the U1 snRNP binding the 5′ss, while the splicing factor SF1 adheres to the BPS. Within the cell, U1 is present in higher numbers than the remaining snRNPs and appears important not only in

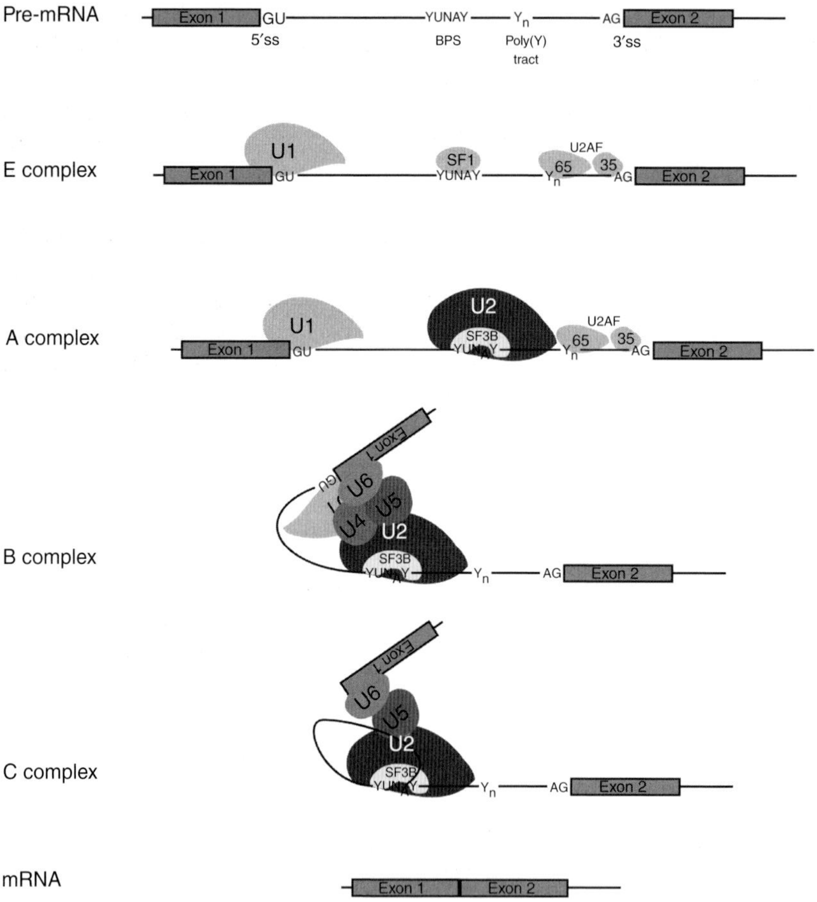

Figure 22.1 *Overview of the splicing process.* The U1 snRNP binds the 5′ss, while SF1 attaches to the branch point sequence (BPS) and U2AF to polypyrimidine tract (Y_n) and 3′ss to form E complex. U2 snRNP replaces SF1 and U2AF, thereby creating A complex. The tri snRNP U4/U5/U6 binds and replaces U1 (B complex) before rearrangement and dissociation of U4 in C complex, the complex that actually carries out the splicing reaction.

splice site selection but also in maintaining mRNA length and polyadenylation [8, 9]. Concurrently, the U2AF65 subunit of the U2AF complex (U2 accessory factor, not to be confused with the U2 snRNP) will attach to the PPT. The other U2AF subunit U2AF35 binds the 3′ss. This first assembly of factors is known as the *E complex*. It marks initiation of the splicing process, but displays no catalytic activity. In the next step, the U2 snRNP replaces SF1 and base pairs with the BPS, which results in the A complex. This likely occurs in a multistep manner, as we will see later. The tri-snRNP complex composed of U4, U5, and U6 then binds, forming the B complex. Finally U1 and U4 dissociate, leaving C complex behind, the catalytically active spliceosome.

The actual catalytic splicing process begins with a nucleophilic attack of the 2′ hydroxyl group of the branch point adenosine (within the BPS) onto the 5′ss, thereby separating the 5′ exon from the 3′ portion of the pre-mRNA and forming a lariat structure within the BPS adenosine holding phosphodiester bonds on its 2′, 3′, and 5′ hydroxyl groups. The 3′ OH group of the 5′ exon then proceeds with a nucleophilic attack on the phosphodiester bond on the 3′ss, thereby connecting the 5′ and 3′ exons with each other, while removing the intron lariat and completing the splicing process.

It appears that the splicing machinery does not freely diffuse throughout the nucleus but concentrates in the nuclear speckles [10]. Usually, each nucleus contains some 20–50 of these small, irregularly shaped aggregates of snRNP and protein complexes. Inhibition of splicing, for instance, with short morpholino oligonucleotides against sequences of the U ribonucleic acids (RNAs), results in speckle reorganization into fewer and larger aggregates.

22.2.2 Alternative Splicing

Besides regulating and maintaining the splicing process itself, the choice of exons for the final transcript is also of vital importance [11, 12].

The inclusion of unwanted exons in a particular transcript can have far-reaching consequences as the alternative splice variants of several genes have opposing functions. For instance, the apoptotic regulator Bcl-X (B-cell lymphoma) contains two possible splice sites within its second intron [13]. Using the upstream 5′ss results in a pro-apoptotic protein called *Bcl-X*$_S$. However, using another 5′ss further downstream yields a longer anti-apoptotic product (Bcl-X_L), which contributes to increased cell survival, rather than controlled apoptosis [14]. Not surprisingly the Bcl-X_L transcript appears upregulated in many cancer cells [15, 16], while Bcl-X_S usually gets repressed [17].

Naturally, a whole cornucopia of accessory factors regulates splicing and splice site selection [18]. In addition to cis-elements within the RNA, several families of RNA factors control which alternative splicing path is chosen. Well-studied examples include serine-rich (SR) proteins and heterologous nuclear ribonuclear particles (hnRNPs) [19, 20]. SR proteins bind to exonic splicing enhancers (ESEs) or intronic splicing enhancers (ISEs) and aid in recruiting the spliceosome itself.

In contrast, hnRNPs mainly interact with silencer sequences to prevent splicing or promote exon skipping.

22.2.3
mRNA Processing

Beyond the mechanics of splicing itself, the splicing machinery closely interacts with other macromolecular assemblies governing mRNA processing, including capping and polyadenylation, and possibly transcription itself [21]. Even after splicing is complete, protein factors deposited during the splicing process, such as exon junction complexes (EJCs) still affect the fate of mRNA.

Besides errors yielding frame-shifted transcripts, other splicing problems can easily result in a functional transcript facing degradation instead of expression. The standard model of nonsense-mediated mRNA decay (NMD) assumes that newly processed transcripts undergo a pioneer round of translation, in which the ribosome displaces EJCs, from the transcript [22]. Transcripts still containing EJCs after this first round of translation become summarily degraded [23]. Hence, a premature stop codon or an error in splicing resulting in an EJC 3′ of the stop codon can prevent gene expression.

Until fairly recently, splicing had not attracted much attention within the community of chemical biologists as useful inhibitors were lacking. Because researchers working on splicing itself had successfully used molecular genetics and small RNA tools for their studies, it seemed nobody was actively looking for inhibitors either.

22.3
The Chemical Approach

22.3.1
The First Splicing Inhibitors

In 2007, two papers were published back to back, each reporting the identification of a small-molecule inhibitor of the spliceosome [3, 4]. Since the initial description of SSA and pladienolide B, a number of further splicing inhibitors have been discovered and opened a new and active field of investigation (Figure 22.2). Besides novelty, these molecules have proved to be potent probes into cellular processes, while also holding clinical promise.

SSA constitutes a methyl ketal derivative of a natural product named FR901464 and was originally isolated from a *Pseudomonas sp.* fermentation broth [24, 25]. Before identification of the molecule's true mechanism, FR901464 had already attracted attention as a potent activator of viral promoters, as well as having visible antitumor activity. It inhibited the growth of various cancer cell lines at low nanomolar concentrations and extended the life span of tumor-bearing mice.

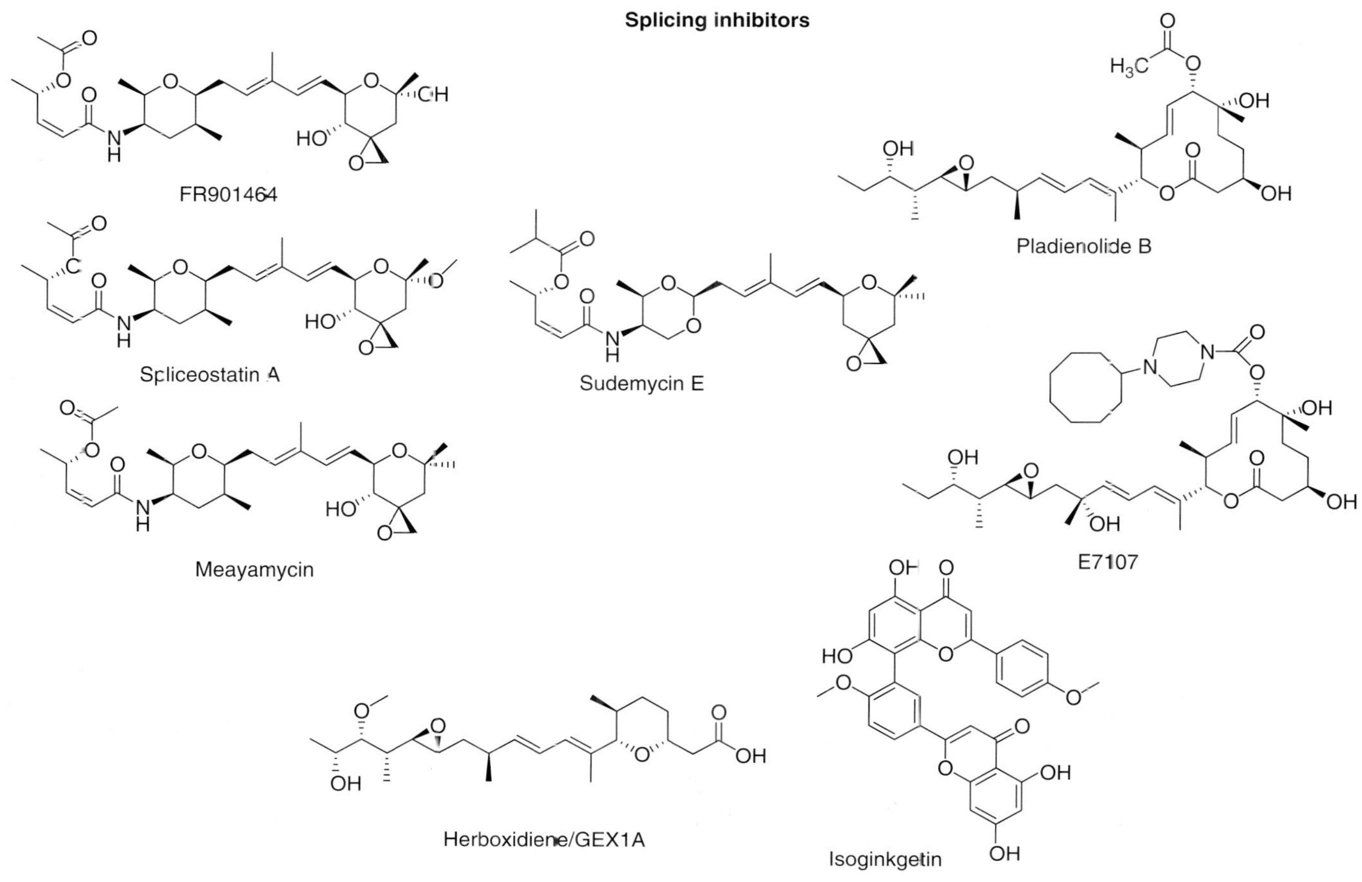

Figure 22.2 *Overview of the splicing inhibitors discussed.* FR901464 and derivatives are listed on the left, pladienolide and relatives on the right with the artificial inhibitor containing elements of both FR901464 and pladienolide, sudemycin in the middle. Herboxidiene/GEX1A and isoginkgetin are at the bottom.

However, the activation of viral promoters hinted more at an involvement in cell signaling and gave no indication of the molecule's actual mechanism.

It had been observed that FR901464 treatment led to cell cycle arrest in G1 and G2/M phases and buildup of a truncated form of the cyclin dependent kinase (CDK) inhibitor p27, dubbed p27*. The derivative SSA remains as potent as its parent compound but is chemically significantly more stable, thus making for a much more convenient research tool.

Concurrently, pladienolide also garnered interest as an antitumor compound. Pladienolide B was initially isolated from *Streptomyces platensis* Mer-11107 and its structure clearly reflects its polyketide origin [26–28]. Pladienolide B caught the attention as a potent suppressor of hypoxia-induced expression of vascular endothelial growth factor (VEGF) and displayed very potent antitumor activity both in tissue culture as well as in mouse xenograft models. Owing to its potency, a synthetic pladienolide derivative, E7107, has entered clinical trials [29, 30].

Pladienolide B and SSA do not share much molecular identity, although they are composed of similar functional groups, including a diene linker, several hydroxyls, a carbonyl, and an epoxy group. A computational side-by-side comparison of their likely lowest energy conformation in solution hinted that the molecules may adopt a shape, which could conceivably bring the same functional groups, namely, the epoxy and carbonyloxy groups into the same relative arrangement with each other [31]. On the basis of this insight, a series of simplified derivatives of FR901464 have been designed, the most potent of which appears to be sudemycin [32]. In a similar vein, the Koide group created an array of FR901464 derivatives, dubbed meayamycins, which allowed detailed structure activity relationships, as well as optimizing potency [33]. Reportedly, the inhibitor concentration 50 (IC_{50}) for cell killing could be pushed into the low picomolar range over a 5- to 10-day period.

As for pladienolide, a related *Streptomyces* natural product, herboxidiene (also called *GEX1A*) had been identified previously and proved amenable to full synthesis [34–36].

22.3.2
Inhibition

Both SSA and pladienolide present a *bona fide* example for target identification in chemical biology. Both molecules proved amenable to derivatization, without losing too much activity or going off-target.

Pladienolide B itself proved very accommodating to chemical modifications including a tritium-labeled form, a biotin conjugate with an additional photoreactive group, as well as a fluorescently tagged version [3]. Investigating subcellular localization, it was discovered that the tritium label consistently concentrated in the nucleus. Tracking the fluorescently tagged variant, the molecule appeared to locate to nuclear speckles. Relying again on the radioactively labeled probe combined with immunoprecipitation of nuclear-speckle-associated proteins identified the U2 snRNP as the most likely binding partner of pladienolide B. Because an antibody against the SAP155 protein precipitated the radio-label most efficiently, the SF3b subcomplex of the U2 snRNP seemed the most likely target. Using the photoreactive biotin conjugate allowed identifying the SAP130 subunit of the SF3b complex as the probable target protein. Involvement of the SAP145 and SAP155 subunits could not be ruled out.

After identification of the spliceosome as the pladienolide target, splicing inhibition under pladienolide B treatment could be confirmed by polymerase chain reaction (PCR)-based assays against spliced and unspliced isoforms of several genes. Furthermore, it was observed that pladienolide application led to rearrangement and enlargement of nuclear speckles.

The discovery of the SSA target followed a very similar route. Intrigued by the presence of a constitutively active truncated p27 protein in SSA-treated cells, the origin of this p27* was investigated. The complementary deoxyribonucleic acid (cDNA) expression of p27 did not yield p27* in presence of SSA and proteasome inhibition did not have any effect on this new form of p27 either, ruling out proteolysis as a source of the truncated protein [4].

Using a FR901464-biotin conjugate, the spliceosomal SF3b complex was also identified as a specific SSA target. The initial report could not narrow down a particular binding protein within the complex, but confirmed *in vitro* that SSA did indeed inhibit pre-mRNA splicing. This finding finally allowed solving the p27* conundrum as SSA treatment resulted in nonspliced pre-mRNA leaking into the cytoplasm and getting translated. The truncated form of p27 resulted from expression of such an improperly processed pre-mRNA. It lacked Thr187, a residue required for cdk-mediated phosphorylation and subsequent ubiquitin-mediated proteasomal degradation. Thus, it constitutively inhibited cdk2, thereby blocking cell-cycle progression. In a sense, this outcome of splicing inhibition presents a special case as one would expect most unspliced RNAs to either remain retained inside the nucleus or to face degradation *v* the NMD pathway.

Following the identification of the first spliceosome inhibitors, research focused on two main questions. First, to more closely explain the mechanism of the new probes and second, to find the SSA and pladienolide binding site.

The original report suggested pladienolide B would bind SAP130; however, more recently, a pladienolide resistance mutation in SAP155 (also called *SF3B1*) has been discovered [37]. Furthermore, a photoreactive herboxidiene analog cross-linked to SAP155 [36]. In light of this data, it appears quite plausible that the drug-binding site spans more than one protein subunit or lies at a subunit–subunit interface.

It appears that most laboratories are either in possession of SSA or pladienolide B. Hence, comparative studies have not been carried out so far, but considering that the two molecules and their relatives act by virtually identical mechanisms, it would seem plausible that they also share the same binding site.

In case of SSA, it was reported that drug treatment prevents formation of the pre-spliceosomal A complex [38]. The U2 snRNP does still bind the pre-mRNA but appears more weakly attached, as the complex can be removed from the RNA by increasing the buffer's heparin concentration *in vitro*. Furthermore, it appeared that U2 snRNP binding became less specific in the presence of SSA with portions other than the U2 RNAs branch point recognition sequence (bprs) interacting with the pre-mRNA and binding within a few nucleotides 5′ of the BPS. This interaction mirrored the behavior of U2 snRNPs lacking the bprs altogether. Furthermore, the authors reported that SSA treatment did not lead to general splicing inhibition in treated cells but affected only a subset of genes, while altering alternative splicing as well.

Using only a low heparin concentration, another group reported that SSA allowed formation of A complex, but stalled the progression to B complex [39]. Together, these findings suggest that SSA application prevents productive interaction between the U2 snRNP and the pre-mRNA, but may still permit assembly of a weak form of A complex. As the authors described, the sequence of the BPS could influence SSA sensitivity with BPS better matching the U2 bprs, still allowing tight U2 binding in the presence of SSA. Unfortunately, the study did not demonstrate whether these tightly bound complexes still had splicing activity or not. Because the U2 snRNP plays a pivotal role in splice site selection, impaired U2 binding may not only inhibit splicing per se but also enable alternative splicing events. In this case, one would expect introns with BPS well matching the U2 bprs still getting spliced normally in the presence of SSA, while weaker matching BPS would less likely receive proper processing.

Considering the very low doses of SSA necessary to take effect *in vivo*, it would be unlikely that all and every U2 complex in treated cells will be inhibited. Rather, the cell will have to contend with several stalled U2 snRNPs on pre-mRNAs with other U2 species not experiencing inhibition, while yet others may simply fall off the template. Together with a possible preference for strong BPS, this might also help explain the observed changes in alternative splicing patterns.

Similar experiments using pladienolide analog E7107 yielded results consistent with those obtained with SSA [40]. In the presence of E7107 A, complex formation appeared compromised. U2 binding to the pre-mRNA template only occurred at low heparin concentrations. The authors observed E7107 inhibiting an adenosine triphosphate (ATP)-dependent remodeling step in the SF3b complex, which would normally expose the bprs and allow tight interaction with the BPS. Unfortunately, to date no comparable data is available for SSA. Therefore, it is too early to say whether SSA, too, inhibits ATP-dependent remodeling, or whether the two compounds' mechanisms differ in this respect.

While SSA and pladienolide B seem to have near-identical properties, a few more splicing inhibitors that actually differ in mechanism have been reported in recent years. One of the first screens looking specifically for splicing inhibitors added the biflavonoid isoginkgetin to the list of active molecules [41, 42]. Similar to SSA and pladienolide, the molecule had been previously identified for possible medical purposes, among others as an antioxidant and neuroprotective agent. Isoginkgetin also proved active in stopping tumor cell invasion [43–45]. It did allow formation of the A complex, but seemed to interfere with B complex formation; that is, binding of the U4/U5/U6 tri-snRNP. Unfortunately, no binding protein has been identified so far and to date no direct comparison to SSA or pladienolide has been undertaken. Considering the reported medical utility of isoginkgetin, it would be interesting to investigate whether splicing inhibition at two subsequent steps in the splicing pathway results in qualitatively different cellular outcomes. Studies on isoginkgetin also reported enlargement of nuclear speckles, reminiscent of the change observed in the presence of SSA or pladienolide.

22.4 Chemical Biological Research/Evaluation

22.4.1 Cellular Effect

Splicing inhibition has a wide range of consequences for a cell. Besides remodeling of the nuclear speckles and leakage of improperly processed pre-mRNA into the cytoplasm, it appears that treatment with a splicing inhibitor also influences transcription and even the chromatin state. It had been observed at a fairly early stage that SSA treatment greatly reduced expression of VEGF [46]. While the VEGF transcript was indeed not properly spliced, expression seemed furthermore reduced on a transcriptional level. In addition, the phosphorylation level of the ribonucleic acid polymerase II (RNAP II) C-terminal domain on Ser2 and Ser5 appeared significantly reduced during splicing inhibition, which would hint at reduced transcription initiation as well as elongation.

A further study presented evidence that splicing does influence transcription. In this case, SSA derailed proper pre-mRNA processing, by abolishing RNAPII stopping at the 3′ end of the message, which in turn enhanced pre-mRNA leakage into the cytoplasm [47].

In addition, SSA treatment led to a significant shift of the histone H3 lysine 36 trimethyl mark (H3K36m3), usually associated with alternative splicing from 5′ to 3′ [48]. While the details of how splicing affects transcription and chromatin state require further study, a picture emerges in which splicing inhibition has much farther consequences for a cell than simply generating insufficiently processed mRNAs.

22.4.2 Clinical Utility

As mentioned before, E7107 has progressed into clinical trials, although currently most of their outcomes have not been reported and several trials were prematurely terminated. A limited study on thyroid cancer reported stable disease or delayed progression in a subset of patients [30]. As the antitumor activity of both SSA and pladienolide had been reported long before the molecule's mechansims were discovered, the general medical utility of splicing inhibitors received increasing attention. Indeed, splicing inhibition or the induction of alternative splicing could prove beneficial in patient treatment. As mentioned earlier, the generation of constitutively active p27* leads to cdk2 inhibition and consequent cell cycle arrest. Furthermore, SSA-induced downregulation of VEGF expression had visible anti-angiogenic activity in animal models [46].

The importance of splicing, especially alternative splicing, in tumorigenesis and tumor progression is well appreciated. Erroneous upregulation of SFs, such as the SR protein SRSF1 (serine/arginine splicing factor) in several cancers has been established. In particular, SRSF1 changes the splicing pattern of several known oncogenes and tumor suppressors, resulting in increased cell growth and protection from apoptosis [20]. For instance, in the case of BIN1, a negative regulator of the MYC proto-oncogene, alternative splicing can incorporate an exon 12A into the final transcript, which reduces BIN1s affinity to MYC [49, 50]. As several splice isoforms can even have opposing effects, such as the above-mentioned case of Bcl-X, chemical manipulation of alternative splicing could also have a therapeutic benefit in certain tumor types. In this case, the most promising targets may lie outside the core spliceosome but among the splice regulators, such as SR proteins or hnRNPs.

While pladienolide B and SSA represent the only molecules for which direct binding to the spliceosome could be demonstrated (and isoginkgetin being a good third candidate), several other compounds affect splicing, especially alternative splicing. These compounds primarily target signaling molecules, which influence splice site choice. TG003 represents one of the earliest identified alternative splicing modulators [51]. It actually inhibits cdc2-like kinase (clks) clk1/sty, which in turn acts upon the SR protein SF2/ASF (anti-silencing function). Consequently, reduced phosphorylation of SF2/ASF changes the splicing pattern of several endogenous genes.

The list of small molecules affecting alternative splicing runs much longer with several compounds affecting SR protein function. These molecules either work on signaling pathways that regulate SR protein phosphorylation or, in the case of some indole derivatives, interacting with SR proteins directly.

Similar to TG003, the inhibitor SRPIN340 blocks kinases acting on SR proteins, while cardiontonic steroids used to treat heart conditions modulate alternative splicing by decreasing the level of SR protein SRSF3 [52]. Meanwhile, amiloride, a drug used against hypertension, is known to decrease phosphorylation of SRSF1 [53, 54]. It would be interesting to know whether patients treated with existing medications that also influence splicing behavior experience different risks or clinical outcomes for various types of tumors. Early estimates suggested that 15% of disease-related mutations affect splicing [55]. More recent predictions have placed that percentage at around 50% [56]. In this light, a directed screen of known splicing modulators against tissue cultures from splicing defect-associated tumors might prove worthwhile.

22.5 Conclusion

In the barely 6 years since the first description of splicing inhibitors, the number of identified molecules has increased. Synthetic derivatives of known compounds have helped generate more potent derivatives, as well as enable detailed structure-activity studies. With E7107, one splicing inhibitor has even entered the clinic. Meanwhile SSA, pladienolide, meayamycin, and others have already aided scientific progress as tool compounds in studies on splicing mechanism and RNA processing. For the next years, we can expect further insight into the complex world of pre-mRNA metabolism through the power of chemical biology.

References

1. Wang, E.T., Sandberg, R., Luo, S., Khrebtukova, I., Zhang, L., Mayr, C., Kingsmore, S.F., Schroth, G.P., and Burge, C.B. (2008) Alternative isoform regulation in human tissue transcriptomes. *Nature*, **456** (7221), 470–476.
2. Pan, Q., Shai, O., Lee, L.J., Frey, B.J., and Blencowe, B.J. (2008) Deep surveying of alternative splicing complexity in the human transcriptome by high-throughput sequencing. *Nat. Genet.*, **40** (12), 1413–1415.
3. Kotake, Y., Sagane, K., Owa, T., Mimori-Kiyosue, Y., Shimizu, H., Uesugi, M., Ishihama, Y., Iwata, M., and Mizui, Y. (2007) Splicing factor SF3b as a target of the antitumor natural product pladienolide. *Nat. Chem. Biol.*, **3** (9), 570–575.
4. Kaida, D., Motoyoshi, H., Tashiro, E., Nojima, T., Hagiwara, M., Ishigami, K., Watanabe, H., Kitahara, T., Yoshida, T., Nakajima, H., Tani, T., Horinouchi, S., and Yoshida, M. (2007) Spliceostatin A targets SF3b and inhibits both splicing and nuclear retention of pre-mRNA. *Nat. Chem. Biol.*, **3** (9), 576–583.
5. Hoskins, A.A. and Moore, M.J. (2012) The spliceosome: a flexible, reversible macromolecular machine. *Trends Biochem. Sci.*, **37** (5), 179–188.
6. Wahl, M.C., Will, C.L., and Lührmann, R. (2009) The spliceosome: design principles of a dynamic RNP machine. *Cell*, **136** (4), 701–718.
7. Staley, J.P. and Woolford, J.J.L. (2009) Assembly of ribosomes and spliceosomes: complex ribonucleoprotein machines. *Curr. Opin. Cell Biol.*, **21** (1), 109–118.
8. Kaida, D., Berg, M.G., Younis, I., Kasim, M., Singh, L.N., Wan, L., and Dreyfuss, G. (2010) U1 snRNP protects pre-mRNAs from premature cleavage and polyadenylation. *Nature*, **468** (7324), 664–668.
9. Berg, M.G., Singh, L.N., Younis, I., Liu, Q., Pinto, A.M., Kaida, D., Zhang, Z., Cho, S., Sherrill-Mix, S., Wan, L., and Dreyfuss, G. (2012) U1 snRNP determines mRNA length and regulates isoform expression. *Cell*, **150** (1), 53–64.
10. Spector, D.L. and Lamond, A.I. (2011) Nuclear speckles. *Cold Spring Harbor Perspect. Biol.*, **3** (2).
11. Kalsotra, A. and Cooper, T.A. (2011) Functional consequences of developmentally regulated alternative splicing. *Nat. Rev. Genet.*, **12** (10), 715–729.
12. David, C.J. and Manley, J.L. (2010) Alternative pre-mRNA splicing regulation in cancer: pathways and programs unhinged. *Genes Dev.*, **24** (21), 2343–2364.
13. Boise, L.H., González-García, M., Postema, C.E., Ding, L., Lindsten, T., Turka, L.A., Mao, X., Nuñez, G., and Thompson, C.B. (1993) Bcl-x, a bcl-2-related gene that functions as a dominant regulator of apoptotic cell death. *Cell*, **74** (4), 597–608.

14. Minn, A.J., Boise, L.H., and Thompson, C.B. (1996) Bcl-x(S) anatagonizes the protective effects of Bcl-x(L). *J. Biol. Chem.*, **271** (11), 6306–6312.
15. Kühl, J.S., Krajewski, S., Durán, G.E., Reed, J.C., and Sikic, B.I. (1997) Spontaneous overexpression of the long form of the Bcl-X protein in a highly resistant P388 leukaemia. *Br. J. Cancer*, **75** (2), 268–274.
16. Xerri, L., Parc, P., Brousset, P., Schlaifer, D., Hassoun, J., Reed, J.C., Krajewski, S., and Birnbaum, D. (1996) Predominant expression of the long isoform of Bcl-x (Bcl-xL) in human lymphomas. *Br. J. Haematol.*, **92** (4), 900–906.
17. Ma, X., Zhao, Y., Li, Y., Lu, H., and He, Y. (2010) Relevance of Bcl-x expression in different types of endometrial tissues. *J. Exp. Clin. Cancer Res.*, **29**, 14.
18. Dreyfuss, G., Kim, V.N., and Kataoka, N. (2002) Messenger-RNA-binding proteins and the messages they carry. *Nat. Rev. Mol. Cell Biol.*, **3** (3), 195–205.
19. Martinez-Contreras, R., Cloutier, P., Shkreta, L., Fisette, J.F., Revil, T., and Chabot, B. (2007) hnRNP proteins and splicing control. *Adv. Exp. Med. Biol.*, **623**, 123–147.
20. Long, J.C. and Caceres, J.F. (2009) The SR protein family of splicing factors: master regulators of gene expression. *Biochem. J.*, **417** (1), 15–27.
21. Moore, M.J. and Proudfoot, N.J. (2009) Pre-mRNA processing reaches back to transcription and ahead to translation. *Cell*, **136** (4), 688–700.
22. Behm-Ansmant, I., Kashima, I., Rehwinkel, J., Saulière, J., Wittkopp, N., and Izaurralde, E. (2007) mRNA quality control: an ancient machinery recognizes and degrades mRNAs with nonsense codons. *FEBS Lett.*, **581** (15), 2845–2853.
23. Le Hir, H., Gatfield, D., Izaurralde, E., and Moore, M.J. (2001) The exon-exon junction complex provides a binding platform for factors involved in mRNA export and nonsense-mediated mRNA decay. *EMBO J.*, **20** (17), 4987–4997.
24. Nakajima, H., Hori, Y., Terano, H., Okuhara, M., Manda, T., Matsumoto, S., and Shimomura, K. (1996) New antitumor substances, FR901463, FR901464 and FR901465. II. Activities against experimental tumors in mice and mechanism of action. *J. Antibiot. (Tokyo)*, **49** (12), 1204–1211.
25. Nakajima, H., Sato, B., Fujita, T., Takase, S., Terano, H., and Okuhara, M. (1996) New antitumor substances, FR901463, FR901464 and FR901465. I. Taxonomy, fermentation, isolation, physico-chemical properties and biological activities. *J. Antibiot. (Tokyo)*, **49** (12), 1196–1203.
26. Mizui, Y., Sakai, T., Iwata, M., Uenaka, T., Okamoto, K., Shimizu, H., Yamori, T., Yoshimatsu, K., and Asada, M. (2004) Pladienolides, new substances from culture of *Streptomyces platensis* Mer-11107. III. In vitro and in vivo antitumor activities. *J. Antibiot. (Tokyo)*, **57** (3), 188–196.
27. Sakai, T., Asai, N., Okuda, A., Kawamura, N., and Mizui, Y. (2004) Pladienolides, new substances from culture of *Streptomyces platensis* Mer-11107. II. Physico-chemical properties and structure elucidation. *J. Antibiot. (Tokyo)*, **57** (3), 180–187.
28. Sakai, T., Sameshima, T., Matsufuji, M., Kawamura, N., Dobashi, K., and Mizui, Y. (2004) Pladienolides, new substances from culture of *Streptomyces platensis* Mer-11107. I. Taxonomy, fermentation, isolation and screening. *J. Antibiot. (Tokyo)*, **57** (3), 173–179.
29. Kanada, R.M., Itoh, D., Nagai, M., Niijima, J., Asai, N., Mizui, Y., Abe, S., and Kotake, Y. (2007) Total synthesis of the potent antitumor macrolides pladienolide B and D. *Angew. Chem. Int. Ed.*, **46** (23), 4350–4355.
30. Tsimberidou, A.M., Vaklavas, C., Wen, S., Hong, D., Wheler, J., Ng, C., Naing, A., Tse, S., Busaidy, N., Markman, M., Sherman, S.I., and Kurzrock, R. (2009) Phase I clinical trials in 56 patients with thyroid cancer: the M. D. Anderson cancer center experience. *J. Clin. Endocrinol. Metab.*, **94** (11), 4423–4432.
31. Lagisetti, C., Pourpak, A., Goronga, T., Jiang, Q., Cui, X., Hyle, J., Lahti, J.M., Morris, S.W., and Webb, T.R. (2009) Synthetic mRNA splicing modulator compounds with *in vivo* antitumor activity. *J. Med. Chem.*, **52** (22), 6979–6990.

32. Fan, L., Lagisetti, C., Edwards, C.C., Webb, T.R., and Potter, P.M. (2011) Sudemycins, novel small molecule analogues of FR901464, induce alternative gene splicing. *ACS Chem. Biol.*, **6** (6), 582–589.
33. Albert, B.J., McPherson, P.A., O'Brien, K., Czaicki, N.L., Destefino, V., Osman, S., Li, M., Day, B.W., Grabowski, P.J., Moore, M.J., Vogt, A., and Koide, K. (2009) Meayamycin inhibits pre-messenger RNA splicing and exhibits picomolar activity against multidrug-resistant cells. *Mol. Cancer Ther.*, **8** (8), 2308–2318.
34. Sakai, Y., Tsujita, T., Akiyama, T., Yoshida, T., Mizukami, T., Akinaga, S., Horinouchi, S., Yoshida, M., and Yoshida, T. (2002) GEX1 compounds, novel antitumor antibiotics related to herboxidiene, produced by Streptomyces sp. II. The effects on cell cycle progression and gene expression. *J. Antibiot. (Tokyo)*, **55** (10), 863–872.
35. Sakai, Y., Yoshida, T., Ochiai, K., Uosaki, Y., Saitoh, Y., Tanaka, F., Akiyama, T., Akinaga, S., and Mizukami, T. (2002) GEX1 compounds, novel antitumor antibiotics related to herboxidiene, produced by Streptomyces sp. I. Taxonomy, production, isolation, physicochemical properties and biological activities. *J. Antibiot. (Tokyo)*, **55** (10), 855–862.
36. Hasegawa, M., Miura, T., Kuzuya, K., Inoue, A., Won Ki, S., Horinouchi, S., Yoshida, T., Kunoh, T., Koseki, K., Mino, K., Sasaki, R., Yoshida, M., and Mizukami, T. (2011) Identification of SAP155 as the target of GEX1A (Herboxidiene), an antitumor natural product. *ACS Chem. Biol.*, **6** (3), 229–233.
37. Yokoi, A., Kotake, Y., Takahashi, K., Kadowaki, T., Matsumoto, Y., Minoshima, Y., Sugi, N.H., Sagane, K., Hamaguchi, M., Iwata, M., and Mizui, Y. (2011) Biological validation that SF3b is a target of the antitumor macrolide pladienolide. *FEBS J.*, **278** (24), 4870–4880.
38. Corrionero, A., Minana, B., and Valcarcel, J. (2011) Reduced fidelity of branch point recognition and alternative splicing induced by the anti-tumor drug spliceostatin A. *Genes Dev.*, **25** (5), 445–459.
39. Roybal, G.A. and Jurica, M.S. (2010) Spliceostatin A inhibits spliceosome assembly subsequent to prespliceosome formation. *Nucleic Acids Res.*, **38** (19), 6664–6672.
40. Folco, E.G., Coil, K.E., and Reed, R. (2011) The anti-tumor drug E7107 reveals an essential role for SF3b in remodeling U2 snRNP to expose the branch point-binding region. *Genes Dev.*, **25** (5), 440–444.
41. Briancon-Scheid, F., Lobstein-Guth, A., and Anton, R. (1983) HPLC separation and quantitative determination of biflavones in leaves from ginkgo biloba. *Planta Med.*, **49** (12), 204–207.
42. O'Brien, K., Matlin, A.J., Lowell, A.M., and Moore, M.J. (2008) The biflavonoid isoginkgetin is a general inhibitor of Pre-mRNA splicing. *J. Biol. Chem.*, **283** (48), 33147–33154.
43. Ellnain-Wojtaszek, M., Kruczynski, Z., and Kasprzak, J. (2003) Investigation of the free radical scavenging activity of Ginkgo biloba L. leaves. *Fitoterapia*, **74** (1-2), 1–6.
44. Kang, S.S., Lee, J.Y., Choi, Y.K., Song, S.S., Kim, J.S., Jeon, S.J., Han, Y.N., Son, K.H., and Han, B.H. (2005) Neuroprotective effects of naturally occurring biflavonoids. *Bioorg. Med. Chem. Lett.*, **15** (15), 3588–3591.
45. Yoon, S.O., Shin, S., Lee, H.J., Chun, H.K., and Chung, A.S. (2006) Isoginkgetin inhibits tumor cell invasion by regulating phosphatidylinositol 3-kinase/Akt-dependent matrix metalloproteinase-9 expression. *Mol. Cancer Ther.*, **5** (11), 2666–2675.
46. Furumai, R., Uchida, K., Komi, Y., Yoneyama, M., Ishigami, K., Watanabe, H., Kojima, S., and Yoshida, M. (2010) Spliceostatin A blocks angiogenesis by inhibiting global gene expression including VEGF. *Cancer Sci.*, **101** (11), 2483–2489.
47. Martins, S.B., Rino, J., Carvalho, T., Carvalho, C., Yoshida, M., Klose, J.M., de Almeida, S.F., and Carmo-Fonseca, M. (2011) Spliceosome assembly is coupled to RNA polymerase II dynamics at the

3′ end of human genes. *Nat. Struct. Mol. Biol.*, **18** (10), 1115–1123.

48. Kim, S., Kim, H., Fong, N., Erickson, B., and Bentley, D.L. (2011) Pre-mRNA splicing is a determinant of histone H3K36 methylation. *Proc. Natl. Acad. Sci. U.S.A.*, **108** (33), 13564–13569.
49. Sakamuro, D., Elliott, K.J., Wechsler-Reya, R., and Prendergast, G.C. (1996) BIN1 is a novel MYC-interacting protein with features of a tumour suppressor. *Nat. Genet.*, **14** (1), 69–77.
50. Karni, R., de Stanchina, E., Lowe, S.W., Sinha, R., Mu, D., and Krainer, A.R. (2007) The gene encoding the splicing factor SF2/ASF is a proto-oncogene. *Nat. Struct. Mol. Biol.*, **14** (3), 185–193.
51. Muraki, M., Ohkawara, B., Hosoya, T., Onogi, H., Koizumi, J., Koizumi, T., Sumi, K., Yomoda, J., Murray, M.V., Kimura, H., Furuichi, K., Shibuya, H., Krainer, A.R., Suzuki, M., and Hagiwara, M. (2004) Manipulation of alternative splicing by a newly developed inhibitor of Clks. *J. Biol. Chem.*, **279** (23), 24246–24254.
52. Fukuhara, T., Hosoya, T., Shimizu, S., Sumi, K., Oshiro, T., Yoshinaka, Y., Suzuki, M., Yamamoto, N., Herzenberg, L.A., Herzenberg, L.A., and Hagiwara, M. (2006) Utilization of host SR protein kinases and RNA-splicing machinery during viral replication. *Proc. Natl. Acad. Sci. U.S.A.*, **103** (30), 11329–11333.
53. Chang, J.G., Yang, D.M., Chang, W.H., Chow, L.P., Chan, W.L., Lin, H.H., Huang, H.D., Chang, Y.S., Hung, C.H., and Yang, W.K. (2011) Small molecule amiloride modulates oncogenic RNA alternative splicing to devitalize human cancer cells. *PLoS One*, **6** (6), e18643.
54. Chang, W.H., Liu, T.C., Yang, W.K., Lee, C.C., Lin, Y.H., Chen, T.Y., and Chang, J.G. (2011) Amiloride modulates alternative splicing in leukemic cells and resensitizes Bcr-AblT315I mutant cells to imatinib. *Cancer Res.*, **71** (2), 383–392.
55. Krawczak, M., Reiss, J., and Cooper, D.N. (1992) The mutational spectrum of single base-pair substitutions in mRNA splice junctions of human genes: causes and consequences. *Hum. Genet.*, **90** (1-2), 41–54.
56. Wang, G.S. and Cooper, T.A. (2007) Splicing in disease: disruption of the splicing code and the decoding machinery. *Nat. Rev. Genet.*, **8** (10), 749–761.

23
Photochemical Control of Gene Function in Zebrafish Embryos with Light-Activated Morpholinos

Qingyang Liu and Alexander Deiters

23.1
Introduction

Morpholino oligomers (MOs) are commonly employed tools for the study of gene function and embryo development. In this chapter, four approaches to photoregulate MO activity with case studies in zebrafish embryos are discussed, including hairpin-caged morpholino oligomers (cMOs), sense-cMOs, nucleobase-cMOs, and cyclic-cMOs. These light-activated MOs can be designed to selectively silence any target gene with precise spatial and temporal resolution, thus potentially providing insight into embryonic development that cannot be obtained with other tools.

23.2
The Biological Problem

Antisense agents are common tools to sequence specifically target messenger ribonucleic acid (mRNA) and block subsequent protein expression [1], through either ribonuclease H (RNase H)-mediated mRNA cleavage (Figure 23.1A) or steric blocking of the ribosome (Figure 23.1B). Both mechanisms lead to efficient inhibition of protein expression and gene silencing. Thus, antisense agents have been proved to be powerful tools for the study of gene function in cells and multicellular model organisms, especially zebrafish embryos.

Early antisense technology was mostly based on single stranded deoxyribonucleic acid (ssDNA) and modified DNA such as phosphorothioate deoxyribonucleic acid (PS DNA). However, their low cellular stability and potential cell toxicity can interfere with their applicability [2]. Modified RNA oligomers especially 2′-*O*-alkyl RNA, which is resistant to nucleases and shows high affinity to complementary mRNA, were developed and efforts were made to generate oligomer backbones with improved cellular stability and reduced toxicity. To this end, a variety of oligomers have been introduced, including 2′-fluoro-arabino nucleic acid (FANA), peptide nucleic acid (PNA), locked nucleic acid (LNA), and

Concepts and Case Studies in Chemical Biology, First Edition. Edited by Herbert Waldmann and Petra Janning.

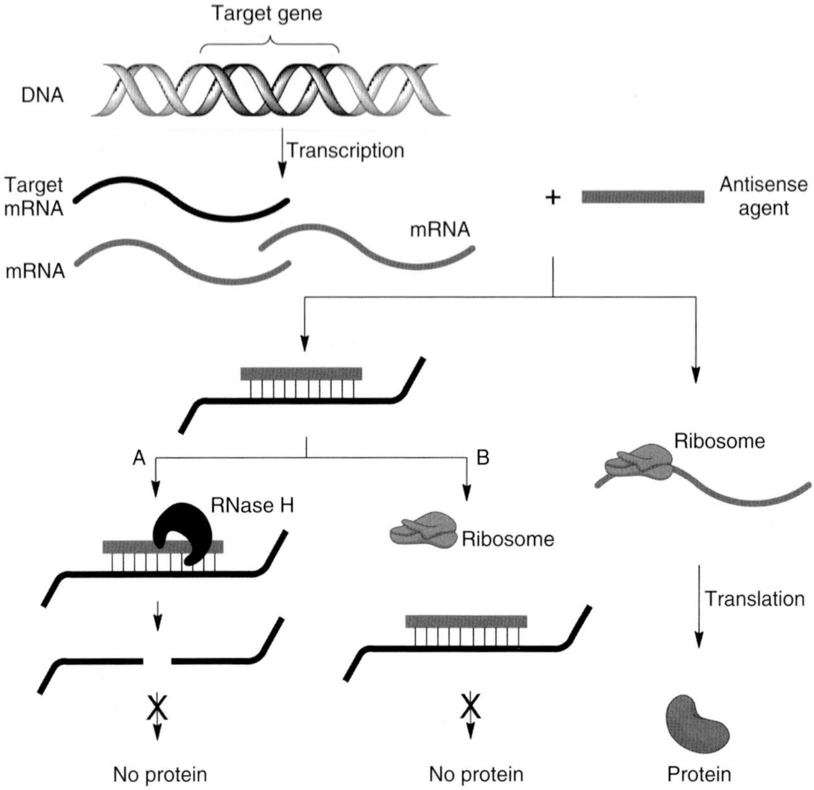

Figure 23.1 *Different mechanisms of gene silencing by antisense agents.* An antisense agent selectively binds to target mRNA (black) and stops mRNA translation through (A) RNase H-mediated mRNA cleavage or (B) blocking of mRNA processing by the ribosome. Meanwhile, other nontargeted mRNAs (gray) are translated into proteins normally.

MOs (Figure 23.2). Owing to the significant structural differences between these novel oligomer backbones and natural oligonucleotides, they do not activate RNase H-mediated degradation with the exception of FANA, but instead, block translation through steric hindrance [1].

MOs are the most commonly used gene-silencing reagents applied in zebrafish embryos [3]. Three strategies targeting different RNA molecules have been reported to trigger gene expression using MOs in zebrafish embryos. The earliest strategy is the use of translation-silencing MOs to target the initiation codon of mature mRNA, thus blocking translation as shown in Figure 23.1. Recently, MOs have been designed to bind to pre-mRNA and inhibit correct pre-mRNA splicing. This strategy requires additional information on intron and exon structure compared to the first strategy, but allows the quantitative measurement of MO efficiency by quantitative real-time polymerase chain reaction (qRT-PCR). MOs targeting microRNAs, which are small noncoding RNAs that are known

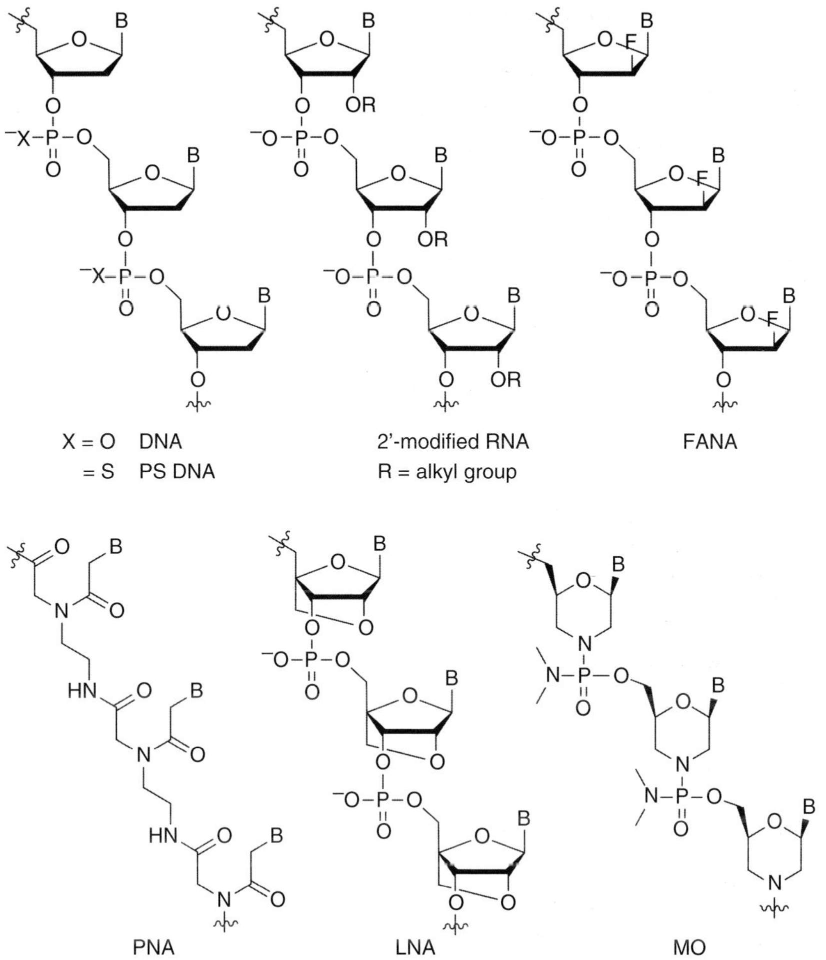

Figure 23.2 Structures of antisense agents. B = nucleobase, including A, T, G, C, or U.

to regulate gene expression, have also been reported, thereby expanding the application of MOs in zebrafish.

Generally, to block translation directly, MOs of 25 base pairs are synthesized and microinjected into embryos at the one- to two-cell stage, followed by distribution throughout the developing embryo and sustained blocking of target gene expression for several days [4]. Usually, within 3–5 days, the embryos injected with antisense MOs show abnormal development and the observed phenotype can be used to elucidate the function of the targeted gene. To further confirm that the phenotype is induced by target gene silencing rather than off-target effects, mRNA encoding target protein but without sequence overlap with antisense MOs can be co-injected with MOs. If the phenotype is recovered, the corresponding gene is responsible for the mutant phenotype. The use of two different sets of

MOs targeting the same gene can also eliminate off-target effects, especially when one translation-silencing MO and one splicing-inhibiting MO are used in combination.

A problem with the MO gene-silencing methodology is a general lack of conditional control. As mentioned previously, after being injected, MOs are distributed evenly throughout the embryo, which leads to gene knockdown throughout the whole embryo right from the injection time point onward. To accurately control embryo development and assign gene function, selective gene perturbation within a certain region and at a specific stage of development can potentially reveal the mechanisms behind embryonic development that cannot be observed otherwise. Several approaches have been developed to accomplish spatial and temporal control over MO activity and these are discussed later.

23.3 The Chemical Approach

Light is noninvasive and orthogonal to most cellular components; thus, it can serve as a regulatory input to biological activities with precise spatiotemporal resolution. Four different approaches have been developed to regulate MO activity with light, based on the installation of light-sensitive groups, termed *caging groups*, within the oligomer (Figure 23.3). The most commonly used caging groups are *o*-nitrobenzyl (ONB) derivatives, which are photolyzed through exposure to 360–365 nm light.

23.3.1 Hairpin-Caged MO

The hairpin-cMO approach was first reported by the Chen Lab in 2007 [5]. A short inhibitor strand was tethered to the antisense MO through a photocleavable dimethoxynitrobenzyl (DMNB) linker (Figure 23.3a). The two linked MO strands form a hairpin duplex that inactivates the antisense MO. Upon light irradiation, the linker is cleaved, releasing the free MO, allowing it to bind to its target mRNA, and thereby silencing mRNA transcription. In this approach, three components are required: the antisense MO, a photocleavable linker, and the inhibitor strand. Both the 3′-amino antisense MO and the 5′-amino inhibitor MO are commercially available and the photocleavable linker, once synthesized, can be applied to various MO strands. The linker contains a DMNB group with an alkynyl functional group at one end and a succinimidyl ester at the other, which is readily reacted with the 5′-amine of the inhibitor strand. The 3′-amine of the antisense MO is treated with 3-azidopropinonic acid succinimidyl ester to afford an azido-modified MO. Then the two MO strands, modified with an azido and an alkynyl group, are linked through a Cu-catalyzed [3+2] cycloaddition reaction. However, the design of the short inhibitor strand is not trivial. If the inhibitor strand is too short, its binding energy is not sufficient to prevent the mRNA from hybridizing

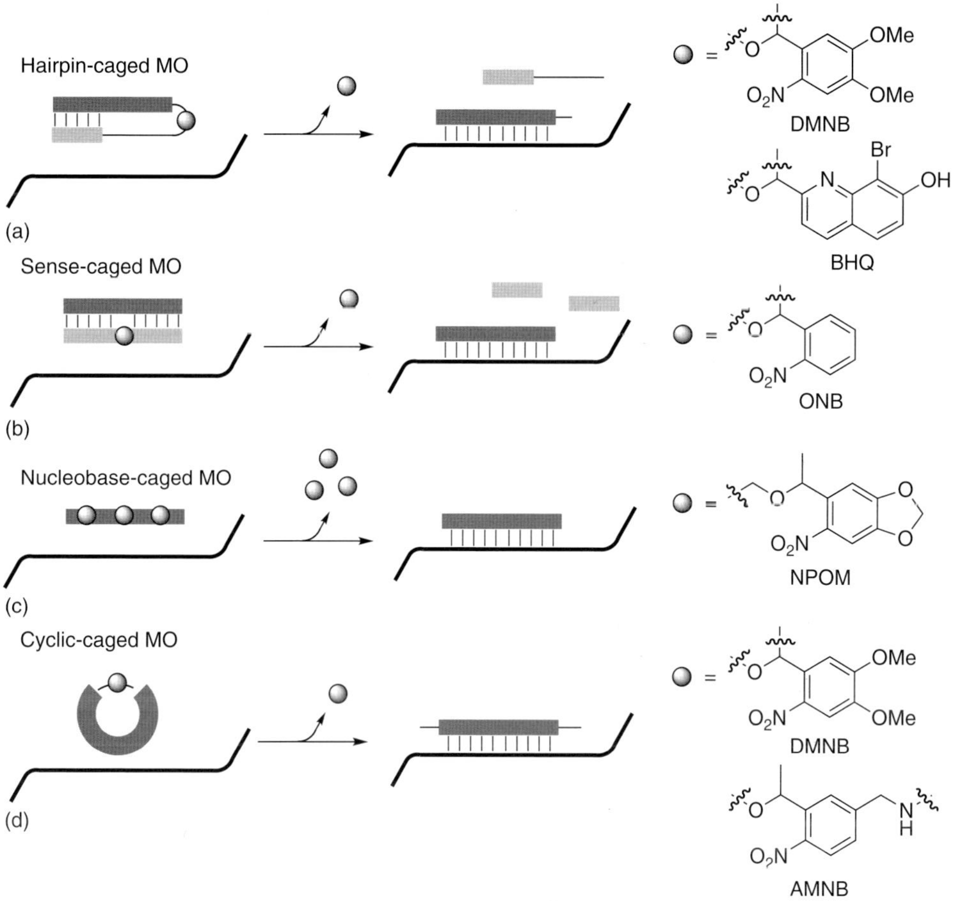

Figure 23.3 *Different strategies to photochemically control MO function.* (a) Hairpin-caged MO with a dimethoxynitrobenzyl (DMNB) or bromohydroxyquinoline (BHQ) linker. (b) Sense-caged MO with an *o*-nitrobenzyl (ONB) linker. (c) Nucleobase-caged MO blocked with 6-nitropiperonyloxymethyl (NPOM) groups. (d) Cyclic-caged MO with a DMNB or 5-aminomethyl-2-nitrobenzyl (AMNB) linker. Black: target mRNA; gray: antisense MO; light gray: inhibitor strand or complementary sense strand; circle: caging group.

to the MO. On the other hand, if its binding affinity to the MO is too high, it will not be efficiently released after linker photolysis. On the basis of melting temperature measurements and biological studies, the Chen Lab derived Equation 23.1 as a general guide for inhibitor design [6]. The equation shows that the melting temperature of the MO duplex (T_m) is related to the number of A/T and G/C base pairs. An efficient inhibitor strand will be generated by designing a hairpin MO with a melting temperature between 41 and 44 °C.

$$T_m = 1.9 \times (A + T) + 5.7 \times (G + C) \tag{23.1}$$

The DMNB hairpin cMOs have been applied to the knockdown of several genes, including *no tail a* (*ntla*), *heart of glass* (*heg*), *floating head* (*flh*), and *ETS1-related protein* (*etv2*) [5, 6]. They also helped reveal an important regulatory mechanism in embryo development, as discussed in Section 23.4.

Apart from UV-cleavable DMNB linkers, a bromohydroxyquinoline (BHQ) linker was also introduced into hairpin cMOs, which enables two-photon activation of MO function at 750 nm [6]. This BHQ linker can provide improved spatial resolution and is less likely to cause any phototoxicity because of UV exposure.

23.3.2
Sense-Caged MO

In the sense-cMO approach, early designs used an MO-blocking RNA sense strand containing a caging group in the center of its phosphate backbone (Figure 23.3b) [7]. The RNA–MO duplex is formed until light cleaves the sense strand and releases the active MO. The caged sense strand can be easily synthesized and is complementary to the MO, which avoids the need to design an inhibitor strand as in the hairpin cMO approach (Section 23.3.1). Among the examples using this RNA-sense-cMO approach, one case study showcases the control of a splicing-inhibitor MO targeting the *ras homolog enriched in brain* (*rheb*) gene [7]. The corresponding gene silencing and decaging time course were monitored by qRT-PCR and the *rheb* levels indicate that the RNA sense cMO effectively cages the antisense MO at early time points, but the activity of the cMO is slowly leaking even without light irradiation due to inhibitor dilution and the degradation of sense RNA in the cell. Also, a large excess of the sense strand is needed to completely inactivate the MO, and high concentrations of RNA can be toxic to the cells.

Thus, the unstable RNA sense strand was replaced by a stable MO sense strand containing a light-cleavable linker, in order to generate a more stable, inactive MO duplex. On the basis of an ONB group, a linker bearing an imidazole carbamate that reacts with the 4-N of morpholine and a trityl protected amine that readily couples to the phosphorodiamidate upon deprotection was incorporated as an MO subunit in the center of the inhibitory sense strand. To avoid potential binding of the cleaved sense MO strand to the antisense MO, it has been shown that for a 25-base antisense MO, either a 21-base sense MO with no mismatch or a 25-base sense MO with four mismatches on both sides of the light-cleavable linker is optimal [8]. In addition, a strict 1 : 1 ratio of antisense MO to sense cMO is crucial for optimal binding and dissociation properties. Several genes have been selectively knocked down by light irradiation with MO sense cMOs, including *ntla*, *sox10*, and *gal4* [8, 9].

One of the advantages of this approach is that the sense cMOs are commercially available and can be readily customized to convert any antisense MO into a light-activated MO. Two recent reports took advantage of the temporal control provided by this light-activation tool. One of these applications was the study of *sox10* function, which encodes protein Sox10 which is known to be a central regulatory factor in neural crest (Box 23.1) development [10]. Bronner and coworkers demonstrated that silencing of *sox10* led to a reduction in olfactory sense neuron formation (Box 23.1) [9]. There are two possible explanations for this: (i) the gene knockdown leads to the lack of neural crest cells, which were shown to be the origin of olfactory sense neurons or (ii) *sox10* directly triggers sense neuron formation. By applying a sense cMO, the temporally controlled light-induced silencing of *sox10* was performed at 17.5 hpf (hours post fertilization) and inhibition of neuron formation was observed even after neuron crest cells have developed. Together with previous results, this proved that *sox10* promotes the formation of the neural crest at an early embryonal stage, and directly regulates the development of olfactory sense neurons later as stated in the second explanation. In the other example, the extrusion of epithelia cells was studied and Rosenblatt and coworkers hypothesized that the ion channel Piezo1 (Box 23.1) is necessary for extrusion formation. Thus, the lack of Piezo1 would reduce cell extrusion and lead to mass growth in the epithelium (Box 23.1). The MO-induced *piezo1* silencing, however, resulted in embryo death at 48 hpf, which prevented the study of cell extrusion. To avoid the lethal phenotype, a sense *piezo1* cMO was applied to selectively knockdown Piezo1 expression at 30 hpf with light and mass growth was observed at 60 hpf as previously hypothesized [11]. Both examples establish the effectiveness of sense cMOs in studying gene function with temporal resolution.

Box 23.1 Definitions of selected terms

Mesoderm. In the early development of zebrafish embryos, gastrulation, a process that lasts from 6 to 10 hpf and involves a variety of cell and tissue movements, results in three germ layers. The mesoderm is the middle layer between the ectoderm (outside layer) and the endoderm (inside layer).

Dorsal–ventral axis. Dorsal refers to the back of the fish and ventral refers to the belly of the fish. The dorsal–ventral axis is the axis from the back to the belly.

Notochord. The notochord is a rod-shaped structure that is derived from the mesoderm close to the neural tube (called *axial mesoderm*) during gastrulation. It extends from head to tail beneath the developing nervous system and physically supports the embryo. It also secretes factors that control the patterning of surrounding tissues.

Neural plate. The neural plate is formed through the thickening of ectodermal tissue on the dorsal side of the embryo. It folds into a tube-shaped structure known as the *neural tube*, representing the precursor to the nervous system.

Medial floor plate. The medial floor plate is a structure located in the ventral midline of the neural tube above the notochord.

Neural crest. The neural crest is a group of multipotent cells at the edge of the neural plate, which, after the formation of the neural tube, migrates throughout the embryo and differentiates into a variety of different cells.

Somite. Somites are paired blocks of mesoderm cells located at the side of the neural tube and are formed after gastrulation.

Microvillous olfactory sensory neurons. Microvillous olfactory sensory neurons are cells that sense smells and bear microscopic protrusions to increase their surface area. They are often found on the surface of the olfactory epithelium.

Epithelium. The epithelium is one of the basic tissue types found in animals and lines the surface of hollow organs and the body.

Ion channel. Ion channels are proteins that gate the flow of ions through cell membranes.

23.3.3 Nucleobase-Caged MO

Instead of inserting light-cleavable linkers into the MO backbone, caging groups have also been installed on nucleobases of the cMO. Proceeding from commercially available 6-nitropiperonal, the 6-nitropiperonyloxymethyl (NPOM) chloromethyl ether caging precursor was synthesized in three steps, which was subsequently reacted with the 3-N on the thymine-MO subunit to provide an NPOM-cMO monomer. After activation via a phosphorodiamidate on the 5′-OH group, the caged monomers were readily incorporated into an MO following standard MO polymerization conditions [12]. With the caging group interfering with the Watson–Crick hydrogen bonding between the MO and target mRNA, the nucleobase-cMO is inactive. After exposure to UV irradiation, which cleaves the caging groups, the MO regains its activity and silences mRNA translation. This approach allows the control over MO activity with the smallest structural change among all light-activated MO approaches. More importantly, after light irradiation, the active MO is the only strand generated, avoiding potential problems that have previously been discussed in the context of photochemically releasing MO oligomers, as in the two previous approaches [3, 8].

In the nucleobase cMO approach, a 25-base oligomer requires four caging groups, distributed evenly if possible, to sufficiently block MO binding. This means that longer irradiation time or higher UV light intensity may be needed to remove multiple caging groups compared to one. The complete removal of all four caging groups is observed *in vitro*, and the efficiency of this approach *in vivo* was demonstrated by the regulation of the *chordin* gene, which is expressed in early embryonic development and triggers dorsal–ventral axis formation (Box 23.1) and brain formation [12]. Embryos injected with a nucleobase *chordin* cMO were either kept in the dark or exposed to UV light (Figure 23.4), and as shown in Figure 23.4d, irradiation resulted in 90% *chordin* mutant phenotype, indicating successful light-induced MO activation and gene silencing. By activating the nucleobase cMO at different time points, it was found that irradiation before 10 hpf mostly led to a severe *chordin* mutant phenotype, while this phenotype was not observed at later irradiations. The importance of *chordin* activity before

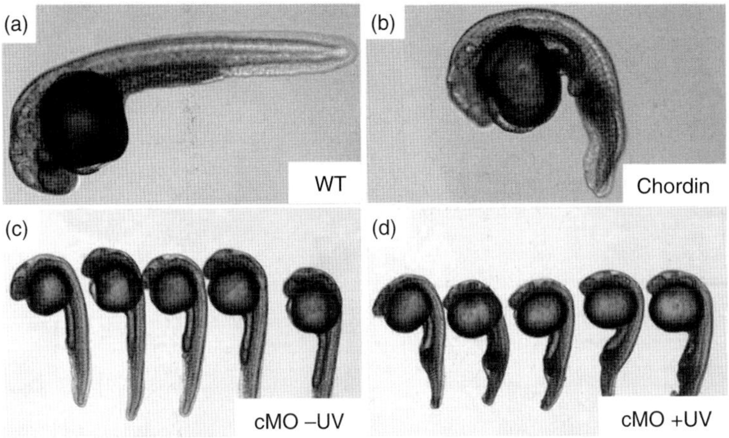

Figure 23.4 *Gene regulation with a nucleobase cMO in zebrafish embryos.* MOs were injected at the one- to four-cell stage and phenotypes were assessed at 24–28 hpf. (a) Wild-type (WT) embryo showing a normal phenotype. (b) Embryos injected with a *chordin* MO show a distinct phenotype, including an abnormal tail fin and a reduced head. (c) Embryos injected with the nucleobase cMO and shielded from light show a normal phenotype. (d) Embryos injected with the nucleobase cMO and exposed to 365 nm light after injection show the *chordin* phenotype. (Adapted with permission from [12]. Copyright (2010) American Chemical Society.)

10 hpf is consistent with known *chordin* function and proves the applicability of nucleobase cMO reagents in temporal gene control [13].

23.3.4
Cyclic-Caged MO

The use of cyclic-cMOs is a recently developed approach to regulate MO activity with light. A photocleavable linker with either a DMNB [14] or a 5-aminomethyl-2-nitrobenzyl (AMNB) group [15] is used to connect the two ends of a linear MO forming a cyclic MO. Owing to the loss of structural flexibility and the induced curvature, the cyclic MO is unable to bind to the target mRNA, enabling conformational gating that is triggered by UV exposure. Upon light irradiation, the linker is cleaved and the MO regains its linear conformation, which enables MO:RNA hybridization and induces transcriptional silencing. The synthesis of the DMNB-containing cyclic cMO takes advantage of commercially available 5′-amine and 3′-disulfide MOs. The synthesis commenced with the coupling of the 5′-amine group on the MO to the succinimidyl ester at one end of the DMNB linker, which was followed by the reduction of the disulfide bond and reaction of the released free thiol group at the 3′-terminus of the MO with the chloroacetamide functionality at the other end of the DMNB linker, thereby delivering the light-activatable cyclic MO. This approach was tested through *ntla* and *pancreas transcription*

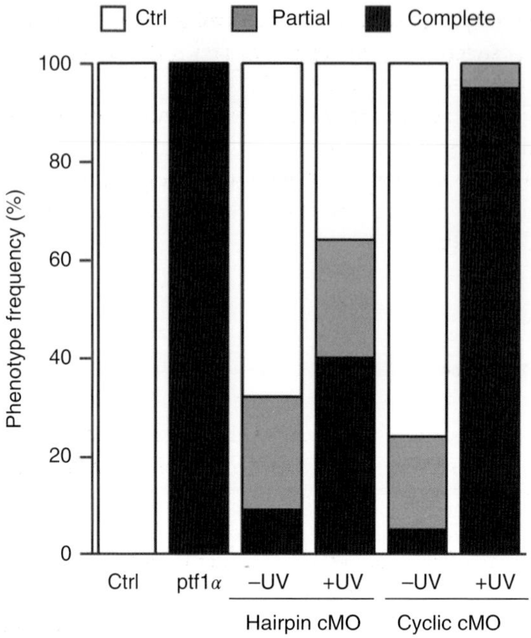

Figure 23.5 *Comparison of the regulation of ptf1α function with a hairpin cMO and a cyclic cMO.* Embryos were injected with either cyclic *ptf1α* cMO or hairpin *ptf1α* cMO at the one-cell stage, irradiated at 3 hpf, and fluorescence were measured at 3 dpf. Ctrl: control without MO injection; partial: some fluorescence observed, indicating partial gene silencing; complete: no fluorescence observed, indicating complete gene silencing. (Adapted with permission from [14]. Copyright © 2012 WILEY-VCH Verlag GmbH & co. KgaA, Weinheim.)

factor 1 alpha (*ptf1α*) gene silencing [14]. In the regulation of *ptf1α,* its expression level was evaluated by measuring the fluorescence intensity of an enhanced green fluorescent protein (EGFP) reporter under control of a *ptf1α* promoter in transgenic zebrafish. As shown in Figure 23.5, cyclic *ptf1α* cMO showed lower background activity than the hairpin *ptf1α* cMO before UV irradiation, possibly due to the more stable linker design and the absence of an equilibrium between the hairpin and the linear MO forms. More importantly, after light irradiation, complete loss of gene function was observed in most embryos injected with cyclic *ptf1α* cMO, which was only obtained in less than half of the hairpin-cMO-injected embryos. This indicates that cyclic cMOs can be more efficient than hairpin cMOs at triggering gene silencing with light.

The AMNB cyclic cMO was synthesized from MOs with the 5′-terminus immobilized on a solid support and the 3′-terminus functionalized with an Fmoc-protected amine. After Fmoc cleavage, the 3′-amine was treated with a disuccinimidyl carbonate-activated AMNB alcohol, followed by two additional steps to install a carboxylic acid function on the linker. Then, the MO was cleaved from the resin and cyclized by coupling a resulting free 5′-amine to the carboxylic acid. The AMNB cyclic cMOs successfully silenced *ntla* and *β-catenin-2* (*cat2*) expression in zebrafish embryos after UV exposure [15].

The cyclic cMO strategy combines several advantages of these three approaches: (i) the MO can be purchased readily; (ii) one caging group linker is sufficient to induce deactivation of the MO; (iii) no byproduct oligomers are generated after photolysis; and (iv) different caging groups, even coumarins which require a low pK_a of the caged target, are tolerated.

23.4
Chemical Biological Research/Evaluation

Having discussed the different approaches that have been developed for the generation of light-activated MOs, an example demonstrating the applicability of these approaches in the assignment of gene function and the investigation of biological problems is summarized here.

Transcription factor Ntla is encoded by the *ntla* gene and promotes the differentiation of axial mesoderm cells into the notochord (Box 23.1) [16]. In zebrafish lacking *ntla* function, these progenitor cells are reprogrammed to become part of the medial floor plate (Box 23.1). Because of the distinct phenotype, *ntla* is often used as a target to test the efficacy of light-activated MOs. Moreover, through temporal activation of a hairpin *ntla* cMO, the Chen Lab demonstrated that the silencing of *ntla* at different time points using light-activated cMOs led to the manifestation of different phenotypes (Figure 23.6a–c). When *ntla* was silenced at an early stage of embryonic development (6 hpf), the embryos failed to assemble a notochord (Figure 23.6b) and lacked the posterior mesoderm, which is consistent with previous reports [17]. Embryos irradiated at later stages (12 hpf) developed an abnormal notochord (Figure 23.6c), indicating that *ntla* is needed for proper assembly of the notochord after the axial mesoderm cells have been committed to notochord cell fates. Thus, further light-activation experiments were designed by Chen and colleagues to explore the role of *ntla* in notochord development [18].

A caged fluorescein-conjugated dextran was co-injected with a hairpin-caged *ntla* MO (discussed in Section 23.3.1) in order to fluorescently mark cells where the MO has been activated after UV exposure. This enabled subsequent fluorescence-activated cell sorting (FACS) to separate irradiated cells from nonirradiated cells in the embryo. Hairpin *ntla* cMOs were spatiotemporally activated within the gray circle at 6 hpf (Figure 23.6d) or 12 hpf (Figure 23.6e), followed by FACS 3–4 h after irradiation. Through transcriptional profiling, it was found that in irradiated tissue 87 genes are downregulated by *ntla* silencing at 6 hpf and an additional 12 genes are downregulated by gene knockdown at 12 hpf. Some of these genes are expressed independently from Ntla in tissues other than the axial mesoderm; thus, it is impossible to differentiate the effect that Ntla has on their expression levels in a whole-embryo analysis, thereby requiring the spatial control of *ntla* silencing and analysis that is enabled by light

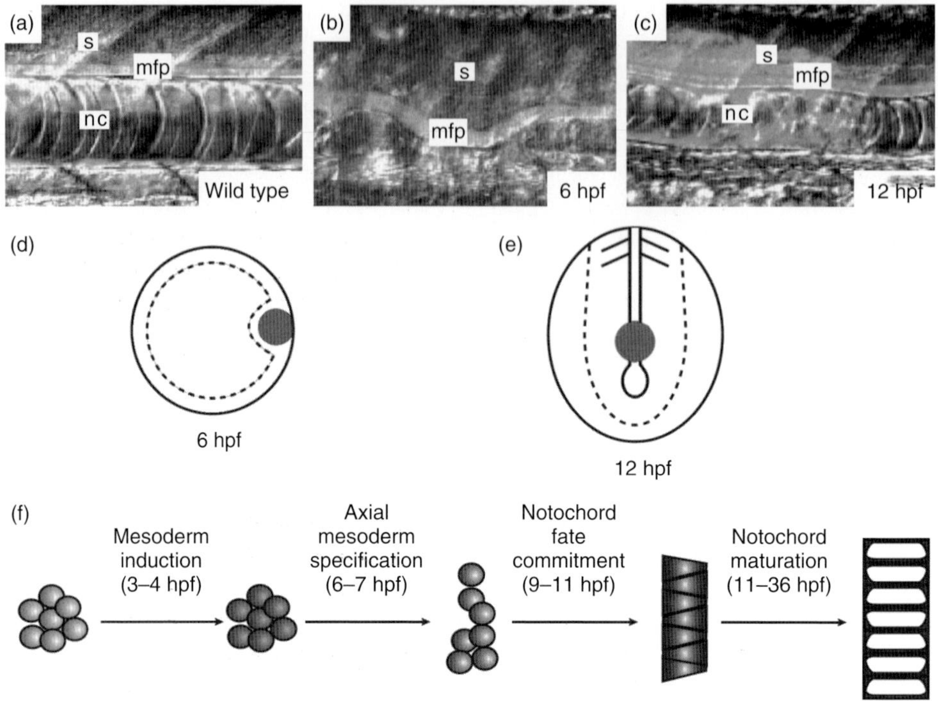

Figure 23.6 *Study of ntla gene function in notochord development with hairpin cMOs.* (a–e) Spatiotemporal control over *ntla* expression. Embryos were injected with hairpin *ntla* cMOs and irradiated at 6 or 12 hpf. (a) Nonirradiated embryos served as a control. (b) Embryos irradiated at 6 hpf lacked the notochord. (c) Embryo irradiated at 12 hpf developed an abnormal notochord. (d) Embryos were irradiated within the gray circle at 6 hpf. (e) Embryos were irradiated within the gray circle at 12 hpf. (f) Model of *ntla*-dependent notochord development. nc: notochord, s: somite (Box 23.1), and mfp: medial floor plate. (Adapted with permission from Macmillan Publishers Ltd: Nature Chemical Biology [18], Copyright (2013).)

activation of cMOs. Several genes in the second subset are known to promote vacuole formation and notochord maturation. Thus, by regulating these genes, *ntla* triggers notochord development after cell fate has been assigned.

Furthermore, the expression of the *flh* gene is known to be Ntla-dependent, but the detailed mechanism of this dependence remains unclear. Two hypotheses regarding the relationship between *flh* and *ntla* have been discussed in the literature: (i) *ntla* directly regulates *flh* expression because an Ntla-binding site is found upstream of *flh* or (ii) *ntla* promotes the formation of notochord progenitors that express *flh* at later cell stages and thus *ntla* indirectly affects *flh* expression. In the previously mentioned transcriptional analysis, *flh* was not among the 99 Ntla-dependent genes. However, light activation of the hairpin *ntla* cMO at 4 hpf downregulated *flh* expression 6 h later, while activation of *ntla* at later stages (5 and 6 hpf) only provided partial or no downregulation of *flh*. Considering that Ntla is depleted in 2 h within the embryo, it is unlikely that *ntla* directly regulates *flh,* but rather that *ntla* promotes cell convergence and/or specification of notochord progenitors and therefore only indirectly affects *flh* expression. Similar results were observed for *fibroblast growth factor 8a* (*fgf8a*), which further supported the indirect regulatory hypothesis. On the basis of these results, an *ntla*-dependent notochord development model was proposed. As shown in Figure 23.6f, *ntla* is first transcribed at 3–4 hpf and promotes the commitment of the axial mesoderm to notochord cell fates by 11 hpf. During notochord maturation (after 11 hpf), *ntla* regulates a set of genes required for proper notochord morphology. Only with the precise spatiotemporal control provided by light-activated MOs was it possible to dissect the multifunctional roles of *ntla* at the cellular and molecular levels.

23.5 Conclusion

Four different types of photoregulated MOs have been developed and used to spatiotemporally trigger MO activity, thereby silencing target mRNAs with spatial and temporal resolution: hairpin cMOs, sense cMOs, nucleobase cMOs, and cyclic cMOs. These approaches allow the study of gene functions at later embryonal stages, which would otherwise be inaccessible as they would be obscured by phenotypes generated at earlier stages through constitutive MO activity. Literature reports have shown that all four approaches are useful tools to study gene function in zebrafish embryos. The programmability to target any sequence and thus any gene of interest gives photoactivated MOs great potential in the study of embryonic development and the detailed mechanisms behind other cellular phenomena.

Acknowledgment

This work was supported in part by the National Institutes of Health (R01GM079114). We thank Dr. James Chen (Stanford University) for helpful discussions.

References

1. Kurreck, J. (2003) Antisense technologies. Improvement through novel chemical modifications. *Eur. J. Biochem.*, **270**, 1628–1644.
2. Eckstein, F. (2000) Phosphorothioate oligodeoxynucleotides: what is their origin and what is unique about them? *Antisense Nucleic Acid Drug Dev.*, **10**, 117–121.
3. Shestopalov, I.A. and Chen, J.K. (2010) Oligonucleotide-based tools for studying zebrafish development. *Zebrafish*, **7**, 31–40.
4. Bill, B.R., Petzold, A.M., Clark, K.J., Schimmenti, L.A., and Ekker, S.C. (2009) A primer for morpholino use in zebrafish. *Zebrafish*, **6**, 69–77.

5. Shestopalov, I.A., Sinha, S., and Chen, J.K. (2007) Light-controlled gene silencing in zebrafish embryos. *Nat. Chem. Biol.*, **3**, 650–651.
6. Ouyang, X.H., Shestopalov, I.A., Sinha, S., Zheng, G.H., Pitt, C.L.W., Li, W.H., Olson, A.J., and Chen, J.K. (2009) Versatile synthesis and rational design of caged morpholinos. *J. Am. Chem. Soc.*, **131**, 13255–13269.
7. Tomasini, A.J., Schuler, A.D., Zebala, J.A., and Mayer, A.N. (2009) PhotoMorphs (TM): a novel light-activated reagent for controlling gene expression in zebrafish. *Genesis*, **47**, 736–743.
8. Tallafuss, A., Gibson, D., Morcos, P., Li, Y., Seredick, S., Eisen, J., and Washbourne, P. (2012) Turning gene function ON and OFF using sense and antisense photo-morpholinos in zebrafish. *Development*, **139**, 1691–1699.
9. Saxena, A., Peng, B.N., and Bronner, M.E. (2013) Sox10-dependent neural crest origin of olfactory microvillous neurons in zebrafish. *Elife*, **2**, e00336.
10. Hong, C.S. and Saint-Jeannet, J.P. (2005) Sox proteins and neural crest development. *Semin. Cell Dev. Biol.*, **16**, 694–703.
11. Eisenhoffer, G.T., Loftus, P.D., Yoshigi, M., Otsuna, H., Chien, C.B., Morcos, P.A., and Rosenblatt, J. (2012) Crowding induces live cell extrusion to maintain homeostatic cell numbers in epithelia. *Nature*, **484**, 546–549.
12. Deiters, A., Garner, R., Lusic, H., Govan, J., Dush, M., Nascone-Yoder, N., and Yoder, J. (2010) Photocaged morpholino oligomers for the light-regulation of gene function in zebrafish and xenopus embryos. *J. Am. Chem. Soc.*, **132** (44), 15644–15650.
13. Crotwell, P.L. and Mabee, P.M. (2007) Gene expression patterns underlying proximal-distal skeletal segmentation in late-stage zebrafish, Danio rerio. *Dev. Dyn.*, **236**, 3111–3128.
14. Yamazoe, S., Shestopalov, I.A., Provost, E., Leach, S.D., and Chen, J.K. (2012) Cyclic caged morpholinos: conformationally gated probes of embryonic gene function. *Angew. Chem. Int. Ed.*, **51**, 6908–6911.
15. Wang, Y., Wu, L., Wang, P., Lv, C., Yang, Z., and Tang, X. (2012) Manipulation of gene expression in zebrafish using caged circular morpholino oligomers. *Nucleic Acids Res.*, **40** (21), 11155–11162.
16. Amacher, S.L., Draper, B.W., Summers, B.R., and Kimmel, C.B. (2002) The zebrafish T-box genes no tail and spadetail are required for development of trunk and tail mesoderm and medial floor plate. *Development*, **129**, 3311–3323.
17. Halpern, M.E., Hatta, K., Amacher, S.L., Talbot, W.S., Yan, Y.L., Thisse, B., Thisse, C., Postlethwait, J.H., and Kimmel, C.B. (1997) Genetic interactions in zebrafish midline development. *Dev. Biol.*, **187**, 154–170.
18. Shestopalov, I.A., Pitt, C.L.W., and Chen, J.K. (2012) Spatiotemporal resolution of the Ntla transcriptome in axial mesoderm development. *Nat. Chem. Biol.*, **8**, 270–276.

24
Life Cell Imaging of mRNA Using PNA FIT Probes

Andrea Knoll, Susann Kummer, Felix Hövelmann, Andreas Herrmann, and Oliver Seitz

24.1
Introduction

Our knowledge about the function of ribonucleic acid (RNA) and RNA-regulated processes has increased enormously during the past years. RNA plays an important role not only in gene regulation and expression but it is also involved in cell signaling or biochemical catalysis within a cell. All these functions offer novel opportunities for the development of gene-expression-based diagnosis as well as therapies of diseases.

Methods to study the biosynthesis of RNA in living cells, their intracellular transport, subcellular localization, and degradation are of great interest for understanding cellular networks and their malfunction [1–3] during diseases. Here, we describe the live cell RNA imaging with peptide nucleic acid-based FIT forced intercalation probes that enabled a simultaneous localization of two viral messenger ribonucleic acid (mRNA) molecules.

24.2
The Biological Problem

In a globalized world, infectious diseases of viral origin such as influenza are a problem, which easily can turn into the threat of pandemia. Therapeutic options require insights into the replication cycle of viruses in host cells. An early consequence of influenza A virus infection is the formation of viral mRNA. An influenza A virion contains eight segmented single-stranded RNA molecules that encode for 11 viral proteins. After endocytotic entry of viruses into the host cell, the segmented RNA genome is released from the virus envelope and transported into the nucleus. Here, viral mRNA molecules are synthesized, which are required for viral protein synthesis upon transport to the cytosol. The newly synthesized viral proteins fulfill various functions during the multistep virus assembly and budding from the plasma membrane. The question arises whether the coordination of the distinct phases of viral replication requires a temporal and spatial regulation of mRNA synthesis and localization.

Concepts and Case Studies in Chemical Biology, First Edition. Edited by Herbert Waldmann and Petra Janning.

To address the various aspects of this question appropriately, imaging at the single host cell is required. For this purpose, we developed fluorogenic hybridization probes that enable imaging of viral mRNA in living infected cells. In order to be useful for imaging, the probes should recognize the target, that is, a specific sequence of the viral mRNA of interest with high specificity and provide emission signals at high signal-to-background ratios. The hybridization probes have to avoid false-positive signals upon target unrelated processes such as binding by RNA/DNA (deoxyribonucleic acid)-binding proteins or nuclease digestion. Further, induction of RNase-H-mediated cleavage of the targeted mRNA complexed with the fluorogenic probe should not occur. To meet these demands, we developed peptide nucleic acid (PNA)-based FIT probes that contain a single hybridization-responsive cyanine dye.

24.2.1 Selection of Biological Targets

For the biological validation of the method, MDCK cells (Madin-Darby canine kidney) were cultivated and infected with influenza A/PR/8 virus (H1N1). The replication cycle includes, among other processes, the RNA-dependent replication and transcription of the viral RNA genome as well as the ribosomal translation into viral proteins such as neuraminidase (NA) and matrix protein M1. These proteins execute distinct functions. The NA protein is required to cleave sialic acid residues from host glycostructures at the budding zone [4]. By comparison, the matrix protein M1 connecting the viral ribonucleoprotein (vRNP) complexes with the viral envelope has upstream and downstream functions. M1 inhibits viral transcription at the late stage of infection and is involved in nuclear export of vRNP molecules but also plays an important role in virus assembly and the budding process [5–9]. Given the difference in function and requirement during the time course of infection, NA and M1 mRNAs are interesting targets for the simultaneous imaging.

The selection of the mRNA segments (NA: nt 625–640; M1: nt 526–550) targeted by the imaging probes was based on previous work, which demonstrated the accessibility and uniqueness within the cellular and viral transcriptome [10].

24.3 The Chemical Approach

24.3.1 Design and Synthesis of PNA FIT Probes

The PNA FIT probes contain a single cyanine dye that belongs to the thiazole orange (TO) family of intercalator dyes. The dye replaces a canonical nucleobase and, thereby, serves as a fluorescent base surrogate. PNA FIT probes respond to changes of the viscosity around the environmentally sensitive TO dye.

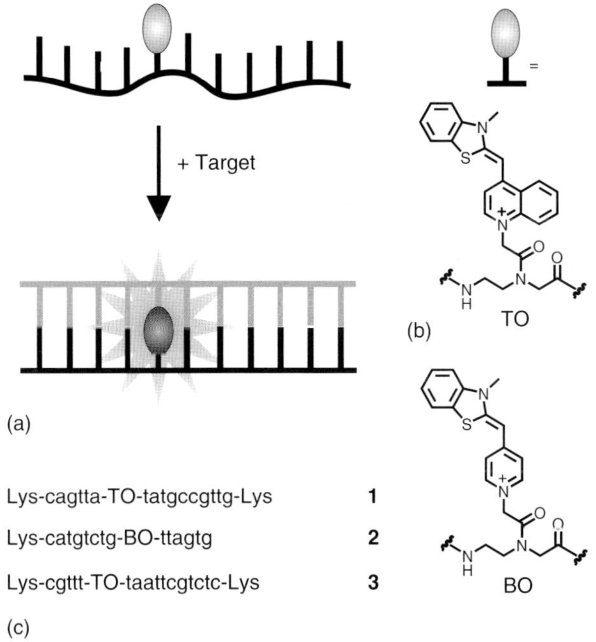

Figure 24.1 (a) Homogeneous detection of nucleic acids (target) with PNA FIT probes. Fluorescence of the single-stranded probe is low, but strongly enhances when the dye intercalates between the helically stacked nucleobases of the probe–target complex. (b) Structure of base surrogates thiazole orange (TO) and pyridinium benzothiazole (BO). (c) FIT probes used for investigation.

Strong enhancement of fluorescent emission is obtained when the dye intercalates between the helically stacked nucleobases of the probe–target complex formed (Figure 24.1a), [11, 12]. By contrast, the fluorescence remains low in the single-stranded state or upon exposure to a complex mixture of biomolecules in cellular media because intercalation of TO in a FIT probe with nucleic acids in *trans* is sterically hindered and requires hybridization of the PNA part. PNA is not subject to nuclease cleavage and cannot induce RNase-H degradation of double-stranded RNA molecules. Therefore, the probe and target remain stable. The high binding affinity of PNA probes for complementary RNA facilitates invasion into folded target segments. PNA FIT probes can be applied at a wide temperature range leading to numerous applications of the same probe such as qualitative live-cell RNA imaging (at 37 °C) as well as quantitative real-time polymerase chain reaction (qPCR) measurements (at 60 °C).

The simultaneous imaging of two viral mRNA targets requires two differently colored probes. This calls for two spectrally resolvable dyes. Among the various members of the TO family of dyes, we selected the quinolinium-based TO (maximal fluorescence emission $F_{em}(\text{max}) = 530$ nm) and the pyridinium-based pyridinium benzothiazole (BO) dye ($F_{em}(\text{max}) = 487$ nm). Carboxymethylated versions of both dyes were introduced as base surrogates in FIT probes

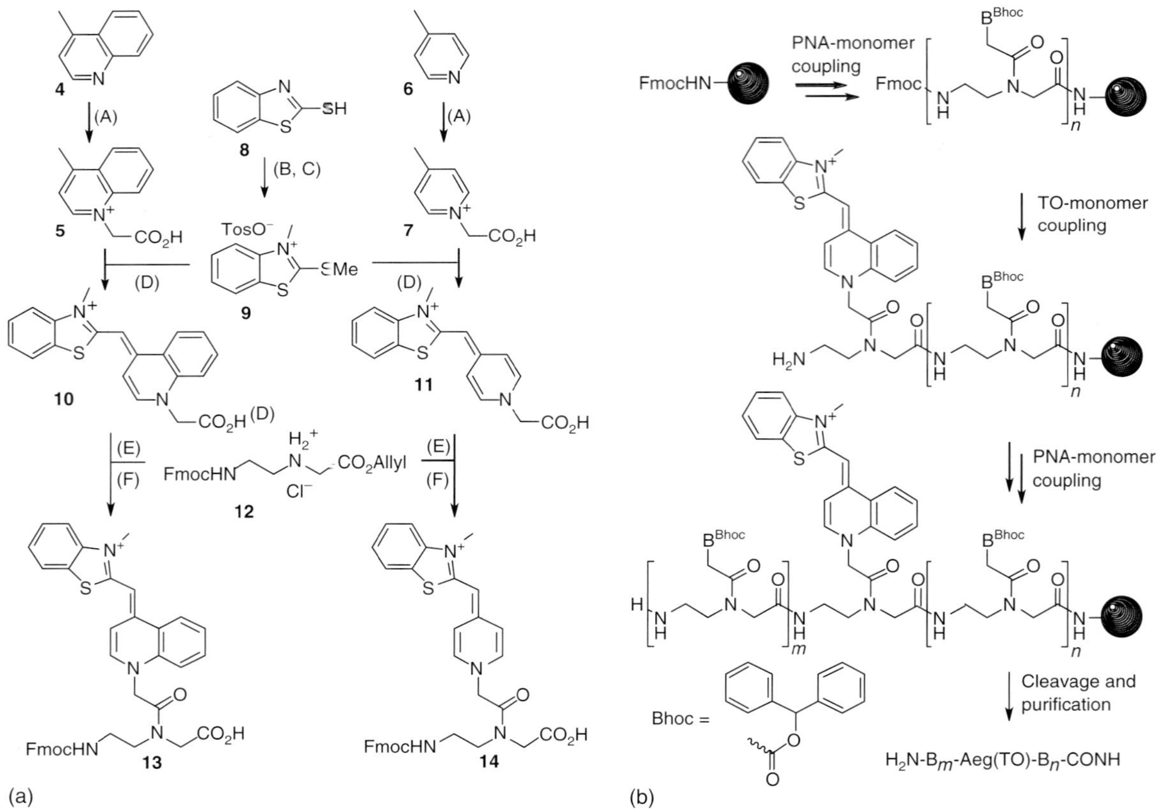

Scheme 24.1 (a) Synthesis of asymmetric cyanine dyes; (A) bromo acetic acid, EtOAc; (B) MeI, K_2CO_3, DMF; (C) TosOMe, neat, 130 °C; (D) NEt_3, CH_2Cl_2; (E) PyBOP, PPTS, NMM, DMF; and (F) MeNHPh, $Pd[PPh_3]_4$, tetrahydrofuran (THF). (b) Solid phase synthesis of TO-modified PNA.

(Figure 24.1b). Owing to the peptide backbone, the PNA oligomers can readily be prepared by automated solid-phase peptide synthesis by using commercially available PNA building blocks (Scheme 24.1a).

The synthesis of the chromophores **10** and **11** was achieved by alkylation of the corresponding *N*-heterocycles (4-methyl quinoline **4** for TO and γ-picoline **6** for BO) with bromoacetic acid. Methylthiobenzothiazolium tosylate **9** was synthesized by double methylation of 2-mercapto benzothiazole **8** and further reacted with **5** or **7**, respectively, according to a method developed by Brooker and coworkers [13]. The resulting carboxymethylated chromophores **10** and **11** were coupled with the aminoethyl-glycine precursor **12** by using benzotriazole-1-yl-oxy-tris-pyrrolidino phosphonium hexafluorophosphate (PyBOP) and *N*-methyl morpholine (NMM) in *N,N*-dimethylformamide (DMF) [14, 15]. The addition of pyridinium *para*-toluene sulfonate (PPTS) was required for solubilizing the chromophores. The final palladium-catalyzed cleavage of the allylester provided TO- and BO-modified building blocks **13** and **14**, respectively, which are isosteric to commercially available PNA monomers and suitable for automated solid-phase synthesis [14] (Scheme 24.1b).

On Fmoc (fluorenylmethoxycarbonyl)-glycin preloaded resin, monomer coupling cycles are carried out consisting of deprotection (piperidine/DMF (1 : 4), coupling (Fmoc-B(Bhoc)-OH (Bhoc = benzhydryloxycarbonyl), NMM, HCTU (2-(6-chloro-1-*H*-benzotriazole-1-yl)-1,1,3,3-tetramethylaminium hexafluorophosphate), NMP (*N*-methylpyrrolidin-2-one)), and capping (acetic anhydride/2,6-dimethyl pyridine, DMF). For coupling of the Fmoc-Aeg(TO)-OH (Aeg, aminoethyl glycine) **13** or Fmoc-Aeg(BO)-OH **14**, PPTS was added to the coupling step. The final probes were obtained by base deprotection and cleavage from the solid support, followed by HPLC purification. For application in cells, polyethylene glycol chains were attached to the probes via a lysine moiety at the N-terminus to increase the water solubility of the probes and prevent segregation [15, 16].

24.4 Chemical Biological Research/Validation

24.4.1 Probe Validation by Fluorescence Measurement

The synthesized probes (Figure 24.1c) are tested in fluorescence measurement in quartz cuvettes by using a fluorescence spectrometer. Fluorescence spectra are recorded before and after hybridization with synthetic target RNA at 37 °C or synthetic target DNA at 60 °C. The latter hybridization experiment emulates the conditions typically used in qPCR analysis. Representative fluorescence spectra of probes providing high fluorescence enhancements (**1** and **2**) are shown in Figure 24.2. The TO probe **1**, which is specific for NA mRNA, furnishes an 11-fold enhancement in fluorescence upon hybridization with the synthetic RNA strand at 37 °C and a 12-fold increase upon hybridization with the DNA target

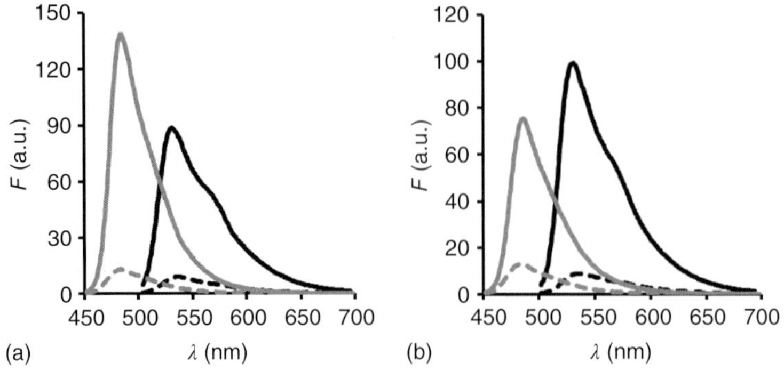

Figure 24.2 *Fluorescence spectra of FIT TO probe **1** (black) and FIT BO probe **2** (gray)* before (dashed line) and after (solid line) addition of perfectly matched (a) RNA at 37 °C or (b) DNA at 60 °C. BO and TO were excited at 440 and 485 nm, respectively.

at 60 °C. The BO-containing probe **2**, which was designed for detection of the mRNA for M1, showed similar properties: a sixfold increase in intensity of BO emission upon hybridization with RNA at 37 °C and an 11-fold intensity increase when DNA was added at 60 °C [15, 16].

24.4.2
Quantitation of Viral mRNA by qPCR

The progression of viral mRNA production during the infection cycle was characterized by means of qPCR measurements. MDCK cells infected with influenza A/PR/8 virus (multiplicity of infection = 100; one cell is virtually infected by 100 virus particles) were harvested at various points post infection. Noninfected cells were collected for control experiments. The total RNA was isolated and purified by using commercially available kits. The optical density (OD) of isolated RNA was measured and aliquots were subjected to *in vitro* transcription. After quantification (by OD measurements), the resulting complementary deoxyribonucleic acid (cDNA) was analyzed in qPCR experiments.

For qPCR, we selected two primers (forward: 5′-GATAATGGAGCAGTGGCT GTATTA-3′, reverse: 5′-GAGGACACAAGAGTCTGAATGTGC-3′) that allowed the amplification of a 101 base-pair long segment (nt 570–671) of the NA mRNA containing the target region of the TO-FIT probe **1**. Another set of primers (forward: 5′-CTGGTATGTGCAACCTGTGAA-3′, reverse: 5′-TCACTCGATCCAGCCATTTG-3′) is used for the amplification of a 155 bp segment (nt 461–615) of the M1 cDNA targeted by the BO probe **2**. As PNA FIT probes become fluorescent upon binding of the complementary strands [17], only the copies of the desired sequences are detected. Figure 24.3 shows amplification curves in which fluorescence measured during the annealing phase is plotted as a function of the number of polymerase chain reaction (PCR) cycles. The increase in fluorescence upon progress of the PCR indicates the presence of NA- (Figure 24.3a) and M1-specific (Figure 24.3b) sequences in infected samples.

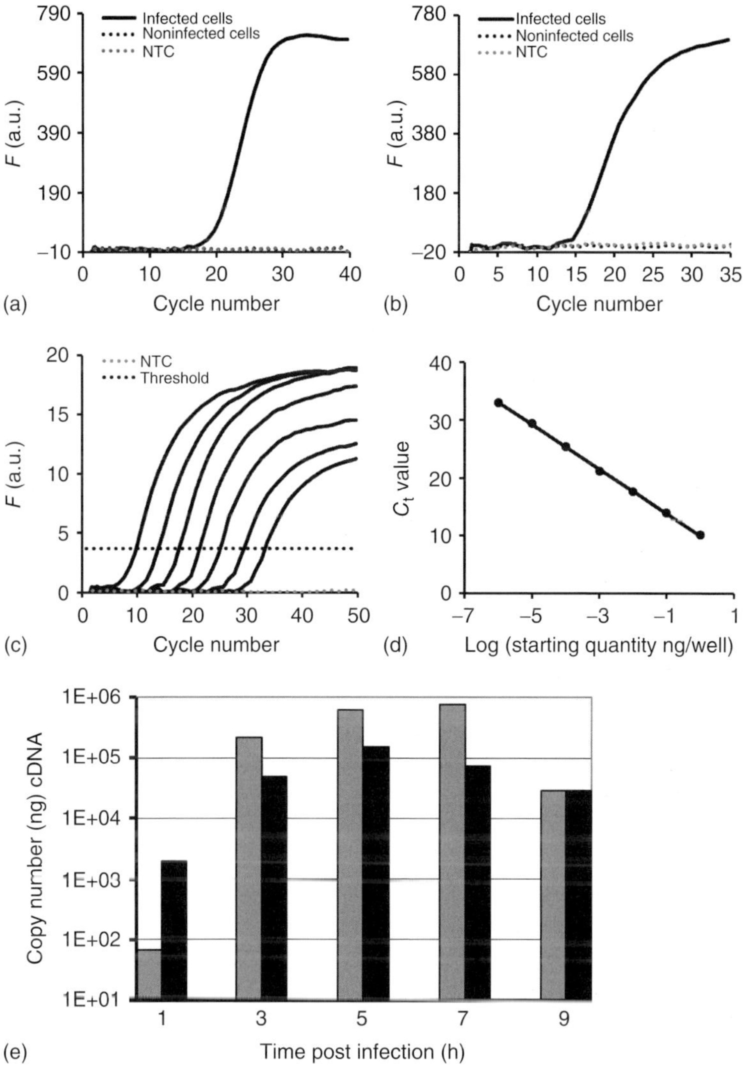

Figure 24.3 *Quantitative PCR analysis.* Amplification curves are obtained by measuring the fluorescence from (a) TO-FIT probe **1** and (b) BO FIT probe **2** in response to amplification of 100 ng cDNA prepared from influenza A/Puerto Rico/8-infected MDCK cells (black, solid), noninfected control cells (black, dashed), or no-template control (NTC, gray). Representative results are shown for a sample prepared 6 h post infection. (c) Amplification curves of a 10-fold dilution series ($1-10^{-6}$ ng) are measured to calculate (d) calibration curves, which allow the estimation of (e) time-dependent expression of NA mRNA (black) and M1 mRNA (gray) as copy numbers/nanogram cDNA starting material. (c,d) Results for a 101-bp DNA target encoding for the neuraminidase. Error bars are not shown. (Adapted with permission from [15]. Copyright © 2012, American Chemical Society.)

The lack of fluorescence increase in measurements carried out in the absence of cDNA (no-template control, NTC) or with cDNA from noninfected MDCK cells attests to the high-sequence specificity of fluorescence signaling by FIT probes.

In preparing for quantitative measurements, calibration curves were measured. For this purpose, PCR is performed by using known amounts of DNA template (Figure 24.3c). The cycle numbers needed to furnish threshold fluorescence (when fluorescence is threefold above the average of NTC fluorescence emission between cycles 3 and 12 and when the sigmoidal is entering a linear phase) are plotted against the logarithm of template quantity (Figure 24.3d). The calibration curves are used to determine the amount of cDNA prepared from cells after different times of infection. It became apparent that maximal levels of NA mRNA (1.5×10^5 copies of NA mRNA per nanogram cDNA used as template in qPCR) were produced 5 h post infection (Figure 24.3e). By comparison, the expression of M1 mRNA reached a maximum (7×10^5 copies of M1 mRNA per nanogram cDNA) 7 h post infection. This corresponds to about 10^4 NA-specific mRNA copies and 10^5 copies of M1 mRNA per infected cell [15, Supporting Information].

24.4.3 Imaging of Viral mRNA in Living Cells

The probes are introduced into living MDCK cells by first permeabilizing the plasma membrane with streptolysine O (SLO, a streptococcal hemolytic exotoxin) in Dulbecco's phosphate buffered saline (DPBS) supplemented with 25 mM 2-[4-(2-hydroxyethyl)piperazin-1-yl]ethanesulfonic acid (HEPES) and 10 mM dithiothreithol (DTT) and then adding an aqueous solution of FIT PNA directly to the culture dish. After 30 min incubation time at 37 °C, the cells were resealed upon addition of fresh Dulbecco's modified Eagle's medium (DMEM) medium. This treatment assures the vitality of the cells. A culture dish with pretreated living cells is mounted in a climate chamber of a confocal laser scanning microscope (CLSM) at 37 °C and the fluorescence image acquisition can be performed.

Before "dual color measurements," cross talk between BO- and TO-containing probes has to be excluded. Therefore, MDCK cells infected with the influenza A virus are stained with the NA-specific TO probe **1**.

The CLSM image measured in the "TO channel" (excitation wavelength (Ex) = 488 nm, emission (Em) = 530–600 nm) reveal a bright fluorescence. Measurements in the "BO channel" (Ex = 440 nm, Em = 470–490 nm) lead to dark images. The complementary behavior is observed when the infected cells are charged with the BO probe **2**. Staining becomes apparent when the "BO channel" is selected, but not when the "TO channel" is used (Figure 24.4). This confirms that probes **1** and **2** allow the independent imaging of NA mRNA and M1 mRNA, respectively. Both probes were added to SLO-permeabilized cells at certain points post infection. The confocal laser scanning microscopic images exposed the differences in the expression pattern (Figure 24.5) [15].

The signal in the TO channel showed that the NA mRNA is detectable 2 h post infection. In this early phase, the NA mRNA predominantly localizes to

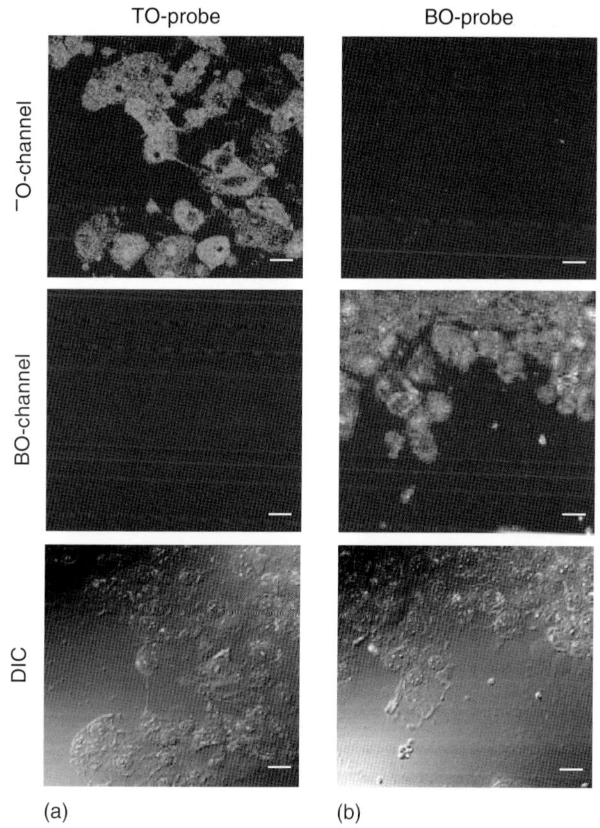

Figure 24.4 *Confocal laser scanning microscope (CLSM) images of living Influenza A/PR/8-infected MDCK cells* stained with (a) NA mRNA specific PNA FIT probe **1** or (b) M1 mRNA specific PNA FIT probe **2**. At 5 h post infection cells were measured in the TO channel (Ex = 488 nm, Em = 530–600 nm) and the BO channel (Ex = 440 nm, Em = 460–490 nm) at 37 °C. White bars correspond to 10 μm. DIC = differential interference contrast. (Adapted with permission from [15]. Copyright © 2012, American Chemical Society.)

the nucleoli, which are clearly visible in the images obtained by differential interference contrast (DIC) microscopy. As the infection proceeds, an increasing TO intensity is measured within the cytosolic region of the cell. Experiments performed with noninfected cells revealed that TO probe **1** does not stain nucleoli *per se*. We infer that the images obtained with TO probe **1** are not perturbed by a localization bias and reflect the localization of available NA mRNA. A different expression pattern is obtained when the BO channel is analyzed. The BO signal appeared 5 h post infection and localized to the nucleoli as well as cytosolic parts as soon as the BO signal exceeded the detection threshold [15].

Several control experiments are required to exclude sources of false-positive signaling. For example, the specificity of the probes can be tested by introducing

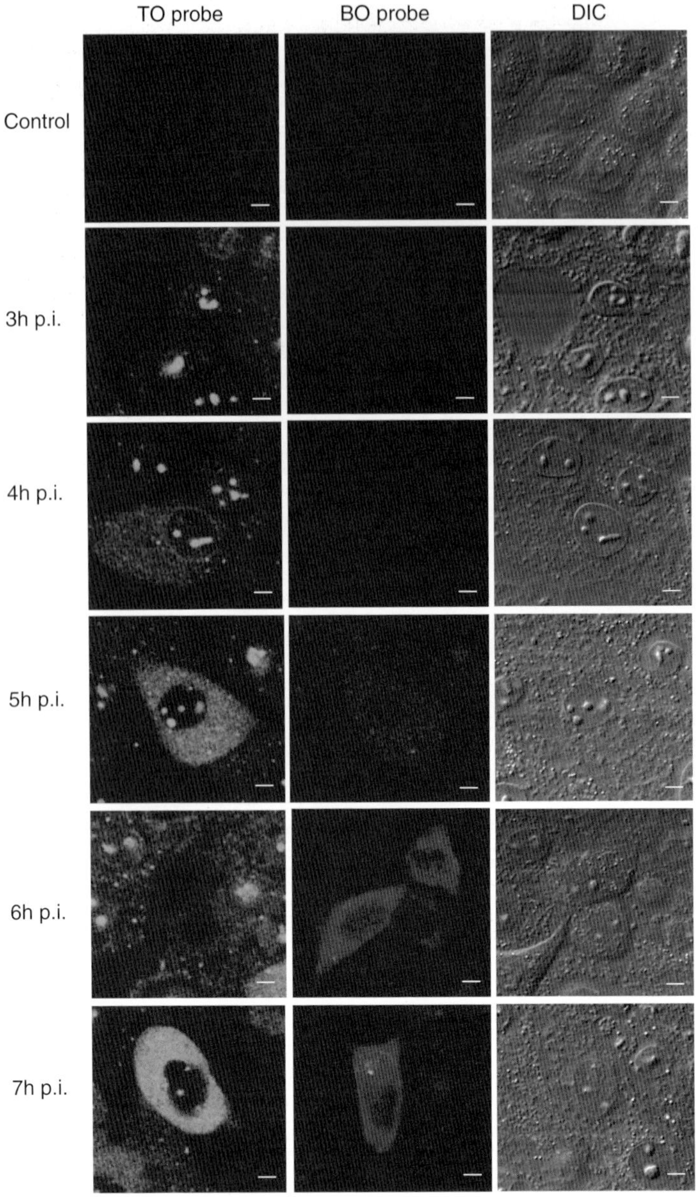

Figure 24.5 *CLSM images of living influenza A-infected MDCK cells* stained simultaneously with NA-specific TO probe **1** and M1-specific BO probe **2** at indicated points post infection and of noninfected (control) MDCK cells. White bars correspond to 10 μm. DIC = differential interference contrast. Measurement conditions, see Figure 24.4. (Adapted with permission from [15]. Copyright © 2012, American Chemical Society.)

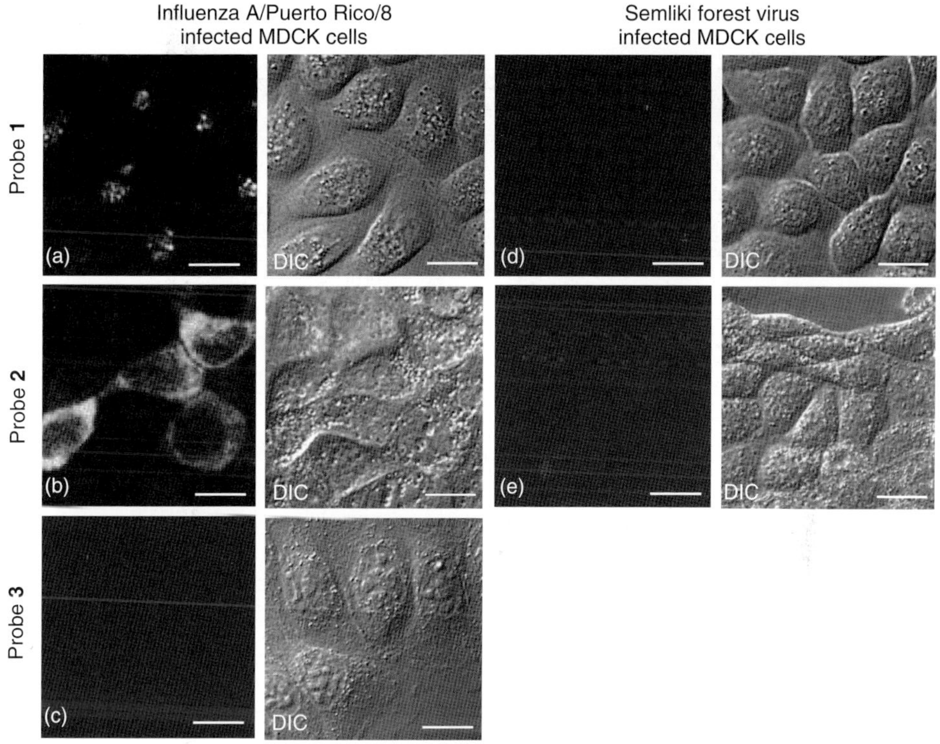

Figure 24.6 *CLSM images of living (a–c) influenza A/PR/8-infected and (d,e) Semliki forest virus-infected MDCK cells* stained with the NA-specific FIT probe **1** (a,d), the M1-specific FIT probe **2** (b,e), or the control probe **3** (c). White bars correspond to 10 μm. DIC = differential interference contrast.

the probes into cells that have been infected by other viruses. Figure 24.6 shows confocal fluorescence images obtained after infection of MDCK cells by the Semliki forest virus (SFV). The viral mRNA molecules produced upon SFV infection bear no resemblance to the sequences targeted by FIT probes **1** and **2**. Consequently, the CLSM images of SFV-infected cells charged with probe **1** or **2** remain dark. An additional control involves the addition of a FIT probe **3** that has no complementarity to the nucleic acid molecules expressed in infected or uninfected cells. Again, attempted staining failed (Figure 24.6). These experiments and the noninfection controls indicate that PNA FIT probes do respond to the mRNA targets but not to other molecules inside a cell [15, 16].

24.5
Conclusion

PNA FIT probes enable studies of gene expression in living cells. The probes show weak fluorescence in the absence of complementary nucleic acid targets. The artificial PNA backbone assures a high biostability. Thus, neither is degradation of the PNA FIT probes by nucleases an issue nor is induction of RNase-H-mediated destruction of target RNA. PNA FIT probes remain dark unless the TO dye is forced to intercalate at a predetermined position upon sequence-specific hybridization. This property prevents false-positive signaling, which occurs when alternative probe technologies such as molecular beacon probes are challenged by DNA-binding proteins.

This account describes experiments in which PNA FIT probes were used to explore the localization pattern of two different viral mRNA molecules expressed upon infection of cells with an H1N1 influenza strain. The study suggests that mRNA encoding for NA passed through nucleoli at an early phase of infection. The "nucleoli-only phase" was absent for mRNA coding for matrix protein M1. This preliminary result confirms the notion that viral gene expression is orchestrated in a spatial and temporal manner. Some mRNA molecules may be needed at later steps of the replication cycle than other molecules. For example, the M1 protein has been described as the master regulator of virus assembly at the budding zone. However, additional experiments are required to study expression of viral genes in more detail. Such studies should involve other mRNA targets as well as color permutations to exclude the possibility of dye-specific bias. We assume that FIT probes will prove useful not only in investigations on mRNA of the influenza virus family (H1N1) but also in other studies aiming for the analysis of live cell RNA dynamics in general.

References

1. Cheung, T.K. and Poon, L.L. (2007) Biology of influenza a virus. *Ann. N.Y. Acad. Sci.*, **1102**, 1–25.
2. Krossoy, B., Hordvik, I., Nilsen, F., Nylund, A., and Endresen, C. (1999) The putative polymerase sequence of infectious salmon anemia virus suggests a new genus within the Orthomyxoviridae. *J. Virol.*, **73**, 2136–2142.
3. Fouchier, R.A.M., Munster, V., Wallensten, A., Bestebroer, T.M., Herfst, S., Smith, D., Rimmelzwaan, G.F., Olsen, B., and Osterhaus, A.D.M.E. (2005) Characterization of a novel influenza A virus hemagglutinin subtype (H16) obtained from black-headed gulls. *J. Virol.*, **79**, 2814–2822.
4. Palese, P., Tobita, K., Ueda, M., and Compans, R.W. (1974) Characterization of temperature sensitive influenza virus mutants defective in neuraminidase. *Virology*, **61**, 397–410.
5. Ruigrok, R.W., Barge, A., Durrer, P., Brunner, J., Ma, K., and Whittaker, G.R. (2000) Membrane interaction of influenza virus M1 protein. *Virology*, **267**, 289–298.
6. Elleman, C.J. and Barclay, W.S. (2004) The M1 matrix protein controls the filamentous phenotype of influenza A virus. *Virology*, **321**, 144–153.
7. Roberts, P.C., Lamb, R.A., and Compans, R.W. (1998) The M1 and M2 proteins of influenza A virus are important determinants in filamentous particle formation. *Virology*, **240**, 127–137.
8. Ye, Z.P., Robinson, D., and Wagner, R.R. (1995) Nucleus-targeting domain of the matrix protein (M1) of influenza virus. *Virology*, **69**, 1964–1970.
9. Martin, K. and Helenius, A. (1991) Role of the influenza virus M1 protein in nuclear export of viral ribonucleoproteins. *Cell*, **67**, 117–130.

10. Wang, W., Cui, Z.Q., Han, H., Zhang, Z.P., Wei, H.P., Zhou, Y.F., Chen, Z., and Zhang, X.E. (2008) Imaging and characterizing influenza A virus mRNA transport in living cells. *Nucleic Acids Res.*, **36**, 4913–4928.
11. Köhler, O., Jarikote, D.V., and Seitz, O. (2005) Forced intercalation probes (FIT Probes): thiazole orange as a fluorescent base in peptide nucleic acids for homogeneous single-nucleotide-polymorphism detection. *ChemBioChem*, **6**, 69–77.
12. Jarikote, D.V., Krebs, N., Tannert, S., Röder, B., and Seitz, O. (2007) Exploring base-pair-specific optical properties of the DNA stain thiazole orange. *Chem. Eur.J.*, **13**, 300–310.
13. Brooker, L.G.S., Keyes, G.H., and Williams, W.W. (1942) Color and constitution. V.1 The absorption of unsymmetrical cyanines. resonance as a basis for a classification of dyes. *J. Am. Chem. Soc.*, **64**, 199–210.
14. Jarikote, D.V., Köhler, O., Socher, E., and Seitz, O. (2005) Divergent and linear solid-phase synthesis of PNA containing thiazole orange as artificial base. *Eur. J. Org. Chem.*, **15**, 3187–3195.
15. Kummer, S., Knoll, A., Socher, E., Bethge, L., Herrmann, A., and Seitz, O. (2012) PNA FIT-probes for the dual color imaging of two viral mRNA targets in influenza H1N1 infected live cells. *Bioconjug. Chem.*, **23**, 2051–2060.
16. Kummer, S., Knoll, A., Socher, E., Bethge, L., Herrmann, A., and Seitz, O. (2011) Fluorescence imaging of influenza H1N1 mRNA in living infected cells using single-chromophore FIT-PNA. *Angew. Chem. Int. Ed.*, **50**, 1931–1934.
17. Socher, E., Jarikote, D.V., Knoll, A., Röglin, L., Burmeister, J., and Seitz, O. (2008) FIT probes: peptide nucleic acid probes with a fluorescent base surrogate enable real-time DNA quantification and single nucleotide polymorphism discovery. *Anal. Biochem.*, **375**, 318–330.

25
Targeting the Transcriptional Hub β-Catenin Using Stapled Peptides

Tom N. Grossmann and Gregory L. Verdine

25.1
Introduction

Inappropriate activation of the Wingless and INT-1 (Wnt) signaling pathway is causally linked to the onset and progression of numerous types of cancer. Owing to the dependence of established tumors on activated Wnt signaling, inhibition of the pathway is considered a promising anticancer strategy. A central hub in Wnt signaling is the protein β-catenin, which regulates pathway activity via involvement in protein–protein interactions (PPIs) with both upstream and downstream signaling components. Interference with these PPIs represents an attractive approach toward suppression of oncogenic Wnt signaling. However, targeting of PPIs is challenging, particularly so when large interaction surfaces are involved. This chapter describes a design-based approach for the development of cell-permeable PPI inhibitors directly targeting β-catenin. The design process described herein relies on a combination of optimization strategies utilizing directed evolution by phage display and synthetic pharmaceuticalization via α-helix stapling. Biochemical, biophysical, and cellular characterization of the stapled peptide inhibitor, plus an X-ray structure of it bound to β-catenin all have provided insights into the molecular basis of Wnt pathway antagonism by this novel agent.

25.2
The Biological Problem

Tumors form in a multistep process whereby wild-type cells evolve into transformed ones. During this process, cells acquire certain biological capabilities that have been defined as the so-called hallmarks of cancer, among which are replicative immortality, sustained proliferation, and increased tissue invasion [1]. Acquisition of these hallmarks is inevitably achieved by activation of certain signaling pathways. Interestingly, several of the most widely usurped pathways are those that are ordinarily active in most cell types only during embryonic

Concepts and Case Studies in Chemical Biology, First Edition. Edited by Herbert Waldmann and Petra Janning.

development, for example, Notch, Hedgehog, and Wnt [2]. Development of pharmacologic inhibitors for these signaling pathways is considered a particularly urgent goal for next-generation therapeutic strategies that target precise cellular and molecular aberrations in cancer. An inhibitor of the Hedgehog pathway protein smoothened was recently approved by the Food and Drug Administration (FDA) [3], and in 2009, our laboratory described the first direct-acting inhibitor of Notch [4]. Before the very recent work described elsewhere and herein [5], agents that selectively counteract the hyperactive Wnt signaling had proved to be elusive indeed.

25.2.1 Canonical Wnt Signaling

The canonical Wnt signal transduction cascade regulates the expression of genes involved in cell survival and proliferation as well as in differentiation. The pathway is regulated via precise control of intracellular levels of the transcriptional hub protein β-catenin [6]. In unstimulated and untransformed cells, β-catenin is recruited into a so-called destruction complex consisting of the proteins axin, adenomatous polyposis coli (APC), glucogen synthase kinase 3β (GSK3β), and casein kinase 1α (CK1α), among others (Figure 25.1). The kinases catalyze or otherwise facilitate phosphorylation of certain key residues in β-catenin, thus leading to ubiquitination by the E3 ligase β-TrCP (transducing repeat-containing protein) and finally proteasomal degradation of β-catenin. The action of this regulatory mechanism results in unstimulated cells maintaining low levels of β-catenin [6]. Activation of Wnt signaling entails engagement of the transmembrane proteins Frizzled and low-density lipoprotein-related receptor (LRP) by diffusible, extracellular Wnt ligand proteins. Ligand-induced dimerization of frizzled and LPR induces the relocalization of the destruction complex to the membrane-embedded Wnt receptor complex via the adaptor protein dishevelled (Dsh), and this in turn leads to inhibition of β-catenin phosphorylation, ubiquitination, and degradation. β-Catenin consequently accumulates in the cytosol and translocates to the nucleus, where it binds directly to transcription factors of the T-cell factor (TCF)/lymphoid enhancer factor (LEF) family, displacing the co-repressor Groucho, and recruiting transcriptional co-activators p300 and CREB-binding protein (CBP) (Figure 25.1) [6]. Transcriptional activation of a large ensemble of genes under the control of TCF/LEF thereby ensues.

25.2.2 Oncogenic Activation of Wnt Signaling

Oncogenic activation of the Wnt pathway can originate from a number of different molecular aberrations in the pathway. Most of these share the common feature of inactivating the destruction complex, leading to high levels of β-catenin. In most cases, this inactivation is caused by mutations in constituents of the destruction complex, such as axin and APC, or by mutations in β-catenin itself

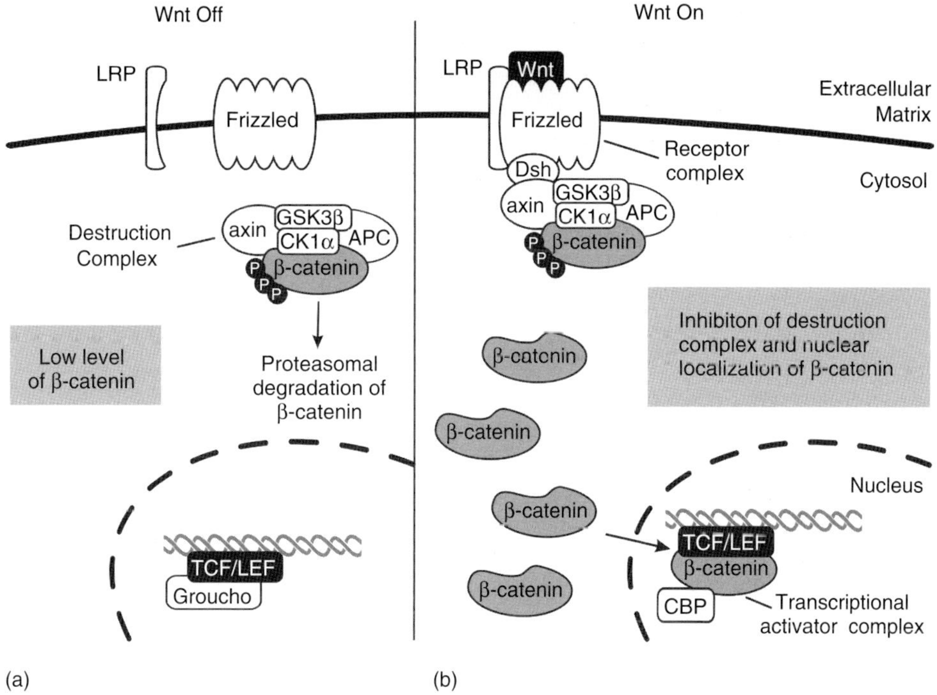

Figure 25.1 *Overview of the canonical Wnt signaling pathway* [6] with its major components in the inactive (a) and active (b) state. The major membrane-bound receptor components are the Wnt ligand receptor frizzled and the low-density lipoprotein-related receptor (LRP); intracellular components are β-catenin, axin, adenomatous polyposis coli (APC), casein kinase 1α (CK1α), glycogen synthase kinase 3β (GSK3β), transcription factors of the T-cell factor (TCF)/lymphoid enhancer factor (LEF) family, and transcriptional co-activators such as CREB-binding protein (CBP).

[6]. Among the various therapeutic options for counteracting constitutive Wnt activation, targeting of pathway components downstream of the destruction complex is considered particularly appealing, as it could diminish the risk of mutational circumvention leading to acquisition of drug resistance [7]. Inhibition of interactions between β-catenin and transcription factors of the TCF/LEF family (Figure 25.2a) has thus emerged as a high-priority target for a next-generation targeted therapy approach. The biological appeal of β-catenin–TCF/LEF interactions is counterbalanced by the chemical intractability of intracellular PPIs as targets, with transcription factors being considered among the most intractable of all PPI targets, owing to their extended interaction interfaces. By way of illustration, Figure 25.2b shows β-catenin (light gray) in complex with TCF4's β-catenin-binding domain (CBD) (black) consisting of an α-helix (site 1) and an extended region (site 2). Very few small molecules have been reported to inhibit the β-catenin–TCF interaction *in vitro* (see Figure 25.2c, for examples), and these have shown a lack of Wnt specificity in cell-based assays and in most cases

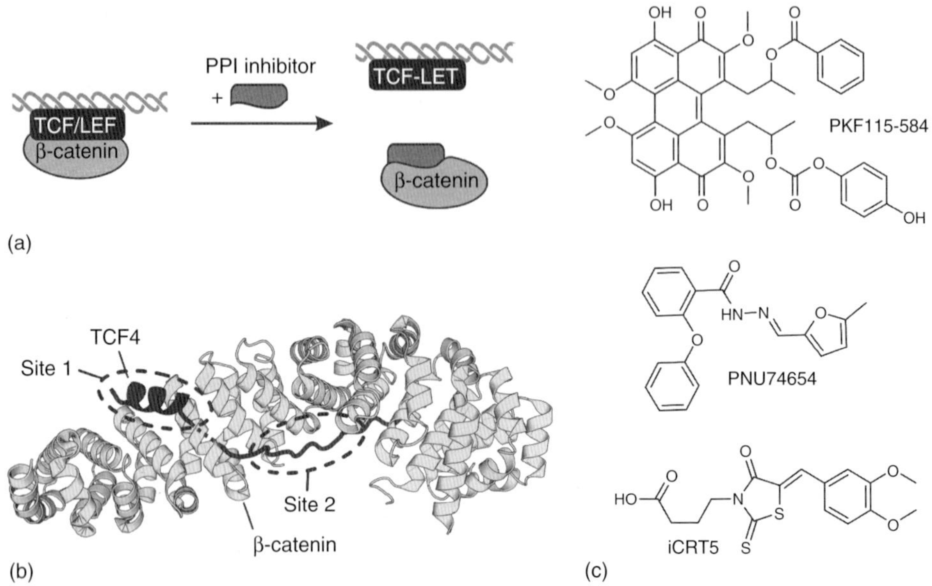

Figure 25.2 (a) Inactivation of canonical Wnt signaling involving inhibition of the β-catenin–TCF/LEF interaction via a direct targeting of β-catenin. (b) Crystal structure (PDB: 2GL7) of β-catenin's armadillo repeat (gray) in complex with the β-catenin-binding domain of TCF4, a member of the TCF/LEF family (black) [11]. (c) Examples for small molecules that potentially inhibit Wnt signaling via direct targeting of β-catenin [7]. PDB: protein data base.

have unelucidated mechanisms of action [7–10]. In an effort to overcome these deficiencies, we explored alternative targeting strategies more suitable for the inhibition of PPIs.

25.3 The Chemical Approach: Hydrocarbon Peptide Stapling

Proteins that engage in PPIs do so using well-ordered interaction surfaces having defined secondary and tertiary structure. Notably, crucial interactions between proteins are frequently mediated by α-helices, suggesting that dominant-negative α-helical peptides might prove particularly useful as PPI antagonists [12]. However, short peptide sequences typically show little or no α-helical character when removed from the stabilizing context of their parent protein, and hence they tend to suffer from poor affinity, poor proteolytic stability, and poor cell permeability. Consequently, strategies capable of enforcing α-helical character upon peptides were used for the design of PPI inhibitors. The peptide stapling technology, which involves introduction of an all-hydrocarbon cross-link into the peptide sequence, efficiently increases the helical character of peptides (Figure 25.3a). In this approach, two α-methyl, α-alkenyl amino acids (Figure 25.3b) are

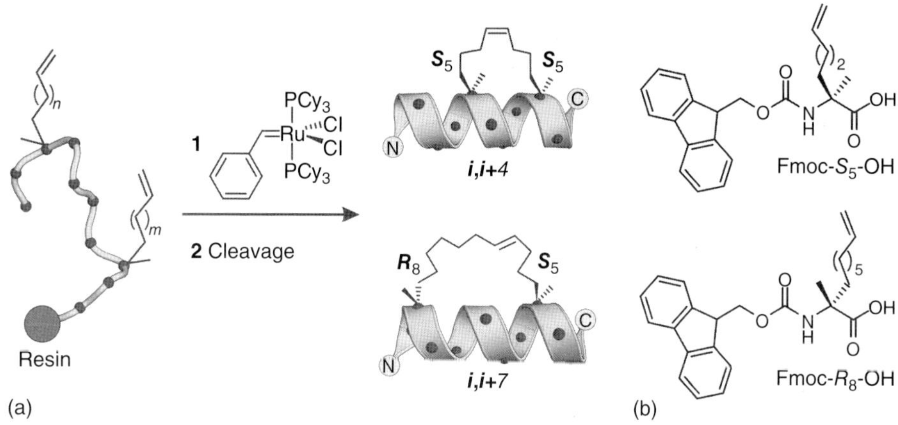

Figure 25.3 *Hydrocarbon stapling approach.* (a) Two olefin-bearing unnatural amino acids are introduced during solid-phase peptide synthesis at two positions of the sequence (i.e., *i*, *i* + 4 and *i*, *i* + 7). Subsequently, ruthenium-mediated ring-closing olefin metathesis is performed on solid support followed by global deprotection and cleavage of the peptide [13]. (b) Olefin-modified Fmoc-protected building blocks used in solid-phase peptide synthesis.

introduced during chain extension via solid-phase peptide synthesis, followed by closure of the macrocyclic bridge using ruthenium-mediated ring-closing olefin metathesis (Figure 25.3a) [13]. The two most successfully applied designs use modified building blocks at amino acid positions i and $i+4$ or $i+7$. The i, $i+4$ arrangement relies on an eight-carbon cross-link connecting the two residues (both with *S*-configuration) spanning one turn of an α-helix. In the i, $i+7$ arrangement, an 11-carbon bridge is used to cross-link the two residues (with *R*-configuration at i and *S*-configuration at $i+7$) spanning two turns of an α-helix. Compared to their unstapled analogs, these hydrocarbon-stapled peptides have been shown to have increased α-helical character, protease resistance as well as cell permeability (e.g., see: [4, 5, 14, 15]). Cell penetration involves an active, endosomal uptake mechanism.

An excellent starting point for the design of stapled peptides is the crystal structure of an α-helical interaction motive in complex with the protein of interest. Considering the increase in affinity that is typically observed upon staple incorporation, peptides that bind the target protein with dissociation constants (K_D) below 100 μM are preferred starting sequences. In order to determine the affinity of short peptides to a protein, fluorescence polarization (FP) assays are usually employed. In these assays, the protein of interest is titrated with a fluorescently labeled ligand peptide. When monitoring FP, the fluorescence of the free peptide is highly depolarized owing to its rapid rotational movements. The large peptide–protein complex, on the other hand, rotates significantly slower, resulting in higher polarization values. On the basis of the concentration-dependent change in FP, the K_D of a peptide–protein complex can be determined.

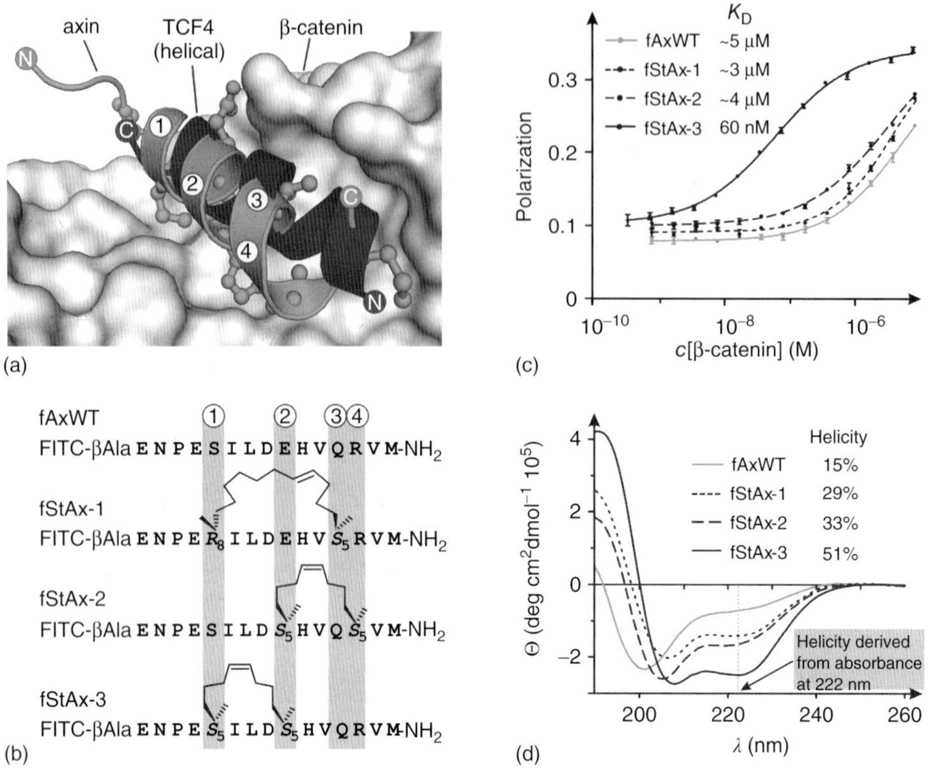

Figure 25.4 (a) Superimposed crystal structures of the β-catenin-binding domains (CBDs) of TCF4 (helical part, black, PDB: 2GL7) [11] and axin (gray, PDB: 1QZ7) [16] in complex with β-catenin (light gray, surface representation). The numbers indicate positions in the axin sequence that were modified for the generation of stapled peptides. (b) Peptides N-terminally labeled with fluorescein isothiocyanate (FITC) including the starting sequence (fAxWT) and three stapled peptides (fStAx-1, -2, and -3). (c) Fluorescence polarization (FP) assay of FITC-labeled peptides binding to full-length β-catenin with corresponding dissociation constants (K_D). (d) Circular dichroism (CD) spectra and derived helicities [5]. ((c) and (d) were adapted from [5].)

Aiming for an inhibition of the β-catenin–TCF/LEF interaction (Figure 25.2b), two potential starting sequences were investigated using FP: the α-helical part of the CBD of TCF4 (black) [11], and the CBD of axin (gray) [16], both binding to site 1 on β-catenin (Figure 25.4a). The α-helical fragment of TCF4's CBD showed very low affinity for β-catenin, and only in combination with the extended region (Figure 25.2b) binding was observed. Extended and unstructured peptide sequences are prone to proteolytic degradation and can hinder cellular uptake. Therefore, hydrocarbon stapling of the TCF peptide was not further pursued. The CBD of axin, on the other hand, which is almost exclusively α-helical, showed moderate affinity ($K_D \sim 5\ \mu M$) and was therefore selected as the starting point for the hydrocarbon stapling approach. The complex structure reveals four amino acids in axin (positions 1–4 in Figure 25.4a) that are not involved in β-catenin recognition. These were selected as potential sites for staple incorporation, and based on the preferred i, $i+4$ and i, $i+7$ arrangement, stapled peptides StAx-1, -2, and -3 (axin-derived stapled peptide) were designed (Figure 25.4b) [5]. Fluorescein-labeled versions of these peptides (fStAx-1, -2, and -3) and of the starting sequence (fAxWT) were synthesized for further evaluation. Using FP, affinities for β-catenin were determined (Figure 25.4c). Compared to fAxWT, both fStAx-1 and -2 did not show improved binding, whereas fStAx-3 exhibited a more than 80-fold increased affinity for β-catenin.

The increased affinity of stapled peptides compared to the unmodified analogs most often originates from their high α-helical character, which is enforced by staple incorporation. In order to determine the degree of α-helicity, all peptides were investigated using circular dichroism (CD) spectroscopy, which can reveal the secondary structure of peptides in solution. A strong negative value at 222 nm, for instance, indicates the presence of an α-helix. This value can also be used to estimate the percentage of α helical content [17]. As expected, the helical character of all three stapled peptides increased when compared to the unmodified starting sequence (Figure 25.4d). Consistent with its highest binding affinity for β-catenin, fStAx-3 also showed the greatest extent of α-helicity (51%).

25.4 The Biological Approach: Phage-Display-Based Optimization

Having identified StAx-3 as the most promising candidate for further optimization, we pursued follow-on experiments aimed at investigating structure–activity relationships for the amino acids that flank the hydrocarbon staple. One focus was to further improve binding to β-catenin. Wishing to access the huge combinatorics provided by directed evolution approaches, we selected phage display technology (Box 25.1) [18] to present an axin-derived peptide library. The library was generated by randomizing different quadruplets of residues within the CBD of axin (Figure 1, left). The resulting phage library was iteratively panned to select for affinity-optimized β-catenin binders. Specifically, applying stringent binding and washing protocols, three selection cycles were performed to provide 32 new sequences. The variations in these selected sequences are summarized in Figure 25.5a, which shows all amino acids that were found at least twice per position. This information was then used in the design of next-generation StAx-3-derived peptides. An analysis of the variations (Figure 25.5a) suggested that incorporation of a hydrophobic residue instead of asparagine N468 and replacement of valine V480 or methionine M481 by tryptophan (W) could result in an increased affinity for β-catenin. Consequently, these changes were incorporated into the next generation of StAx peptides (Figure 25.5b).

Box 25.1 Phage Display

Phage display is a technology that allows the identification of peptide or protein binders for a given target [18]. The starting point is a deoxyribonucleic acid (DNA) library coding for a large number of different amino acid sequences (Figure 1, left: example for axin-derived peptides). These DNA sequences are ligated into a viral vector that is used to generate the phage library. Each phage displays multiple copies of the amino acid sequence encoded in its viral vector. The phage library is then exposed to the immobilized target (e.g., β-catenin, Figure 1) resulting in the binding of a small fraction of phages. After a washing step, the remaining bound phages are eluted for amplification in bacteria. This new phage library is enriched in target binders and can be used in another selection cycle. After each selection cycle, phage plasmids can be collected and sequenced to determine the displayed amino acid sequences. After a certain number of cycles (depending on the nature of the target and the stringency of the washing steps), a sufficient enrichment in affine binders is achieved.

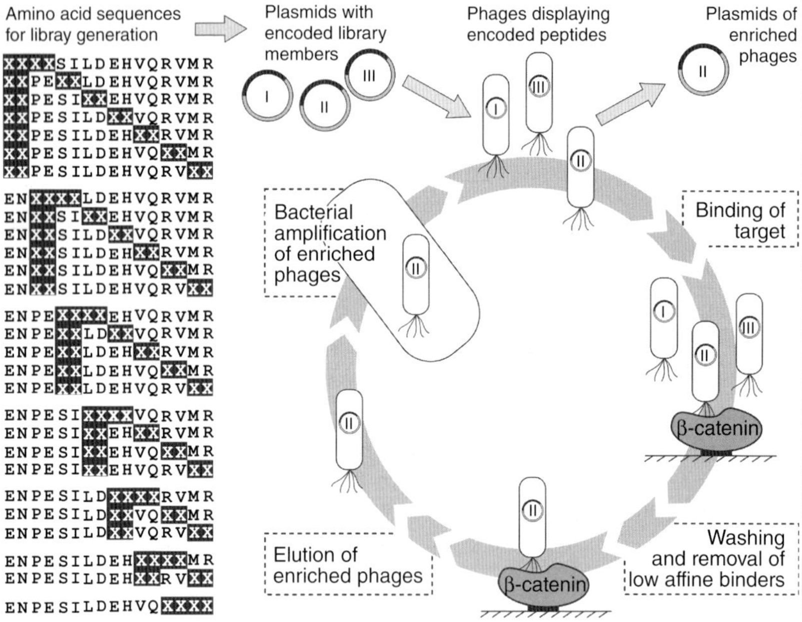

Figure 1 Overview of phage display technology with axin-drived peptide libary and β-catenin as target protein.

In addition to target binding, cellular uptake of StAx peptides is crucial for their biological activity. So far, strict rules for the design of cell-penetrating peptides have not been identified. Nevertheless, it has been reported that removal of negatively charged residues and introduction of arginine (R) generally axin starting sequence with mutations derived from phage display:

			470					475				480			
E	N	P	E	S	I	L	D	E	H	V	Q	R	V	M	R
6R	3L	3S	6Q								5W	3E	11W	12W	6G
2Q	3W	2W									2L	2L		2F	2P
2W	3Y										2M				
	2V														

(a)

fStAx	Peptide sequence	K_D	Charge	Cellular uptake
3	E N P E S_5 I L D S_5 H V Q R V M	60 nM	−3.0	−
31	P E S_5 I L D S_5 H V Q R V M	70 nM	−2.0	−
33	P Q S_5 I L D S_5 H V R R V W R	16 nM	+1.1	~
34	R W P Q S_5 I L D S_5 H V R R V W R	8 nM	+1.1	+
35	R R W P Q S_5 I L D S_5 H V R R V W R	13 nM	+1.1	++
35R	R R W P Q S_5 I L D S_5 H V R R V W R	53 nM	+1.1	++
41R	R R W P Q S_5 I L H S_5 D V R R V A R	> 10^4 nM	+1.1	++

(b)

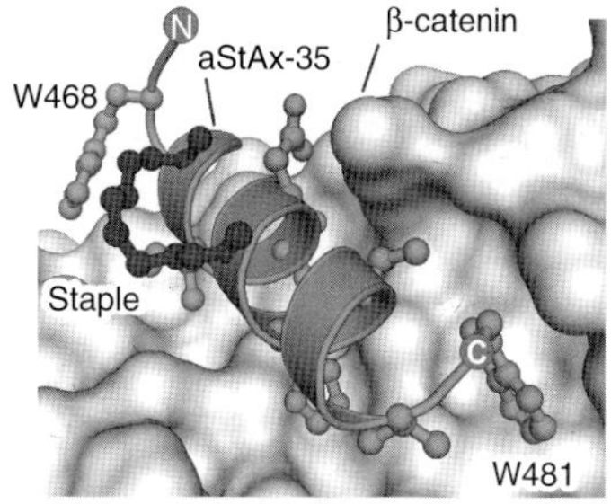

(c)

Figure 25.5 (a) Starting sequence in phage-display-based affinity optimization with a summary of variations found at least twice in the 32 selected phage sequences (numbers indicate frequency of occurrence). (b) A series of stapled peptide sequences (varied amino acids highlighted in black) including their dissociation constant (K_D) with β-catenin, overall charge (calculated with Marvin 5.2.3, 2009, ChemAxon for FITC-labeled peptides at pH 7.5) and performance in cell-permeability tests (incubation for 24 h at 7.5 μM using DLD1 cells, readout: confocal microscopy; cellular uptake: not detectable (−), very low (~), high (+), very high (++)). (c) Crystal structure (PDB: 4DJS) of acetylated StAx-35 (gray cartoon with black staple) bound to β-catenin (light gray, surface representation).

improve cell penetration by stapled peptides. On the basis of these observations, the phage-derived variations were inspected for residues that were likely to support cell penetration. The observed variations suggest a substitution of the two glutamates (D) at position 467 and 470 by arginine (R) and glutamine (Q), respectively. Therefore, these changes were also incorporated into the StAx peptides. Figure 25.5b gives examples for StAx-3-derived peptides including their affinities for β-catenin. In addition, calculated values for the total charge at pH 7.5 and the classification of their cellular uptake by a cancer cell line (DLD1) are given. The elimination of glutamate D467 and glutamine Q468 in fStAx-31 does not affect binding. Overall, the substitution of glutamate D470 by glutamine (Q), of glutamine Q478 by arginine (R), and of methionine M481 by tryptophan (W) result in improved binding of fStAx-33. Further increase in affinity was observed upon addition of arginine (R) and tryptophan (W) at positions 467 and 468, respectively, leading to fStAx-34. With the intention to improve cell permeability, the positive charge of these peptides was increased through incorporation of arginine (R) residues, yielding peptides fStAx-35 and -35R. Cell permeability was investigated with fluorescein-labeled peptides (fStAx) using confocal microscopy (Box 25.2). Colon cancer cells (DLD1) were incubated with 7.5 μM peptide for 24 h. After cell fixation, intracellular levels of fStAx peptides were investigated. Negatively and slightly positively charged peptides did not show significant cellular uptake (fStAx-31 and -33), whereas peptides with an overall charge of +2.1 and more (fStAx-34 to -41R) efficiently penetrated cells. In particular, incubation with fStAx-35 and -35R as well as their related negative control fStAx-41R resulted in intensive homogeneously distributed cytosolic and nuclear fluorescence.

Box 25.2 Confocal Microscopy

Confocal microscopy is characterized by increased optical resolution relative to standard microscopy employing lenses and a fixed focal length. In contrast to classical wide-field microscopy, confocal microscopy employs point illumination, which in combination with a spatial pinhole eliminates out-of-focus fluorescent light. A combination of a large number of these single-point measurements allows the reconstruction of high-resolution 2D or 3D images. Stacked images representing slices through the cell in the Z plane are particularly informative, and can clearly differentiate surface-bound from intracellular peptide, while also revealing the subcellular localization.

To elucidate the molecular basis of the interaction between StAx peptides and β-catenin, the crystal structure of N-terminally acetylated axin-derived stapled peptide-35 (aStAx-35) in complex with residues 134–665 of β-catenin was determined (Figure 25.5c). As expected, aStAx-35 binds to the same binding site as the starting sequence (AxWT), with a nearly complete root-mean-square (RMS) overlay of their backbones (data not shown). The inclusion of two tryptophan (W) residues at positions 468 and 481 within the StAx peptide sequence, as indicated by phage display, resulted in a significant increase in binding affinity (Figure 25.5b). The crystal structure suggests that these tryptophans (W) form additional interactions with β-catenin, in particular, tryptophan W481, which binds in a hydrophobic pocket on β-catenin. As mentioned, arginines (R) were added to StAx-35 at the N- and C-terminal positions 466 and 482, respectively, as well as at position 467 to increase cell permeability. Arginines R476 and R482 are disordered in the crystal structure and R467 does not contact β-catenin. In addition, the α,α-methylated amino acids, forming the all-hydrocarbon staple, are oriented away from the interaction face with β-catenin (Figure 25.5c) and do not contribute directly to the protein–peptide interaction.

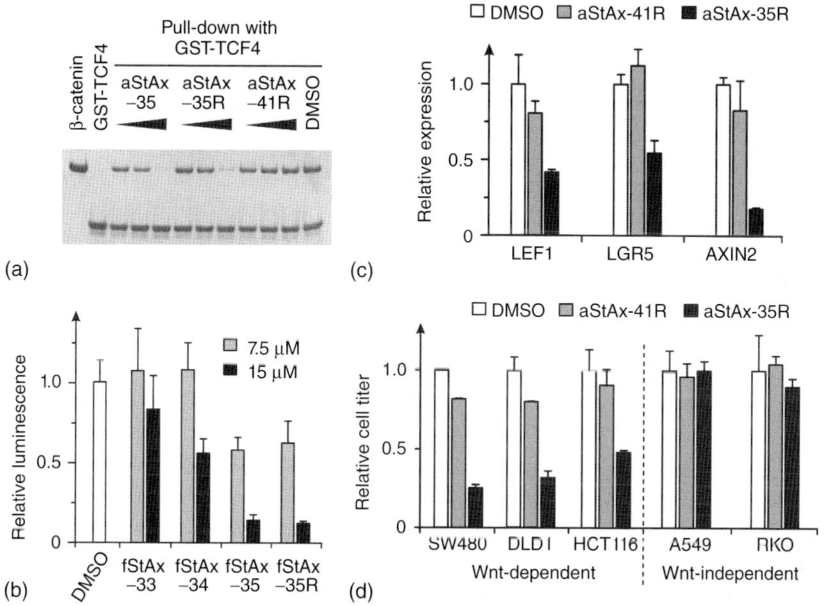

Figure 25.6 (a) *In vitro* competition of aStAx peptides (0.1, 0.5, 2.5 μM) with bead-immobilized GST-TCF4(1-52) for β-catenin (0.5 μM). (b) fStAx peptides inhibit TOP flash luciferase reporter activity in Wnt3a-stimulated HeLa cells. (c) aStAx-35R reduces mRNA level of Wnt/β-catenin target genes in SW480 cells. Relative mRNA level was normalized using the mRNA level of the housekeeping gene β-actin (treatments with 10 μM peptide for 24 h). (d) aStAx-35R inhibits cell proliferation of Wnt-dependent cancer cell lines SW480, DLD1, and HCT116, leaving Wnt-independent cell lines A549 and RKO unaffected (treatments with 10 μM peptide for 5 days, cell titer was determined via cellular ATP level). ((a, b, c and d) Adapted from ref [5].)

25.5 Biochemical and Biological Evaluation

Using an *in vitro* pull-down assay, the StAx peptides showing the highest levels of cell penetration were investigated with respect to the efficiency with which they compete with TCF4 for β-catenin binding. Briefly, the glutathione-S-transferase (GST)-tagged CBD of TCF4(1-52) was immobilized on glutathione-labeled agarose beads and used to precipitate β-catenin (Figure 25.6a). In the presence of acetylated peptides aStAx-35 and -35R, the binding of β-catenin to GST-TCF4(1-52) was inhibited competitively, whereas in the presence of negative control aStAx-41R, β-catenin pull down was not affected. To explore the effects of StAx peptides on Wnt-dependent transcriptional activity, a reporter gene assay was performed using cells that were stimulated with Wnt3a. These cells were transfected with two plasmids, one containing a 10-tandem repeat of the TCF4-binding element with a promoter that is located upstream of the firefly luciferase gene, and the other containing a *renilla* luciferase gene for normalization. In the presence of Wnt3a, cells were treated with fStAx peptides for 24 h followed by luciferase activity measurements. From a panel of peptides, the most cell-permeable stapled peptides, fStAx-35 and fStAx-35R, were identified as potent inhibitors of β-catenin/TCF4-driven firefly luciferase activity (Figure 25.6b). At 15 μM, fStAx-35 and -35R were found to be three times more active in inhibiting luciferase activity than fStAx-34, which showed a higher affinity for β-catenin but a reduced cellular uptake (Figure 25.5b). This observation confirms the importance of efficient cell penetration for potent cellular activity of StAx peptides. Next, the effect of StAx-35R on the messenger ribonucleic acid (mRNA) expression level of Wnt/β-catenin-driven target genes was investigated. Colorectal cancer cells (DLD1), known to have elevated β-catenin levels, were treated with StAx peptides for 24 h followed by total RNA extraction and quantitative real-time polymerase chain reaction (qRT-PCR) of known β-catenin target genes including *LEF1, LGR5,* and *AXIN2.* In agreement with the reporter gene assay, aStAx-35R incubation reduced the mRNA level of these target genes (Figure 25.6c).

Finally, the inhibitory effect of aStAx-35R on the proliferation of cancer cells was investigated. Colorectal cancer cell lines that depend on active Wnt signaling for growth were used: DLD1 and SW480 cells harbor deletions of APC, whereas HCT116 harbors both APC deletion and a mutation in β-catenin blocking its degradation. The treatment of these cells with 10 μM of peptide for 5 days showed significantly reduced cellular adenosine triphosphate (ATP) levels with aStAx-35R as compared to dimethyl sulfoxide (DMSO) and negative control aStAx-41R (Figure 25.6d). To exclude a general cytotoxicity of aStAx-35R, the peptide was tested with cell lines reported to grow independent of Wnt signaling (colorectal cancer cell line, RKO and A549). After 5 days of treatment with 10 μM aStAx-35R, no reduction in A549 and RKO cancer cell proliferation was observed (Figure 25.6d), verifying a Wnt-specific mode-of-action for aStAx-35R.

25.6 Conclusions

Owing to its involvement in the onset and progression of numerous types of cancers, the Wnt/β-catenin pathway is considered a high-priority target for next-generation precision medicine approaches toward the treatment of cancer. This chapter has described an approach in which the formation of the transcriptional activator complex between β-catenin and TCF4 is competitively inhibited by direct targeting of β-catenin. Through a systematic approach that involved screening of different stapling positions, affinity optimization via phage display, and introduction of residues that promote cell penetration, StAx-peptides were discovered. These stapled peptides were shown to function as direct β-catenin antagonists *in vitro* and in cultured cells. A crystal structure of the StAx-35–β-catenin complex has verified the proposed interaction site and confirmed a nearly identical overlay of StAx-35 and the parent axin sequence. Wnt-driven reporter gene assays and the analysis of direct β-catenin/TCF target gene levels confirm inhibition of β-catenin-mediated transcriptional activities. In addition, StAx-35R was demonstrated to selectively reduce the proliferation of Wnt-dependent cancer cells. These stapled peptides represent promising leads for the development of first-in-class β-catenin antagonists.

References

1. Hanahan, D. and Weinberg, R.A. (2011) Hallmarks of cancer: the next generation. *Cell*, **144** (5), 646–674.
2. Katoh, M. (2007) Networking of WNT, FGF, notch, BMP, and hedgehog signaling pathways during carcinogenesis. *Stem Cell Rev.*, **3** (1), 30–38.
3. Von Hoff, D.D., LoRusso, P.M., Rudin, C.M., Reddy, J.C., Yauch, R.L., Tibes, R., Weiss, G.J., Borad, M.J., Hann, C.L., Brahmer, J.R., Mackey, H.M., Lum, B.L., Darbonne, W.C., Marsters, J.C., de Sauvage, F.J., and Low, J.A. (2009) Inhibition of the hedgehog pathway in advanced basal-cell carcinoma. *N. Engl. J. Med.*, **361** (12), 1164–1172.
4. Moellering, R.E., Cornejo, M., Davis, T.N., Del Bianco, C., Aster, J.C., Blacklow, S.C., Kung, A.L., Gilliland, D.G., Verdine, G.L., and Bradner, J.E. (2009) Direct inhibition of the NOTCH transcription factor complex. *Nature*, **462** (7270), 182–188.
5. Grossmann, T.N., Yeh, J.T.H., Bowman, B.R., Chu, Q., Moellering, R.E., and Verdine, G.L. (2012) Inhibition of oncogenic Wnt signaling through direct targeting of beta-catenin. *Proc. Natl. Acad. Sci. U.S.A.*, **109** (44), 17942–17947.
6. Clevers, H. and Nusse, R. (2012) Wnt/beta-catenin signaling and disease. *Cell*, **149** (6), 1192–1205.
7. Hahne, G. and Grossmann, T.N. (2013) Direct targeting of β-catenin: Inhibition of protein–protein interactions for the inactivation of Wnt signaling. *Bioorg. Med. Chem.*, **21**, 4020–4026.
8. Lepourcelet, M., Chen, Y.N.P., France, D.S., Wang, H.S., Crews, P., Petersen, F., Bruseo, C., Wood, A.W., and Shivdasani, R.A. (2004) Small-molecule antagonists of the oncogenic Tcf/beta-catenin protein complex. *Cancer Cell*, **5** (1), 91–102.
9. Gonsalves, F.C., Klein, K., Carson, B.B., Katz, S., Ekas, L.A., Evans, S., Nagourney, R., Cardozo, T., Brown, A.M.C., and DasGupta, R. (2011) An RNAi-based chemical genetic screen identifies three small-molecule inhibitors of the Wnt/wingless signaling pathway. *Proc. Natl. Acad. Sci. U.S.A.*, **108** (15), 5954–5963.
10. Trosset, J.Y., Dalvit, C., Knapp, S., Fasolini, M., Veronesi, M., Mantegani, S., Gianellini, L.M., Catana, C., Sundstrom, M., Stouten, P.F.W., and Moll, J.K. (2006) Inhibition of protein-protein interactions: the discovery of druglike beta-catenin inhibitors by combining virtual and biophysical screening. *Proteins*, **64** (1), 60–67.
11. Sampietro, J., Dahlberg, C.L., Cho, U.S., Hinds, T.R., Kimelman, D., and Xu, W.Q. (2006) Crystal structure of a beta-catenin/BCL9/Tcf4 complex. *Mol. Cell*, **24** (2), 293–300.
12. Azzarito, V., Long, K., Murphy, N.S., and Wilson, A.J. (2013) Inhibition of alpha-helix-mediated protein protein interactions using designed molecules. *Nat. Chem.*, **5** (3), 161–173.
13. Kim, Y.W., Grossmann, T.N., and Verdine, G.L. (2011) Synthesis of all-hydrocarbon stapled alpha-helical peptides via ring-closing olefin metathesis. *Nat. Protoc.*, **6**, 761–771.
14. Schafmeister, C.E., Po, J., and Verdine, G.L. (2000) An all-hydrocarbon cross-linking system for enhancing the helicity and metabolic stability of peptides. *J. Am. Chem. Soc.*, **122** (24), 5891–5892.
15. Walensky, L.D., Kung, A.L., Escher, I., Malia, T.J., Barbuto, S., Wright, R.D., Wagner, G., Verdine, G.L., and

Korsmeyer, S.J. (2004) Activation of apoptosis in vivo by a hydrocarbon-stapled BH3 helix. *Science*, **305** (5689), 1466–1470.

16. Xing, Y., Clements, W.K., Kimelman, D., and Xu, W.Q. (2003) Crystal structure of a beta-catenin/axin complex suggests a mechanism for the beta-catenin destruction complex. *Genes Dev.*, **17** (22), 2753–2764.

17. Chen, Y.H., Yang, J.T., and Chau, K.H. (1974) Determination of helix and beta-form of proteins in aqueous-solution by circular-dichroism. *Biochemistry*, **13** (16), 3350–3359.

18. Smith, G.P. (1985) Filamentous fusion phage – novel expression vectors that display cloned antigens on the virion surface. *Science*, **228** (4705), 1315–1317.

26
Diversity-Oriented Synthesis: Developing New Chemical Tools to Probe and Modulate Biological Systems

Warren R. J. D. Galloway, David Wilcke, Feilin Nie, Kathy Hadje-Georgiou, Luca Laraia, and David R. Spring

26.1
Introduction

Small molecular mass chemical entities (the so-called small molecules) are capable of interacting with biological macromolecules and exerting profound effects upon their function [1]. The use of small molecules to selectively perturb biological systems underpins the field of chemical biology and forms the basis of modern medicine [1, 2]. Put simply, humanity has a significant dependence on biologically active small molecules [3]. It is therefore unsurprising that significant effort is directed toward the identification of new small molecules with specific biological activity.

26.2
The Biological Problem

26.2.1
How to Discover New Chemical Modulators of Biological Function?

In situations where the biological target is structurally well defined, it is often possible to use this information to rationally design or select small-molecule-binding partners. Similarly, if the structure of the natural ligand is known, this can be used as a "template" to guide compound synthesis or selection [1, 4]. However, in cases where the precise nature of the biological target is unknown (e.g., a phenotypic screen), natural ligands are unidentified, or a novel mode of binding to a particular target is desired, a "rational" design, or selection process is clearly not possible. In such situations, the discovery of bioactive small molecules relies upon the biological assessment (screening) of collections (or "libraries") of small molecules to identify those with the desired properties (the so-called "hits"). The success of any such screening endeavor will clearly be inherently dependent upon the molecular composition of the library [1, 4].

Concepts and Case Studies in Chemical Biology, First Edition. Edited by Herbert Waldmann and Petra Janning.

26.2.2 Sources of Small Molecules for Screening

The molecules comprising small-molecule screening collections may be obtained from natural (natural products) or nonnatural (chemical synthesis) sources [4].

26.2.2.1 Natural Products

Natural products show a wide range of biological activities and have been used medicinally throughout the course of human history [4]. However, there are problems associated with using natural products in screening experiments; they can be difficult to source, and the identification, purification, and chemical modification (to improve properties, e.g., potency) of the bioactive components can be very challenging [1]. In addition, there may be areas of chemical space (Box 26.1) that nature has ignored, which nonetheless contain compounds with biologically interesting properties; such compounds would not be detected if screening was limited to natural products only. Thus, it is not always realistic, or indeed desirable, to generate, and screen libraries consisting solely of natural products [4].

Box 26.1 Chemical Space

Chemical compounds can be characterized by a wide variety of molecular "descriptors" such as physiochemical properties (e.g., their lipophilicity) and topological features (e.g., degree of branching) [5, 6]. The term *chemical space* is commonly used in place of "multi-dimensional descriptor space": it is a region defined by a particular selection of molecular descriptors and the upper and lower values (limits) placed upon them [5, 6]. In the context of small-molecule collections, chemical space is generally defined as the total descriptor space that encompasses all small carbon-based molecules that could, in principle, be created [5, 6]. Each chemically unique small molecule will have a unique set of molecular properties and thus molecular descriptor values, and will therefore reside at a discrete point in chemical space [5].

26.2.2.2 Chemical Synthesis and the Need for Structural Diversity

Deliberate chemical synthesis represents an important alternative means of obtaining small-molecule libraries for biological screening [4]. But what molecules should be synthesized? Between the late 1980s and mid-1990s, a "brute force" approach was adopted; libraries of large numbers of compounds (literally millions in some cases) could be efficiently produced by combinatorial-type methods and these were routinely screened using high-throughput methods [7]. However, libraries of this sort have had limited success in the discovery of novel biologically active small molecules [4]. This has been largely attributed to the lack of structural variation between the compounds within such collections. It is now widely acknowledged that the success of any screening process (in terms of the hit frequency) is inherently dependent upon the structural diversity of the library used; the size of the library is not everything [1, 8, 9]. There is a direct correlation between the overall structural diversity of a small-molecule library and its functional diversity (i.e., the range of biological activities displayed by the library). High functional diversity is clearly valuable in screens where the precise nature of the biological target is unknown or ill defined (e.g., a phenotypic screen) [1]. The presence of multiple structural classes within a molecular collection being screened against a single, specific target also increases the probability of discovering a molecule capable of binding in a novel manner [4].

Why should the functional diversity of a small-molecule library be related to its overall structural diversity? Biological macromolecules interact with each other in a three-dimensional (3D) manner [2]. On a molecular scale, biomolecules can be thought of as large 3D environments with certain defined potential binding regions. Consequently, they will only interact with small molecules that display a complementary 3D binding surface [4]. That is, a given biomolecule imposes a degree of shape selection for binding partners [1, 5, 8, 9]. Thus, the 3D shape of a small molecule is the most important factor controlling its biological effects [1]. Molecular shape is dictated by molecular structure [1, 7]. Structurally diverse libraries should therefore contain compounds with a diverse range of distinct molecular shapes; consequently, the library as a whole would cover a broad range of potential biological binding partners [4, 10].

There are four principal components of structural diversity that are typically identified [1]:

1) **Appendage diversity** – variation in structural moieties around a common scaffold
2) **Functional group diversity** – variation in the functional groups present
3) **Stereochemical diversity** – variation in the orientation of potential macromolecule-interacting elements
4) **Scaffold diversity** – presence of many distinct molecular scaffolds.

Scaffold diversity is the most crucial of these aspects in terms of the functional diversity of a library [10]. The shape–space coverage of any compound set (and thus its functional diversity) stems mainly from the nature and 3D geometries of the central scaffolds, with the peripheral substituents being of considerably less importance in this regard [7, 10]. Traditional combinatorial libraries typically possessed low levels of scaffold diversity; the molecules in such collections were broadly similar structures, with structural variation restricted to the presence of different appendages around a common scaffold. This explains their poor performance in many biological screens, especially those where the precise nature of the biological target was poorly defined or unknown [7].

Many commercially available and proprietary compound collections are synthesized in a combinatorial-type manner and so suffer from a lack of structural (principally scaffold) diversity [4, 11]. Another drawback of such collections stems from the nature of medicinal chemistry research over the course of the past few decades, which has focused upon a limited set of biological targets [7]. As a result, commercially available and proprietary compound libraries are often heavily biased toward compounds that satisfy predefined criteria for the modulation of such "traditional" targets (e.g., the Lipinski "rule of 5" criteria for oral bioavailability [1, 7, 12]). Consequently, these libraries are intrinsically biased toward *known* bioactive chemical space (the chemical space spanned by known biologically active molecules), leaving potentially large swathes of biologically relevant chemical space underexplored. There is a widespread acknowledgement that the targets of the current pharmacopeia represent only a small fraction of potential targets that could impact on health [7, 11]. There are many other human-disease-related targets (such as protein–protein interactions), which have traditionally been termed *undruggable* as they have proved difficult, if not impossible, to address with small molecules [7, 11, 13]. However, it is becoming increasingly evident that these targets are indeed tractable to small molecule modulation; it is simply that they have traditionally been challenged with the wrong types of molecules [2, 11]. The molecules comprising typical commercially available and proprietary compound collections seem to be well suited to modulating "traditional" medicinal chemistry targets, but lack the structural features necessary to affect other processes [7, 11, 13].

26.3
The Chemical Approach

26.3.1
Diversity-Oriented Synthesis

Diversity-oriented synthesis (DOS) was developed over the past decade in order to address the need for new small-molecule collections with higher levels of structural, and thus functional, diversity [2, 14, 15]. DOS libraries aim to efficiently interrogate large areas of chemical space simultaneously. This includes known biologically relevant chemical space (by definition, a fruitful region for the discovery of useful small-molecule modulators of biological function) *and* under-explored (and, indeed, completely novel) regions of chemical space, which may contain molecules with unusual or exciting biological properties (e.g., the capability to modulate classically "undruggable" targets) [1, 7]. The screening of such libraries should provide hits against a broad range of biological targets with increased frequency and decreased cost relative to less diverse libraries, facilitating the discovery of new agents for therapeutic intervention and novel probes for biological research [1, 11].

26.3.1.1 DOS and Scaffold Diversity
Ideally, a DOS should address all four of the principal types of structural diversity mentioned previously. However, the ideal synthesis of a structurally diverse small molecule collection is one in which the diversity is achieved in the most efficient manner possible [1]. As alluded to previously, it is the scaffold diversity of the library that is the key parameter in this regard. It is generally acknowledged

that increasing the scaffold diversity in a small-molecule library is one of the most effective ways of increasing its overall structural diversity (and, consequently, its shape, and thus functional diversity) [1, 10]. Thus, the efficient incorporation of multiple molecular scaffolds in a single library is of central importance to the success of a DOS. This is undoubtedly the most challenging facet of any DOS program [1, 2].

There are two basic strategies for generating scaffold diversity in a DOS context (Figure 26.1). The first is a "branching" approach, where divergent reactions are carried out on a substrate to produce compounds with distinct molecular scaffolds. The second is a "folding" approach, where intramolecular reactions are used to "pair up" strategically positioned functional groups. This could involve either the use of different starting materials and common reaction conditions, such that each starting material yields a product containing a different molecular scaffold, or a densely functionalized molecule where different functional groups can be reacted together under distinct reaction conditions and so create a number of different scaffolds. These strategies are not orthogonal to each other and many

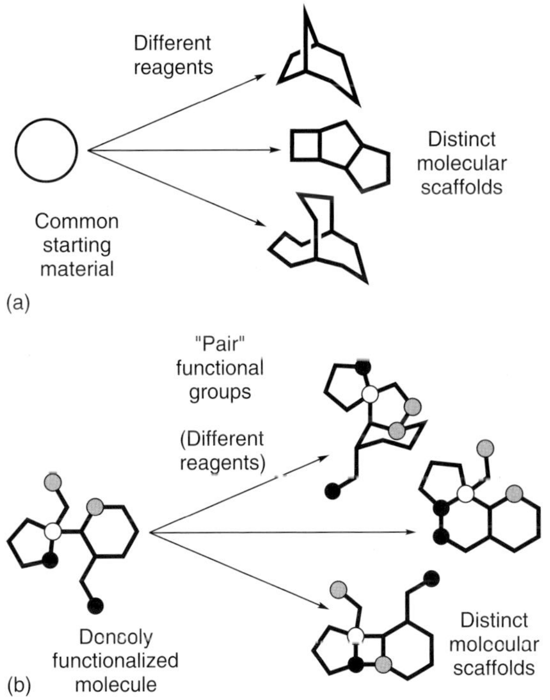

Figure 26.1 *Examples of strategies for generating scaffold diversity in DOS.* (a) An example of the "branching" approach. Here, the exposure of a given starting material to different reagents results in the generation of different molecular scaffolds. (b) An example of the "folding" approach. Here, different "complementary" functional groups (indicated by different colored circles) of a densely functionalized molecule are reacted together (the "pair" process) in functional group-specific intramolecular reactions, to yield different scaffolds [8].

DOS programmers will utilize both [1, 4]. The resulting products should ideally contain synthetic handles for further transformations, thereby providing scope for additional diversification [5]. Variation in starting materials and/or reagents allows for the introduction of appendage, functional group, and stereochemical diversity (the latter of which may also be incorporated through the use of stereoselective reactions) [1].

26.4
Chemical Biology Research

26.4.1
DOS as a Tool for Identifying New Modulators of Mitosis

Antimitotic compounds are used clinically for the treatment of cancer, and this target class is widely regarded to still hold great promise for anticancer therapy. How can new, structurally distinct antimitotic agents be identified? Recently, Spring and coworkers described the discovery of new small-molecule modulators of mitosis using DOS, illustrating the utility of this synthetic approach for the identification of new biologically relevant chemical entities [16].

26.4.1.1 DOS Library Synthesis

Diazoacetates represent attractive starting units for the branching DOS pathways [5, 17]. The diazoacetate functionality exhibits enormous synthetic versatility, permitting the use of a wide variety of different synthetic transformations. Thus, diazoacetate compounds have the potential to be converted into several products with different scaffolds, which should themselves be suitable for further diversification [5].

Spring and coworkers recently reported the use of two different, readily accessible, phenyldiazoacetate compounds (**1** and **2**) as starter units for two different branching DOS pathways (Schemes 26.1 and 26.2) [16]. The second pathway (Scheme 26.2) utilized the highly functionalized derivative **3** as a key branch-point intermediate. The presence of both an electron deficient and an electron neutral olefin, coupled with the proximal aryl bromide and a carboxylic ester, afforded the opportunity to regioselectively modify the scaffold of **3** in a multidirectional approach. Overall, these two DOS pathways combined generated a library totaling 35 small molecules, with 10 distinct molecular scaffolds, comprising complex fused ring systems of varying sizes and a multiplicity of stereocenters present. Cheminformatic analysis of the library indicated that it accessed biologically relevant areas of chemical space and had a good level of shape diversity.

26.4.1.2 Biological Studies: Phenotypic Screening for Antimitotic Effects

The DOS library compounds were screened for their ability to induce mitotic arrest in human osteosarcoma cells (U2OS line) [18, 19]. Cells were incubated with compounds at a range of concentrations and then stained for the mitotic

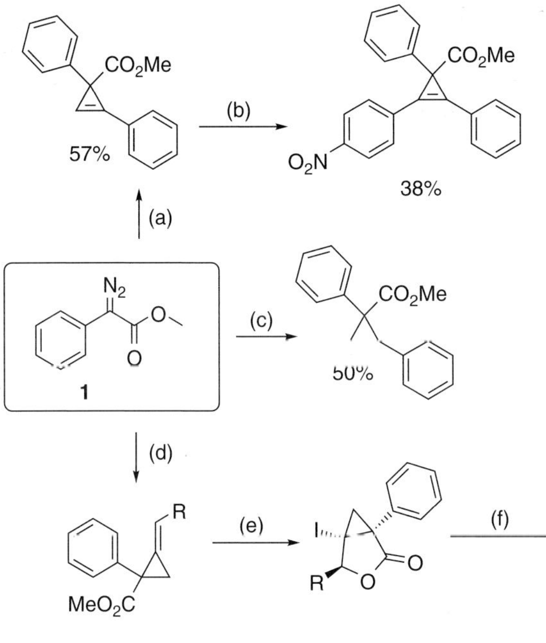

Scheme 26.1 Synthesis of subset of DOS library from compound **1**; (a) phenylacetylene, $Rh_2(OAc)_4$ (1 mol%), CH_2Cl_2; (b) *p*-nitroiodobenzene, $Pd(OAc)_2$ (10 mol%), K_2CO_3, DMF; (c) styrene, $Rh_2(OAc)_4$ (1 mol%), CH_2Cl_2; (d) allene, $Rh_2(OAc)_4$ (1 mol%), CH_2Cl_2; (e) *N*-iodosuccinimide, MeCN-H_2O (2 : 1), 50 °C; and (f) Bu_3SnH, AIBN, PhH, 80 °C. DMF: *N,N*-dimethylformamide, AIBN: 2,2′-azobis(2-methylpropionitrile).

marker phosphohistone H3 and imaged on a Cellomics Arrayscan high-content microscope. The percentage of cells arrested in mitosis following compound treatment was then calculated by image analysis. The most potent compound (**4**) gave a large (35–40%) mitotic arrest (Figure 26.2 for structure, Table 26.1 for screening data). On the basis of this result, the partially saturated analog of **4**, compound **5**, was prepared in a racemic form (Figure 26.2). Compound **5**, subsequently termed *dosabulin*, was also found to give a mitotic arrest in U2OS cells, with a twofold increase in potency compared to **4**. Treatment with dosabulin also resulted in growth inhibition in the low micromolar range over a period of 72 h (Figure 26.3, Table 26.1). Separation of both enantiomers of dosabulin by preparative chiral high-performance liquid chromatography (HPLC) and subsequent retesting, revealed that all the activity resided in the *(S)*-enantiomer (Figure 26.3, Table 26.1). It was found that *(S)*-dosabulin treated cells died through apoptosis while cells treated with *(R)*-dosabulin did not.

26.4.1.3 **Biological Studies: Target Identification**

While phenotypic screening allows for the rapid identification of biologically active molecules from a library, subsequent target identification (identification of the biological target(s) that interact with a compound of interest) is notoriously

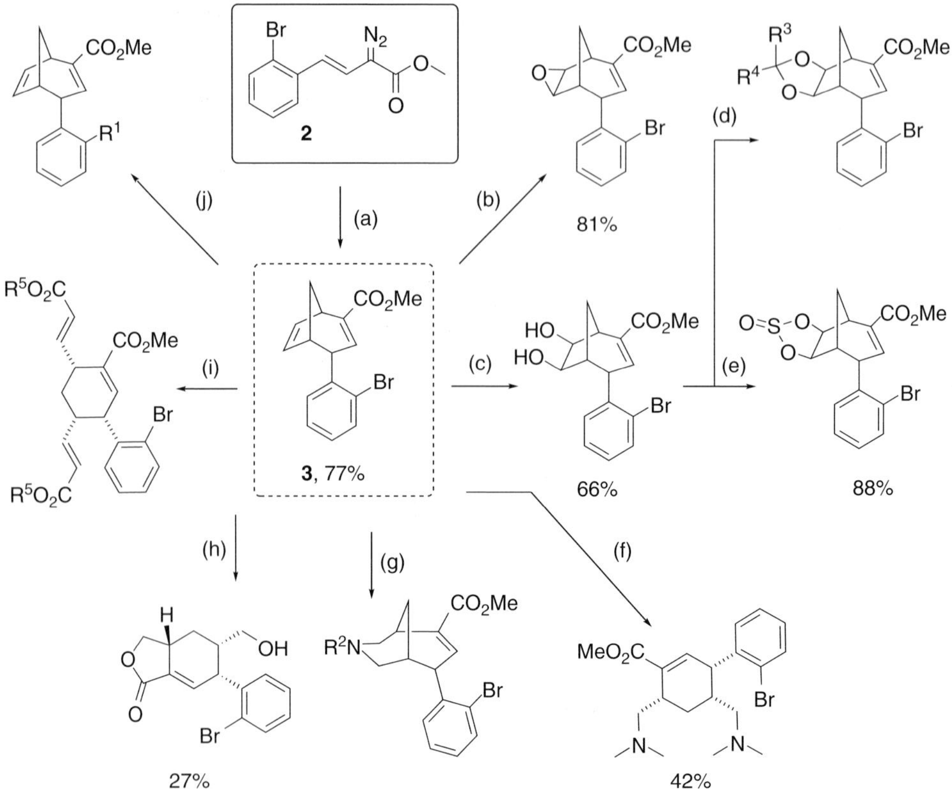

Scheme 26.2 Synthesis of subset of DOS library from compound **2**; (a) cyclopentadiene, $Rh_2(OAc)_4$ (1 mol%), CH_2Cl_2; (b) *m*-CPBA, CH_2Cl_2; (c) OsO_4 (2.5 mol%), NMO, acetone-H_2O (9 : 1); (d) aldehyde/ketone, CSA (10 mol%), 3 Å molecular sieves, CH_2Cl_2; (e) $SOCl_2$, CH_2Cl_2; (f) 2,6-lutidine, NMO, OsO_4 (2.5 mol%), $PhI(OAc)_2$, acetone-H_2O (10 : 1), then dimethylamine, $NaBH(OAc)_3$, CH_2Cl_2; (g) 2,6-lutidine, NMO, OsO_4 (2.5 mol%), $PhI(OAc)_2$, acetone-H_2O (10 : 1), then primary amine, $NaBH(OAc)_3$, CH_2Cl_2; (h) 2,6-lutidine, NMO, OsO_4 (2.5 mol%), $PhI(OAc)_2$, acetone-H_2O (10 : 1), then $NaBH_4$, MeOH; (i) alkene, Hoveyda-Grubbs (II) catalyst (10 mol%), ethylene, PhMe, 100 °C; and (j) $Pd(OAc)_2$ (10 mol%), boronic acid, PPh_3 (15 mol%), 2N K_2CO_3, PhMe, 90 °C. NMO: *N*-methylmorpholine-*N*-oxide, CSA: camphorsulfonic acid.

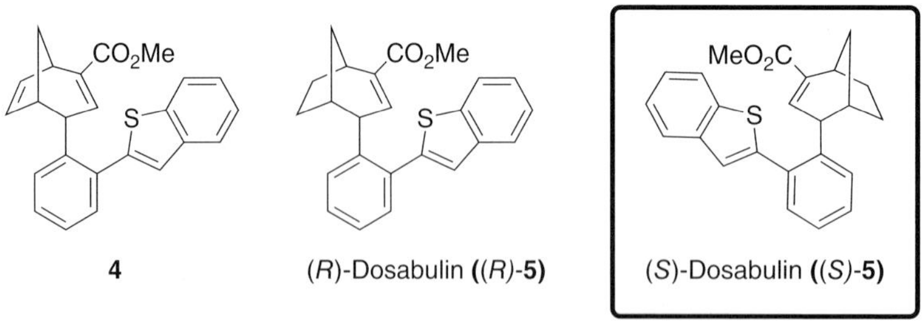

4 (*R*)-Dosabulin (**(*R*)-5**) (*S*)-Dosabulin (**(*S*)-5**)

Figure 26.2 Structures of some antimitotics from the DOS library.

Table 26.1 Mitotic arrest (EC_{50}) and growth inhibition (IC_{50}) values for selected compounds from the DOS library.

Compound	Mitotic index (MI) EC_{50} (μM)	Growth inhibition $(GI)_{50}$ (μM)
4	6.25 ± 0.91	3.70 ± 0.71
(*rac*)-Dosabulin	3.13 ± 0.32	1.47 ± 0.03
(*R*)-Dosabulin	N/A	N/A
(*S*)-Dosabulin	1.23 ± 0.10	0.81 ± 0.37

EC_{50}, effective concentration 50; IC_{50}, inhibitor concentration 50.
Growth inhibition assessed by sulforhodamine B colorimetric assay for cytotoxic effects. All values are mean ± standard deviation. N/A = not active.

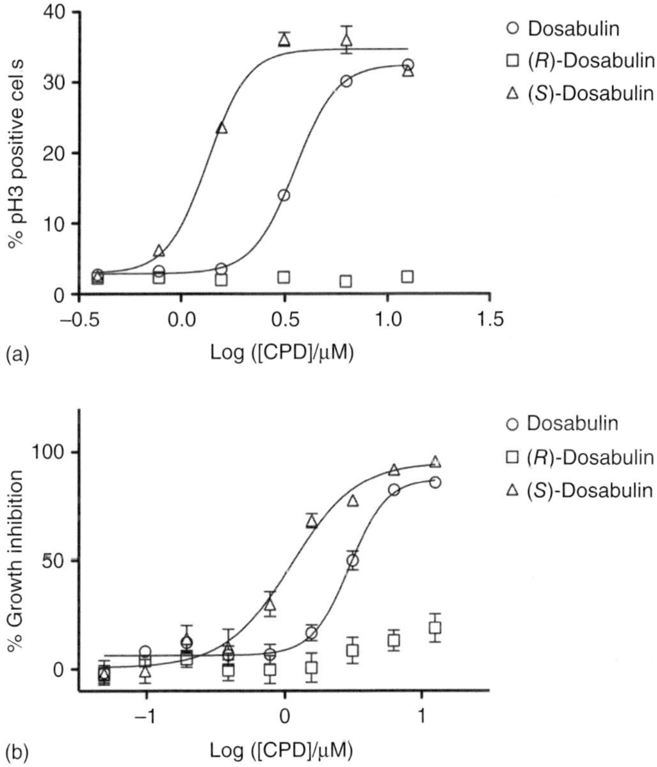

Figure 26.3 (a) Representative mitotic index assay data for racemic dosabulin and its purified enantiomers. Data points are mean ± standard error in the mean of an experiment conducted in triplicate. "% PH3 positive cells" refers to the proportion of cells stained with an antibody against phosphor-histone H3. "CPD" = compound under investigation. (b) Growth inhibition curves assessed by sulforhodamine B assay for racemic dosabulin and its purified enantiomers. Data points are mean ± standard error in the mean of an experiment conducted in triplicate. "CPD" = compound under investigation.

difficult [5, 19]. However, careful observation of the phenotype may sometimes offer clues [20]. Spring and coworkers used confocal microscopy to look at the key mitotic protein, tubulin; it was found that the tubulin network was heavily disrupted upon treatment with *(S)*-dosabulin. This led to the hypothesis that *(S)*-dosabulin was targeting tubulin itself. This phenotype was recapitulated by nocodazole, a known tubulin depolymerizer, providing indirect evidence for this assertion. Existing agents targeting tubulin suffer from administration and resistance problems; thus, new antimitotics-targeting tubulin are of significant therapeutic interest [21]. An *in vitro* tubulin polymerization assay established that *(S)*-dosabulin acts as a tubulin depolymerizing agent. Several small-molecule-binding sites are known to exist in the tubulin polymer [22]. For example, vinblastine binds the β-tubulin subunit, while colchicine binds at the α/β interface. Further work demonstrated that *(S)*-dosabulin partially inhibits the binding of colchicine to tubulin, suggesting that it may bind in a site vicinal or allosteric to colchicine.

26.5 Conclusion

Over the course of the past decade, DOS has established itself as a powerful tool for the efficient *de novo* creation of structurally, and thus functionally, diverse small molecule collections. Many ingenious DOS strategies have been reported, which have enabled the efficient synthesis of libraries based on tens of different molecular scaffolds; and the screening of DOS libraries has led to the identification of numerous bioactive small molecules (including modulators of a range of undruggable targets) [7]. For example, Spring and coworkers used DOS to discover dosabulin, a novel small molecule that causes mitotic arrest and cancer cell death by apoptosis at submicromolar concentrations.

A key challenge for future DOS campaigns is to improve the balance between broad chemical space coverage and biological relevance [1]. A DOS should aim to specifically and efficiently access *both* known and unknown *biologically relevant* chemical space, rather than regions that are not going to provide biologically useful small molecules [1]. Toward this end, future years may witness the emergence of more "constrained" DOS campaigns that seek to generate maximum structural diversity within preselected limits, such that a better balance between structural (scaffold) diversity (which is valuable for broad bioactive chemical space coverage) and target relevance and/or drug likeness is achieved.

References

1. Galloway, W.R.J.D., Isidro-Llobet, A., and Spring, D.R. (2010) Diversity-oriented synthesis as a tool for the discovery of novel biologically active small molecules. *Nat. Commun.*, **1**, 80.
2. O'Connor, C.J., Beckmann, H.S.G., and Spring, D.R. (2012) Diversity-oriented synthesis: producing chemical tools for dissecting biology. *Chem. Soc. Rev.*, **41**, 4444–4456.
3. O'Connor, C.J., Laraia, L., and Spring, D.R. (2011) Chemical genetics. *Chem. Soc. Rev.*, **40**, 4332–4345.
4. O'Connell, K.M.G., Galloway, W.R.J.D., Ibbeson, B.M., Isidro-Llobet, A., O'Connor, C.J., and Spring, D.R.

(2012) in *Solid-Phase Organic Synthesis:Conepts, Strategies and Applications* (eds P.H. Toy and Y. Lam), John Wiley & Sons, Inc., Hoboken, NJ, pp. 131–150.
5. Galloway, W.R.J.D., Bender, A., Welch, M., and Spring, D.R. (2009) The discovery of antibacterial agents using diversity-oriented synthesis. *Chem. Commun.*, 2446–2462.
6. Dobson, C.M. (2004) Chemical space and biology. *Nature*, **432**, 824–828.
7. Galloway, W.R.J.D. and Spring, D.R. (2013) Towards drugging the undruggable: enhancing the scaffold diversity of synthetic small molecule screening collections using diversity-oriented synthesis. *Divers. Oriented Synth.*, **1**, 21–28.
8. Burke, M.D. and Schreiber, S.L. (2004) A planning strategy for diversity-oriented synthesis. *Angew. Chem. Int. Ed.*, **43**, 46–58.
9. Burke, M.D., Berger, E.M., and Schreiber, S.L. (2003) Generating diverse skeletons of small molecules combinatorially. *Science*, **302**, 613–618.
10. Sauer, W.H. and Schwarz, M.K. (2003) Molecular shape diversity of combinatorial libraries: a prerequisite for broad bioactivity. *J. Chem. Inf. Comput. Sci.*, **43**, 987–1003.
11. Schreiber, S.L. (2009) Organic chemistry: molecular diversity by design. *Nature*, **457**, 153–154.
12. Lipinski, C. and Hopkins, A. (2004) Navigating chemical space for biology and medicine. *Nature*, **432**, 855–861.
13. Altmann, K.H., Buchner, J., Kessler, H., Diederich, F., Krautler, B., Lippard, S., Liskamp, R., Muller, K., Nolan, E.M., Samori, B., Schneider, G., Schreiber, S.L., Schwalbe, H., Toniolo, C., van Boeckel, C.A., Waldmann, H., and Walsh, C.T. (2009) The state of the art of chemical biology. *ChemBioChem*, **10**, 16–29.
14. Lee, D., Sello, J.K., and Schreiber, S.L. (2000) Pairwise use of complexity-generating reactions in diversity-oriented organic synthesis. *Org. Lett.*, **2**, 709–712.
15. Schreiber, S.L. (2000) Target-oriented and diversity-oriented organic synthesis in drug discovery. *Science*, **287**, 1964–1969.
16. Ibbeson, B.M., Laraia, L., Alza, E., O'Connor, C.J., Tan, Y., Davies, H.M.L., McKenzie, G., Venkitaraman, A.R., and Spring, D.R. (2014) Diversity-oriented synthesis as a tool for identifying new modulators of mitosis. *Nat. Commun.*, **5**, 3155.
17. Wyatt, E.E., Fergus, S., Galloway, W.R.J.D., Bender, A., Fox, D.J., Plowright, A.T., Jessiman, A.S., Welch, M., and Spring, D.R. (2006) Skeletal diversity construction *via* a branching synthetic strategy. *Chem. Commun.*, 3296–3298.
18. Young, D.W., Bender, A., Hoyt, J., McWhinnie, E., Chirn, G.W., Tao, C.Y., Tallarico, J.A., Labow, M., Jenkins, J.L., Mitchison, T.J., and Feng, Y. (2008) Integrating high-content screening and ligand-target prediction to identify mechanism of action. *Nat. Chem. Biol.*, **4**, 59–68.
19. Ziegler, S., Pries, V., Hedberg, C., and Waldmann, H. (2013) Target identification for small bioactive molecules: finding the needle in the haystack. *Angew. Chem. Int. Ed.*, **52**, 2744–2792.
20. Neumann, B., Walter, T., Heriche, J.K., Bulkescher, J., Erfle, H., Conrad, C., Rogers, P., Poser, I., Held, M., Liebel, U., Cetin, C., Sieckmann, F., Pau, G., Kabbe, R., Wunsche, A., Satagopam, V., Schmitz, M.H.A., Chapuis, C., Gerlich, D.W., Schneider, R., Eils, R., Huber, W., Peters, J.M., Hyman, A.A., Durbin, R., Pepperkok, R., and Ellenberg, J. (2010) Phenotypic profiling of the human genome by time-lapse microscopy reveals cell division genes. *Nature*, **464**, 721–727.
21. Dumontet, C. and Jordan, M.A. (2010) Microtubule-binding agents: a dynamic field of cancer therapeutics. *Nat. Rev. Drug Discovery*, **9**, 790–803.
22. Gigant, B., Wang, C., Ravelli, R.B.G., Roussi, F., Steinmetz, M.O., Curmi, P.A., Sobel, A., and Knossow, M. (2005) Structural basis for the regulation of tubulin by vinblastine. *Nature*, **435**, 519–522.

27
Scaffold Diversity Synthesis with Branching Cascades Strategy

Kamal Kumar

27.1
Introduction

Chemical biology plays significant roles in unraveling diverse and unknown functions of different proteins and as a consequence contributes immensely to drug discovery research. Small molecules with greater structural diversity are required in order to identify the best candidates that interact with protein targets and modulate biological functions. A synthetic challenge therefore is posed for chemists to devise new ways to generate compound collections covering not only large areas of chemical space but also beholding unknown and novel molecular architectures. In this chapter, we discuss one of these synthetic approaches targeting scaffold diversity, which is termed as *branching pathways*, and as a case study a branching cascades strategy that utilizes cascade reaction sequences as key transformations is illustrated.

27.2
The Biological/Pharmacological Problem: Discovering Small Bioactive Molecules

The emergence of medium- to high-throughput screening (HTS) in the pharmaceutical industry and its slow but steady entry into academia has drawn great attention to the compound collections that undergo these screenings [1, 2]. The quality of a compound collection greatly influences the hit rates of any screening endeavor and therefore "*quality*" of molecules has become the key criterion in design and synthesis of compound collections [3, 4]. One source of inspiration for this quality is the structural diversity and complexity of natural products, which remain the major source of bioactive hit and lead structures in drug discovery till date [5–7]. Traditionally, natural products and their analogs are synthesized in what is famously termed as *total synthesis or target oriented synthesis* (*TOS*), wherein simple building blocks are assembled stepwise in a long and often tedious synthesis route to generate a complex natural product [8–10]. For years, TOS has been a highly appreciated but equally hard synthesis discipline to follow that delivers in successful cases small amounts of few natural product derivatives.

Concepts and Case Studies in Chemical Biology, First Edition. Edited by Herbert Waldmann and Petra Janning.

Although, recently, many successful "scale-up" TOS endeavors [11–13] have been reported, yet, TOS remains an impractical method when it comes to providing a large number of molecules for screening purposes. Synthesis of compound libraries that are either natural-product-derived/inspired molecules [14] or based on "privileged scaffolds" [15] are endeavors to fill up the natural-product-like chemical space. Diversity-oriented synthesis (DOS) [16–20] and biology-oriented synthesis (BIOS) [21–24] have emerged as two principles for the design and synthesis of compound collections for chemical biology and medicinal chemistry research and quite often generate natural-product-like molecules (Box 27.1).

Box 27.1 Synthesis Approaches to Explore Biologically Relevant Chemical Space

TOS, DOS, and BIOS are three major synthetic approaches that aim to provide biologically active molecules for discovery research. TOS remains an elegant, yet multistep and tedious approach to synthesize natural products and their analogs in the laboratory. In most cases, this practice assembles simpler building blocks in a convergent manner to build a complex natural product. For instance, synthesis of complex anticancer epothilone analogs [25] from simpler building blocks is depicted in Figure 1. The complexity of natural products obviously limits the synthesis of analog libraries to a small size, thereby limiting the exploration of their potential in biomedical research and drug discovery. To provide natural product-like and thus biologically relevant compound collections of relatively bigger size, DOS [16–20] and BIOS [21–24] approaches have been introduced. DOS aims to populate the chemical space by generating structurally diverse and complex molecular architectures in a combinatorial manner. DOS follows the principle of forward synthetic analysis in which each chemical reaction step adds complexity and structural diversity (skeletal, functional, and stereochemical) in a large pool of small molecules. For instance, Thomas *et al.* used a DOS strategy to generate a series of diverse scaffolds employing different cycloaddition reactions of solid supported α,β-unsaturated compounds (Figure 1). Thus, a library of 242 small molecules embodying 18 natural-product-like scaffolds was generated, which yielded several hits in a screen for activity against pathogenic strains of methicillin-resistant *Staphylococcus aureus* (MRSA) [26]. BIOS [27] builds on the biological relevance and prevalidation of natural products and other bioactive compound classes and develop focused compound libraries based on scaffolds present in those structural classes [21–24]. A major goal for BIOS is to identify ligands for proteins that are functionally diverse but share similarity in their ligand-binding cores [28]. In a BIOS application, Koch and coworkers synthesized natural-product Dysidiolide-inspired libraries consisting of γ-hydroxybutenolides and closely related α,β-unsaturated five-membered lactones as well as the decalines. Biological screening of the library yielded hits for Cdc25A (cell division control protein), AChE (acetylcholinesterase), or 11β-HSD1/2 (11β-hydroxysteroid dehydrogenase). These proteins are functionally very different from one another but share structural similarities in their ligand-sensing cores.

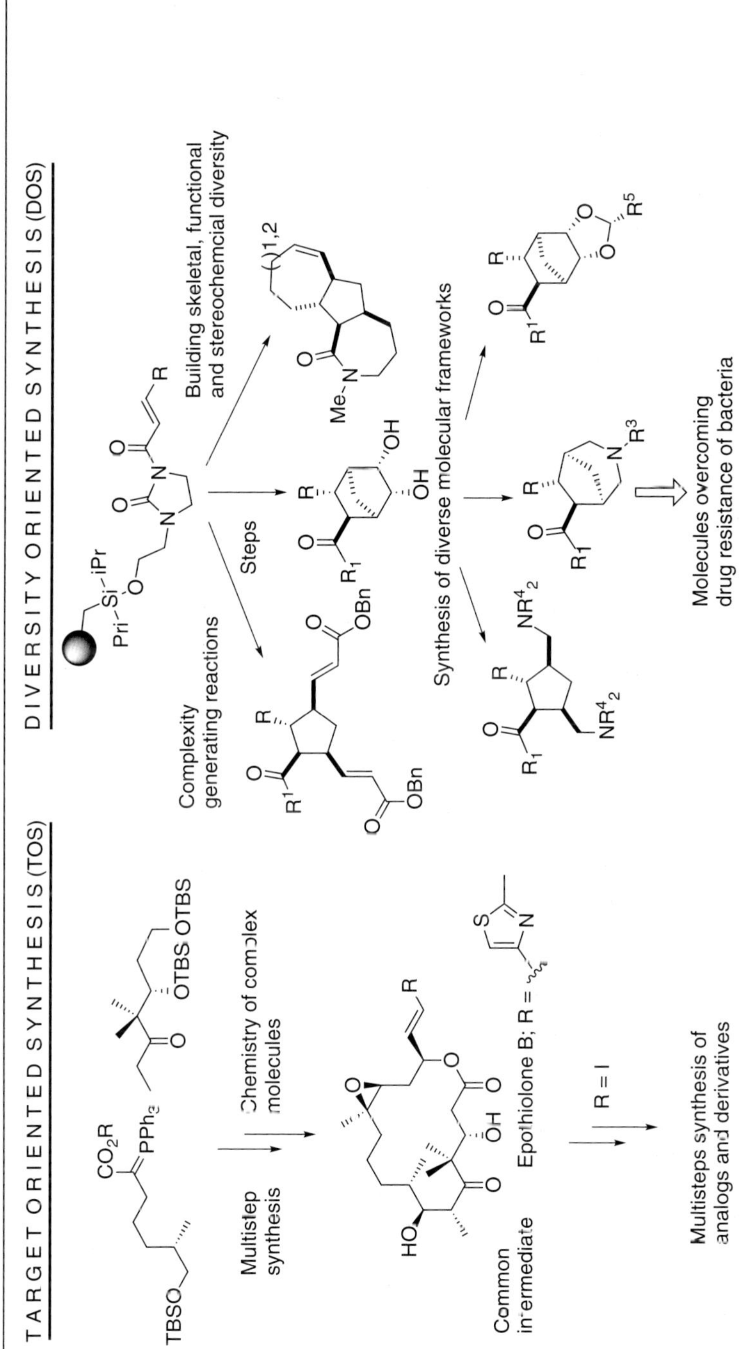

Figure 1 *Different synthetic approaches to identify potent biologically active small molecules.* (Reproduced with permission from [24]. Copyright American Chemical Society.)

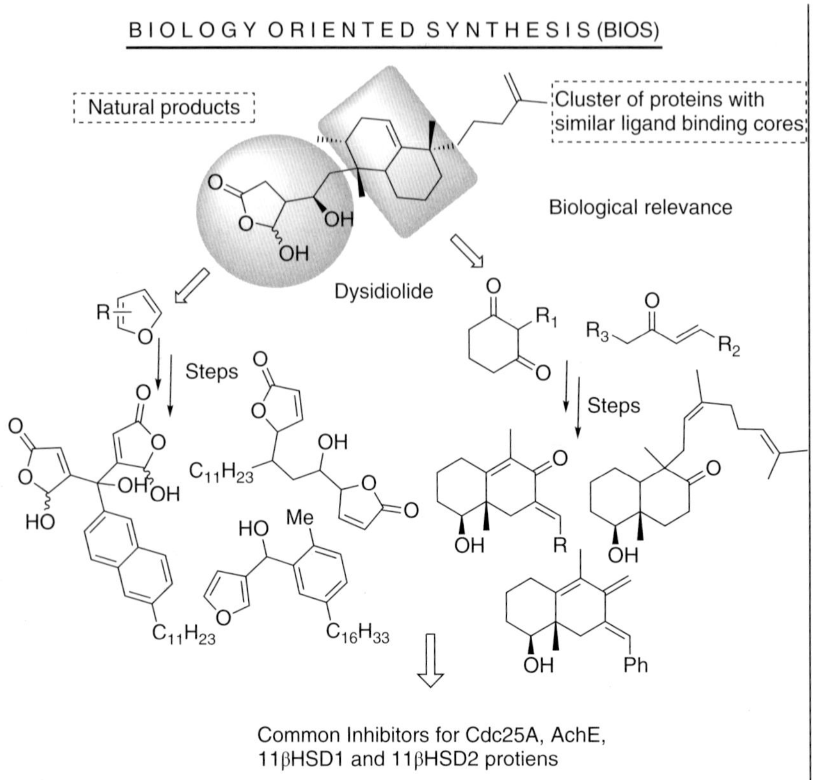

Figure 1 (*Continued*)

Over the years, it has been realized that organic synthesis has merely focused only on a small percentage of vast chemical space in its structural diversity in the synthetic molecules [29] and therefore a large potential of unknown and novel molecular entities remain untapped. Considering the fact that there exist only small islands of biological activities in chemical space [4, 30–32], there is a huge demand to cover as much broader areas of structural diversity in the compound libraries as possible. Therefore, advances in the organic synthesis methods that can provide efficient access to diverse structural classes of small molecules are highly desired.

27.3
The Chemical Approach: Scaffold Diversity

27.3.1
Beyond the Biased Exploration of Chemical Space

The scaffold is the core molecular framework [33] that gives a molecule its basic shape, provides rigidity or flexibility to the molecule, and respective positions to various substitutions over its periphery in three-dimensional space (Box 27.2). Often, scaffolds are directly involved in interactions with protein targets or receptors, either by hydrogen bonds or hydrophobic interactions. Scaffold diversity is therefore a very important parameter to characterize compound libraries and to identify diverse ligands that could interact with diverse protein targets [34, 35]. However; a balance between the diversity of core frameworks or scaffolds within a library and the density of representation of each scaffold is nevertheless required. While the dense representation over small numbers of scaffolds is often applicable in libraries focused on a particular biological target class, a sparse representation of a large number of scaffolds may not provide the desired structure–activity relationship and the hit confirmations, in particular, for molecules that are single exemplars of a particular scaffold. Thus, library design must take into consideration a fine balance between scaffold diversity and scaffold representation.

Box 27.2 Defining a Scaffold

The meaning of the term *scaffold* is quite subjective and changes both with the chemists and the context in which it is explained. For instance, a medicinal chemist would define a scaffold as the core structure required for a given pharmacological activity, while a synthetic chemist may define it as a particular ring system or framework considering the synthetic planning that generates it. However, a generally acceptable term as outlined by Bemis and Murcko [36] defines scaffold as a core structure or framework derived from molecules by removing side chain atoms while preserving the atoms in ring systems or linking ring systems and sp^2 atoms directly bonded to these atoms. For instance, removing the side chains of the medicinally important aminoquinolines, as presented in Figure 2a, provides the quinoline as a Murcko scaffold. A complex Murcko scaffold can be further dissected into more than one ring system by cleaving linker bonds between the rings [37]. Each molecule has $n + 1$ levels, which are numbered sequentially from level 0 (the single remaining ring) up to level n (the whole molecule). The level $n - 1$ is termed *Murcko framework* in this case. For instance, the Hsp90 inhibitor NVP AUY922 depicted in Figure 2b is reduced to different simpler scaffolds [38].

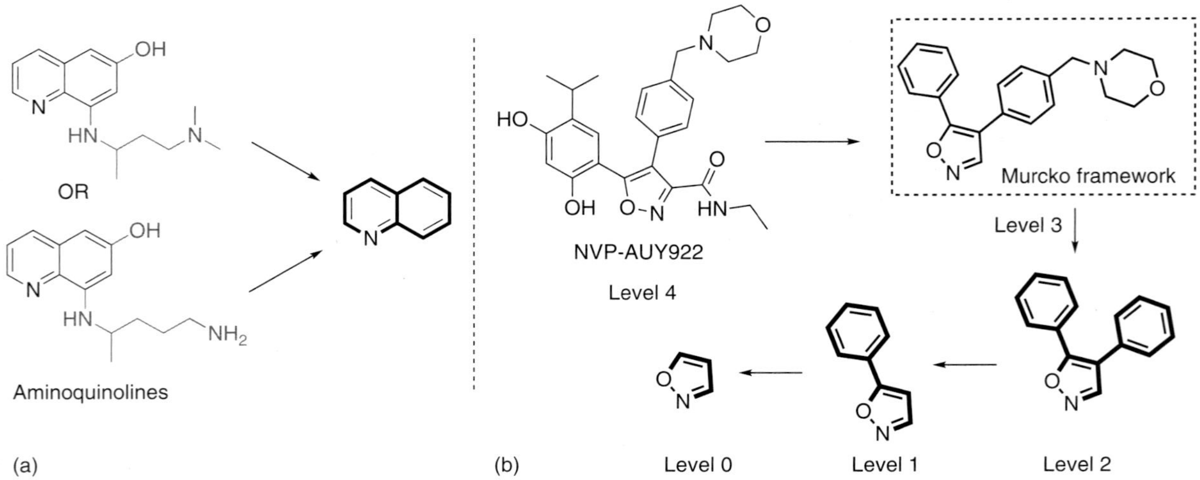

Figure 2 *Murcko structures for bioactive molecules.* (a) Quinoline as Murcko framework of bioactive aminoquinolines after removing side chains and (b) different levels of structural simplification of Hsp90 inhibitor NVP-AUY922 provide different structural levels of scaffolds.

The morpholine-benzyl, phenyl isoxazole is the Murcko framework for this bioactive molecule.

Earlier cases of a compound library synthesis had mostly kept the core scaffold constant throughout all compounds of a given library. By the end of the past century, there was a paradigm shift in the way that the pharmaceutical industry has applied combinatorial chemistry to drug discovery. To acquire more quality-based compound libraries, synthesis for primary screening has steered away from numerically large libraries based on limited number of scaffolds toward collections of small libraries comprising many diverse chemotypes. This is a highly significant move, as synthesis validation and optimization for each given scaffold type is the most time-consuming step in library production. While the synthesis of several small libraries around different scaffolds requires more time and resources than to build one large library based on a single chemotype, there had to be solid reasons to supersede the practicality and feasibility arguments. Among them, a major concern in HTS of early combinatorial libraries against a number of biological targets was the "sporadic" biological results obtained for a given scaffold, that is, the hit rates observed for different targets were either very high or essentially zero [39–41]. In contrast, compound libraries possessing diverse and complex chemotypes, such as natural products, exhibited more consistent hit rates across a variety of targets. That clearly hints toward the inherent inability of a combinatorial library derived from a single scaffold, irrespective of the library size, to demonstrate structural diversity to interact with a number of different biological targets. Therefore, in the past decade, organic synthesis has been focusing on the chemical transformations that can be adapted to suit the demand of generating scaffold diversity in rapid and efficient ways.

27.3.2
Scaffold Diversity Synthesis

Among the three approaches depicted in Box 27.1 guiding the synthesis of compound libraries, DOS strategy, as the name suggests, offers a divergent pathway to generate collections of a large number of diverse small molecules. As depicted in Chapter 26 by Spring and coworkers, the most important aspect of structural diversity in DOS of a compound collection is the scaffold diversity. To achieve efficient access to diverse scaffolds, the following two pathways are often considered in DOS [17].

1) The first approach is termed *folding pathway* [42, 43] and uses a common set of reaction conditions to transform a range of substrates into products with distinct and diverse molecular skeletons (Figure 27.1a). The substrates are encoded to "fold" into the alternative scaffolds through strategically embodied functionalities, known as *σ-elements*. Each σ-element thus dictates the formation of diverse molecular framework.

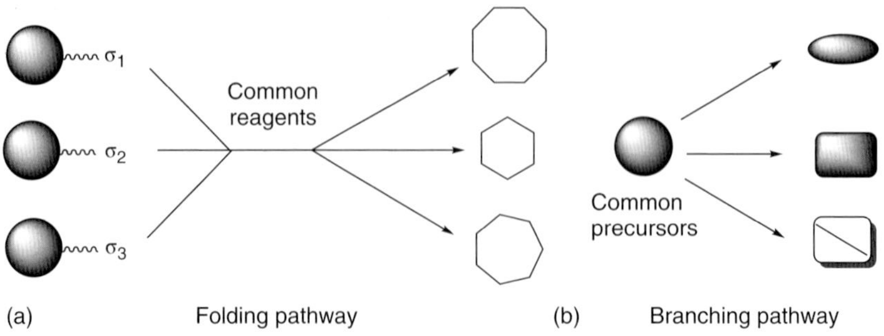

Figure 27.1 (a,b) DOS pathways to generate scaffold diversity.

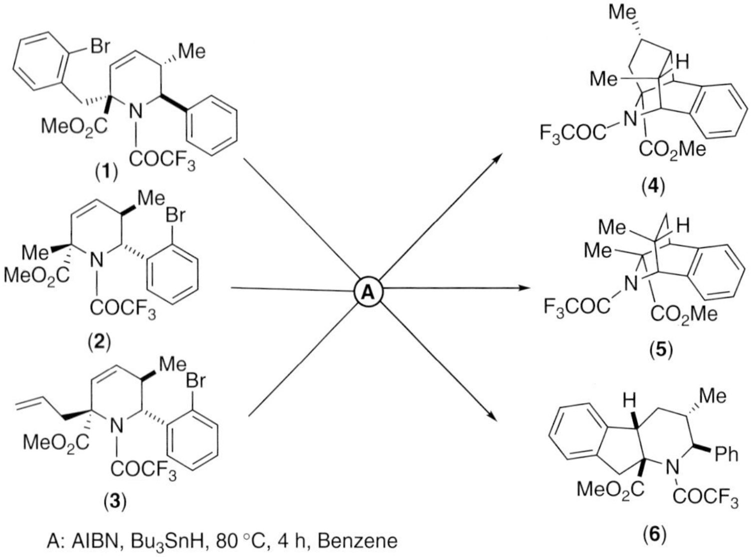

Scheme 27.1 *Folding pathways in DOS.* AIBN = azobisisobutyronitrile.

Exploring radical cyclizations, Panek and Porco designed a *folding* approach to generate skeletal diversity (Scheme 27.1), employing a set of tetrahydropyridines (**1**–**3**) as substrates [44]. The skeletons of the products **4**–**6** were pre-encoded in the substrates **1**–**3** by the location of the radical-initiating sites and the unsaturated groups. The folding processes were triggered by treatment of the tetrahydropyridines **1**–**3** with tributyltin hydride and a substoichiometric amount of 2,2′-azobis(2-methylpropionitril) (AIBN) at 80 °C. The bromine atom on the phenyl group was strategically placed in substrates in order to generate the site-specific radical and consequently selective cyclization reaction to yield a range of distinct polycyclic alkaloid-like frameworks.

2) The "branching pathway" strategy involves the conversion of common precursors into a range of distinct molecular scaffolds (Figure 27.1b). This strategy is more challenging because substrates that can be flexibly transformed into distinct molecular scaffolds need careful synthetic design and planning. Moreover, the synthesis of the molecules should remain more or less combinatorial [43, 45, 46].
A branching pathway based on the chemistry of Michael adducts (**7**) was developed by Porco (Scheme 27.2) [47]. Reduction of the nitro group triggered lactamization to yield γ-lactams such as **8**. In contrast, with appropriately positioned alkenyl and alkynyl substituents, cyclization via ring-closing metathesis or Pauson–Khand reaction was possible. With R^1 = allyl and R^2=C≡CCH$_2$OMe, enyne metathesis yielded the cyclic diene **9**. In contrast, with R^1=C≡CH and R^2 = allyl, a Pauson–Khand reaction allowed the remarkable bridged cyclopentenone **10** to be obtained.

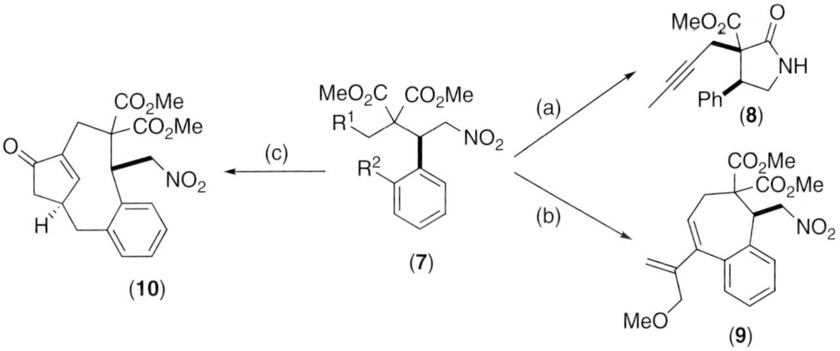

(**10**) (**7**) (**8**) (**9**)

Scheme 27.2 *Porco's branching pathway.* (a) Zn, AcOH-THF, then Na$_2$CO$_3$ (aq.) [R^1 = C≡CMe, R^2 = H]; (b) Grubb's first generation catalyst, ethylene, microwave, 150 W, 50 °C, CH$_2$Cl$_2$ [R^1 = allyl, R^2 = C≡CMCH$_2$OMe]; and (c) Co$_2$(CO)$_8$, microwave, 150 W, 80 °C, CH$_2$Cl$_2$ [R^1 = C≡CH, R^2 = allyl].

27.4
Chemical/Biological Evaluation – Branching Cascades Strategy in Scaffold Diversity Synthesis

Diversity and molecular complexity are two important criteria that enrich a compound collection in biological activity [48, 49]. Therefore, strategies that efficiently build up diverse and relatively complex molecular architectures, in particular based on natural product frameworks, are highly desired. Branching pathways in DOS pose formidable challenge of incorporating scaffold diversity in a compound collection. Cascade or domino reaction sequences [50, 51] wherein more than one reaction happens consecutively in a one-pot strategy and molecular complexity is rapidly built up can immensely improve the efficiency of diversity syntheses endeavors (Box 27.3). Kumar and coworkers introduced a branching pathway strategy employing cascade or domino reaction sequences, termed *branching cascades*, wherein a common precursor facilitates diverse cascade reactions on its surface in order to yield diverse and complex molecular frameworks [52]. To design a common substrate whereupon different cascade reactions can take place, that too in an efficient manner and leading to a diverse set of molecules, is a challenging task. Obviously, such a common precursor needs to have multiple reactive sites and functionalities to initiate and propagate different reactions and/or reaction sequences. To the common precursor, one can add additional substrates or reagents as cascade-initiating molecules or cascade triggers. Chemo-physical properties (e.g., hard or soft nucleophilicity) of different cascade triggers, say X, Y, or Z, would direct the initiation and propagation of different cascade reaction sequences, exploiting the diverse functional sites (F_1, F_2, F_3) on the common multifunctionalized precursor and thereby yielding structurally diverse, complex, and functionalized scaffolds (Figure 27.2). Each of these functionalized scaffolds could be further modified to generate diverse focused compound collections.

Box 27.3 Multistep Versus Cascade or Domino Synthesis

Complex molecules are traditionally synthesized by assembling simpler building blocks in a stepwise manner. This multistep synthesis generally makes one reaction at a time followed by its workup and purification of the product, which is then again employed in a further reaction. In a simple form, say, substrate A makes B, which is purified after the reaction workup and B undergoes another reaction to make C. The latter is again purified and used to make the final product D. Cascade or domino reactions, however, use substrates where multiple reactions happen one after another in a sequence and one does not need multiple workups and purifications. Thus, in a cascade reaction sequence, conversion of A into D happens via conversion of A to E and E to F, however without any need to isolate either E or F and thereby shortening the synthesis and avoiding a number of tedious and time-consuming workups and purifications. To further illustrate this, both stepwise and a cascade synthesis of natural product Gravelliferone is depicted in Figure 3. While the stepwise synthesis begins with benzyloxycoumarin (**11**) and takes three synthetic steps to reach an intermediate (**16**), which is transformed into natural product (**21**) in a further three steps in an overall 9.4% yield [53]. A synthesis step here means a reaction that follows the usual reaction workup and purification of the reaction product for the next step. Obviously, it took at least six workups and purifications in this process. The cascade synthesis of Gravelliferone begins with 2,4-bis-prenylated benzaldehyde (**22**), which undergoes a cascade reaction sequence when treated with phosphorane (**23**) and yields the natural product (**21**) in *one synthetic step* and 10% yield [54]. However, this one synthesis step contained seven individual reaction steps happening in a sequence to finally

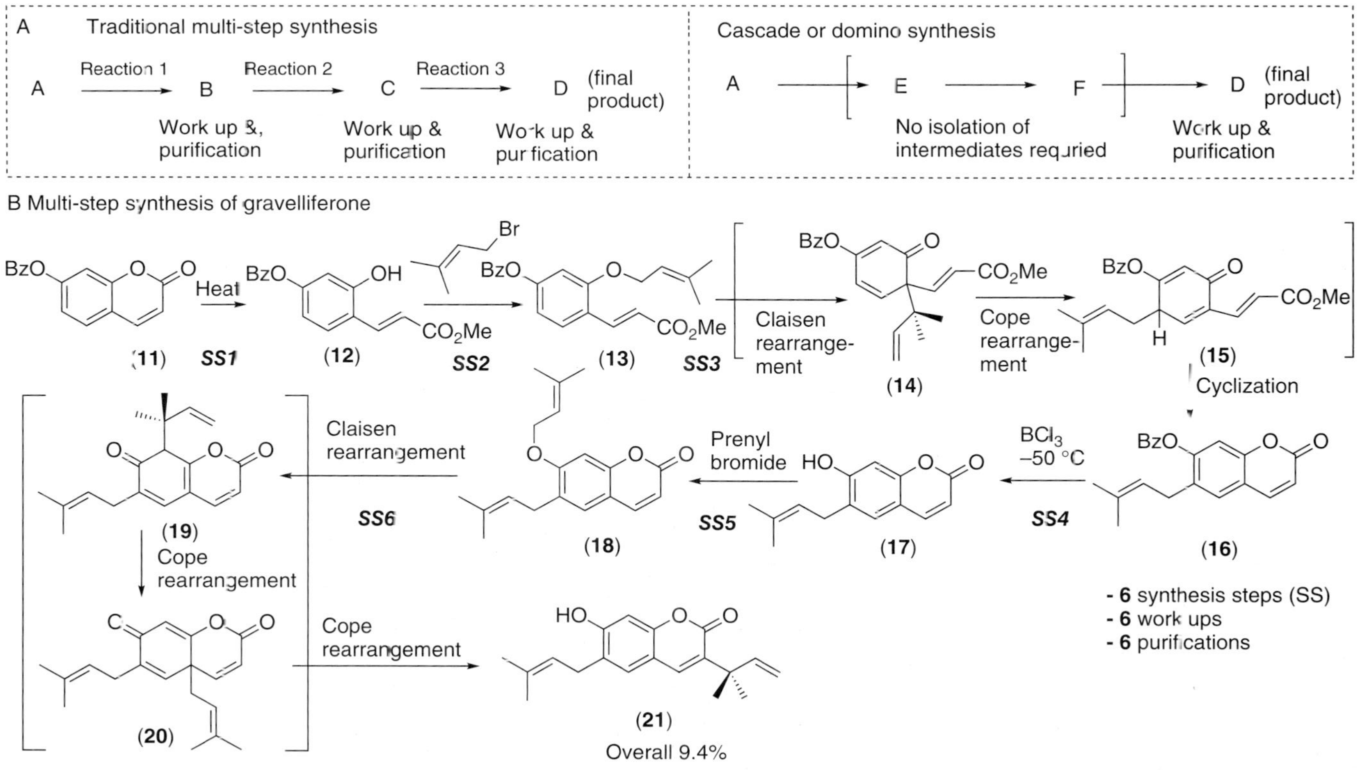

Figure 3 Stepwise versus cascade synthesis.

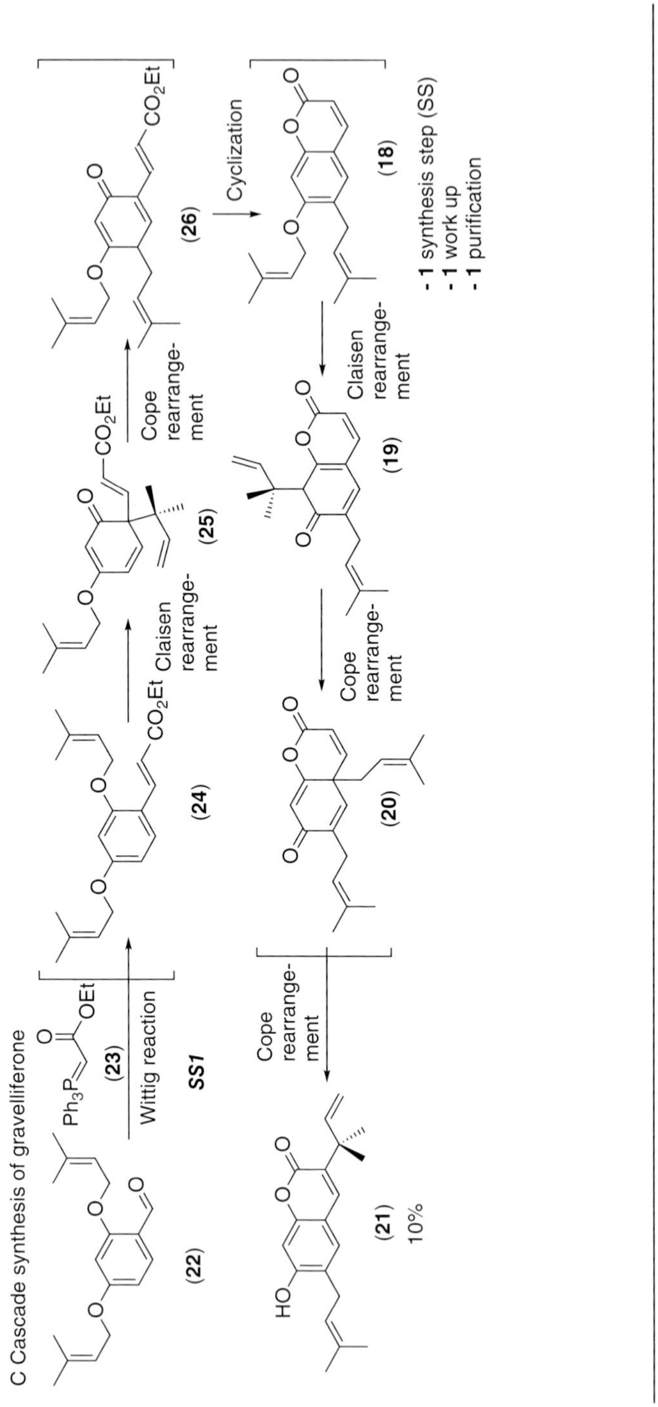
C Cascade synthesis of gravelliferone
(22)
Ph3P=
OEt
(23)
Wittig reaction
SS1
(24)
CO2Et
Claisen rearrange-ment
(25)
Cope rearrange-ment
(26)
Cyclization
(18)
- 1 synthesis step (SS)
- 1 work up
- 1 purification
Claisen rearrange-ment
(19)
Cope rearrange-ment
(20)
Cope rearrange-ment
HO
(21)
10%

yield the natural product. Although the yields of the two syntheses are similar, the time and energy spent on the workups and purifications make a big difference. The cascade synthesis obviously is easier and efficient in this case.

Kumar and coworkers synthesized the substrate **27** in quantitative yields from the commercially available 3-formylchromones and acetylene carboxylates in two steps [55]. Substrate **27** embodies multiple electrophilic sites (δ+), a vinylogous ester (a leaving group) and a couple of pronucleophilic sites (2 and 2′) to facilitate the interplay of diverse cascade reaction sequences (Figure 27.3).

To exploit the dispersed electrophilicity over the surface of **27** toward yielding diverse scaffolds, a reaction screening with different nucleophiles was planned. Nucleophiles that could generate different and complex scaffolds were carefully chosen. Thus, a batch of 23 nucleophiles that included 12 bisnucleophiles (*N–N*, *N–O*, *N–C*, and *O–C* bis-nucleophiles), 6 mononucleophiles, 4 zwitterionic species, and *tert*-butylmalonate (Figure 27.4), was screened to observe the substrates conversions, product profiles, and optimal reaction conditions. Initial reaction screening was performed using a Radley's Carousel Reaction Station with 12 reaction tubes under an argon atmosphere at room temperature.

Reaction conditions employed included the reaction in the presence or absence of external bases for the bisnucleophiles; different concentrations of mononucleophiles and nucleophilic zwitterions. After a quick aqueous workup, a liquid chromatography coupled to mass spectrometry (LC-MS) analysis of the crude reaction mixture was performed. Reactions with high conversions (consumption of **27**) and relatively cleaner product profiles were then separately optimized (when required) for better results. The reaction screening revealed that **27**-ketoesters were better

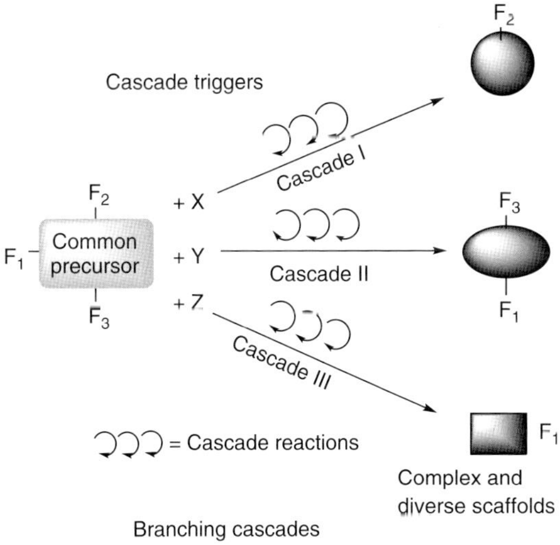

Figure 27.2 Branching cascades strategy.

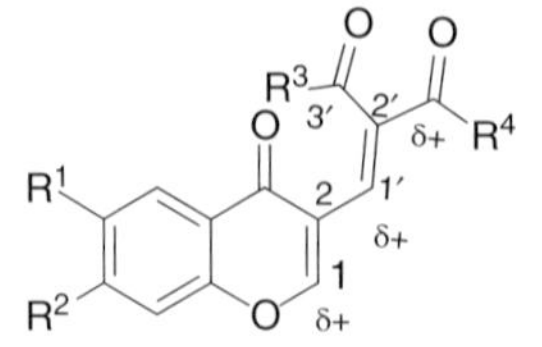

27
Common precursor

(*E*)-ketoesters

27a; $R^1 = R^2 = H$, $R^3 = CO_2Me$, $R^4 = OMe$;
27b; $R^1 = R^2 = H$, $R^3 = CO_2Et$, $R^4 = OEt$;
27c; $R^1 = Me$, $R^2 = H$, $R^3 = CO_2Me$, $R^4 = OMe$;
27d; $R^1 = Cl$, $R^2 = H$, $R^3 = CO_2Et$, $R^4 = OEt$;
27e; $R^1 = Cl$, $R^2 = H$, $R^3 = CO_2Me$, $R^4 = OMe$;
27f; $R^1 = i$Pr, $R^2 = H$, $R^3 = CO_2Me$, $R^4 = OMe$;

(*Z*)-aldehydes

27g; $R^1 = R^2 = H$, $R^3 = OMe$, $R^4 = H$;
27h; $R^1 = Me$, $R^2 = H$, $R^3 = OMe$, $R^4 = H$;
27i; $R^1 = i$Pr, $R^2 = H$, $R^3 = OMe$, $R^4 = H$;
27j; $R^1 = Cl$, $R^2 = Me$, $R^3 = OMe$, $R^4 = H$.

Figure 27.3 A common substrate for branching cascades.

substrates in branching cascade strategy than the corresponding **27**-aldehydes, which often yielded the mixtures of products and with varying conversions.

Among the unsuccessful nucleophiles were diaminoalkane **28**, *t*-butylmalonate **40** (yielded a complex mixture of products); amino alcohols **33** and **34** (displayed sluggish reactivities and unstable products); mononucleophiles **41–44** (yielded mixtures of inseparable products); and zwitterions **48–49** (no reaction). Although amino alcohol **35** reacted well and provided clean reaction, the structures for the diastereoisomers of the adducts could not be assigned. The scaffolds generated by the other 11 nucleophiles are depicted in Scheme 27.3.

Four of the *N*,*N*-bisnucleophiles successfully provided diverse and complex scaffolds. The cascade reaction of 4-aminopiperidine (**29**), the *N*,*N*-bisnucleophile with **27** led to the formation of complex tetrahydro-1,4-ethanopyrido[1,2-*a*] pyrimidine ring-system **58** in high yields (Scheme 27.3). Another cascade reaction initiated by *N*,*N*-bisnucleophile 2-(2-aminophenyl)indole **30** with **27**-ketoesters in dichloromethane at room temperature led to the formation of novel chromone substituted benzo[2, 3]azocino[4,5-*b*]indoles **61** along with a minor product with a different molecular architecture that is dihydroindolo-[3,2-*c*]pyrido-[1,2-*a*]quinolone **51** in varying yields (Scheme 27.3). Interestingly, under basic reaction conditions (equimolar triethylamine), the same nucleophile (**30**) followed another domino reaction to cleanly transform **27** into another ring-system **62**, an indolo[1,2-*c*]pyrido[1,2-*a*]quinazoline in excellent yields (Scheme 27.3). 2-(1*H*-benzo[d]imidazol-2-yl)ethanamine **31**, another *N*,*N*-bisnucleophile, provided novel pyridinium salts **60** following yet another cascade reaction sequence. Reaction of tryptamine with **27**-ketoesters in dichloromethane followed by treatment with 10% trifluoroacetic acid (TFA) led to clean synthesis of biologically active *centrocountins* – the tetrahydroindoloquinolizines (**53**, see Chapter 16 for chemical biology of centrocountins) [56, 57].

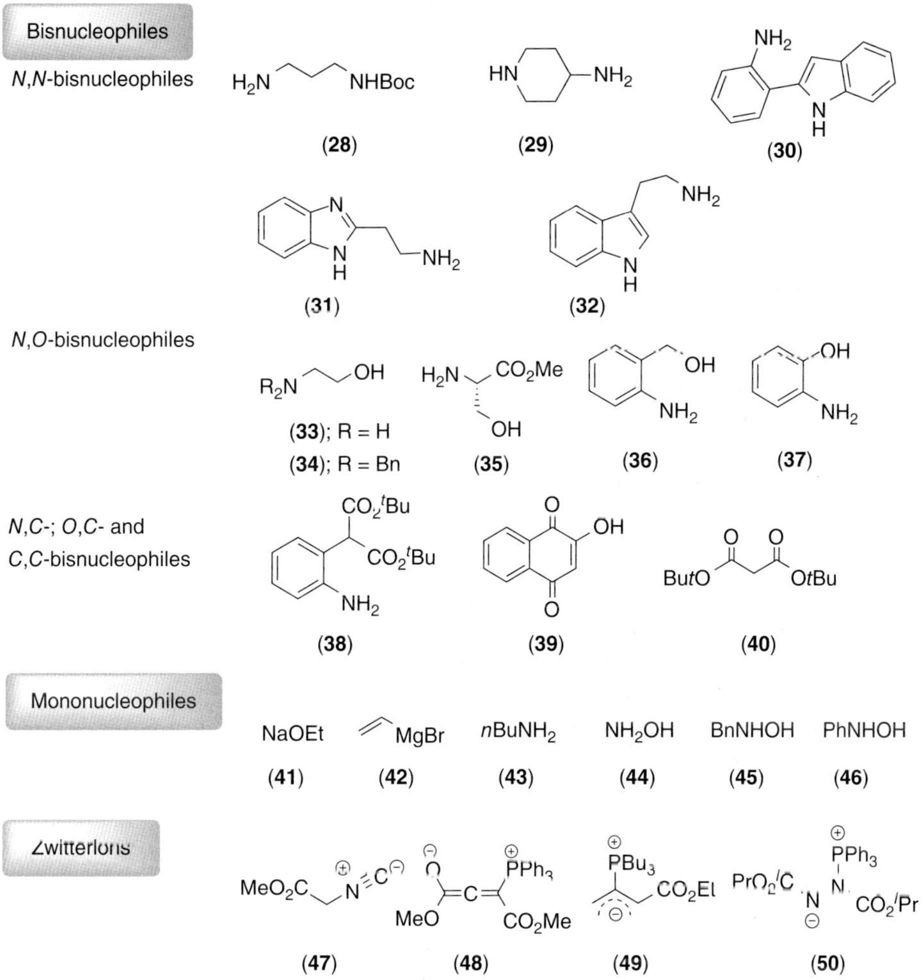

Figure 27.4 Nucleophiles used in reaction screening in branching cascade strategy.

2-aminobenzyl alcohol (**36**) the *N,O*-bisnucleophile could add to the scaffold diversity by transforming **27**-ketoesters into dihydrobenzo[*d*]pyrido[2,1-*b*][1,3]oxazine rings (**59**) in excellent yields (Scheme 27.3).

N,C- and *O,C*-bisnucleophiles **38** and **39**, respectively, had displayed very sluggish reactivities in the initial screening. After further reaction condition optimization, **38** led to the formation of benzoindolizine **52** (Scheme 27.3); a scaffold that is part of many biologically active natural products. Similarly, **39** could provide the natural-product-based pyranonaphthoquinones scaffold **57** in good yields and diastereoselectivity.

N-Benzyl and phenyl hydroxylamines **45** and **46**, the mononucleophiles triggered a new cascade reaction sequence (Scheme 27.4) with both **27** leading to

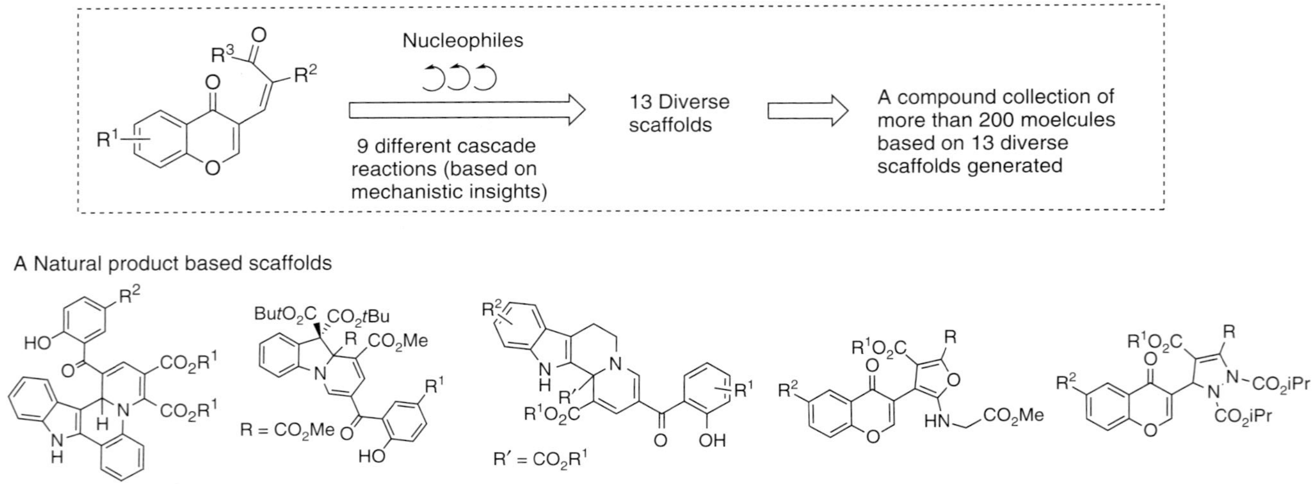

51 (31 – 49%) from 30 (Cascade IV)

52 (79 – 87%) from 38 (Cascade I)

53 (30 – 85%) from 32 (Cascade I)

54 (91 – 95%) from 47 (Cascade VIII)

55 (73 – 91%) from 50 (Cascade VI)

Scheme 27.3 *Reaction conditions to synthesize* **51**: **27**-ketoesters (1 mmol), **30** (1.1 mmol), CH_2Cl_2, r t, 3–6 h; **52**: **27**-ketoesters (1 mmol), **38** (1.2 mmol), EtOH (10 ml), 60 °C, 2,6-lutidine (1.5 mmol), AgOTf (0.1 mmol), 8 h; **53**: **27**-ketoesters (1 mmol), **32** (1 mmol), $HC(OMe)_3/CH_2Cl_2$ (1 : 2, 5 ml), r t, 3–4 h, then 10% TFA, r t, 4–8 h; **54**: **27** (1 mmol), **47** (1.1 mmol), r t, 3–6 h; **55**: **27** (1 mmol), DIAD (1.2 mmol), PPh_3 (1.3 mmol), THF, r t, 2–5 h; **56**: **27** (1 mmol), **45/46** (1.2 mmol), CH_2Cl_2, r t, overnight; **57**: **27**-ketoesters (1 mmol), **39** (1.2 mmol), Ac_2O/AcOH (1 : 3; 4 ml), 150 °C, 5 min; **58**: **27**-ketoesters (1 mmol), **29** (1.2 mmol), CH_2Cl_2, r t, 4 h; **59**: **27**-ketoesters (1 mmol), **36** (1.2 mmol), CH_2Cl_2, r t, 3 h; **60**: **27**-ketoesters (1 mmol), **31** (1 mmol), CH_2Cl_2, r t, Et_3N (3.0 mmol), 4 h; **61**: same as for **51**; **62**: **27**-ketoesters/aldehydes (1 mmol), **30** (1.2 mmol), CH_2Cl_2, r t, Et_3N (3.0 mmol), 6 h; **63**: same as for **50**. r t = room temperature; DIAD = diisopropyldiazadicarboxylate; THF = tetrahydrofuran.

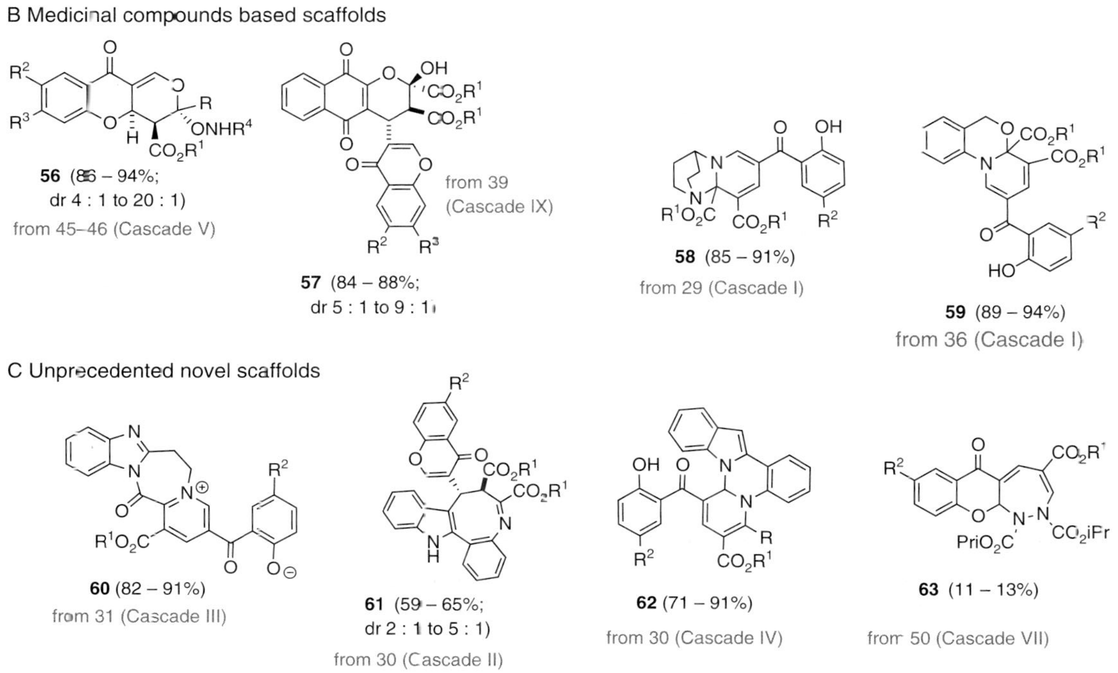

Scheme 27.3 *(Continued)*

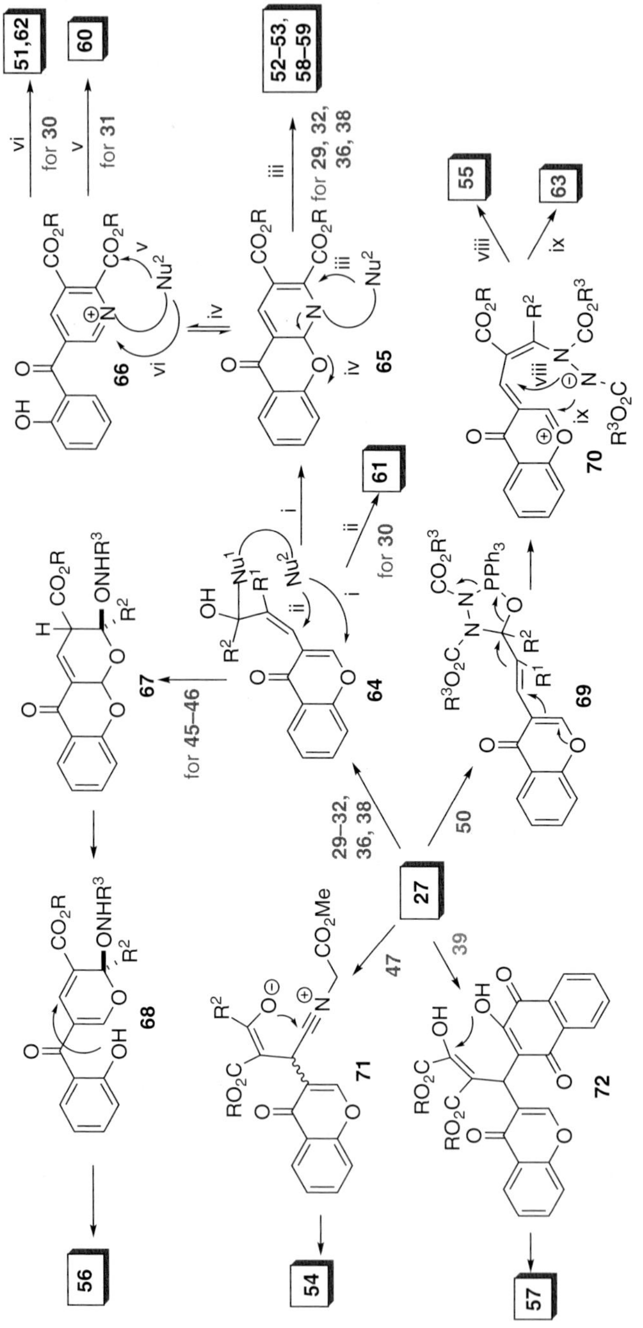

Cascade I: via 64-i-65-iii; Cascade II: via 64-ii;
Cascade III: via 64-65-iv-v; Cascade IV: via 64-65-iv-vi;
Cascade V: via 64-67-68; Cascade VI: via 69-70-viii;
Cascade VII: via 69-70-ix; Cascade VIII:via 71; Cascade IX:via 72

Scheme 27.4 Mechanistic insights into diverse cascade directing scaffold diversity synthesis.

natural-product-based and highly substituted tricyclic benzopyrones **56** in excellent yields and with appreciable diastereoselectivities (Scheme 27.3).

Among the nucleophilic zwitterions, Huisgen's zwitterions (**50**) reacted with both **27**-ketoesters and **27**-aldehydes to provide chromone-substituted pyrrole ring-system **55** as the major product. Only in some cases of **27**-aldehydes, **63** a novel ring system was obtained as minor products. Zwitterion **47**, that is, methyl isocyanoacetate displayed unbiased reactivity toward **27**-ketoesters/aldehydes and yielded benzopyrones supporting substituted furan ring in high yields (**54**, Schemes 27.3 and 27.4).

Mechanistic insights as proposed by Kumar and coworkers supported the very design of the branching cascade strategy because the successful nucleophiles have utilized the dispersed electrophilic sites over the surface of common precursors **27** beautifully to provide different products via different cascade reactions (Scheme 27.4). In many cases, the reaction sequence apparently begins with the addition of nucleophiles to aldehyde or ketoester moiety and followed by further cyclization on to the chromones moiety (cascades I, III, IV, VI) before the second nucleophilic addition or rearrangement leading to diverse scaffolds. Only in a few cases, the nucleophiles preferred to add on the unsaturated side chain appended to chromone ring in **27** and thus providing isoflavaone-based ring systems (cascades II, VII–IX).

In summary, branching cascades could successfully utilize 11 nucleophiles to transform the common precursor **27** into 13 diverse and complex ring systems or functionalized scaffolds (**51–63**, Scheme 27.3). Nine different cascade reaction sequences were triggered by the nucleophiles employed and each cascade led to a different scaffold (Scheme 27.4). The structural diversity obtained covers both biologically relevant and novel chemical space as the molecules behold both natural product and medicinal-chemistry-based ring systems as well as unprecedented molecular frameworks (Scheme 27.3). The scaffolds obtained were highly functionalized and thus could be used for further modifications to have enough representation of each scaffold in a library rich in scaffold diversity to provide novel and interesting small molecules for medicinal chemistry and chemical biology research.

27.5 Conclusions

Scaffold diversity is an important parameter to characterize any compound collection. Organic synthesis is expected to provide efficient approaches to create three-dimensionally complex and structurally diverse small molecules. Branching pathways in DOS is an elegant approach to access scaffold diversity that could lead to small, yet diverse, focused compound libraries. Branching cascades strategy has successfully demonstrated the power and application of cascade or domino reactions in compound collection synthesis to build rapidly complex scaffolds for library synthesis. Further synthesis endeavors toward branching pathways are appearing in the literature [58–60]. Biological results of the molecules generated by these strategies are anticipated and may provide interesting lead or probe molecules. Organic synthesis has embraced new responsibilities in this century to provide both the *quality and the quantity* of small and relatively complex molecules for screening endeavors. To help aid this cause, reaction discovery approaches too need to change and unravel the number of chemical transformations that can be utilized in scaffold diversity syntheses.

References

1. Fox, S., Farr-Jones, S., Sopchak, L., Boggs, A., and Comley, J. (2004) High-throughput screening: searching for higher productivity. *J. Biomol. Screening*, **9**, 354–358.
2. Macarron, R. (2006) Critical review of the role of HTS in drug discovery. *Drug Discovery Today*, **11**, 277–279.
3. Piggott, A.M. and Karuso, P. (2004) Quality, not quantity: the role of natural products and chemical proteomics in modern drug discovery. *Comb. Chem. High Throughput Screening*, **7**, 607–630.
4. Shelat, A.A. and Guy, R.K. (2007) Scaffold composition and biological relevance of screening libraries. *Nat. Chem. Biol.*, **3**, 442–446.
5. Ganesan, A. (2008) The impact of natural products upon modern drug discovery. *Curr. Opin. Chem. Biol.*, **12**, 306–317.
6. Newman, D.J. and Cragg, G.M. (2009) Natural product scaffolds as leads to drugs. *Future Med. Chem.*, **8**, 1415–1427.
7. Beghyn, T., Poulain, R.D., Willand, N., Folleas, B., and Deprez, B. (2008) Natural compounds: leads or ideas? Bioinspired molecules for drug discovery. *Chem. Biol. Drug Des.*, **72**, 3–15.
8. Wilson, R.M. and Danishefsky, S.J. (2006) Small molecule natural products in the discovery of therapeutic agents: the synthesis connection. *J. Org. Chem.*, **71**, 8329–8351.
9. Nicolaou, K.C. and Snyder, S.A. (2004) The essence of total synthesis. *Proc. Natl. Acad. Sci. U.S.A.*, **101**, 11929–11936.
10. Schreiber, S.L. (2000) Target-oriented and diversity-oriented organic synthesis in drug discovery. *Science*, **287**, 1964–1969.
11. Dixon, D.D., Lockner, J.W., Zhou, Q., and Baran, P.S. (2012) Scalable, divergent synthesis of meroterpenoids *via* "borono-sclareolide". *J. Am. Chem. Soc.*, **134**, 8432–8435.
12. Mendoza, A., Ishihara, Y., and Baran, P.S. (2012) Scalable enantioselective total synthesis of taxanes. *Nat. Chem.*, **4**, 21–25.
13. Su, S., Rodriguez, R.A., and Baran, P.S. (2011) Scalable, stereocontrolled total synthesis of (±)-axinellamines A and B. *J. Am. Chem. Soc.*, **133**, 13922–13925.
14. Kumar, K. and Waldmann, H. (2009) Synthesis of natural product inspired compound collections. *Angew. Chem. Int. Ed.*, **48**, 3224–3242.
15. Welsch, M.E., Snyder, S.A., and Stockwell, B.R. (2010) Privileged scaffolds for library design and drug discovery. *Curr. Opin. Chem. Biol.*, **14**, 347–361.
16. Burke, M.D., Berger, E.M., and Schreiber, S.L. (2004) A synthesis strategy yielding skeletally diverse small molecules combinatorially. *J. Am. Chem. Soc.*, **126**, 14095–14104.
17. Burke, M.D. and Schreiber, S.L. (2004) A planning strategy for diversity-oriented synthesis. *Angew. Chem. Int. Ed.*, **43**, 46–58.
18. Tan, D.S. (2005) Diversity-oriented synthesis: exploring the intersections between chemistry and biology. *Nat. Chem. Biol.*, **1**, 74–84.

19. Galloway, W.R.J.D., Isidro-Llobet, A., and Spring, D.R. (2010) Diversity-oriented synthesis as a tool for the discovery of novel biologically active small molecules. *Nat. Commun.*, **1**, 80.
20. O'Connor, C.J., Beckmann, H.S.G., and Spring, D.R. (2012) Diversity-oriented synthesis: producing chemical tools for dissecting biology. *Chem. Soc. Rev.*, **41**, 4444–4456.
21. Wetzel, S., Bon, R.S., Kumar, K., and Waldmann, H. (2011) Biology-oriented synthesis. *Angew. Chem. Int. Ed.*, **50**, 10800–10826.
22. Bon, R.S. and Waldmann, H. (2010) Bioactivity-guided navigation of chemical space. *Acc. Chem. Res.*, **43**, 1103–1114.
23. Kaiser, M., Wetzel, S., Kumar, K., and Waldmann, H. (2008) Biology-inspired synthesis of compound libraries. *Cell. Mol. Life Sci.*, **65**, 1186–1201.
24. Lachance, H., Wetzel, S., Kumar, K., and Waldmann, H. (2012) Charting, navigating, and populating natural product chemical space for drug discovery. *J. Med. Chem.*, **55**, 5989–6000.
25. Feyen, F., Cachoux, F., Gertsch, J., Wartmann, M., and Altmann, K.-H. (2008) Epothilones as lead structures for the synthesis-based discovery of new chemotypes for microtubule stabilization. *Acc. Chem. Res.*, **41**, 21–31.
26. Thomas, G.L., Spandl, R.J., Glansdorp, F.G., Welch, M., Bender, A., Cockfield, J., Lindsay, J.A., Bryant, C., Brown, D.F.J., Loiseleur, O., Rudyk, H., Ladlow, M., and Spring, D.R. (2008) Anti-MRSA agent discovery using diversity-oriented synthesis. *Angew. Chem. Int. Ed.*, **47**, 2808–2812.
27. Nören-Müller, A., Corrêa, I.R., Prinz, H., Rosenbaum, C., Saxena, K., Schwalbe, H.J., Vestweber, D., Cagna, G., Schunk, S., Schwarz, O., Schiewe, H., and Waldmann, H. (2006) Discovery of protein phosphatase inhibitor classes by biology-oriented synthesis. *Proc. Natl. Acad. Sci. U.S.A.*, **103**, 10606–10611
28. Koch, M.A., Wittenberg, L.-O., Basu, S., Jeyaraj, D.A., Gourzoulidou, E., Reinecke, K., Odermatt, A., and Waldmann, H. (2004) Compound library development guided by protein structure similarity clustering and natural product structure. *Proc. Natl. Acad. Sci. U.S.A.*, **101**, 16721–16726.
29. Lipkus, A.H., Yuan, Q., Lucas, K.A., Funk, S.A., Bartelt, W.F. III,, Schenck, R.J., and Trippe, A.J. (2008) Structural diversity of organic chemistry. A scaffold analysis of the CAS registry. *J. Org. Chem.*, **73**, 4443–4451.
30. Dow, M., Fisher, M., James, T., Marchetti, F., and Nelson, A. (2012) Towards the systematic exploration of chemical space. *Org. Biomol. Chem.*, **10**, 17–28.
31. Hert, J., Irwin, J.J., Laggner, C., Keiser, M.J., and Shoichet, B.K. (2009) Quantifying biogenic bias in screening libraries. *Nat. Chem. Biol.*, **5**, 479–483.
32. Dobson, C.M. (2004) Chemical space and biology. *Nature*, **432**, 824–828.
33. Ertl, P., Jelfs, S., Mühlbacher, J., Schuffenhauer, A., and Selzer, P. (2006) Quest for the rings. *In silico* exploration of ring universe to identify novel bioactive heteroaromatic scaffolds. *J. Med. Chem.*, **49**, 4568–4573.
34. Galloway, W.R.J.D. and Spring, D.R. (2013) Towards drugging the undruggable: enhancing the scaffold diversity of synthetic small molecule screening collections using diversity-oriented synthesis. *Diversity Oriented Synth.*, **1**, 21–28.
35. MacLellan, P. and Nelson, A. (2013) A conceptual framework for analysing and planning synthetic approaches to diverse lead-like scaffolds. *Chem. Commun.*, **49**, 2383–2393.
36. Bemis, G.W. and Murcko, M.A. (1996) The properties of known drugs. 1. Molecular frameworks. *J. Med. Chem.*, **39**, 2887–2893.
37. Lewell, X.Q., Jones, A.C., Bruce, C.L., Harper, G., Jones, M.M., Mclay, I.M., and Bradshaw, J. (2003) Drug rings database with web interface. A tool for identifying alternative chemical rings in lead discovery programs. *J. Med. Chem.*, **46**, 3257–3274.
38. Brough, P.A., Aherne, A., Barril, X., Borgognoni, J., Boxall, K., Cansfield, J.E., Cheung, K.-M.J., Collins, I., Davies, N.G.M., Drysdale, M.J., Dymock, B., Eccles, S.A., Finich, H., Fink, A., Hayes, A., Howes, R., Hubbard, R.E.,

James, K., Jordan, A.M., Lockie, A., Martins, V., Massey, A., Matthews, T.P., McDonald, E., Northfield, C.J., Pearl, L.H., Prodromou, C., Ray, S., Raynaud, F.I., Roughley, S.D., Sharp, S.Y., Surgenor, A., Walmsley, D.L., Webb, P., Wood, M., Workman, P., and Wright, L. (2008) 4,5-Diarylisoxazole Hsp90 chaperone inhibitors: potential therapeutic agents for the treatment of cancer. *J. Med. Chem.*, **51**, 196–218.

39. Roth, B., Sheffler, D., and Kroeze, W. (2004) Magic shotguns versus magic bullets: selectively non-selective drugs for mood disorders and schizophrenia. *Nat. Rev. Drug Discovery*, **3**, 353–359.
40. Paolini, G., Shapland, R., van Hoorn, W., Mason, J., and Hopkins, A. (2006) Global mapping of pharmacological space. *Nat. Biotechnol.*, **24**, 805–815.
41. Munos, B. (2009) Lessons from 60 years of pharmaceutical innovation. *Nat. Rev. Drug Discovery*, **8**, 959–968.
42. Oguri, H. and Schreiber, S.L. (2005) Skeletal diversity *via* a folding pathway: synthesis of indole alkaloid-like skeletons. *Org. Lett.*, **7**, 47–50.
43. Spandl, R.J., Diaz-Gavilan, M., O'Connell, K.M.G., Thomas, G.L., and Spring, D.R. (2008) Diversity-oriented synthesis. *Chem. Rec.*, **8**, 129–142.
44. Dandapani, S., Duduta, M., Panek, J.S., and Porco, J.A. (2007) Skeletal diversity through radical cyclization of tetrahydropyridine scaffolds. *Org. Lett.*, **9**, 3849–3852.
45. Liao, Y., Hu, Y., Wu, J., Zhu, Q., Donovana, M., Fathia, R., and Yang, Z. (2003) Diversity oriented synthesis and branching reaction pathway to generate natural product-like compounds. *Curr. Med. Chem.*, **10**, 2285–2316.
46. Wyatt, E.E., Fergus, S., Galloway, W.R.J.D., Bender, A., Fox, D.J., Plowright, A.T., Jessiman, A.S., Welch, M., and Spring, D.R. (2006) Skeletal diversity construction *via* a branching synthetic strategy. *Chem. Commun.*, **31**, 3296–3298.
47. Comer, E., Rohan, E., Deng, L., and Porco, J.A. (2007) An approach to skeletal diversity using functional group pairing of multifunctional scaffolds. *Org. Lett.*, **9**, 2123–2126.
48. Haggarty, S.J. (2005) The principle of complementarity: chemical versus biological space. *Curr. Opin. Chem. Biol.*, **9**, 296–303.
49. Schreiber, S.L. (2005) Small molecules: the missing link in the central dogma. *Nat. Chem. Biol.*, **1**, 64–66.
50. Tietze, L.F. (1996) Domino reactions in organic synthesis. *Chem. Rev.*, **96**, 115–136.
51. Enders, D., Grondal, C., and Hüttl, M.R.M. (2007) Asymmetric organocatalytic domino reactions. *Angew. Chem. Int. Ed.*, **46**, 1570–1581.
52. Liu, W., Khedkar, V., Baskar, B., Schürmann, M., and Kumar, K. (2011) Branching cascades: a concise synthetic strategy targeting diverse and complex molecular frameworks. *Angew. Chem. Int. Ed.*, **50**, 6900–6905.
53. Cairns, N., Harwood, L.M., and Astles, D.P. (1994) Tandem thermal Claisen–Cope rearrangements of coumarate derivatives. Total syntheses of the naturally occurring coumarins: suberosin, demethylsuberosin, ostruthin, balsamiferone and gravelliferone. *J. R. Soc. Chem., Perkin Trans. 1*, 3101–3107.
54. Patre, R.E., Shet, J.B., Parameswaran, P.S., and Tilve, S.G. (2009) Cascade wittig reaction-double claisen and cope rearrangements: one-pot synthesis of diprenylated coumarins gravelliferone, balsamiferone, and 6,8-diphenylumbelliferone. *Tetrahedron Lett.*, **50**, 6488–6490.
55. Khedkar, V., Liu, W., Dückert, H., and Kumar, K. (2010) Efficient and atom-economic synthesis of α-substituted β-chromonyl-α,β-unsaturated carbonyls through molecular rearrangement. *Synlett*, 403–406, erratum 1576.
56. Dückert, H., Pries, V., Khedkar, V., Menninger, S., Bruss, H., Bird, A.W., Maliga, Z., Brockmeyer, A., Janning, P., Hyman, A., Grimme, S., Schürmann, M., Preut, H., Hübel, K., Ziegler, S., Kumar, K., and Waldmann, H. (2012) Natural product–inspired cascade synthesis yields modulators of centrosome integrity. *Nat. Chem. Biol.*, **8**, 179–184.
57. Eschenbrenner-Lux, V., Dückert, H., Khedkar, V., Bruss, H., Waldmann, H., and Kumar, K. (2013) Cascade syntheses

routes to the centrocountins. *Chem. Eur. J.*, **19**, 2294–2304.

58. Patil, N.T., Shinde, V.S., and Sridhar, B. (2013) Relay catalytic branching cascade: a technique to access diverse molecular scaffolds. *Angew. Chem. Int. Ed.*, **52**, 2251–2255.

59. Baskar, B., Wittstein, K., Sankar, M.G., Khedkar, V., Schürmann, M., and Kumar, K. (2012) Stereoselective cascade double-annulations provide diversely ring-fused tetracyclic benzopyrones. *Org. Lett.*, **14**, 5924–5927.

60. Waldmann, H., Kühn, M., Liu, M., and Kumar, K. (2008) Reagent-controlled domino synthesis of skeletally-diverse compound collections. *Chem. Commun.*, 1211–1213.

Index

q

r

s

t